Universitext

Series Editors

Nathanaël Berestycki, Universität Wien, Vienna, Austria

Carles Casacuberta, Universitat de Barcelona, Barcelona, Spain

John Greenlees, University of Warwick, Coventry, UK

Angus MacIntyre, Queen Mary University of London, London, UK

Claude Sabbah, École Polytechnique, CNRS, Université Paris-Saclay, Palaiseau, France

Endre Süli, University of Oxford, Oxford, UK

Universitext is a series of textbooks that presents material from a wide variety of mathematical disciplines at master's level and beyond. The books, often well class-tested by their author, may have an informal, personal, or even experimental approach to their subject matter. Some of the most successful and established books in the series have evolved through several editions, always following the evolution of teaching curricula, into very polished texts.

Thus as research topics trickle down into graduate-level teaching, first textbooks written for new, cutting-edge courses may find their way into *Universitext*.

Hannah Geiss • Stefan Geiss

Measure, Probability and Functional Analysis

Springer

Hannah Geiss
Department of Mathematics and Statistics
University of Jyväskylä
Jyväskylä, Finland

Stefan Geiss
Department of Mathematics and Statistics
University of Jyväskylä
Jyväskylä, Finland

ISSN 0172-5939   ISSN 2191-6675 (electronic)
Universitext
ISBN 978-3-031-84066-1   ISBN 978-3-031-84067-8 (eBook)
https://doi.org/10.1007/978-3-031-84067-8

© The Editor(s) (if applicable) and The Author(s), under exclusive license to Springer Nature Switzerland AG 2025

This work is subject to copyright. All rights are solely and exclusively licensed by the Publisher, whether the whole or part of the material is concerned, specifically the rights of translation, reprinting, reuse of illustrations, recitation, broadcasting, reproduction on microfilms or in any other physical way, and transmission or information storage and retrieval, electronic adaptation, computer software, or by similar or dissimilar methodology now known or hereafter developed.
The use of general descriptive names, registered names, trademarks, service marks, etc. in this publication does not imply, even in the absence of a specific statement, that such names are exempt from the relevant protective laws and regulations and therefore free for general use.
The publisher, the authors and the editors are safe to assume that the advice and information in this book are believed to be true and accurate at the date of publication. Neither the publisher nor the authors or the editors give a warranty, expressed or implied, with respect to the material contained herein or for any errors or omissions that may have been made. The publisher remains neutral with regard to jurisdictional claims in published maps and institutional affiliations.

This Springer imprint is published by the registered company Springer Nature Switzerland AG
The registered company address is: Gewerbestrasse 11, 6330 Cham, Switzerland

If disposing of this product, please recycle the paper.

To the memory of our parents:

Margot and Ernst Stiewe
Hannah Geiss

Christel and Siegfried Geiß
Stefan Geiss

To the memory of our teachers in probability theory at the Friedrich Schiller University Jena:

Hans-Jürgen Engelbert, Johannes Kerstan, and Ulrich Zähle

Preface

Measure theory and probability live from the interaction with functional analysis, where both sides benefit from each other. We start with a self-contained introduction to probability based on measure and integration theory and continue presenting related concepts from functional analysis. The book is designed to serve as lecture notes, as a self-study course, and also as a reference.

The first topic of the book, the chapters until Sect. 15.1, evolved to a large extent from courses given at the *University of Jyväskylä* (Finland) and the *University of Innsbruck* (Austria) with the aim to provide the necessary knowledge and tools in probability that are required for further studies in stochastic process theory and stochastic analysis. Here we follow the axiomatic approach intended for students from mathematics, but explicitly also for other students who need more mathematical background for their further studies. We do not assume any knowledge of measure and integration theory, instead we introduce the necessary parts. In this way, we offer a somewhat demanding but rewarding access to rigorous probability theory.

The second and third topics of the book (see below) are intertwined and cover advanced limit theorems and functional analysis. Regarding functional analysis we treat two aspects: Firstly we consider real-valued random variables using Banach function spaces. Banach function spaces are Banach spaces which are used here to describe certain properties of random variables. For example, the tail behaviour of random variables can be expressed in terms of Lorentz and Orlicz spaces. Using properties of those Banach function spaces one gains new insights and, at the same time, existing techniques become more transparent. Secondly, we consider random variables with values in Banach spaces. For example, the sample paths of the Brownian motion can be interpreted as elements of the space of continuous functions as we will do in this book. More generally, solutions to stochastic differential equations with trajectories having certain regularity properties, like the Hölder continuity, can be treated as Banach space valued objects.

Topic I: Measure Theory and Basics of Probability Theory

In Chap. 2 we introduce measure and probability spaces, treat basic properties of measures, and continue with the lemmas of Fatou and Borel-Cantelli.

Chapter 3 is devoted to the construction of measure spaces, where we start with discrete measure spaces. The main part of this chapter concerns the extension theorem (also named after Carathéodory) that enables us to extend σ-additive measures on algebras to σ-additive measures on σ-algebras. This statement is fully proven in Appendix B. By this extension theorem we construct, among others, the fundamental Lebesgue measure on the real line.

In Chap. 4 necessary concepts and statements on metric spaces and Banach spaces are collected which will be used throughout the book.

In Chap. 5 close relations between the topology of a metric space and measures on its Borel σ-algebra become already visible. We prove inner and outer regularity of measures with respect to closed and open sets, respectively, and Ulam's theorem about the inner regularity with respect to compact sets. The notion of the support of a measure is one more connection between the topology of a metric space and its Borel σ-algebra.

Chapter 6 introduces measurable maps between measurable spaces, which is the abstract approach that is very beneficial in various situations. Besides standard distribution functions for scalar valued random variables, we also consider the multivariate framework.

Independence belongs to the most important probabilistic concepts. In Chap. 7 we introduce independence on three different abstraction levels, ranging from the independence of events to the independence of σ-algebras. A main result of this chapter is to verify the existence of countable products of probability spaces. By this result we ensure the existence of sequences of independent random variables. The concept of independence leads also to the phenomena of 0-1-laws. We will prove the Hewitt-Savage 0-1-law and apply it to a risk-model from non-life insurance in Chap. 22.

Chapter 8 presents the full integration theory for general measure spaces including the main theorems about integration named after Fatou, Lebesgue, Fubini, and Tonelli. The more analytical part concerns the inequalities due to Markov, Jensen, Hölder, and Minkowski. We also add the Hoeffding inequality which will be useful when inspecting the speed of convergence of Monte Carlo methods. We end by Wald's identity that we need in our application to non-life insurance.

Chapter 9 covers all types of convergence of random variables and their relations to each other that are relevant for us. One of the main results is Vitali's convergence theorem connecting uniform integrability to the convergence in L_p.

In Chap. 10 we prove the Radon-Nikodym theorem and deduce the existence of conditional expectations.

Chapter 11 introduces Fourier transforms of measures and characteristic functions of random variables. These concepts are central in probability as they allow to prove the existence of measures, to investigate their properties, and to study distributional properties of random variables in terms of complex analysis. So we can introduce Gaussian random vectors in an elegant way without taking care whether the distribution is degenerated or not.

In Chap. 12 we consider the weak convergence of measures on $\mathbb{R}^d$ which is also based on the Fourier analysis methods from Chap. 11. The so-called *Portmanteau Theorem*, Lévy's continuity theorem, and the Prokhorov theorem are presented and fully proven in an extra section for further reading. The *Portmanteau Theorem* summarizes the most important properties of weak convergence. Finally, we close the circle by proving the Dudley-Skorohod theorem that takes us back from the weak convergence to the almost sure convergence introduced in Chap. 9.

We continue with two important limit theorems, the *Strong Law of Large Numbers* and the *Central Limit Theorem* in Chap. 13 and Sect. 15.1. The strong law of large numbers is proven in the modern form of Etemadi.

Topic II: Advanced Limit Theorems

The *Ergodic Theorem of Birkhoff and Khintchin* is a statement about iterative applications of a measure preserving map and is a central result in ergodic theory. In Chap. 14 we prove this statement and deduce a generalized version of the strong law of large numbers in the Kolmogorov form.

Normal approximation refers to limit theorems that approximate a Gaussian distribution with a control on the speed of convergence. In Chap. 15 we present with proofs two fundamental results for this approximation, the *Berry-Esseen Theorem* and *Stein's Method*.

The *Law of Iterated Logarithm* is a fascinating result about the asymptotic behaviour of random walks. In Chap. 21 we prove this result by *Berry-Esseen's Theorem*. This chapter comes last as we need a maximal inequality provided in Chap. 20 belonging to the class of Lévy-Ottaviani-inequalities, where we also include a recent result of Szewczak giving a quantitative improvement in one of these inequalities.

Topic III: Functional Analysis

In Chap. 16 we study inversion formulas for the Fourier transform. We continue with the proof of the *Bochner-Khinchin Theorem* for measures on $\mathbb{R}^d$ which turns out to be extremely useful in various situations.

The Fourier transform of measures is an essential tool to study the existence and properties of measures. Specializing this transform to measures with a density, we face the Fourier transform of functions, that is a subject of harmonic analysis. So in Chap. 17 we prove Plancherel's equality and its generalization, the Hausdorff-Young inequality—two fundamental statements regarding the Fourier transform of functions. To prove the Hausdorff-Young inequality we need to verify the Riesz-Thorin theorem in Appendix A.8. We continue with the *Hilbert transform* which is used, for example, in signal processing to transform signals into signals with a non-negative Hermitian spectrum. The Hilbert transform is a prime example from the theory of singular integral operators.

Chapter 18 deals with *Representation Theorems* based on the concept of duality from Banach space theory. More precisely, we will find the dual spaces of the L_p-spaces by the Riesz-Fréchet representation theorem, and in terms of the Riesz-

Kakutani representation theorem the dual space of $C(K)$, the space of continuous functions on the compact metric space K. By the Riesz-Kakutani theorem we will learn that the extension theorem for measures from Chap. 3 becomes a statement about duality in functional analysis.

In Chap. 19 we introduce and investigate Banach function spaces. Banach function spaces are based on the nonincreasing rearrangements of random variables. The nonincreasing rearrangement has the disadvantage not to share certain convexity properties. So one replaces non-increasing rearrangements by maximal functions. Based on this, we introduce Banach function spaces and prove the Hardy-Littlewood inequality. Special cases of Banach function spaces are Lorentz spaces and Orlicz spaces. Furthermore, we introduce the K-functional, a central tool in interpolation theory. As an application we provide formulas to compute the p-th mean of the supremum of independent random variables and to describe and handle distributional properties of random variables.

In Chap. 20 we consider random variables with values in Banach spaces. They already appear if one considers stochastic processes with values in $\mathbb{R}$ and wants to interpret their paths as random functions and to find their distribution on a function space. For a separable Banach space E we consider for random variables

$$f : \Omega \to E$$

- the Pettis measurability theorem,
- the Bochner integral,
- the strong law of large numbers,
- Lévy-Ottaviani maximal inequalities,
- the Itô-Nisio theorem about the convergence of sums of independent Banach space valued random variables,
- the Ciesielski's theorem and its connection to the Brownian motion.

In Chap. 22 we consider a risk-model from non-life insurance. By Wald's identity from Chap. 8 and the Hewitt-Savage 0-1-law from Chap. 7 we prove that the ruin probability is one if the Net-Profit-Condition does not hold. Moreover, we apply results from Chap. 20 to get information about maximal waiting times between the first n claims in dependence on n.

Jyväskylä, Finland

Hannah Geiss
Stefan Geiss

Acknowledgements

We thank the anonymous referees and Dirk Werner for their contributions which lead to numerous substantial improvements of the manuscript. Moreover, we would like to thank our colleagues for their comments and helpful support: Sarah Geiss, Friedrich Hubalek, Tero Kilpeläinen, Jesse Koivu, Thuan Nguyen, Stefan Perko, Markus Riedle, Alexander Steinicke, Jan van Neerven, and Mark Veraar.

Competing Interests The authors have no competing interests to declare that are relevant to the content of this manuscript.

How Is the Book Arranged?

The sections **up to (including) Sect. 15.1** are intended for standard probability courses in mathematics but also for interested students from other fields that need a solid foundation in measure and probability theory. The material can be typically divided into a basic course and one or two advanced courses. Sections that are marked with * are intended for extended reading.

The sections **following Sect. 15.1** are intended for topic classes, seminars, and self-study, and are for master's and PhD students and for postdoctoral researchers. Here no sections are marked by *.

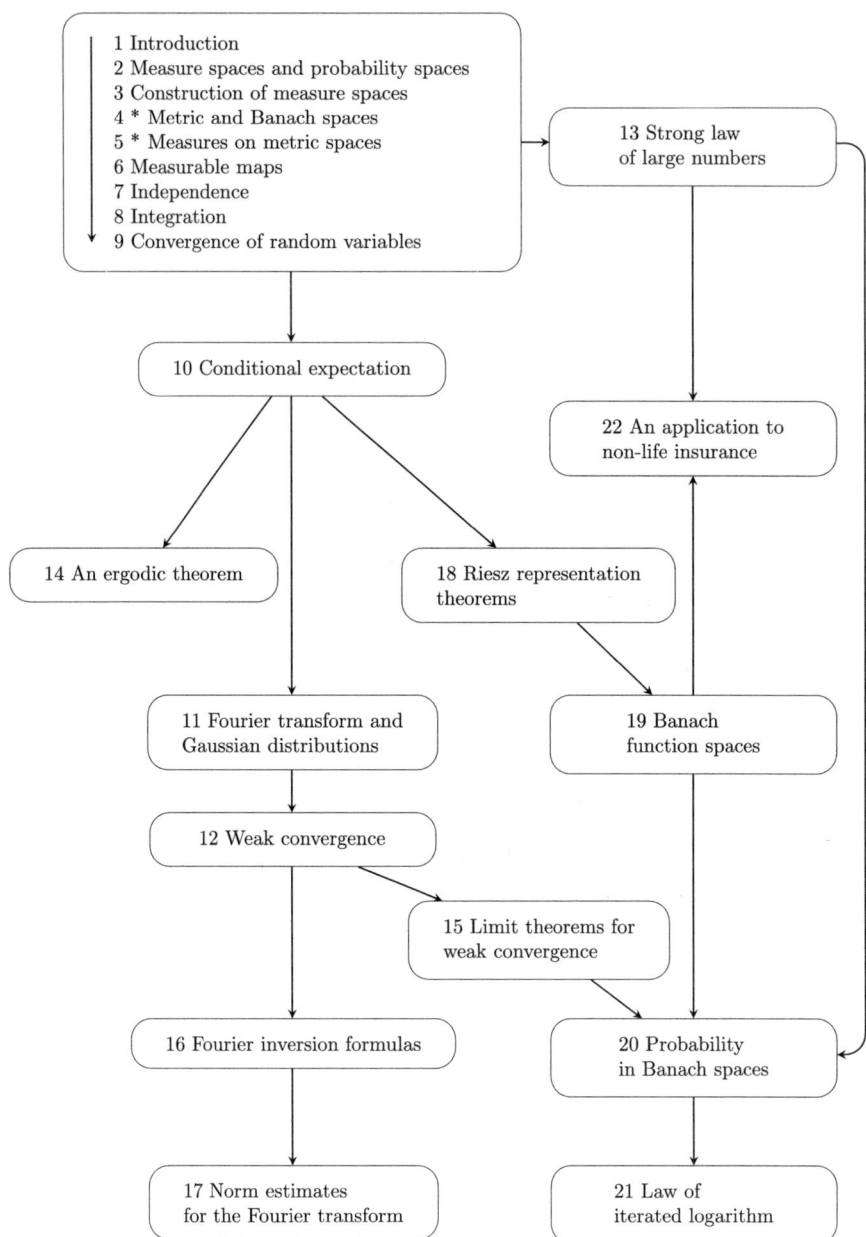

Contents

1 Introduction: With Two Examples 1

2 Measure Spaces and Probability Spaces 7
    2.1   The Elementary Events 7
    2.2 σ-Algebras 8
    2.3   Definition and First Properties of Measures 11
    2.4*   The Completion of Measure Spaces 14
    2.5   Infinitely Often Occurring Events and a First Look at
            Independence: The Lemmas of Fatou and Borel-Cantelli 14
    2.6   A First Look at Conditional Probabilities and Bayes' Rule 19
    2.7   Exercises 21
    2.8   Comments 24

3 Construction of Measure Spaces 25
    3.1   Measures on $\{0, \ldots, n\}$ and $\mathbb{N}_0$ 25
            3.1.1   Binomial Distribution with Parameters $n \in \mathbb{N}$
                        and $0 < p < 1$ 26
            3.1.2   Poisson Distribution with Parameter $\lambda > 0$ 28
            3.1.3   Geometric Distribution with Parameter $0 < p < 1$ 29
    3.2   The Extension Theorem and Consequences 30
    3.3   Measures on $(\mathbb{R}, \mathcal{B}(\mathbb{R}))$ 31
            3.3.1   The Borel σ-Algebra $\mathcal{B}(\mathbb{R})$ 31
            3.3.2   The Extension Theorem for $(\mathbb{R}, \mathcal{B}(\mathbb{R}))$ 34
            3.3.3   Lebesgue Measure on $\mathbb{R}$ 37
            3.3.4   Gaussian Distribution on $\mathbb{R}$ 38
            3.3.5   Exponential Distribution on $\mathbb{R}$ 39
            3.3.6   The Trace σ-Algebra 40
            3.3.7   Uniform Distribution on $[c, d]$ 42
    3.4   Exercises 42
    3.5   Comments 44

4	***Metric and Banach Spaces**		45
	4.1	Metric Spaces	45
	4.2	Banach Spaces and Quasi-Banach Spaces	48
	4.3	Linear Operators Between Banach Spaces	51
	4.4	Main Theorems	53
	4.5	Comments	54
5	***Measures on Metric Spaces**		57
	5.1	The Borel σ-Algebra $\mathcal{B}(M)$	57
	5.2	Regularity of Measures on Metric Spaces	58
	5.3	The Support of a Measure	63
	5.4	Exercises	64
	5.5	Comments	65
6	**Random Variables and Measurable Maps**		67
	6.1	Measurable Maps with Values in $\mathbb{R}$ and Random Variables	68
	6.2	Measurable Maps Between Measurable Spaces	71
	6.3*	Measurable Maps Between Metric Spaces	73
	6.4	Summary	74
	6.5	Image Measures and Distribution Functions	74
	6.6	Discrete and Continuous Random Variables	77
	6.7*	Distribution Functions on $\mathbb{R}^d$	77
	6.8	Exercises	80
	6.9	Comments	81
7	**Independence**		83
	7.1	The Basic Concept	84
	7.2	Kolmogorov's Zero-One Law	86
	7.3	Products of Measure Spaces	88
	7.4*	A Remark on Non σ-Finite Measure Spaces	95
	7.5	Construction of Independent Random Variables	95
	7.6	Product Spaces and Independence	96
	7.7*	Hewitt-Savage's 0-1 Law	98
	7.8	Exercises	102
	7.9	Comments	103
8	**Integration**		105
	8.1	Definition of the Lebesgue Integral	106
	8.2	Theorem About Monotone Convergence	109
	8.3	Basic Properties of the Lebesgue Integral	112
	8.4	Extended Measurable Maps	114
	8.5	Fatou's Lemma and Lebesgue's Theorem	116
	8.6	Examples and Relations to Other Integrals	117
		8.6.1 Countable Measure Spaces	117
		8.6.2* Lebesgue's Criterion for Riemann Integrability	118
		8.6.3 The Lebesgue-Stieltjes Integral	122
	8.7	Change of Variable Formula	122

	8.8	Affine Images of the Lebesgue Measure λ_d	123
	8.9	The Theorems of Tonelli and Fubini	125
	8.10	Applications of Tonelli's and Fubini's Theorem	130
	8.11	The Variance of a Random Variable	135
	8.12	Inequalities	136
		8.12.1 Markov's Inequality	136
		8.12.2 Jensen's Inequality	137
		8.12.3 Hölder's and Minkowski's Inequality	138
		8.12.4* Hoeffding's Inequality	142
	8.13*	Wald's Identity	143
	8.14*	Atomless Probability Spaces	145
	8.15	Exercises	146
	8.16	Comments	149
9	**Convergence of Random Variables**		151
	9.1	Almost Sure Convergence	152
	9.2	Convergence in Probability	157
		9.2.1 Probabilistic Formulation	157
		9.2.2* Analytical Formulation	161
	9.3	Convergence in L_p	162
		9.3.1 Probabilistic Formulation	162
		9.3.2* Analytical Formulation	164
	9.4	Uniform Integrability	165
	9.5	Summary	171
	9.6	Exercises	172
10	**The Theorem of Radon-Nikodym and Conditional Expectation**		175
	10.1	Signed Measures	176
	10.2	Theorem of Radon-Nikodym	181
	10.3	Conditional Expectation	184
	10.4	A Representation of Conditional Expectations	189
	10.5	Exercises	194
	10.6	Comments	196
11	**Fourier Transform and Gaussian Distributions**		197
	11.1	Preliminaries	197
	11.2	Definition and Basic Properties of Characteristic Functions	199
	11.3	Convolutions of Measures	205
	11.4	Densities and Convolutions of Functions	206
	11.5	The theorem of Riemann-Lebesgue and the Uniqueness Theorem	209
	11.6	Moments of Measures	213
	11.7	Gaussian Distribution on $\mathbb{R}^d$	216
	11.8	Gaussian $\mathbb{R}^d$-Valued Random Variables	221
	11.9	A Characterization of Independent Random Variables	221
	11.10	Exercises	223
	11.11	List of Characteristic Functions	225

12 Weak Convergence 227
  12.1 The Main Theorems 227
  12.2 Proofs of the Main Theorems 231
  12.3 From Weak Convergence Back to Almost Sure Convergence 237
  12.4 Exercises 240
  12.5 Comments 240

13 Strong Law of Large Numbers 243
  13.1 Formulation of the Strong Law of Large Numbers 243
  13.2 Proof of the SLLN in Etemadi Form 244
  13.3 Theorem About Normal Numbers 246
  13.4 Speed of Convergence in Monte-Carlo Methods 247
  13.5 Buffon's Needle Problem 249
  13.6 Exercises 250
  13.7 Comments 251

14 An Ergodic Theorem 253
  14.1 Birkhoff-Khinchin Ergodic Theorem 253
  14.2 A Generalized SLLN 256
  14.3 Exercises 258
  14.4 Comments 258

15 Limit Theorems for Weak Convergence 259
  15.1 Central Limit Theorems 259
  15.2 The Berry-Esseen Theorem 267
  15.3 Stein's Method 274
  15.4 The Poisson Limit Theorem 283
  15.5 Exercises 286
  15.6 Comments 288

16 Fourier Inversion Formulas 289
  16.1 Inversion Formulas 289
  16.2 The Theorem of Bochner and Khinchin 293
  16.3 Cauchy Distribution 296
  16.4 Exercises 299
  16.5 Comments 300

17 Norm Estimates for the Fourier Transform 301
  17.1 Plancherel's Equality and Hausdorff-Young Inequality 301
  17.2 Hilbert Transform 308
  17.3 Comments 310

18 Riesz Representation Theorems 311
  18.1 L_p Spaces Over σ-Finite Measures 311
  18.2 The Dual Space of $L_p(\Omega, \mathcal{F}, \mu)$ 312
  18.3 The Spaces $C(K)$ 317
  18.4 The Dual Space of $C(K)$ 320

	18.5 Exercises	324
	18.6 Comments	325
19	**Banach Function Spaces**	**327**
	19.1 Nonincreasing Rearrangement and Maximal Functions	327
	19.2 Banach Function Spaces	342
	19.3 Exercises	351
	19.4 Comments	353
20	**Probability in Banach Spaces**	**355**
	20.1 Banach Space Valued Measurable Maps	355
	20.2 The Bochner Integral	359
	20.3 The Dunford and the Pettis Integral	363
	20.4 Strong Law of Large Numbers in Separable Banach Spaces	367
	20.5 Maximal Inequalities	370
	20.6 The Itô-Nisio Theorem	372
	20.7 Ciesielski's Theorem and the Wiener Process	374
	20.8 Exercises	380
	20.9 Comments	382
21	**Law of Iterated Logarithm**	**383**
	21.1 The Law of Iterated Logarithm	383
	21.2 Berry-Esseen Meets the LIL	384
	21.3 Exercises	390
	21.4 Comments	390
22	**An Application to Non-life Insurance**	**391**
	22.1 The Sparre-Anderson Model	391
	22.2 Necessity of the NPC Condition and Expected Waiting Times	392
	22.3 Comments	395
A	**Analysis**	**397**
	A.1 Complex Numbers	397
	A.2 Limit Inferior and Limit Superior	398
	A.3 Theorem of Stone and Weierstrass	399
	A.4 Oscillatory Integrals and Dirichlet Integral	400
	A.5 Schur's Product Theorem	402
	A.6 The Axiom of Choice and the Lemma of Zorn-Kuratowski	402
	A.7 Equivalence Classes	403
	A.8 Riesz–Thorin Theorem	403
	A.9 Ciesielski's Theorems	407
B	**Measure and Integration Theory**	**413**
	B.1 Monotone Class Theorems and π-λ-Theorem	413
	B.2 Outer Measures	417
	B.3 Countably Generated σ-Algebras	421
	B.4 The Lebesgue σ-Algebra	422

C	**R-code**		427
	C.1	monte-carlo.r	427
	C.2	monte-carlo_for_pi.r	428
	C.3	slln_vs_clt.r	429
	C.4	lil.r	430

Bibliography 431

Index 439

Chapter 1
Introduction: With Two Examples

Probability theory can be understood as a mathematical model for the intuitive notion of *uncertainty*. Some parts of probability theory belong to pure mathematics, other parts to applied mathematics, whereas the division between pure- and applied mathematics is not unique and might change over time. Without probability theory all the stochastic models in physics, biology, and economics would either not have been developed or would not be rigorous. In order to be rigorous, a solid mathematical foundation is needed to assist us with various tools, like set theory, functional analysis, complex analysis, and the theory of special functions. On the other hand, probability is used in many branches of pure mathematics itself, even in branches where one does not expect this, like in convex geometry.

Probability theory based on an axiomatic approach came into being with Kolmogorov's[1] [105] monograph from 1933. The approach of Kolmogorov is based on the strict abstract mathematical formulation of the notion of an event and its probability as transparent as possible, an approach that was different from more empirical approaches at that time, as of S.N. Bernstein[2] [12] and R.v. Mises[3] [132]. The development of the Lebesgue measure in a geometrical context and its generalization by Fréchet have been important steps done before Kolmogorov's abstract approach.

On the one hand side this means that modern probability is a field of mathematics based on measure theory. On the other side it might be misleading to understand modern probability simply as a sub-branch of measure theory as own notions, techniques, and tools were and are developed.

---

[1] Andrey Nikolaevich Kolmogorov, 25/04/1903 (Tambov, Tambov province, Russia)–20/10/1987 (Moscow, USSR).

[2] Sergei Natanovich Bernstein, 5/03/1880 (Odessa, Ukraine)–26/10/1968 (Moscow, USSR).

[3] Richard von Mises, 19/04/1883 (Lemberg, Galicia, Austrian Empire (now Lviv, Ukraine))–14/07/1953 (Boston, Massachusetts, USA).

The axiomatic approach to probability theory enables the usage of analytical tools leading to intrinsic connections between probability theory, measure theory, and several areas in analysis. In this book we present some of these connections and conclude the introduction by two examples that will guide us through some parts of the book:

First Example Let us start with a classical example from 1733, Buffon's[4] needle experiment (see [27] and [8] for the historical background): We take needles of a fixed length $L > 0$ and a plane area with parallel lines, where the distance between two lines equals the length L of the needles. We throw one needle after the other on the plane area, without caring about the direction and position of the needles, and count how many needles have a common point with one or two of the parallel lines. If we did throw n needles and counted S_n cases where a needle intersects one line or touches two lines, then we will observe for large n that $S_n/n \sim 2/\pi$, or more formally that

$$\lim_{n \to \infty} \frac{S_n}{n} = \frac{2}{\pi}.$$

So we obtain by a simple experiment the number π. Besides this surprising fact, the basic question is how to model this experiment mathematically and how to confirm within this model the above observation. Intuitively, the randomness comes from two sources: the position of one fixed end point of the needle and the angle against the parallel lines. It turns out that we need to consider only one stripe, so that we have the state space $\Omega := [0, L) \times [0, 2\pi)$, where the parameters $\omega = (l, \alpha)$ model the position of the needle in one stripe. Then we consider the map

$$f : \Omega = [0, L) \times [0, 2\pi) \to \{0, 1\}$$

---

[4] Georges Louis Leclerc Comte de Buffon, 07/09/1707 (Montbard, Côte d'Or, France)–16/04/1788 (Paris, France).

1 Introduction

such that $f(\omega) := 1$ if the needle intersects one line or touches two lines, and $f(\omega) := 0$ otherwise. Now we proceed as follows:

Step 1: We model the randomness of our experiment by considering the probability space $(\Omega, \mathcal{F}, \mathbb{P})$, where $\mathcal{F}$ will be the Borel σ-algebra on $[0, L) \times [0, 2\pi)$ and $\mathbb{P}$ be the normalized Lebesgue measure, which is the uniform distribution. The knowledge about this will be provided in Chaps. 2 and 3.

Step 2: We interpret the map f as a measurable map or random variable. These concepts are introduced in Chap. 6.

Step 3: We model the repeated throwing of a needle by taking *independent copies* $f_1, f_2, \ldots : \overline{\Omega} \to \mathbb{R}$ of f, where $(\overline{\Omega}, \overline{\mathcal{F}}, \overline{\mathbb{P}})$ is an appropriate 'larger' probability space. The concept of independence is introduced in Chap. 7.

Step 4: In the final step we use the *Strong Law of Large Numbers* discussed in Chap. 13 and obtain for $S_n := f_1 + \cdots + f_n$ that

$$\lim_{n \to \infty} \frac{S_n}{n} = \mathbb{E} f_1$$

in an almost sure sense. The various notions of convergence are treated in Chap. 9, the *expected value* $\mathbb{E} f_1$ in Chap. 8. It turns out that $\mathbb{E} f_1 = \frac{2}{\pi}$ which confirms the experiment.

Second Example Our second example is about the movement of pollen in water, observed by the botanist Robert Brown. We fix one particle of the pollen that is in position zero at time $t = 0$ and observe its vertical movement over time. To model this by a discretization, we consider a time-stepping of mesh-size $1/2^n$ and independent movements up or down by $\varepsilon_n > 0$ with equal probability $1/2$. So the position of the particle at time $t = i/2^n$ with $i = 0, 1, 2, \ldots$ is described by

$$\varepsilon_n f_1 + \cdots + \varepsilon_n f_i$$

with independent random variables $f_1, f_2, \ldots : \Omega \to \mathbb{R}$ such that each f_i takes the values -1 and 1 with probability $1/2$ (Fig. 1.1).

What is the right choice of the scaling factor $\varepsilon_n > 0$? One choice is that the variance at time $t = 1$ is the same for all approximations, in other words

$$\mathrm{var}_n^2 := \mathbb{E} |\varepsilon_n f_1 + \cdots + \varepsilon_n f_{2^n}|^2$$

does not depend on n. Normalizing this to $\mathrm{var}_n^2 = 1$, we get that $\varepsilon_n := 2^{-n/2}$. So for a time $t = i 2^{-n}$, $i = 0, 1, \ldots$, we get the model

$$W_t^n := \frac{f_1 + \cdots + f_i}{2^{n/2}} = \frac{f_1 + \cdots + f_{t 2^n}}{2^{n/2}}.$$

Fig. 1.1 One path of $t \mapsto W_t^3$

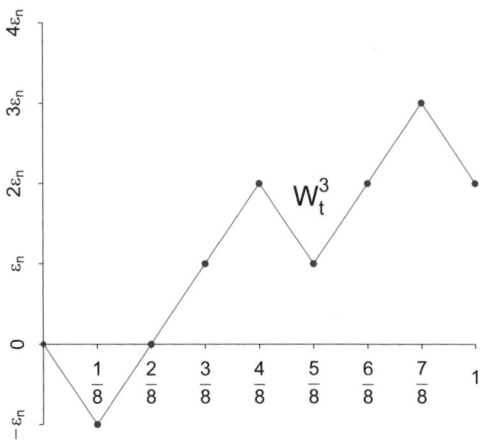

To understand the processes $t \mapsto W_t^n(\omega)$ as $n \to \infty$ we take $n_0 \in \mathbb{N}$ and consider dyadic time points $0 = t_0 < t_1 < \cdots < t_L = 1$ with $t_l := i_l 2^{-n_0}$ and $0 \leqslant i_0 < \cdots < i_L \leqslant 2^{n_0}$. What happens with the random vector $(W_{t_0}^n, \ldots, W_{t_L}^n)$, that stands for the vertical positions of the particle at the times $0 = t_0 < t_1 < \cdots < t_L = 1$, if $n \geqslant n_0$ and $n \to \infty$?

Step 1: In Chap. 12 we show by the *Central Limit Theorem* that the laws of the random vectors $\omega \mapsto (W_{t_0}^n(\omega), \ldots, W_{t_L}^n(\omega))$ converge weakly to a Gaussian measure $\mu_{t_0,\ldots,t_L}$ on $\mathbb{R}^{L+1}$ as $n \to \infty$. Here it is remarkable that replacing the scaling factor $2^{n/2}$ in the denominator by 2^n would yield to the Strong Law of Large Numbers from the first example.

Step 2: We obtain a family of Gaussian measures $(\mu_{t_0,\ldots,t_L})$ indexed by the gridpoints $0 = t_0 < t_1 < \cdots < t_L = 1$, $L = 1, 2, \ldots$. In Sect. 20.7 we show with Ciesielski's theorem by means of functional analysis that this family uniquely defines a Gaussian probability measure, the Wiener measure on the Banach space $C([0, 1])^5$. In this way the Wiener measure gives a direct interpretation of the continuous movement of the particle we did start from. How do we proceed to obtain this Wiener measure? We will use the results from the chapter about probability in Banach spaces (Chap. 20) as follows:

- First we have to choose a σ-algebra on $E := C([0, 1])$. As the space E is a *separable* Banach space we can choose the σ-algebra generated by the norm-open sets. Here the separability of E is important as we learn from Theorem 20.1.6 that there are bad surprises in the case the Banach space E is not separable.
- Using random Schauder function expansions and Ciesielski's results about the representation of Hölder functions (Appendix A.9) we prove in Theo-

---

[5] The space $C([0, 1])$ stands for the space of continuous functions on the interval $[0, 1]$.

1 Introduction

rem 20.7.3 that there is an infinite dimensional Gaussian random variable $W : \Omega \to C[0, 1]$ such that, almost surely,

$$|W_t(\omega) - W_s(\omega)| \leqslant D(\omega)\sqrt{|t - s|\left(1 + \log_2 \frac{2}{|t - s|}\right)}$$

with $D : \Omega \to [0, \infty) \in L^{\Phi_2}$ and L^{Φ_2} is an Orlicz space considered in Proposition 19.2.7. The law of the random variable W is the Wiener measure.

- As the proof of Theorem 20.7.3 reveals that the random variable $W : \Omega \to C[0, 1]$ is the limit of a *sum of independent Gaussian random variables*, we have the connection to the celebrated Itô and Nisio Theorem 20.6.1 about the equivalence of different types of convergences of these sums.

Historical information about mathematicians can be found in the *MacTutor History of Mathematics Archive* under

https://mathshistory.st-andrews.ac.uk/,

and this source has been also used throughout this book.

Some Notation We close with some notation used in the sequel.

Set of all subsets of Ω	2^Ω	=	$\{A : A \subseteq \Omega\}$
Empty set	$\emptyset$	=	set, without any element
For $A, B \subseteq \Omega$:			
Intersection	$A \cap B$	=	$\{\omega \in \Omega : \omega \in A \text{ and } \omega \in B\}$
Union	$A \cup B$	=	$\{\omega \in \Omega : \omega \in A \text{ or } \omega \in B\}$
Set-theoretical minus	$A \setminus B$	=	$\{\omega \in \Omega : \omega \in A \text{ and } \omega \notin B\}$
Complement	A^c	=	$\{\omega \in \Omega : \omega \notin A\}$
Symmetric difference	$A \Delta B$	=	$(A \cup B) \setminus (A \cap B)$
Real numbers	$\mathbb{R}$		
Rational numbers	$\mathbb{Q}$		
Integers	$\mathbb{Z}$	=	$\{\ldots, -2, -1, 0, 1, 2, \ldots\}$
Natural numbers with 0	$\mathbb{N}_0$	=	$\{0, 1, 2, \ldots\}$
Natural numbers	$\mathbb{N}$	=	$\{1, 2, 3, \ldots\}$
Indicator function	$\mathbb{1}_A(x)$	=	$\begin{cases} 1 & \text{if } x \in A \\ 0 & \text{if } x \notin A \end{cases}$
Extended positive half-line	$[0, \infty]$	=	$[0, \infty) \cup \{\infty\}$
Extended real line	$[-\infty, \infty]$	=	$(-\infty, \infty) \cup \{-\infty, \infty\}$
Minimum	$\alpha \wedge \beta$	=	$\min\{\alpha, \beta\}$
Signum	$\text{sgn}(x)$	=	$\begin{cases} 1 & \text{if } x > 0 \\ 0 & \text{if } x = 0 \\ -1 & \text{if } x < 0 \end{cases}$
Ceiling function	$\lceil x \rceil$	=	smallest integer greater than or equal to $x \in \mathbb{R}$
Cardinality of a set	$\#A$	=	number of elements in A
$A, B \geqslant 0$ and $c \geqslant 1$	$A \sim_c B$	means	$\frac{1}{c} A \leqslant B \leqslant cA$

If nothing contrary is stated, we also use the

Convention A set A is said to be **countable** if either $\#A < \infty$ or A has the cardinality of $\mathbb{N}$. In other words, countable means finite or countably infinite.

Chapter 2
Measure Spaces and Probability Spaces

In this chapter we introduce measure spaces and, as a special case, the probability space—the basis of probability theory. Here we build on the abstract framework finally formulated by Kolmogorov in 1933. A measure space $(\Omega, \mathcal{F}, \mu)$ consists of three components:

- A non-empty set Ω which is called in probability theory the set of elementary events.
- The σ-algebra $\mathcal{F}$ which is a set of subsets of Ω. The elements of $\mathcal{F}$ are called events.
- The measure μ that assigns to each set of the σ-algebra $\mathcal{F}$ a non-negative number. In probability theory this models how likely it is that an $\omega \in \Omega$ belongs to an $A \in \mathcal{F}$.

Next we explain the three components in detail. We start with examples of Ω, the σ-algebras are introduced in Sect. 2.2, and measures in Sect. 2.3. We conclude with the lemmas of Fatou and Borel-Cantelli and with Bayes' rule.

2.1 The Elementary Events

In general, Ω is simply a non-empty set. In probability theory the elements $\omega \in \Omega$ are called **elementary events** or **states**. The set Ω is chosen according to the application one has in mind. Let us give some examples:

Example 2.1.1

(1) If we roll a dice, then the possible outcomes are the numbers between 1 and 6. We write

$$\Omega = \{1, 2, 3, 4, 5, 6\}.$$

(2) If we flip a coin, then we have either 'heads' or 'tails' showing, so we put

$$\Omega := \{H, T\}.$$

If we have two coins, then the set of all possible outcomes is

$$\Omega := \{(H, H), (H, T), (T, H), (T, T)\}.$$

(3) If we wish to model the life-time of an electric bulb, then one canonical choice is $\Omega := [0, \infty)$, where $\omega \in \Omega$ models the life-time.

2.2 σ-Algebras

The σ-algebra is a basic tool in measure and probability theory. It serves as the domain of definition of measures. For example, the concept of a σ-algebra is needed to introduce the Lebesgue[1] measure and Gaussian[2] measures on the real line, or the exponential distribution on the positive half-line.

Definition 2.2.1 (σ-Algebra, Algebra, Measurable Space) Let Ω be a non-empty set. A collection $\mathcal{F}$ of subsets $A \subseteq \Omega$ is called σ-**algebra** on Ω if

(1) $\emptyset, \Omega \in \mathcal{F}$,
(2) $A \in \mathcal{F}$ implies that $A^c = \Omega \setminus A \in \mathcal{F}$,
(3) $A_1, A_2, \ldots \in \mathcal{F}$ implies that $\bigcup_{i=1}^{\infty} A_i \in \mathcal{F}$.

The pair $(\Omega, \mathcal{F})$, where $\mathcal{F}$ is a σ-algebra on Ω, is called **measurable space**. If one replaces (3) by

(3') $A, B \in \mathcal{F}$ implies that $A \cup B \in \mathcal{F}$,

then $\mathcal{F}$ is called an **algebra**.

Remark 2.2.2

(1) In probability theory and statistics the sets $A \in \mathcal{F}$ are called **events**. One says that an event A occurs if $\omega \in A$ and that it does not occur if $\omega \notin A$. The idea behind this is the intuition that one usually does not know in which particular state $\omega \in \Omega$ a system is, but given an event $A \in \mathcal{F}$, one can decide at least whether $\omega \in A$ or $\omega \notin A$.

---

[1] Henri Léon Lebesgue, 28/06/1875 (Beauvais, Oise, Picardie, France)–26/07/1941 (Paris, France).
[2] Johann Carl Friedrich Gauss, 30/04/1777 (Brunswick, Duchy of Brunswick, now Germany)–23/02/1855 (Göttingen, Hanover, now Germany).

2.2 σ-Algebras

This way of thinking is in accordance with the properties of a σ-algebra: For example, if $\omega \in A$ can be decided, then $\omega \notin A$ can be decided as well, and if $\omega \in A_1, \omega \in A_2, \ldots$ can be decided, then $\omega \in \bigcup_{i=1}^{\infty} A_i$ can be also decided.

(2) One of the main aspects in Definition 2.2.1 is that we use an *infinite* union in (3), but do *not* allow uncountably many sets, only *countably many* sets.

(3) Sometimes, the terms σ-field and field are used instead of σ-algebra and algebra, respectively.

We state De Morgan's[3] rule which we will use frequently in the following:

Lemma 2.2.3 (De Morgan's Rule) *If Ω is non-empty, I is an arbitrary index set, and $A_i \subseteq \Omega$ for all $i \in I$, then it holds that*

$$\left(\bigcup_{i \in I} A_i\right)^c = \bigcap_{i \in I} A_i^c.$$

The proof is the subject of Exercise 1. The next proposition follows directly from Definition 2.2.1:

Proposition 2.2.4

(1) Every σ-algebra is an algebra.
(2) Given a measurable space $(\Omega, \mathcal{F})$ and $A, B, A_1, A_2, \cdots \in \mathcal{F}$, then

$$\bigcap_{i=1}^{\infty} A_i \in \mathcal{F} \quad \text{and} \quad A \setminus B \in \mathcal{F}.$$

Proof (1) In Definition 2.2.1 we have that (3) implies (3'). This can be seen by taking $A_1 = A$ and $A_2 = A_3 = \cdots = B$. (2) is an exercise (Exercise 2). □

We continue with some examples.

Example 2.2.5 (Smallest and Largest σ-Algebra) For any σ-algebra $\mathcal{F}$ on Ω one has

$$\{\emptyset, \Omega\} \subseteq \mathcal{F} \subseteq 2^{\Omega},$$

where 2^{Ω} stands for the power set, the set of all subsets of Ω. It is easy to see that 2^{Ω} and $\{\Omega, \emptyset\}$ are σ-algebras as well. Consequently, $\{\Omega, \emptyset\}$ and 2^{Ω} are the smallest and largest σ-algebras on Ω. A σ-algebra, different from those two, is simply obtained as follows: We assume $\#\Omega > 2$, choose $\emptyset \neq A \subsetneq \Omega$, and let

$$\mathcal{F} := \{\Omega, \emptyset, A, A^c\}.$$

Then $\{\emptyset, \Omega\} \subsetneq \mathcal{F} \subsetneq 2^{\Omega}$.

---

[3] Augustus De Morgan, 27/06/1806 (Madura, India)–18/03/1871 (London, England).

We list some more examples:

Example 2.2.6 (Rolling a Dice)
(1) Assume first the model for one dice, i.e. $\Omega := \{1, \ldots, 6\}$ and $\mathcal{F} := 2^\Omega$. Then, for example, the event 'the dice shows an even number' is described by $A = \{2, 4, 6\}$.
(2) Assume now a model with two dice, i.e. $\Omega := \{(k, l) : k, l = 1, \ldots, 6\}$. We want to describe the event 'the sum of the two dice equals four', i.e. $A = \{(1, 3), (2, 2), (3, 1)\}$. There are at least two canonical options to take an appropriate σ-algebra: Either the largest one $\mathcal{F}_1 := 2^\Omega$ or a smaller one

$$\mathcal{F}_2 := \left\{ A \subseteq \Omega : A = \bigcup_{m \in J} A_m, \ J \subseteq \{2, \ldots, 12\} \right\},$$

where $A_m := \{(k, l) : k + l = m\}$, and the convention $\bigcup_{m \in \emptyset} A_m := \emptyset$ is used. The σ-algebra $\mathcal{F}_2$ corresponds to the following model: One player throws two dice but only tells the other player the sum of the outcomes, not the values of the particular dice.

Example 2.2.7 (Heads and Tails)
(1) Assume a model for two coins, i.e. $\Omega := \{(H, H), (H, T), (T, H), (T, T)\}$ and $\mathcal{F} := 2^\Omega$. 'Exactly one of two coins shows heads' is modelled by

$$A = \{(H, T), (T, H)\}.$$

(2) Assume now that we have a game with infinitely many trials. As set of elementary events we take the set of all possible outcomes, i.e. $\Omega := \{(a_1, a_2, a_3, \ldots) : a_k \in \{T, H\}\}$. Imagine a game where player I wins, if there are 10 heads more than tails for the first time, and player II wins if there are 10 tails more than heads for the first time.

If Ω has finitely many elements, $\Omega = \{\omega_1, \ldots, \omega_n\}$, then any algebra $\mathcal{F}$ on Ω is automatically a σ-algebra. However, in general this is not the case as shown by the next example:

Example 2.2.8 (An Algebra Which Is Not a σ-Algebra) Let $\mathcal{A}$ be the collection of subsets $A \subseteq \mathbb{R}$ such that A can be written as

$$A = (a_1, b_1] \cup (a_2, b_2] \cup \cdots \cup (a_n, b_n]$$

or

$$A = (a_1, b_1] \cup (a_2, b_2] \cup \cdots \cup (a_n, \infty)$$

where $-\infty \leqslant a_1 \leqslant b_1 \leqslant \cdots \leqslant a_n \leqslant b_n < \infty$ with the convention that $(a,a] = \emptyset$. Then $\mathcal{A}$ is an algebra, but not a σ-algebra (Exercise 3).

2.3 Definition and First Properties of Measures

Before we introduce the notion of a measure, we define what we mean by a partition of Ω.

Definition 2.3.1 (Partition) Let $\Omega \neq \emptyset$ be a set and let $I \neq \emptyset$ be an index-set. A family $(\Omega_i)_{i \in I}$ with $\Omega_i \subseteq \Omega$ is called **partition** of Ω if

(1) $\Omega_i \cap \Omega_j = \emptyset$ for $i \neq j$,
(2) $\Omega = \bigcup_{i \in I} \Omega_i$.

Definition 2.3.2 (Measure Space, Probability Space) Let $(\Omega, \mathcal{F})$ be a measurable space.

(1) A map $\mu : \mathcal{F} \to [0, \infty]$ is called **measure** if

   (a) $\mu(\emptyset) = 0$,
   (b) σ-**additivity** holds, that is for all $A_1, A_2, \ldots \in \mathcal{F}$ with $A_i \cap A_j = \emptyset$ for $i \neq j$ one has

$$\mu\left(\bigcup_{i=1}^{\infty} A_i\right) = \sum_{i=1}^{\infty} \mu(A_i). \tag{2.1}$$

   The triplet $(\Omega, \mathcal{F}, \mu)$ is called **measure space**.
(2) A measure space $(\Omega, \mathcal{F}, \mu)$ or a measure μ is called **finite** if $\mu(\Omega) < \infty$.
(3) A measure space $(\Omega, \mathcal{F}, \mu)$ or a measure μ is called σ-**finite** provided that there exists a partition $(\Omega_k)_{k=1}^{\infty} \subseteq \mathcal{F}$ of Ω such that for all $k = 1, 2, \ldots$ it holds $\mu(\Omega_k) < \infty$.
(4) A finite measure μ on $(\Omega, \mathcal{F})$ is called **probability measure** or **distribution** if $\mu(\Omega) = 1$. Then the triplet $(\Omega, \mathcal{F}, \mu)$ is called **probability space**.

Remark 2.3.3

(1) If μ is a probability measure, then we usually write $\mathbb{P}$ instead of μ.
(2) In the definition of a measure μ the assumption $\mu(\emptyset) = 0$ can be omitted if there is at least one set $A \in \mathcal{F}$ with $\mu(A) < \infty$: indeed, we have the disjoint union $A = A \cup \bigcup_{i=1}^{\infty} \emptyset$ so that the σ-additivity (2.1) implies

$$\infty > \mu(A) = \mu(A) + \sum_{i=1}^{\infty} \mu(\emptyset)$$

and $\mu(\emptyset) = 0$. The only situation, where $\mu(\emptyset) > 0$ would be possible, is when $\mu(A) \equiv \infty$ for all $A \in \mathcal{F}$, so that we would have $\mu(\emptyset) = \infty$.

We continue with our previous examples:

Example 2.3.4 (Rolling a Dice) We proceed with Example 2.2.6 and assume the model of a dice, i.e. $\Omega = \{1, \ldots, 6\}$ and $\mathcal{F} = 2^\Omega$. Assuming that all outcomes for rolling a dice are equally likely, leads to the Ansatz

$$\mathbb{P}(\{\omega\}) := \frac{1}{6}.$$

Then, for example, $\mathbb{P}(\{2, 4, 6\}) = \frac{1}{2}$.

Example 2.3.5 (Heads and Tails) We proceed with Example 2.2.7 and assume two coins, i.e.

$$\Omega = \{(T, T), (H, T), (T, H), (H, H)\}$$

and $\mathcal{F} = 2^\Omega$. Then the intuitive assumption 'fair' leads to

$$\mathbb{P}(\{\omega\}) := \frac{1}{4} \quad \text{for all} \quad \omega \in \Omega.$$

That means, for example, the probability that exactly one of two coins shows heads is $\mathbb{P}(\{(H, T), (T, H)\}) = \frac{1}{2}$.

Other examples for measures are the Dirac and the counting measure.

Example 2.3.6 (Dirac[4] and Counting Measure) Let Ω be an arbitrary non-empty set and $\mathcal{F} = 2^\Omega$.

(1) **Dirac measure**: For a fixed $\omega_0 \in \Omega$ we let

$$\delta_{\omega_0}(A) := \begin{cases} 1 : \omega_0 \in A \\ 0 : \omega_0 \notin A \end{cases}.$$

(2) **Counting measure**: Define

$$\mu(A) := \#A,$$

the cardinality of the set A. This measure is σ-finite if and only if Ω is countable.

---

[4] Paul Adrien Maurice Dirac, 08/08/1902 (Bristol, England)–20/10/1984 (Tallahassee, Florida, USA).

2.3 Definition and First Properties of Measures

(3) Let $A_0 \subseteq \Omega$ with $\#A_0 > 1$, and let

$$\mu(A) := \begin{cases} 1 : A \cap A_0 \neq \emptyset \\ 0 : \text{otherwise} \end{cases}.$$

Then μ is *not* a measure (Exercise 11).

We close this section with some basic properties of measures:

Proposition 2.3.7 *Let $(\Omega, \mathcal{F}, \mu)$ be a measure space. Then the following assertions are true:*

(1) *If $A_1, \ldots, A_n \in \mathcal{F}$ are pairwise disjoint, then $\mu\left(\bigcup_{i=1}^n A_i\right) = \sum_{i=1}^n \mu(A_i)$.*
(2) *If $A, B \in \mathcal{F}$, then $\mu(B \cap A) + \mu(B \setminus A) = \mu(B)$. In particular, $\mu(A) \leq \mu(B)$ if $A \subseteq B$.*
(3) *If $A \in \mathcal{F}$ and μ is a probability measure, then $\mu(A^c) = 1 - \mu(A)$.*
(4) *If $A_1, A_2, \ldots \in \mathcal{F}$ then $\mu\left(\bigcup_{i=1}^\infty A_i\right) \leq \sum_{i=1}^\infty \mu(A_i)$.*
(5) **Continuity from below:** *If $A_1, A_2, \ldots \in \mathcal{F}$ such that $A_1 \subseteq A_2 \subseteq A_3 \subseteq \cdots$, then*

$$\lim_{n \to \infty} \mu(A_n) = \mu\left(\bigcup_{i=1}^\infty A_i\right).$$

(6) **Continuity from above:** *If $A_1, A_2, \ldots \in \mathcal{F}$ such that $A_1 \supseteq A_2 \supseteq A_3 \supseteq \cdots$ and $\mu(A_1) < \infty$, then*

$$\lim_{n \to \infty} \mu(A_n) = \mu\left(\bigcap_{i=1}^\infty A_i\right).$$

Proof (1) We let $A_{n+1} = A_{n+2} = \cdots = \emptyset$ and use $\mu(\emptyset) = 0$, to get

$$\mu\left(\bigcup_{i=1}^n A_i\right) = \mu\left(\bigcup_{i=1}^\infty A_i\right) = \sum_{i=1}^\infty \mu(A_i) = \sum_{i=1}^n \mu(A_i).$$

(2) Since $(B \cap A) \cap (B \setminus A) = \emptyset$, we get that

$$\mu(B \cap A) + \mu(B \setminus A) = \mu((B \cap A) \cup (B \setminus A)) = \mu(B).$$

(3) We apply (2) to $B = \Omega$ and exploit that $\mu(\Omega) = 1$.
(4) Put $B_1 := A_1$ and $B_i := A_i \setminus (A_1 \cup \cdots \cup A_{i-1})$ for $i = 2, 3, \ldots$ Obviously, $\mu(B_i) \leq \mu(A_i)$ for all $i \geq 1$ because $B_i \subseteq A_i$. Since the sets B_i are pairwise disjoint and $\bigcup_{i=1}^\infty A_i = \bigcup_{i=1}^\infty B_i$ it follows that

$$\mu\left(\bigcup_{i=1}^\infty A_i\right) = \mu\left(\bigcup_{i=1}^\infty B_i\right) = \sum_{i=1}^\infty \mu(B_i) \leq \sum_{i=1}^\infty \mu(A_i).$$

(5) We define $B_1 := A_1$, $B_2 := A_2 \setminus A_1$, $B_3 := A_3 \setminus A_2$, $B_4 := A_4 \setminus A_3$, ... and get that

$$\bigcup_{i=1}^{\infty} B_i = \bigcup_{i=1}^{\infty} A_i \quad \text{and} \quad B_i \cap B_j = \emptyset \quad \text{for} \quad i \neq j.$$

Consequently,

$$\mu\left(\bigcup_{i=1}^{\infty} A_i\right) = \mu\left(\bigcup_{i=1}^{\infty} B_i\right) = \sum_{i=1}^{\infty} \mu(B_i) = \lim_{n \to \infty} \sum_{i=1}^{n} \mu(B_i) = \lim_{n \to \infty} \mu(A_n)$$

since $\bigcup_{i=1}^{n} B_i = A_n$.
(6) This is the subject of Exercise 4. □

2.4* The Completion of Measure Spaces

In certain applications one needs the following additional property of a measure space:

Definition 2.4.1 (Complete Measure Spaces) A measure space $(\Omega, \mathcal{F}, \mu)$ is called **complete** provided that $A \in \mathcal{F}$, $\mu(A) = 0$, and $B \subseteq A$ imply that $B \in \mathcal{F}$.

A given arbitrary measure space $(\Omega, \mathcal{F}, \mu)$ can be completed by the following procedure:

Proposition 2.4.2 *For a measure space* $(\Omega, \mathcal{F}, \mu)$ *define* $\overline{\mathcal{F}}^\mu$ *to be the collection of all sets* $B \subseteq \Omega$ *such that there are* $A, C \in \mathcal{F}$ *with* $A \subseteq B \subseteq C$ *and* $\mu(C \setminus A) = 0$. *Moreover, define* $\overline{\mu}(B) := \mu(A) = \mu(C)$. *Then the following assertions hold true:*

(1) $\overline{\mu}$ *is well-defined.*
(2) $(\Omega, \overline{\mathcal{F}}^\mu, \overline{\mu})$ *is a complete measure space.*

The proof is the subject of Exercise 15.

2.5 Infinitely Often Occurring Events and a First Look at Independence: The Lemmas of Fatou and Borel-Cantelli

We start with the notion of the *limit superior* and *limit inferior* of sets (for the limit inferior and superior of a sequence of real numbers see Appendix A.2) that are used to describe infinitely often occurring events:

2.5 Infinitely Often Occurring Events and a First Look at Independence: The...

Definition 2.5.1 ($\liminf_{n\to\infty} A_n$ **and** $\limsup_{n\to\infty} A_n$) Let Ω be non-empty and $A_1, A_2, \ldots$ be subsets of Ω. Then

$$\liminf_{n\to\infty} A_n := \bigcup_{n=1}^{\infty}\left(\bigcap_{i=n}^{\infty} A_i\right) \quad \text{and} \quad \limsup_{n\to\infty} A_n := \bigcap_{n=1}^{\infty}\left(\bigcup_{i=n}^{\infty} A_i\right),$$

where we will also use the shorter notation $\bigcup_{n=1}^{\infty} \bigcap_{i=n}^{\infty} A_i$ and $\bigcap_{n=1}^{\infty} \bigcup_{i=n}^{\infty} A_i$.
One can understand these expressions as

$$\omega \in \bigcup_{n=1}^{\infty}\left(\bigcap_{i=n}^{\infty} A_i\right) \iff \exists n \in \mathbb{N} \text{ such that } \omega \in \bigcap_{i=n}^{\infty} A_i$$
$$\iff \exists n \in \mathbb{N} \ \forall i \geq n \text{ one has } \omega \in A_i$$

and

$$\omega \in \bigcap_{n=1}^{\infty}\left(\bigcup_{i=n}^{\infty} A_i\right) \iff \forall n \geq 1 \text{ one has } \omega \in \bigcup_{i=n}^{\infty} A_i$$
$$\iff \forall n \geq 1 \ \exists i \geq n \text{ such that } \omega \in A_i.$$

The following lemma gives the interpretation of $\liminf_{n\to\infty} A_n$ and $\limsup_{n\to\infty} A_n$:

Lemma 2.5.2

(1) We have that $\omega \in \liminf_{n\to\infty} A_n$ if and only if there is an $n \geq 1$ such that for all $i \geq n$ we have $\omega \in A_i$.
(2) We have that $\omega \in \limsup_{n\to\infty} A_n$ if and only if ω belongs to infinitely many of the A_n.
(3) $(\liminf_{n\to\infty} A_n)^c = \limsup_{n\to\infty} A_n^c$.

Proof (1) follows directly from the definition and (3) by the De Morgan's rules. Regarding (2) we observe that, again by the definition, we have that

$$\omega \in \limsup_{n\to\infty} A_n \iff \text{for all } n \geq 1 \text{ there is an } i \geq n \text{ such that } \omega \in A_i.$$

Hence, if ω belongs to infinitely many of the A_n, then this condition is satisfied. Conversely, assume that

$$\text{for all } n \geq 1 \text{ there is an } i \geq n \text{ such that } \omega \in A_i.$$

We start by $n_1 = 1$ and find an $i_1 \geq 1$ with $\omega \in A_{i_1}$. Then, letting $n_2 := i_1 + 1$, we find an $i_2 \geq n_2$ with $\omega \in A_{i_2}$. Continuing this construction, we find $1 \leq i_1 < i_2 < \cdots$ with $\omega \in A_{i_k}$ for $k \geq 1$. □

Theorem 2.5.3 (Lemma of Fatou[5]) *Let $(\Omega, \mathcal{F}, \mathbb{P})$ be a probability space and $A_1, A_2, \ldots \in \mathcal{F}$. Then*

$$\mathbb{P}\left(\liminf_{n\to\infty} A_n\right) \leqslant \liminf_{n\to\infty} \mathbb{P}(A_n) \leqslant \limsup_{n\to\infty} \mathbb{P}(A_n) \leqslant \mathbb{P}\left(\limsup_{n\to\infty} A_n\right).$$

Proof From Proposition 2.3.7(5) we conclude

$$\mathbb{P}\left(\liminf_{n\to\infty} A_n\right) = \mathbb{P}\left(\bigcup_{n=1}^{\infty}\bigcap_{i=n}^{\infty} A_i\right) = \lim_{n\to\infty} \mathbb{P}\left(\bigcap_{i=n}^{\infty} A_i\right), \tag{2.2}$$

where we use that $\bigcap_{i=n}^{\infty} A_i \subseteq \bigcap_{i=n+1}^{\infty} A_i$. Then the first inequality of the assertion follows from

$$\lim_{n\to\infty} \mathbb{P}\left(\bigcap_{i=n}^{\infty} A_i\right) = \liminf_{n\to\infty} \mathbb{P}\left(\bigcap_{i=n}^{\infty} A_i\right) \leqslant \liminf_{n\to\infty} \mathbb{P}(A_n).$$

The second inequality is the general relation between $\liminf_{n\to\infty} a_n$ and $\limsup_{n\to\infty} a_n$ from (A.2). The last inequality follows if we apply the first relation to the sequence of complements $(A_n^c)_{n=1}^{\infty}$ and use Lemma 2.5.2(3), relation (A.1), and $\mathbb{P}(A^c) = 1 - \mathbb{P}(A)$ for $A \in \mathcal{F}$. □

A natural question is whether we might have a strict inequality

$$\limsup_{n\to\infty} \mathbb{P}(A_n) < \mathbb{P}\left(\limsup_{n\to\infty} A_n\right).$$

For this we give an example after Corollary 2.5.8 below. A first consequence of the Lemma of Fatou is the following statement:

Corollary 2.5.4 *For a probability space $(\Omega, \mathcal{F}, \mathbb{P})$ and $A_1, A_2, \ldots \in \mathcal{F}$ one has that*

(1) $\liminf_{n\to\infty} \mathbb{P}(A_n) = 0$ *implies* $\mathbb{P}\left(\liminf_{n\to\infty} A_n\right) = 0$,
(2) $\limsup_{n\to\infty} \mathbb{P}(A_n) = 1$ *implies* $\mathbb{P}\left(\limsup_{n\to\infty} A_n\right) = 1$.

Now we turn to the fundamental notion of independence.

Definition 2.5.5 (Independence of Events) Let $(\Omega, \mathcal{F}, \mathbb{P})$ be a probability space and $I \neq \emptyset$ be an arbitrary index set. The events $(A_i)_{i \in I} \subseteq \mathcal{F}$ are called **independent**, provided that for all $n \geqslant 2$ and distinct $i_1, \ldots, i_n \in I$ one has that

$$\mathbb{P}\left(A_{i_1} \cap A_{i_2} \cap \cdots \cap A_{i_n}\right) = \mathbb{P}\left(A_{i_1}\right)\mathbb{P}\left(A_{i_2}\right) \cdots \mathbb{P}\left(A_{i_n}\right).$$

---

[5] Pierre Joseph Louis Fatou, 28/02/1878 (Lorient, France)–09/08/1929 (Pornichet, France).

2.5 Infinitely Often Occurring Events and a First Look at Independence: The...

The above definition is a probabilistic approach to model one aspect of the intuitive notion of independence from daily life. For example, inviting two friends A and B, where A comes with the probability $p \in (0, 1)$ and B with the probability $q \in (0, 1)$, which is already a model to express the fact how likely it is that they come, and assuming that both friends do not have any contact nor are connected by outer circumstances, i.e. they are 'independent', the probability that both come is intuitively $pq \in (0, 1)$.

Remark 2.5.6

(1) It follows directly from the definition that the sets $(A_i)_{i \in J}$ are independent if $\emptyset \neq J \subseteq I$ provided that the sets $(A_i)_{i \in I}$ are independent.
(2) The definition of independence needs to be handled with care: Assume a probability space $(\Omega, \mathcal{F}, \mathbb{P})$ and $I = \{1, \ldots, n\}$ with $n \geq 3$.

  (a) If any two sets of a collection of sets are independent we call this **pairwise independence**. Pairwise independence is not sufficient for independence: As we will see in Exercise 12, we can find sets $A_1, A_2, A_3 \in \mathcal{F}$ such that A_i and A_j are independent for $i \neq j$ from $\{1, 2, 3\}$, but A_1, A_2, A_3 are not independent.
  (b) It is not sufficient to check only all sets: Assume any sets $A_1, \ldots, A_n \in \mathcal{F}$ with $A_1 = \emptyset$, then we always have that

  $$0 = \mathbb{P}(A_1 \cap A_2 \cap \cdots \cap A_n) = \mathbb{P}(A_1) \cdots \mathbb{P}(A_n)$$

  even if $A_2, \ldots, A_n$ are not independent.

Now we turn to the fundamental Lemma of Borel-Cantelli:

Theorem 2.5.7 (Lemma of Borel-Cantelli[6]) *Let $(\Omega, \mathcal{F}, \mathbb{P})$ be a probability space and $A_1, A_2, \ldots \in \mathcal{F}$. Then one has the following:*

(1) *If $\sum_{n=1}^{\infty} \mathbb{P}(A_n) < \infty$, then $\mathbb{P}\left(\limsup_{n \to \infty} A_n\right) = 0$.*
(2) *If $A_1, A_2, \ldots$ are independent and $\sum_{n=1}^{\infty} \mathbb{P}(A_n) = \infty$, then*

$$\mathbb{P}\left(\limsup_{n \to \infty} A_n\right) = 1.$$

Consequently, if $A_1, A_2, \ldots$ are independent, then

$$\sum_{n=1}^{\infty} \mathbb{P}(A_n) < \infty \iff \mathbb{P}\left(\limsup_{n \to \infty} A_n\right) = 0$$

---

[6] Francesco Paolo Cantelli, 20/12/1875 (Palermo, Sicily, Italy)–1/07/1966 (Rome, Italy).

and

$$\sum_{n=1}^{\infty} \mathbb{P}(A_n) = \infty \iff \mathbb{P}\left(\limsup_{n\to\infty} A_n\right) = 1.$$

The conditions $\sum_{n=1}^{\infty} \mathbb{P}(A_n) < \infty$ and $\sum_{n=1}^{\infty} \mathbb{P}(A_n) = \infty$, respectively, are conditions on how quickly the probabilities of the events A_n get small and to what extent there are sufficiently many events with a large probability, respectively.

As a by-product of the Lemma of Borel-Cantelli we obtain a zero-one law:

Corollary 2.5.8 (Zero-One Law) *For a probability space $(\Omega, \mathcal{F}, \mathbb{P})$ and independent events $A_1, A_2, \cdots \in \mathcal{F}$ one has*

$$\mathbb{P}\left(\limsup_{n\to\infty} A_n\right) \in \{0, 1\}.$$

Using this corollary we get a simple family of examples for the strict inequality

$$\limsup_{n\to\infty} \mathbb{P}(A_n) < \mathbb{P}\left(\limsup_{n\to\infty} A_n\right)$$

as follows: If we take independent events $A_1, A_2, \ldots \mathcal{F}$ such that $\mathbb{P}(A_n) = \varepsilon \in (0, 1)$, then

$$\limsup_{n\to\infty} \mathbb{P}(A_n) = \varepsilon < 1 = \mathbb{P}\left(\limsup_{n\to\infty} A_n\right).$$

Here the question arises whether such an infinite sequence of independent events $A_1, A_2, \ldots$ exists. The existence will be shown in Corollary 7.5.1 by constructing suitable probability spaces.

Proof of Theorem 2.5.7 (1) The *continuity from above* (Proposition 2.3.7(6)) allows to conclude that

$$\mathbb{P}\left(\limsup_{n\to\infty} A_n\right) = \mathbb{P}\left(\bigcap_{n=1}^{\infty}\bigcup_{i=n}^{\infty} A_i\right) = \lim_{n\to\infty} \mathbb{P}\left(\bigcup_{i=n}^{\infty} A_i\right) \leqslant \lim_{n\to\infty} \sum_{i=n}^{\infty} \mathbb{P}(A_i) = 0,$$

where the last inequality follows from Proposition 2.3.7(4).

(2) It holds that

$$\left(\limsup_{n\to\infty} A_n\right)^c = \liminf_{n\to\infty} A_n^c = \bigcup_{n=1}^{\infty}\bigcap_{i=n}^{\infty} A_i^c.$$

From (2.2) we have that

$$\mathbb{P}\left(\bigcup_{n=1}^{\infty}\bigcap_{i=n}^{\infty} A_i^c\right) = \lim_{n\to\infty}\mathbb{P}\left(\bigcap_{i=n}^{\infty} A_i^c\right).$$

Hence to get $\mathbb{P}\left((\limsup_{n\to\infty} A_n)^c\right) = 0$ it suffices to show that

$$\mathbb{P}\left(\bigcap_{i=n}^{\infty} A_i^c\right) = 0.$$

Since the independence of $A_1, A_2, \ldots$ implies the independence of $A_1^c, A_2^c, \ldots$ (Exercise 7), we get (setting $p_n := \mathbb{P}(A_n)$) that

$$\mathbb{P}\left(\bigcap_{i=n}^{\infty} A_i^c\right) = \lim_{N\to\infty, N\geqslant n}\mathbb{P}\left(\bigcap_{i=n}^{N} A_i^c\right) = \lim_{N\to\infty, N\geqslant n}\prod_{i=n}^{N}\mathbb{P}\left(A_i^c\right)$$

$$= \lim_{N\to\infty, N\geqslant n}\prod_{i=n}^{N}(1-p_i) \leqslant \lim_{N\to\infty, N\geqslant n}\prod_{i=n}^{N} e^{-p_i}$$

$$= \lim_{N\to\infty, N\geqslant n} e^{-\sum_{i=n}^{N} p_i} = e^{-\sum_{i=n}^{\infty} p_i}$$

$$= 0$$

because of $\sum_{i=n}^{\infty} p_i = \infty$, where we have used that $1 - x \leqslant e^{-x}$ for $x \in \mathbb{R}$. □

2.6 A First Look at Conditional Probabilities and Bayes' Rule

Independence can be also expressed in terms of conditional probabilities. A conditional probability describes the probability of an event B given the event A occurs:

Definition 2.6.1 (Conditional Probability) Let $(\Omega, \mathcal{F}, \mathbb{P})$ be a probability space and $A \in \mathcal{F}$ with $\mathbb{P}(A) > 0$. Then

$$\mathbb{P}(B|A) := \frac{\mathbb{P}(B \cap A)}{\mathbb{P}(A)} \quad \text{for} \quad B \in \mathcal{F},$$

is called **conditional probability of B given A**.

The construction is justified by the following properties:

Proposition 2.6.2 *Let $(\Omega, \mathcal{F}, \mathbb{P})$ be a probability space and $A \in \mathcal{F}$ with $\mathbb{P}(A) > 0$. Then the following holds:*

(1) *For $\mathbb{P}_A(B) := \mathbb{P}(B|A)$, $B \in \mathcal{F}$, the triplet $(\Omega, \mathcal{F}, \mathbb{P}_A)$ is a probability space.*
(2) *For $B \in \mathcal{F}$ we have that A is independent from B if and only if $\mathbb{P}(B|A) = \mathbb{P}(B)$.*

The proof is the subject of Exercise 10. Note that there is no requirement that the event A occurs somehow before B in time. For example, for rolling two dice one can consider

$$\mathbb{P}(B|A) = \mathbb{P}(\text{first dice shows } 3 \mid \text{outcome of the sum of the two dice is } 7).$$

Now we formulate Bayes' formula:

Theorem 2.6.3 (Bayes'[7] Formula) *Let $(\Omega, \mathcal{F}, \mathbb{P})$ be a probability space, $\Omega = \bigcup_{i=1}^{n} B_i$ with $n \geq 2$, $B_i \in \mathcal{F}$, $B_i \cap B_j = \emptyset$ for $i \neq j$, and $\mathbb{P}(B_i) > 0$ for $i = 1, \ldots, n$. Then, for $A \in \mathcal{F}$ with $\mathbb{P}(A) > 0$ one has*

$$\mathbb{P}(B_i|A) = \frac{\mathbb{P}(A|B_i)\mathbb{P}(B_i)}{\sum_{k=1}^{n} \mathbb{P}(A|B_k)\mathbb{P}(B_k)}.$$

Proof The formula follows from

$$\frac{\mathbb{P}(A|B_i)\mathbb{P}(B_i)}{\sum_{k=1}^{n} \mathbb{P}(A|B_k)\mathbb{P}(B_k)} = \frac{\mathbb{P}(A \cap B_i)}{\sum_{k=1}^{n} \mathbb{P}(A \cap B_k)} = \frac{\mathbb{P}(A \cap B_i)}{\mathbb{P}(A)}.$$

□

Before we explain some terminology related to Bayes' formula let us consider an example:

Example 2.6.4 We apply Theorem 2.6.3 for $n = 2$, $B_1 = B$, and $B_2 = B^c$, and get that

$$\mathbb{P}(B|A) = \frac{\mathbb{P}(A|B)\mathbb{P}(B)}{\mathbb{P}(A|B)\mathbb{P}(B) + \mathbb{P}(A|B^c)\mathbb{P}(B^c)}.$$

A laboratory blood test is 95% effective in detecting a certain disease when it is, in fact, present. However, the test also yields a 'false positive' result for 1% of the healthy persons tested. If 0.5% of the population actually has the disease, what is the probability a person has the disease given his test result is positive? We set

$$B := \text{'the person has the disease'},$$
$$A := \text{'the test result is positive'}.$$

---
[7] Thomas Bayes, 1702 (London, England)–17/04/1761 (Tunbridge Wells, Kent, England).

2.7 Exercises

Hence we have

$$\mathbb{P}(A|B) = \mathbb{P}(\text{'a positive test result'}|\text{'person has the disease'}) = 0.95,$$
$$\mathbb{P}(A|B^c) = 0.01,$$
$$\mathbb{P}(B) = 0.005.$$

Applying the above formula gives that

$$\mathbb{P}(B|A) = \frac{0.95 \times 0.005}{0.95 \times 0.005 + 0.01 \times 0.995} \approx 0.323.$$

That means only 32% of the persons whose test results are positive actually have the disease.

In Theorem 2.6.3 an event B_j is also called **hypothesis** (which one wishes to test), the probabilities $\mathbb{P}(B_j)$ are called the **prior probabilities** (or *a priori probabilities*), and the probabilities $\mathbb{P}(B_j|A)$ the **posterior probabilities** (or *a posteriori probabilities*) of B_j.

2.7 Exercises

Ex 1: Let Ω be a non-empty set.

(a) For $A, B \subseteq \Omega$ show that $(A \setminus B) \cup (B \setminus A) = (A \cup B) \setminus (A \cap B)$. The expression $(A \cup B) \setminus (A \cap B) =: A \Delta B$ is called the symmetric difference of A and B.

(b) Prove De Morgan's law $\left(\bigcup_{i \in I} A_i \right)^c = \bigcap_{i \in I} A_i^c$ where $A_i \subseteq \Omega$ and I is an arbitrary index set.

Ex 2: Verify Proposition 2.2.4(2): For a σ-algebra $\mathcal{F}$ and $A, B, A_1, A_2, \ldots \in \mathcal{F}$ prove that $\bigcap_{i=1}^{\infty} A_i \in \mathcal{F}$ and $A \setminus B \in \mathcal{F}$. What if $\mathcal{F}$ is only an algebra?

Ex 3: Verify Example 2.2.8: Let $\mathcal{A}$ be the collection of subsets $A \subseteq \mathbb{R}$ such that A can be written as

$$A = (a_1, b_1] \cup (a_2, b_2] \cup \cdots \cup (a_n, b_n]$$

or

$$A = (a_1, b_1] \cup (a_2, b_2] \cup \cdots \cup (a_n, \infty),$$

where $n \in \mathbb{N}$ and $-\infty \leqslant a_1 \leqslant b_1 \leqslant \cdots \leqslant a_n \leqslant b_n < \infty$ with the convention that $(a, a] = \emptyset$. Show that $\mathcal{A}$ is an algebra but not a σ-algebra.

Ex 4: Prove Proposition 2.3.7(6): Let $(\Omega, \mathcal{F}, \mu)$ be a measure space and assume that $A_1, A_2, \ldots \in \mathcal{F}$ are such that $A_1 \supseteq A_2 \supseteq A_3 \supseteq \cdots$.

(a) Assume there exists some $n_0 \in \mathbb{N}$ with $\mu(A_{n_0}) < \infty$. Show that

$$\lim_{n \to \infty} \mu(A_n) = \mu\left(\bigcap_{n=1}^{\infty} A_n\right).$$

Why is this equivalent with the formulation in Proposition 2.3.7(6)?

(b) Find an example of $A_1, A_2, \ldots \in \mathcal{F}$ with $A_1 \supseteq A_2 \supseteq A_3 \supseteq \cdots$ and $\mu(A_n) = \infty$ for all $n \in \mathbb{N}$ such that

$$\lim_{n \to \infty} \mu(A_n) \neq \mu\left(\bigcap_{n=1}^{\infty} A_n\right).$$

Ex 5: Let $(\Omega, \mathcal{F}, \mathbb{P})$ be a probability space. For $A_1, A_2 \in \mathcal{F}$ we know the relation

$$\mathbb{P}(A_1 \cup A_2) = \mathbb{P}(A_1) + \mathbb{P}(A_2) - \mathbb{P}(A_1 \cap A_2).$$

Complete and prove the formula for the probability of the union of three sets $A_1, A_2, A_3 \in \mathcal{F}$:

$$\mathbb{P}(A_1 \cup A_2 \cup A_3) = \mathbb{P}(A_1) + \cdots + \mathbb{P}(A_1 \cap A_2 \cap A_3).$$

Find such a formula for the probability of the union of $A_1, \ldots, A_n \in \mathcal{F}$ where $n \in \mathbb{N}$ is arbitrary. This is known as the *inclusion-exclusion principle*.

Ex 6: (a) Let $A_1, A_2, A_3, \ldots \subseteq \Omega$. Prove that $\liminf_{n \to \infty} A_n \subseteq \limsup_{n \to \infty} A_n$.

(b) For $n = 1, 2, \ldots$ let $A_n := [(-1)^n, 1 + 1/n)$. Find $\liminf_{n \to \infty} A_n$ and $\limsup_{n \to \infty} A_n$.

(c) For $n = 1, 2, \ldots$ let $A_n := [(-1)^n, 2 - 1/n)$. Find $\liminf_{n \to \infty} A_n$ and $\limsup_{n \to \infty} A_n$.

Ex 7: Let $(\Omega, \mathcal{F}, \mathbb{P})$ be a probability space and $A_1, \ldots, A_n \in \mathcal{F}$ be independent events. Prove that $A_1^c, \ldots, A_n^c$ are independent.

Ex 8: Verify the inequality $1 + x \leq e^x$, $x \in \mathbb{R}$, used in the proof of Theorem 2.5.7.

Ex 9: Assume that a self-learning robot has to try to carry out certain tasks. Denote for $i \in \mathbb{N}$ the event

$$A_i := \text{'the robot has done the task } i \text{ successfully'}.$$

Since the robot is self-learning we assume that

$$\lim_{i \to \infty} \mathbb{P}(A_i) = 1.$$

2.7 Exercises

Can one compute the probability that the robot carries out infinitely many tasks successfully?

Ex 10: Prove Proposition 2.6.2: Let $(\Omega, \mathcal{F}, \mathbb{P})$ be a probability space and $A \in \mathcal{F}$ with $\mathbb{P}(A) > 0$. Define $\mathbb{P}_A(B) := \mathbb{P}(B|A)$ for $B \in \mathcal{F}$.

(a) Prove that $(\Omega, \mathcal{F}, \mathbb{P}_A)$ is a probability space.
(b) Prove that A is independent from B if and only if $\mathbb{P}(B|A) = \mathbb{P}(B)$.

Ex 11: The 'wrong' Dirac measure: Assume a set $\Omega \neq \emptyset$ and a subset $A_0 \subseteq \Omega$ that contains at least two elements. For $\mathcal{F} := 2^\Omega$ and $A \in \mathcal{F}$ define $\mu(A) := 1$ if $A \cap A_0 \neq \emptyset$ and $\mu(A) := 0$ if $A \cap A_0 = \emptyset$. Is $\mu : \mathcal{F} \to [0, 1]$ a measure?

Ex 12: Pairwise independence does not imply independence:

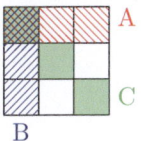

Let $\Omega := \{1, \ldots, 9\}$ and $\mathcal{F} := 2^\Omega$. Identify Ω with the 9 sub-squares in the picture. Find a probability measure on $(\Omega, \mathcal{F})$ such that

- A, B, C are pairwise independent,
- but A, B, C are not independent.

Ex 13: A probability space which is too small for independence: We assume the discrete uniform distribution on $\Omega = \{1, 2, 3, 4\}$, i.e. for any $A \in \mathcal{F} = 2^\Omega$ we have $\mathbb{P}(A) = \frac{\#A}{4}$. Then $B := \{1, 2\}$ and $C := \{2, 3\}$ are independent. Is there a set $D \in 2^\Omega$ such that $D \notin \{\emptyset, \Omega\}$ and B, C, D are independent?

Ex 14: Let $(\Omega, \mathcal{F}, \mathbb{P})$ be a probability space, $A_1, A_2, A_3, \ldots \in \mathcal{F}$ be independent with $\mathbb{P}(A_n) = p_n \in (0, 1)$.

(a) Prove that
$$\mathbb{P}(\liminf_{n \to \infty} A_n) = \lim_{N \to \infty} e^{-\sum_{n=N}^{\infty} \log \frac{1}{p_n}}.$$

(b) Deduce that $\mathbb{P}(\liminf_{n \to \infty} A_n) \in \{0, 1\}$ and that the probability is 1 if and only if $\sum_{n=1}^{\infty} \log \frac{1}{p_n} < \infty$.

Ex 15: Prove Proposition 2.4.2: For a measure space $(\Omega, \mathcal{F}, \mu)$ define $\overline{\mathcal{F}}^\mu$ to be the collection of all sets $B \subseteq \Omega$ such that there are $A, C \in \mathcal{F}$ with $A \subseteq B \subseteq C$ and $\mu(C \setminus A) = 0$, and set $\overline{\mu}(B) := \mu(A) = \mu(C)$. Verify the following:

(a) $\overline{\mu}$ is well-defined.
(b) $(\Omega, \overline{\mathcal{F}}^\mu, \overline{\mu})$ is a complete measure space.

2.8 Comments

Sections 2.2 and 2.3: The axiomatic approach to probability spaces is due to Kolmogorov [105] from 1933. He also refers to earlier approaches due to Bernstein [12] and von Mises [132].

Section 2.5: Regarding the Lemma of Fatou for sets (Theorem 2.5.3) see the remarks for Theorem 8.5.1. The Borel-Cantelli Lemma Theorem 2.5.7 is a result of several contributions: Borel [24, page 252], Hausdorff [82, page 419–421] (within the proof of a strong law of large numbers for dyadic expansions due to Borel), and finally Cantelli [29, 30]. We learnt this from Dudley [50, page 276] and the reader is referred for further information to this beautiful book.

Chapter 3
Construction of Measure Spaces

Distributions or, more general, measures which can be modelled on $\Omega \subseteq \mathbb{N}_0$ are often called discrete. We start this chapter with a list of common examples of discrete measures. The main part of the chapter is devoted to the rigorous definition of measures and distributions where $\Omega = \mathbb{R}$, for example the Lebesgue measure and the Gaussian measure. We will learn that those measures need a smaller domain of definition than $2^\mathbb{R}$. This requires the introduction of the Borel σ-algebra and the usage of the extension theorem stating a unique extension of a suitable pre-measure on an algebra to a measure on the σ-algebra generated by the algebra.

3.1 Measures on $\{0, \ldots, n\}$ and $\mathbb{N}_0$

The basis for the construction of measures is the following simple fact whose proof we leave to the reader.

Lemma 3.1.1 *Let $\Omega \neq \emptyset$ be countable (finite or countably infinite), let $\mathcal{F} := 2^\Omega$, and assume weights $p_\omega \in [0, \infty]$. For $B \subseteq \Omega$ define*

$$\mu(B) := \sum_{\omega \in B} p_\omega = \sum_{\omega \in \Omega} p_\omega \delta_\omega(B),$$

where

$$\delta_\omega(B) = \begin{cases} 1 : \omega \in B \\ 0 : \omega \notin B \end{cases}$$

is the Dirac measure introduced in Example 2.3.6. Then $(\Omega, \mathcal{F}, \mu)$ is a measure space.

3.1.1 Binomial Distribution with Parameters $n \in \mathbb{N}$ and $0 < p < 1$

(1) $\Omega := \{0, 1, \ldots, n\}$.
(2) $\mathcal{F} := 2^{\Omega}$.
(3) $\mathbb{P}(B) = \text{Bin}_{n,p}(B) := \sum_{k=0}^{n} \binom{n}{k} p^k (1-p)^{n-k} \delta_k(B)$.

Proposition 3.1.2 *For $0 < p < 1$ the triplet $(\{0, \ldots, n\}, 2^{\{0,\ldots,n\}}, \text{Bin}_{n,p})$ is a probability space.*

Proof In view of Lemma 3.1.1 we only need to verify $\text{Bin}_{n,p}(\Omega) = 1$. But this follows from the binomial theorem as

$$\text{Bin}_{n,p}(\Omega) = \sum_{k=0}^{n} \binom{n}{k} p^k (1-p)^{n-k} \delta_k(\Omega)$$

$$= \sum_{k=0}^{n} \binom{n}{k} p^k (1-p)^{n-k} = (p + (1-p))^n = 1.$$

□

The function $\{0, 1, \ldots, n\} \ni k \mapsto \text{Bin}_{n,p}(\{k\})$ is called *probability mass function* (Fig. 3.1).

Interpretation The binomial distribution $\text{Bin}_{n,p}(\{k\})$ describes the probability of k successes out of n independent trials, where the probability of success of each trial is $p \in (0, 1)$.

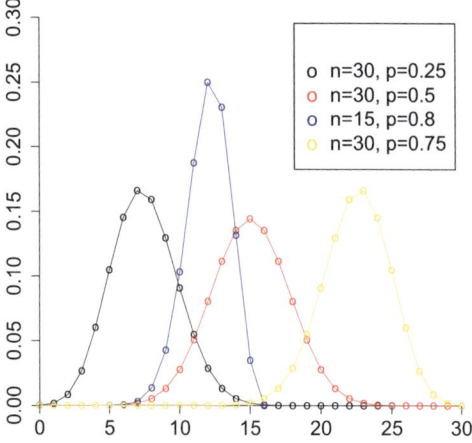

Fig. 3.1 Binomial distribution: probability mass functions

3.1 Measures on $\{0, \ldots, n\}$ and $\mathbb{N}_0$

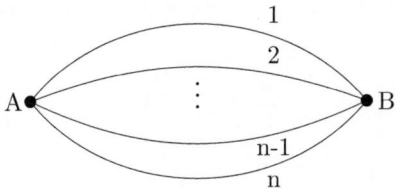

As an example we consider n communication channels between locations A and B. Each channel has a communication rate $\rho > 0$ (say ρ bits per second), which is assumed to yield to the rate ρk, in case k channels are working. Each channel independently works with the probability $p \in (0, 1)$ and fails with a probability $1 - p \in (0, 1)$, so that we have a random communication rate $R \in \{0, \rho, \ldots, n\rho\}$.

What is an appropriate mathematical model for this? We start with

$$\Omega := \{\omega = (\varepsilon_1, \ldots, \varepsilon_n) : \varepsilon_1 \in \{0, 1\}, \ldots, \varepsilon_n \in \{0, 1\}\}$$

and interpret $\varepsilon_i = 1$ that the i-th event occurs (success), and $\varepsilon_i = 0$ that the i-th event does not occur (failure). As σ-algebra we take $\mathcal{F} := 2^\Omega$. Finally, we let, for $\omega = (\varepsilon_1, \ldots, \varepsilon_n)$,

$$p_\omega := p^{\#\{i \in \{1,\ldots,n\} : \varepsilon_i = 1\}} (1 - p)^{\#\{i \in \{1,\ldots,n\} : \varepsilon_i = 0\}}.$$

The number of $\omega = (\varepsilon_1, \ldots, \varepsilon_n) \in \Omega$ such that $\varepsilon_1 + \cdots + \varepsilon_n = k$ equals $\binom{n}{k}$, $k \in \{0, \ldots, n\}$, so that

$$\sum_{\omega \in \Omega} p_\omega = \sum_{k=0}^{n} \binom{n}{k} p^k (1-p)^{n-k} = (p + (1-p))^n = 1.$$

Therefore, with

$$\mathbb{P}(B) := \sum_{\omega \in B} p_\omega$$

we obtain a probability space $(\Omega, \mathcal{F}, \mathbb{P})$. For $i = 1, \ldots, n$, we can describe the success and failure of the i-th event using $A_i := \{(\varepsilon_1, \ldots, \varepsilon_n) \in \Omega : \varepsilon_i = a_i\}$ with $a_i = 1$ and $a_i = 0$, respectively. Fixing $a_1, \ldots, a_n \in \{0, 1\}$ one can verify by a computation the following:

(1) The events $A_1, \ldots, A_n$ are independent.
(2) $\mathbb{P}(A_i) = p$ if $a_i = 1$ and $\mathbb{P}(A_i) = 1 - p$ if $a_i = 0$.

That means that we have the right model for our problem. The link to the binomial distribution is given by

$$\mathbb{P}\left(\left\{\omega \in \Omega : \sum_{i=1}^{n} \varepsilon_i = k\right\}\right) = Bin_{n,p}(\{k\}).$$

3.1.2 Poisson Distribution with Parameter $\lambda > 0$

(1) $\Omega := \mathbb{N}_0$.
(2) $\mathcal{F} := 2^{\Omega}$.
(3) $\mathbb{P}(B) = \text{Pois}_\lambda(B) := \sum_{k=0}^{\infty} e^{-\lambda} \frac{\lambda^k}{k!} \delta_k(B)$ (Fig. 3.2).

Proposition 3.1.3 *For $\lambda > 0$ the triplet $(\mathbb{N}_0, 2^{\mathbb{N}_0}, \text{Pois}_\lambda)$ is a probability space.*

Proof Again we use Lemma 3.1.1 and observe that

$$\text{Pois}_\lambda(\Omega) = \sum_{k=0}^{\infty} e^{-\lambda} \frac{\lambda^k}{k!} \delta_k(\Omega) = \sum_{k=0}^{\infty} e^{-\lambda} \frac{\lambda^k}{k!} = e^{-\lambda} e^\lambda = 1.$$

□

Application The Poisson[1] distribution occurs, for example, in connection with the Poisson process, a stochastic processes with a continuous time parameter and

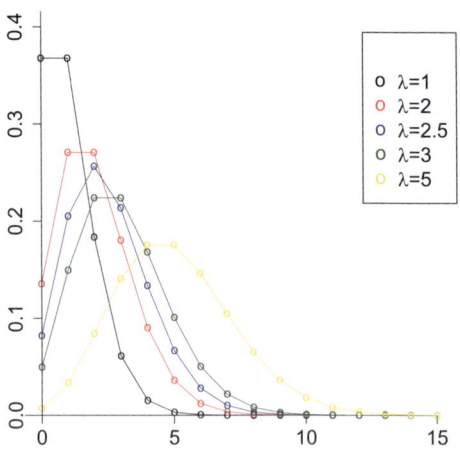

Fig. 3.2 Poisson distribution: probability mass functions

---

[1] Siméon Denis Poisson, 21/06/1781 (Pithiviers, France)–25/04/1840 (Sceaux, France).

jumps, which is used, for example, in insurance mathematics. The probability that a homogeneous Poisson process with intensity $\lambda > 0$ jumps k times in the time interval $(s, t]$ with $0 \leqslant s < t < \infty$ is equal to $\text{Pois}_{\lambda(t-s)}(\{k\})$.

3.1.3 Geometric Distribution with Parameter $0 < p < 1$

(1) $\Omega := \mathbb{N}_0$.
(2) $\mathcal{F} := 2^\Omega$.
(3) $\mathbb{P}(B) = \text{Geom}_p(B) := \sum_{k=0}^\infty (1-p)^k p \delta_k(B)$ (Fig. 3.3).

Proposition 3.1.4 *For $0 < p < 1$ the triplet $(\mathbb{N}_0, 2^{\mathbb{N}_0}, \text{Geom}_p)$ is a probability space.*

Proof Again we use Lemma 3.1.1 and observe

$$\text{Geom}_p(\Omega) = \sum_{k=0}^\infty (1-p)^k p \delta_k(\Omega) = \sum_{k=0}^\infty (1-p)^k p = \frac{p}{1-(1-p)} = 1.$$

□

Application The probability that an electric light bulb burns out per unit of time is $p \in (0, 1)$. We assume that the bulb does not have a 'memory', that means its failure is independent of the time the bulb has been already switched on. So, we get the following model: At day 0 the probability of failure is p. If the bulb has not burned out at day 0, it will fail again with probability p at the first day so that the total probability of failure at day 1 is $(1-p)p$. If we continue in this way, then we get that failure at day k has the probability $(1-p)^k p$.

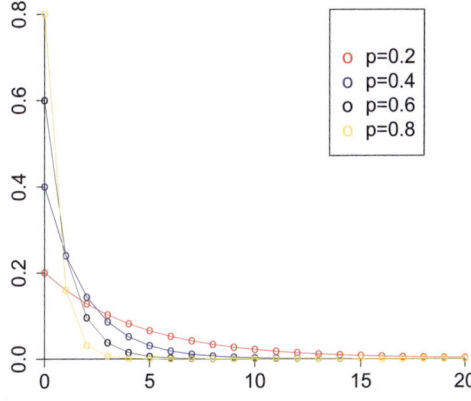

Fig. 3.3 Geometric distribution: probability mass functions

3.2 The Extension Theorem and Consequences

Although the definition of a measure is not difficult, to prove existence and uniqueness of measures may sometimes be demanding. The reason is the fact that the σ-algebra, the domain of definition of a measure, cannot be constructed explicitly in certain cases. To overcome this problem the following extension theorem is used:

Theorem 3.2.1 (Extension Theorem) *Let $\Omega \neq \emptyset$ and $\mathcal{A}$ be an algebra with $\mathcal{F} = \sigma(\mathcal{A})$ (where $\sigma(\mathcal{A})$ is introduced in Proposition 3.3.2 below). Assume $\mu_0 : \mathcal{A} \to [0, \infty]$ such that*

(1) $\mu_0(\Omega_l) < \infty$, $l = 1, 2, \ldots$, *for some partition* $(\Omega_l)_{l=1}^{\infty} \subseteq \mathcal{A}$ *of* Ω,
(2) *for pairwise disjoint* $A_n \in \mathcal{A}$ *it holds that*

$$\bigcup_{n=1}^{\infty} A_n \in \mathcal{A} \quad \text{implies} \quad \mu_0\left(\bigcup_{n=1}^{\infty} A_n\right) = \sum_{n=1}^{\infty} \mu_0(A_n).$$

Then there exists a unique measure $\mu : \mathcal{F} \to [0, \infty]$ such that

$$\mu(A) = \mu_0(A) \quad \text{for all} \quad A \in \mathcal{A}.$$

The obtained measure μ is σ-finite because $\mu(\Omega_l) < \infty$ for all $l = 1, 2, \ldots$. The proof of Theorem 3.2.1 can be found in Appendix B.2. The basic idea is to define in (B.2) the **outer measure**

$$\mu_0^*(A) := \inf\left\{\sum_{n=1}^{\infty} \mu_0(A_n) : A \subseteq \bigcup_{n=1}^{\infty} A_n, A_n \in \mathcal{A}\right\},$$

which can be defined for *all* subsets $A \subseteq \Omega$. This also leads to the following outer regularity with respect to the algebra $\mathcal{A}$ of the measure μ, proven in Appendix B.2 as well:

Theorem 3.2.2 (Outer Regularity) *Assume the setting and the notation from Theorem 3.2.1. Then, for all $A \in \mathcal{F}$ one has that*

$$\mu(A) = \inf\left\{\sum_{n=1}^{\infty} \mu(A_n) : A \subseteq \bigcup_{n=1}^{\infty} A_n, A_n \in \mathcal{A}\right\}.$$

If one is only interested in the uniqueness of measures, then one can also use the following approach:

Definition 3.2.3 (π-System) *A non-empty collection $\mathcal{P}$ of subsets $A \subseteq \Omega$ is called π-system, provided that*

$$A \cap B \in \mathcal{P} \quad \text{for all} \quad A, B \in \mathcal{P}.$$

Any algebra is a π-system, but a π-system is not an algebra in general. For example, take the π-system $\mathcal{P} := \{(a,b) : -\infty < a < b < \infty\} \cup \{\emptyset\}$.

Theorem 3.2.4 (Uniqueness of Measures) *Let $(\Omega, \mathcal{F})$ be a measurable space such that*

(1) $\mathcal{F} = \sigma(\mathcal{P})$, *where $\mathcal{P}$ is a π-system containing Ω,*
(2) μ *and ν are finite measures on $\mathcal{F}$ such that $\mu(A) = \nu(A)$ for all $A \in \mathcal{P}$.*

Then $\mu(B) = \nu(B)$ for all $B \in \mathcal{F}$.

The proof can be found in Appendix B.1.

3.3 Measures on $(\mathbb{R}, \mathcal{B}(\mathbb{R}))$

3.3.1 The Borel σ-Algebra $\mathcal{B}(\mathbb{R})$

Our goal is to introduce the Lebesgue measure λ on the real line which measures the length of an interval as we expect, i.e. we want that

$$\lambda((a,b)) = b - a$$

for $-\infty < a < b < \infty$. So we have the set $\Omega = \mathbb{R}$ of elementary events, the open intervals, and a natural candidate for a measure, the Lebesgue measure. The question arises which other sets can be measured by this type of measure. In particular, can we take the collection of all subsets $\mathcal{F} = 2^\Omega$? Surprisingly it turns out that we cannot use this σ-algebra as it is too large if we wish to have some translation properties (see Remark B.4.3). This leads us to a task in measure theory: the construction of a reasonable σ-algebra for the Lebesgue measure. Clearly, as a minimal requirement this σ-algebra should contain all open intervals. To get this idea working we use Proposition 3.3.2 below, which is based on the following fundamental construction of σ-algebras:

Proposition 3.3.1 (Intersection of σ-Algebras Is a σ-Algebra) *Let Ω and J be arbitrary non-empty sets, and let $\mathcal{F}_j$, $j \in J$, be a family of σ-algebras on Ω. Then*

$$\mathcal{F} := \bigcap_{j \in J} \mathcal{F}_j$$

is a σ-algebra as well.

Proof First we notice that $\emptyset, \Omega \in \mathcal{F}_j$ for all $j \in J$, so that $\emptyset, \Omega \in \bigcap_{j \in J} \mathcal{F}_j$. Now let $A, A_1, A_2, \ldots \in \bigcap_{j \in J} \mathcal{F}_j$. Hence $A, A_1, A_2, \ldots \in \mathcal{F}_j$ for all $j \in J$, so that

$$A^c = \Omega \setminus A \in \mathcal{F}_j \quad \text{and} \quad \bigcup_{i=1}^{\infty} A_i \in \mathcal{F}_j$$

for *all* $j \in J$. Consequently,

$$A^c \in \bigcap_{j \in J} \mathcal{F}_j \quad \text{and} \quad \bigcup_{i=1}^{\infty} A_i \in \bigcap_{j \in J} \mathcal{F}_j,$$

and $\bigcap_{j \in J} \mathcal{F}_j$ is a σ-algebra. □

Proposition 3.3.2 (Smallest σ-Algebra Containing a Collection of Sets) Let $\Omega \neq \emptyset$, $\mathcal{G}$ be a non-empty collection of subsets of Ω, and

$$\sigma(\mathcal{G}) := \bigcap_{\mathcal{F}\sigma\text{-algebra with } \mathcal{F} \supseteq \mathcal{G}} \mathcal{F}.$$

Then $\sigma(\mathcal{G})$ is the smallest σ-algebra containing $\mathcal{G}$, i.e. we have the following:

(1) The collection $\sigma(\mathcal{G})$ is a σ-algebra containing all sets from $\mathcal{G}$.
(2) If $\mathcal{F}_0$ is any σ-algebra on Ω containing all $G \in \mathcal{G}$, then $\sigma(\mathcal{G}) \subseteq \mathcal{F}_0$.

Proof We let

$$J := \left\{ \mathcal{F} \subseteq 2^\Omega : \mathcal{F} \text{ is a } \sigma\text{--algebra and } \mathcal{F} \supseteq \mathcal{G} \right\}.$$

One has $J \neq \emptyset$, because $\mathcal{G} \subseteq 2^\Omega$ and 2^Ω is a σ-algebra. Hence

$$\sigma(\mathcal{G}) := \bigcap_{\mathcal{F} \in J} \mathcal{F}$$

is a σ-algebra according to Proposition 3.3.1 and it holds $\mathcal{G} \subseteq \sigma(\mathcal{G})$ by construction. It remains to show that $\sigma(\mathcal{G})$ is the smallest σ-algebra containing $\mathcal{G}$. Assume another σ-algebra $\mathcal{F}_0$ with $\mathcal{G} \subseteq \mathcal{F}_0$. By definition of J we have that $\mathcal{F}_0 \in J$ so that

$$\sigma(\mathcal{G}) = \bigcap_{\mathcal{F} \in J} \mathcal{F} \subseteq \mathcal{F}_0.$$

□

Remark 3.3.3 It is easy to see that the definition of $\sigma(\mathcal{G})$ implies:

(1) If $\mathcal{G}$ is a σ-algebra, then $\sigma(\mathcal{G}) = \mathcal{G}$.
(2) If $\mathcal{G} \subseteq \mathcal{H}$ are non-empty collections of subsets of Ω, then $\sigma(\mathcal{G}) \subseteq \sigma(\mathcal{H})$.

The above construction of the smallest σ-algebra containing a given collection of sets is elegant and the basis for various fundamental measure spaces. However one needs to notice that this elegance has a price as we do not know explicitly all the elements of this generated smallest σ-algebra. Nevertheless, this approach works. One of the most important σ-algebras generated in this way is the Borel σ-algebra on $\mathbb{R}$. For its definition we introduce open sets.

3.3 Measures on $(\mathbb{R}, \mathcal{B}(\mathbb{R}))$

Definition 3.3.4 (Open and Closed Sets in $\mathbb{R}$)

(1) A subset $A \subseteq \mathbb{R}$ is called **open** if either $A = \emptyset$ or for each $x \in A$ there is an $\varepsilon > 0$ such that $(x - \varepsilon, x + \varepsilon) \subseteq A$.
(2) A subset $B \subseteq \mathbb{R}$ is called **closed** if $A := \mathbb{R} \backslash B$ is open.

Definition 3.3.5 (Borel σ-Algebra on $\mathbb{R}$) The smallest σ-algebra generated by the open sets in $\mathbb{R}$ is called **Borel** σ-algebra and denoted by $\mathcal{B}(\mathbb{R})$.

The following proposition shows that there are several ways to generate the Borel σ-algebra.

Proposition 3.3.6 (Generating the Borel σ-Algebra on $\mathbb{R}$) *We let*

(0) *$\mathcal{G}_0$ be the collection of all open subsets of $\mathbb{R}$,*
(1) *$\mathcal{G}_1$ be the collection of all closed subsets of $\mathbb{R}$,*
(2) *$\mathcal{G}_2$ be the collection of all intervals $(-\infty, b]$, $b \in \mathbb{R}$,*
(3) *$\mathcal{G}_3$ be the collection of all intervals $(-\infty, b)$, $b \in \mathbb{R}$,*
(4) *$\mathcal{G}_4$ be the collection of all intervals $(a, b]$, $-\infty < a < b < \infty$,*
(5) *$\mathcal{G}_5$ be the collection of all intervals (a, b), $-\infty < a < b < \infty$.*

Then $\sigma(\mathcal{G}_0) = \sigma(\mathcal{G}_1) = \sigma(\mathcal{G}_2) = \sigma(\mathcal{G}_3) = \sigma(\mathcal{G}_4) = \sigma(\mathcal{G}_5)$.

Proof We only show that $\sigma(\mathcal{G}_0) = \sigma(\mathcal{G}_1) = \sigma(\mathcal{G}_3) = \sigma(\mathcal{G}_5)$, the remaining parts are intended to be an exercise.

$\sigma(\mathcal{G}_3) \subseteq \sigma(\mathcal{G}_0)$: This follows from $\mathcal{G}_3 \subseteq \mathcal{G}_0$ and Remark 3.3.3.
$\sigma(\mathcal{G}_5) \subseteq \sigma(\mathcal{G}_3)$: For $-\infty < a < b < \infty$ one has that

$$(a, b) = \bigcup_{n=1}^{\infty} \left((-\infty, b) \backslash \left(-\infty, a + \frac{1}{n} \right) \right) \in \sigma(\mathcal{G}_3)$$

so that $\mathcal{G}_5 \subseteq \sigma(\mathcal{G}_3)$. Hence by Remark 3.3.3 we have $\sigma(\mathcal{G}_5) \subseteq \sigma(\mathcal{G}_3)$.

$\sigma(\mathcal{G}_0) \subseteq \sigma(\mathcal{G}_5)$: We assume a non-empty open set $A \subseteq \mathbb{R}$. For all $x \in A$ there is a *maximal* $\varepsilon_x > 0$ such that

$$(x - \varepsilon_x, x + \varepsilon_x) \subseteq A.$$

(Why it is important to work with ε_x becomes clear in Exercise 3.) Hence

$$A = \bigcup_{x \in A \cap \mathbb{Q}} (x - \varepsilon_x, x + \varepsilon_x),$$

which proves $\mathcal{G}_0 \subseteq \sigma(\mathcal{G}_5)$ implying $\sigma(\mathcal{G}_0) \subseteq \sigma(\mathcal{G}_5)$.

$\sigma(\mathcal{G}_1) = \sigma(\mathcal{G}_0)$: If $A \in \mathcal{G}_0$, then $A^c \in \mathcal{G}_1 \subseteq \sigma(\mathcal{G}_1)$ and $A \in \sigma(\mathcal{G}_1)$. Hence $\mathcal{G}_0 \subseteq \sigma(\mathcal{G}_1)$ and $\sigma(\mathcal{G}_0) \subseteq \sigma(\mathcal{G}_1)$. The remaining inclusion $\sigma(\mathcal{G}_1) \subseteq \sigma(\mathcal{G}_0)$ can be shown in the same way. □

Remark 3.3.7 It is not always possible to construct from a generating system the corresponding σ-algebra by applying countable sequences of set-theoretical operations. For example, in [16, Section 1.2, Constructing σ-fields] this is outlined for $\Omega = (0, 1]$ and the generating system of intervals $(a, b], 0 \leqslant a < b \leqslant 1$.

3.3.2 The Extension Theorem for $(\mathbb{R}, \mathcal{B}(\mathbb{R}))$

We consider a function $F : \mathbb{R} \to \mathbb{R}$ and set

$$\mu((a, b]) := F(b) - F(a) \quad \text{for all} \quad -\infty < a < b < \infty.$$

What are the properties of F we need to assume such that μ is a measure on $(\mathbb{R}, \mathcal{B}(\mathbb{R}))$? Firstly, the monotonicity of the measure makes it necessary that F is non-decreasing. Secondly, the continuity of the measure from above shown in Proposition 2.3.7(6) requires right-continuity of F. This is all what we need, so we proceed as follows:

(1) $\Omega := \mathbb{R}$.
(2) $\mathcal{F} := \mathcal{B}(\mathbb{R})$.
(3) As generating algebra $\mathcal{A}$ for $\mathcal{B}(\mathbb{R})$ we take the algebra from Example 2.2.8, i.e. the collection of subsets $A \subseteq \mathbb{R}$ such that A can be written as

$$A = (a_1, b_1] \cup (a_2, b_2] \cup \cdots \cup (a_n, b_n]$$

or

$$A = (a_1, b_1] \cup (a_2, b_2] \cup \cdots \cup (a_n, \infty)$$

where $n \in \mathbb{N}$ and $-\infty \leqslant a_1 \leqslant b_1 \leqslant \cdots \leqslant a_n \leqslant b_n < \infty$.

(4) We assume a non-decreasing function $F : \mathbb{R} \to \mathbb{R}$ which is right-continuous and define for a set A given in item (3)

$$\mu_0(A) := \sum_{i=1}^{n} (F(b_i) - F(a_i)) \tag{3.1}$$

with $F(\infty) := \lim_{b \to \infty} F(b)$ and $F(-\infty) := \lim_{a \to -\infty} F(a)$.

The following theorem is the counterpart to Lemma 3.1.1 from the discrete setting:

Theorem 3.3.8 (Existence of Borel-Measures)

(1) *The collection $\mathcal{A}$ is an algebra.*

3.3 Measures on $(\mathbb{R}, \mathcal{B}(\mathbb{R}))$

(2) *The map $\mu_0 : \mathcal{A} \to [0, \infty]$ given in (3.1) is well-defined and satisfies the assumptions of the extension Theorem 3.2.1.*

(3) *The unique σ-additive extension $\mu : \mathcal{B}(\mathbb{R}) \to [0, \infty]$ is σ-finite.*

(4) *If ν is a measure on $(\mathbb{R}, \mathcal{B}(\mathbb{R}))$ such that*

$$\nu((a, b]) = \mu((a, b]) \quad \text{for all } -\infty < a < b < \infty,$$

then $\nu = \mu$.

Remark 3.3.9 The half-open intervals in the algebra $\mathcal{A}$ are chosen to guarantee the properties of an algebra. Here also the choice $[a, b)$ would have been possible, but the choice $(a, b]$ fits better the notion of the distribution function introduced in Definition 6.5.4 below.

Proof of Theorem 3.3.8 (1) was considered in Example 2.2.8.

(2) The map μ_0 is well-defined, that means $\mu_0(A)$ does not depend on the representation $A = (a_1, b_1] \cup (a_2, b_2] \cup \cdots \cup (a_n, b_n]$. We need to prove that assumption Theorem 3.2.1(2) is satisfied. For this we start with pairwise disjoint $A_1, A_2, \ldots \in \mathcal{A}$ such that

$$(a, b] = \bigcup_{n=1}^{\infty} A_n \quad \text{with} \quad -\infty < a < b < \infty.$$

As each A_n is a finite union of half open intervals, it is sufficient to show the following: If we have pairwise disjoint intervals $(a_n, b_n]$ with

$$(a, b] = \bigcup_{n=1}^{\infty} (a_n, b_n], \tag{3.2}$$

then we have that

$$F(b) - F(a) = \sum_{n=1}^{\infty} (F(b_n) - F(a_n)). \tag{3.3}$$

It is easy to see that for all $N \in \mathbb{N}$ we have

$$\sum_{n=1}^{N} (F(b_n) - F(a_n)) \leq F(b) - F(a)$$

so that

$$\sum_{n=1}^{\infty} (F(b_n) - F(a_n)) \leq F(b) - F(a).$$

We now show the opposite inequality. Let $\varepsilon > 0$ and assume $\tilde{b}_n \in (b_n, \infty)$ such that $F(\tilde{b}_n) \leq F(b_n) + 2^{-n}\varepsilon$, where we exploit the right continuity of F. For $\tilde{a} \in (a, b)$ such that $F(\tilde{a}) - \varepsilon \leq F(a)$, where again the right continuity of F is exploited, we get

$$[\tilde{a}, b] \subseteq \bigcup_{n=1}^{\infty} \left(a_n, \tilde{b}_n\right).$$

Hence we have an open covering of a compact set, and by the Heine[2]-Borel theorem there exists a subcover

$$[\tilde{a}, b] \subseteq \bigcup_{n \in I(\varepsilon)} \left(a_n, \tilde{b}_n\right)$$

for some *finite* set $I(\varepsilon)$. This implies that

$$(\tilde{a}, b] \subseteq \bigcup_{n \in I(\varepsilon)} \left(a_n, \tilde{b}_n\right]$$

and

$$F(b) - F(a) - \varepsilon \leq F(b) - F(\tilde{a}) \leq \sum_{n \in I(\varepsilon)} (F(\tilde{b}_n) - F(a_n))$$

$$\leq \sum_{n=1}^{\infty} \left[(F(b_n) - F(a_n)) + \frac{\varepsilon}{2^n}\right] = \varepsilon + \sum_{n=1}^{\infty} (F(b_n) - F(a_n)).$$

By $\varepsilon \downarrow 0$ we get $F(b) - F(a) \leq \sum_{n=1}^{\infty}(F(b_n) - F(a_n))$. If in (3.2) $a = -\infty$ or $b = \infty$, or both, then on the right hand side of (3.2) we have a union of finite $(a_n, b_n]$ and intervals $(-\infty, B]$ or (A, ∞) or both (but at most one from each type). We can subtract these terms in (3.3) and are in the situation we have considered.
(3) Because $\mathbb{R} = \bigcup_{n=-\infty}^{\infty} (n-1, n]$ and $\mu((n-1, n]) = \mu_0((n-1, n]) = F(n) - F(n-1) < \infty$ the measure μ is σ-finite.
(4) Let ν be a measure on $(\mathbb{R}, \mathcal{B}(\mathbb{R}))$ such that $\nu((a, b]) = \mu((a, b])$ for all $-\infty < a < b < \infty$. Then by the σ-additivity we have $\nu((-\infty, b]) = \mu((-\infty, b])$ and $\nu((a, \infty)) = \mu((a, \infty))$ for all $-\infty < a < b < \infty$. Therefore, μ and ν coincide on $\mathcal{A}$. The uniqueness Theorem 3.2.1 gives $\nu = \mu$. □

In the remainder of this chapter we have a closer look at some examples of measures which appear frequently in applications. Let $p : \mathbb{R} \to [0, \infty)$ be a function such that p is continuous on $\mathbb{R}$ or continuous on $\mathbb{R}$ except at finitely many

---

[2] Heinrich Eduard Heine, 16/03/1821 (Berlin, Germany)–21/10/1881 (Halle, Germany).

3.3 Measures on $(\mathbb{R}, \mathcal{B}(\mathbb{R}))$

points $-\infty < x_1 < \cdots < x_L < \infty$ and such that in these points we assume that

$$\lim_{x \uparrow x_l} p(x) \quad \text{and} \quad \lim_{x \downarrow x_l} p(x)$$

exist and are finite. Therefore, p is Riemann integrable, and the Riemann integrals (**denoted by** dx **in contrast to the integral w.r.t. the Lebesgue measure which we will denote by** $d\lambda(x)$)

$$\int_a^b p(x)dx$$

can be defined for all $-\infty < a < b < \infty$. Then μ given by

$$\mu((a, b]) = \int_a^b p(x)dx, \tag{3.4}$$

is a measure on $(\mathbb{R}, \mathcal{B}(\mathbb{R}))$. The function p is called the **density** of μ. We fix parameters $m \in \mathbb{R}$, $\sigma > 0$, $\lambda > 0$, and $-\infty < c < d < \infty$, and consider the following often used densities:

Function p	Name of measure μ	Probability measure
$p(x) = 1$	Lebesgue measure	No, it is σ-finite
$p(x) = \frac{1}{\sqrt{2\pi\sigma^2}} e^{-\frac{(x-m)^2}{2\sigma^2}}$	Gaussian measure	Yes
$p(x) = \mathbb{1}_{[0,\infty)}(x)\lambda e^{-\lambda x}$	Exponential distribution	Yes
$p(x) = \mathbb{1}_{[c,d]}(x)$	Uniform distribution	If $d - c = 1$, otherwise finite

3.3.3 Lebesgue Measure on $\mathbb{R}$

Proposition 3.3.10 (Existence of the Lebesgue Measure)

(1) *There exists a unique measure λ on $(\mathbb{R}, \mathcal{B}(\mathbb{R}))$ such that $\lambda((a, b]) = b - a$ for all $-\infty < a < b < \infty$.*
(2) *λ is a σ-finite measure.*

Proof (1) and (2) follows from Theorem 3.3.8 by taking $F(x) = x$. □

Definition 3.3.11 (Lebesgue Measure) The measure λ on $(\mathbb{R}, \mathcal{B}(\mathbb{R}))$ from Proposition 3.3.10 is called **Lebesgue measure**.

3.3.4 Gaussian Distribution on $\mathbb{R}$

For $m \in \mathbb{R}$ and $\sigma > 0$ we let

$$p_{m,\sigma^2}(x) := \frac{1}{\sqrt{2\pi\sigma^2}} e^{-\frac{(x-m)^2}{2\sigma^2}}.$$

Proposition 3.3.12 (Existence of the Gaussian Distribution)

(1) *There exists a unique measure $\mathcal{N}_{m,\sigma^2}$ on $(\mathbb{R}, \mathcal{B}(\mathbb{R}))$ such that*

$$\mathcal{N}_{m,\sigma^2}((a,b]) = \int_a^b p_{m,\sigma^2}(x)\,\mathrm{d}x \quad \textit{for all} \quad -\infty < a < b < \infty.$$

(2) $\mathcal{N}_{m,\sigma^2}$ *is a probability measure.*

Definition 3.3.13 The measure $\mathcal{N}_{m,\sigma^2}$ on $(\mathbb{R}, \mathcal{B}(\mathbb{R}))$ from Proposition 3.3.12 is called **Gaussian measure (Gaussian distribution, normal distribution)** with mean $m \in \mathbb{R}$ and variance $\sigma^2 > 0$, and the function p_{m,σ^2} is called **Gaussian density** (Fig. 3.4).

Proof of Proposition 3.3.12 (1) follows from Theorem 3.3.8 by taking

$$F(x) := \int_{-\infty}^x p_{m,\sigma^2}(y)\,\mathrm{d}y.$$

(2) The proof of $\mathcal{N}_{m,\sigma^2}(\mathbb{R}) = 1$ will be given in Corollary 8.10.2. □

Fig. 3.4 Gaussian densities

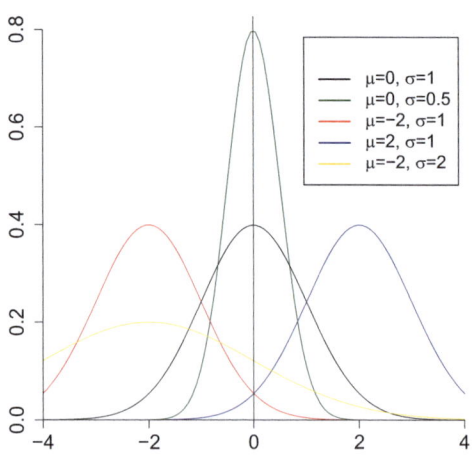

3.3.5 Exponential Distribution on $\mathbb{R}$

For $\lambda > 0$ we let

$$p_\lambda(x) := \mathbb{1}_{[0,\infty)}(x)\lambda e^{-\lambda x}.$$

Proposition 3.3.14 (Existence of the Exponential Distribution)

(1) *There exists a unique measure* Exp_λ *on* $(\mathbb{R}, \mathcal{B}(\mathbb{R}))$ *such that*

$$\mathrm{Exp}_\lambda((a, b]) = \int_a^b p_\lambda(x)\mathrm{d}x \quad \text{for all} \quad -\infty < a < b < \infty.$$

(2) Exp_λ *is a probability measure.*
(3) *The probability measure* Exp_λ *does not have a memory in the sense that for* $a, b \geqslant 0$ *we have*

$$\mathrm{Exp}_\lambda([a+b, \infty)|[a, \infty)) = \mathrm{Exp}_\lambda([b, \infty)),$$

where the conditional probability on the left-hand side was defined in Definition 2.6.1.

Definition 3.3.15 The measure Exp_λ on $(\mathbb{R}, \mathcal{B}(\mathbb{R}))$ from Proposition 3.3.14 is called **exponential distribution** with parameter $\lambda > 0$, and the function p_λ is the density of the exponential distribution (Fig. 3.5).

Proof of Proposition 3.3.14 (1) follows from Theorem 3.3.8 with $F(x) := 0$ for $x \leqslant 0$ and $F(x) := 1 - e^{-\lambda x}$ for $x \geqslant 0$.

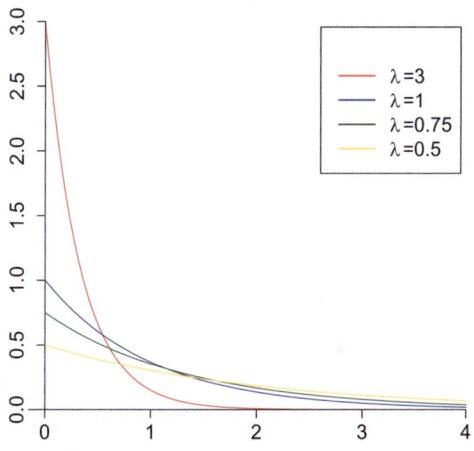

Fig. 3.5 Exponential distribution: densities

(2) We have that

$$\operatorname{Exp}_\lambda(\mathbb{R}) = \lim_{n\to\infty} \operatorname{Exp}_\lambda((-n, n]) = \lim_{n\to\infty} (1 - e^{-\lambda n}) = 1.$$

(3) follows from

$$\operatorname{Exp}_\lambda([a+b, \infty)|[a, \infty)) = \frac{\operatorname{Exp}_\lambda([a+b, \infty) \cap [a, \infty))}{\operatorname{Exp}_\lambda([a, \infty))} = \frac{\lambda \int_{a+b}^\infty e^{-\lambda x} dx}{\lambda \int_a^\infty e^{-\lambda x} dx}$$

$$= \frac{e^{-\lambda(a+b)}}{e^{-\lambda a}} = \operatorname{Exp}_\lambda([b, \infty)).$$

□

Example 3.3.16 Suppose that the amount of time one spends waiting in the queue at a particular supermarket checkout is exponentially distributed with $\lambda = \frac{1}{10}$. We ask: (a) What is the probability, that a customer will spend more than 15 minutes in the queue? (b) What is the probability, that one spends more than 15 minutes from the beginning in the queue, given that one already spent at least 10 minutes in the queue?

The answer for (a) is $\operatorname{Exp}_\lambda([15, \infty)) = e^{-15\frac{1}{10}} \approx 0.223$. For (b) we get $\operatorname{Exp}_\lambda([15, \infty)|[10, \infty)) = \operatorname{Exp}_\lambda([5, \infty)) = e^{-5\frac{1}{10}} \approx 0.607$.

3.3.6 The Trace σ-Algebra

We need the concept of trace σ-algebras to introduce the uniform distribution on an interval $[c, d]$ given we know the Lebesgue measure on $(\mathbb{R}, \mathcal{B}(\mathbb{R}))$. As trace σ-algebras and the corresponding restrictions of measures are of general relevance, and we shall use them at different places, we devote a separate section to this subject.

Given a probability space $(\Omega, \mathcal{F}, \mathbb{P})$ and a set $A \in \mathcal{F}$ with $\mathbb{P}(A) > 0$, we introduced in Definition 2.6.1 the conditional probability

$$\mathbb{P}_A(B) = \mathbb{P}(B|A) = \frac{\mathbb{P}(B \cap A)}{\mathbb{P}(A)}.$$

This did lead us to the probability space $(\Omega, \mathcal{F}, \mathbb{P}_A)$. Because it is natural to restrict also Ω to A one needs to find out what the corresponding restriction of $\mathcal{F}$ will be. And there is one more point one needs to understand: If $\mathcal{G}$ is a collection of subsets from Ω that generates $\mathcal{F}$, what is the collection of subsets of A that generates the restriction of $\mathcal{F}$? This is clarified by the following lemma:

Lemma 3.3.17 Let $(\Omega, \mathcal{F})$ be a measurable space and $\emptyset \neq A \subseteq \Omega$.

(1) The collection $\mathcal{F}_A := \{F \cap A : F \in \mathcal{F}\} \subseteq 2^A$ is a σ-algebra on A.

3.3 Measures on $(\mathbb{R}, \mathcal{B}(\mathbb{R}))$

(2) If $A \in \mathcal{F}$, then $\mathcal{F}_A = \{F \in \mathcal{F} : F \subseteq A\}$.
(3) If $\mathcal{G}$ is a non-empty set of sub-sets of Ω such that $\sigma(\mathcal{G}) = \mathcal{F}$, then the collection $\mathcal{G}_A := \{G \cap A : G \in \mathcal{G}\}$ generates the σ-algebra $\mathcal{F}_A$ on A, i.e. it holds $\sigma(\mathcal{G})_A = \sigma(\mathcal{G}_A) = \mathcal{F}_A$.
(4) If $(\Omega, \mathcal{F}, \mu)$ is a measure space, $A \in \mathcal{F}$, and $\mu_A(B) := \mu(B)$ for $B \in \mathcal{F}_A$, then $(A, \mathcal{F}_A, \mu_A)$ is a measure space.

Proof of Proposition 3.3.14 (1) Choosing $F = \emptyset$ and $F = \Omega$ implies $\emptyset \in \mathcal{F}_A$ and $A \in \mathcal{F}_A$, respectively. Any element $G \in \mathcal{F}_A$ can be written as $G = F \cap A$ for some $F \in \mathcal{F}$. Therefore $A \setminus G \in \mathcal{F}_A$ follows from $F^c \in \mathcal{F}$ and

$$A \setminus G = A \setminus (F \cap A) = F^c \cap A.$$

Assume that for $i = 1, 2, \ldots$ the G_i are given by $G_i = F_i \cap A$ with $F_i \in \mathcal{F}$. Then

$$\bigcup_{i=1}^{\infty} G_i = \bigcup_{i=1}^{\infty}(F_i \cap A) = \left(\bigcup_{i=1}^{\infty} F_i\right) \cap A \in \mathcal{F}_A$$

since $\bigcup_{i=1}^{\infty} F_i \in \mathcal{F}$.

(2) follows directly from the definition.

(3) We have that $\mathcal{G} \subseteq \sigma(\mathcal{G})$ so that $\mathcal{G}_A \subseteq \sigma(\mathcal{G})_A$. Hence part (1) implies that

$$\sigma(\mathcal{G}_A) \subseteq \sigma(\mathcal{G})_A.$$

For the inclusion $\sigma(\mathcal{G}_A) \supseteq \sigma(\mathcal{G})_A$ we use the *principle of good sets* and define

$$\mathcal{G}_0 := \{G \in \sigma(\mathcal{G}) : G \cap A \in \sigma(\mathcal{G}_A)\}.$$

The *principle of good sets* means that one collects all 'good sets' which satisfy a desired property and shows that this collection is large enough, for example a σ-algebra. Indeed, since

$$\mathcal{G} \subseteq \mathcal{G}_0 \subseteq \sigma(\mathcal{G}),$$

if $\mathcal{G}_0$ would be a σ-algebra, then we would have $\mathcal{G}_0 = \sigma(\mathcal{G})$ and $G \cap A \in \sigma(\mathcal{G}_A)$ would hold for any $G \in \sigma(\mathcal{G})$, i.e. $\sigma(\mathcal{G})_A \subseteq \sigma(\mathcal{G}_A)$. So it remains to show that $\mathcal{G}_0$ is a σ-algebra:

(a) Since $A, \emptyset \in \sigma(\mathcal{G}_A)$ we conclude from the definition of $\mathcal{G}_0$ that $\Omega, \emptyset \in \mathcal{G}_0$.
(b) Assume $G \in \mathcal{G}_0$. Then we have $G \cap A \in \sigma(\mathcal{G}_A)$, and since $\sigma(\mathcal{G}_A)$ is a σ-algebra on A, also $G^c \cap A = A \setminus (G \cap A) \in \sigma(\mathcal{G}_A)$. Hence $G^c \in \mathcal{G}_0$.
(c) Similarly as in the proof of part (1) one derives from $G_1, G_2, \ldots \in \mathcal{G}_0$ that $\bigcup_{n=1}^{\infty} G_n \in \mathcal{G}_0$.

(4) The statement is similar to Proposition 2.6.2(1). $\square$

Definition 3.3.18 The σ-algebra $\mathcal{F}_A$ on A constructed in Lemma 3.3.17(1) is called the **trace of $\mathcal{F}$ on A**.

3.3.7 Uniform Distribution on $[c, d]$

Now we obtain the uniform distribution on $[c, d]$ for $-\infty < c < d < \infty$ from the Lebesgue measure on $\mathbb{R}$ constructed before:

(1) $\Omega := [c, d]$.
(2) $\mathcal{F}$ is the trace σ-algebra of $\mathcal{B}(\mathbb{R})$ on $[c, d]$ which is denoted by $\mathcal{B}([c, d])$.
(3) $\mathcal{U}_{[c,d]}(A) := \frac{\lambda(A)}{d-c}$ for $A \in \mathcal{B}([c, d])$, where λ is the Lebesgue measure on $(\mathbb{R}, \mathcal{B}(\mathbb{R}))$.

By Lemma 3.3.17 any of the collections $\mathcal{G}_0, \ldots, \mathcal{G}_5$ from Proposition 3.3.6 leads to a generating system of $\mathcal{B}([c, d])$ by choosing

$$\{[c, d] \cap G : G \in \mathcal{G}_i\}.$$

Please be aware that, for example, the collection

$$\{(a, b) : c \leqslant a < b \leqslant d\}$$

is *not* a generating system of $\mathcal{B}([c, d])$ (see Exercise 7). The uniform distribution on $[c, d]$ can be seen as a measure on $(\mathbb{R}, \mathcal{B}(\mathcal{R}))$ as well by defining

$$\widetilde{\mathcal{U}}_{[c,d]}(A) := \frac{\lambda([c, d] \cap A)}{d - c},$$

which yields to $\widetilde{\mathcal{U}}_{[c,d]}((a, b]) = \int_a^b p(x) dx$ for $-\infty < a < b < \infty$ with $p(x) := \frac{\mathbb{1}_{[c,d]}(x)}{d-c}$.

3.4 Exercises

Ex 1: Verify Lemma 3.1.1. More generally, assume a countable set $\Omega \neq \emptyset$, $\mathcal{F} := 2^\Omega$, and assume $p_\omega \in [0, \infty]$ for $\omega \in \Omega$. Define a set function on $(\Omega, 2^\Omega)$ by

$$\mathbb{P}(A) := \sum_{\omega \in A} p_\omega \quad \text{for} \quad A \in 2^\Omega \quad \text{with} \quad \sum_{\omega \in \emptyset} p_\omega := 0.$$

(a) Does one get always a measure space?
(b) Find the conditions on the weights p_ω such that the measure space is a probability space, a finite measure space, or a σ-finite measure space.

3.4 Exercises

Ex 2: In Proposition 3.3.1 we showed that the intersection of σ-algebras is again a σ-algebra. Find two σ-algebras on a set Ω such that their union is not a σ-algebra.

Ex 3: (a) Let $\emptyset \neq A \subseteq \mathbb{R}$ be a bounded open set and for $x \in A$ let $\varepsilon_x > 0$ denote the maximal real value such that $(x - \varepsilon_x, x + \varepsilon_x) \in A$. Show that the maximum is really attained and that

$$A = \bigcup_{x \in A}(x - \varepsilon_x, x + \varepsilon_x) = \bigcup_{x \in A \cap \mathbb{Q}}(x - \varepsilon_x, x + \varepsilon_x).$$

(b) Deduce that $\emptyset \neq G \subseteq \mathbb{R}$ is open if and only if there are countably many pairwise disjoint intervals (a_i, b_i), $i \in I$, with $-\infty \leq a_i < b_i \leq \infty$ such that $G = \bigcup_{i \in I}(a_i, b_i)$.

(c) Now we see what can happen, if one drops the requirement 'maximal possible size' in (a): Assume $(0, 1) \cap \mathbb{Q} = \{q_1, q_2, \ldots\}$, which is a dense subset of $(0, 1)$. Fix positive radii $r_i > 0$ and consider

$$A = \bigcup_{i=1}^{\infty}(q_i - r_i, q_i + r_i).$$

Is it always true that $(0, 1) \subseteq A$?

Hint: Choose, for example $r_i = \frac{1}{4^i}$ and estimate $\lambda(A) = \lambda\left(\bigcup_{i=1}^{\infty}(q_i - r_i, q_i + r_i)\right)$ in a suitable way from above.

Ex 4: Finish the proof of Proposition 3.3.6 by showing that $\sigma(\mathcal{G}_2) = \sigma(\mathcal{G}_4) = \sigma(\mathcal{G}_i)$, for some $i \in \{0, 1, 3, 5\}$.

Ex 5: While the elements of a Borel σ-algebra cannot be constructed from a generating system in general in the sense of Remark 3.3.7, an example where the elements are explicitly known is the smallest σ-algebra on $\mathbb{R}$ which is generated by

$$\mathcal{G} := \{\{q\} : q \in \mathbb{Q}\}.$$

Describe explicitly the sets in $\sigma(\mathcal{G})$.

Ex 6: Similarly to Exercise 5, define on $\mathbb{R}$ the smallest σ-algebra generated by

$$\mathcal{G} := \{\{x\} : x \in \mathbb{R}\}.$$

Describe explicitly the sets in $\mathcal{G}$, and show that $\sigma(\mathcal{G}) \subsetneq \mathcal{B}(\mathbb{R})$.

Ex 7: The σ-algebra $\mathcal{B}([a, b])$ is the trace σ-algebra of $\mathcal{B}(\mathbb{R})$ (see Definition 3.3.18). Show that $\sigma((c, d) : a \leq c < d \leq b)$ is strictly smaller than $\mathcal{B}([a, b])$.

Hint: $\mathcal{B}([a, b])$ contains the set $\{a\}$. Check whether $\{a\}$ is also an element of the other σ algebra.

Ex 8: Assume a set $A \in \mathcal{B}([0, 1])$ which is not dense in $[0, 1]$ with respect to the Euclidean metric. Prove that $\mathcal{U}_{[0,1]}(A) < 1$.

3.5 Comments

Section 3.2: The extension Theorem 3.2.1 appears under different names in the literature as a result of various contributions to the theory of set functions at this time. A starting point is the work of Fréchet [67]. The name *Hahn extension* is due to Hahn's work [76]. As the extension theorem can be proved using outer measures, as we do this in this book, it is also called *Carathéodory extension theorem* because of Carathéodory's work [31]. The extension theorem appears in Kolmogorov's work [105] as well. The reader is also referred to the presentation in Dunford-Schwartz [52, Section III.5].

Chapter 4
*Metric and Banach Spaces

This chapter assembles the necessary concepts and statements about metric spaces and Banach spaces which will be used throughout the book. Separability of these spaces will play an important role for their usage in probability theory. It turns out to be useful to consider not only Banach spaces but also quasi-Banach spaces and p-normed Banach spaces. Linear operators between Banach spaces are introduced, and the equivalence of continuity and boundedness is shown. The dual space of a Banach space is defined, and for the convenience of the reader, the Hahn-Banach theorem, a separation theorem, and the closed graph theorem are listed.

The reader might also go straight to the next chapter and use this chapter as a reference whenever necessary.

4.1 Metric Spaces

Let us start with the definition of a metric space:

Definition 4.1.1 (Metric Space) Let $M \neq \emptyset$. The pair $[M, d]$ is called **metric space** and d a **metric** on M if $d : M \times M \to [0, \infty)$ satisfies

(1) $d(x, y) = 0$ if and only if $x = y$ (reflexivity),
(2) $d(x, y) = d(y, x)$ (symmetry),
(3) $d(x, z) \leq d(x, y) + d(y, z)$ (triangle inequality).

The notion of a metric space generalizes various concepts known for the Euclidean space. We recall those concepts that are relevant for us:

Definition 4.1.2 Let $[M, d]$ be a metric space.

(1) A set $G \subseteq M$ is **open** if either $G = \emptyset$ or for all $x \in G$ there is an $\varepsilon > 0$ such that for the open ball around x with radius $\varepsilon > 0$ it holds
$$U_\varepsilon(x) := \{y \in M : d(x, y) < \varepsilon\} \subseteq G.$$

(2) A set $F \subseteq M$ is **closed** if its complement $F^c = \{x \in M : x \notin F\}$ is open.

(3) A set $K \subseteq M$ is **pre-compact** if for all $\varepsilon > 0$ there are $x_1, \ldots, x_n \in M$ with
$$K \subseteq \bigcup_{i=1}^n U_\varepsilon(x_i).$$

(4) A sequence $(x_n)_{n=1}^\infty \subseteq M$ converges to $x \in M$ if $\lim_{n \to \infty} d(x_n, x) = 0$.

(5) A sequence $(x_n)_{n=1}^\infty \subseteq M$ is called **Cauchy**[1] **sequence** if for all $\varepsilon > 0$ there exists an $n(\varepsilon) \in \mathbb{N}$ with $d(x_m, x_n) \leqslant \varepsilon$ whenever $m, n \geqslant n(\varepsilon)$.

(6) The metric space is **complete** if for all Cauchy sequences $(x_n)_{n=1}^\infty \subseteq M$ there is a limit $x \in M$, i.e. $\lim_{n \to \infty} d(x_n, x) = 0$.

(7) A non-empty set $K \subseteq M$ is **compact** if it is pre-compact and all Cauchy sequences $(x_n)_{n=1}^\infty \subseteq K$ have a limit $x \in K$.

(8) A set $A \subseteq M$ is called **dense** in M if for all $\varepsilon > 0$ and all $x \in M$ there is an $a \in A$ such that $d(x, a) < \varepsilon$. A metric space $[M, d]$ is called **separable** if there is a countable dense set $A \subseteq M$.

(9) A metric space is called a **Polish space** if it is complete and separable.

(10) Given $A \subseteq M$, then $\overline{A} \subseteq M$ is the smallest closed set such that $A \subseteq \overline{A}$, i.e. $\overline{A} = \bigcap F$, where the intersection is taken over all closed sets $F \supseteq A$. The set $\overline{A}$ is called **closure** of A.

We summarize (without proof) some properties of metric spaces:

Proposition 4.1.3 Let $[M, d]$ be a metric space.

(1) The set M and the empty set $\emptyset$, both are open and closed.
(2) Each set consisting of finitely many points is closed.
(3) For any collection of open sets $(G_i)_{i \in I} \subseteq M$, $I \neq \emptyset$, one has that $G := \bigcup_{i \in I} G_i$ is open.
(4) For any collection of closed sets $(F_i)_{i \in I} \subseteq M$, $I \neq \emptyset$, one has that $F := \bigcap_{i \in I} F_i$ is closed.
(5) The Heine-Borel property holds. Let $\emptyset \neq K \subseteq M$. Then K is compact if and only if the following holds: If $K \subseteq \bigcup_{i \in I} G_i$, where the sets $G_i \subseteq M$ are open, then there are $i_1, \ldots, i_n \in I$ such that $K \subseteq G_{i_1} \cup \cdots \cup G_{i_n}$.
(6) If (M, d) is complete and $\emptyset \neq A \subseteq M$ is pre-compact, then the closure $\overline{A}$ is compact.

---

[1] Augustin Louis Cauchy, 21/08/1789 (Paris, France)–23/05/1857 (Sceaux, France).

4.1 Metric Spaces

The property that a metric space is separable will be of importance for us.

Proposition 4.1.4 *Let $[M, d]$ be a separable metric space and let $(x_l)_{l \in I} \subseteq M$ be dense, where I is countable. Define the countable system of open balls*

$$\mathcal{B} := \{U_r(x_l) : l \in I, r \in \mathbb{Q} \cap (0, \infty)\}.$$

Let $G \subseteq M$ be a non-empty open set. Then $G = \bigcup U$, where the union is taken over all $U \in \mathcal{B}$ with $U \subseteq G$.

Proof The relation $G \supseteq \bigcup_{U \in \mathcal{B}, U \subseteq G} U$ is obvious. So we only need to show that $G \subseteq \bigcup_{U \in \mathcal{B}, U \subseteq G} U$. For this assume that $x \in G$. As G is open, we find an $\varepsilon > 0$ such that $U_\varepsilon(x) \subseteq G$. Because the system $(x_l)_{l \in I}$ is dense in M, we find an $l_0 \in I$ such that $d(x, x_{l_0}) < \varepsilon/2$. Then we choose a rational number $r \in (d(x, x_{l_0}), \varepsilon/2)$ and get that for $y \in U_r(x_{l_0})$ that

$$d(x, y) \leqslant d(x, x_{l_0}) + d(x_{l_0}, y) < \frac{\varepsilon}{2} + r < \varepsilon$$

so that $x \in U_r(x_{l_0}) \subseteq U_\varepsilon(x) \subseteq G$. □

Proposition 4.1.4 says that there is a countable system $\mathcal{B}$ of open balls such that each open set G is a union of elements from $\mathcal{B}$. In the language of topology, $\mathcal{B}$ is a countable basis of the topology generated by the metric, so that a separable metric space is a *second countable topological space*.

Let us turn to continuous functions between metric spaces:

Definition 4.1.5 *Let $[M_1, d_1]$ and $[M_2, d_2]$ be metric spaces. A function $f : M_1 \to M_2$ is* **continuous** *provided that $\lim_{n \to \infty} d_1(x_n, x) = 0$ implies that $\lim_{n \to \infty} d_2(f(x_n), f(x)) = 0$.*

This definition of continuity is the sequential continuity. Although the definition reflects the intuition of continuity, in topology there is an equivalent definition which bridges between topology and the measurability concept in measure theory, and which allows generalizations more easily:

Proposition 4.1.6 *Let $[M_1, d_1]$ and $[M_2, d_2]$ be metric spaces. A function $f : M_1 \to M_2$ is continuous if and only if $f^{-1}(G)$ is open in M_1 for all open sets $G \subseteq M_2$.*

Finally, we shall need the following two results:

Lemma 4.1.7 (Urysohn's[2] Lemma) *Let $[M, d]$ be a metric space and let F_0 and F_1 be non-empty closed sets such that $F_0 \cap F_1 = \emptyset$. Then there exists a continuous function $f : M \to [0, 1]$ such that $f(x) = i$ for $x \in F_i$.*

---

[2] Pavel Samuilovich Urysohn, 03/02/1898 (Odessa, Ukraine)–17/08/1924 (Batz-sur-Mer, France).

Theorem 4.1.8 (Partition of Unity) *Assume a compact metric space $[M, d]$, a compact non-empty subset $K \subseteq M$, and open non-empty sets $G_1, \ldots, G_N \subseteq M$ such that $K \subseteq G_1 \cup \cdots \cup G_N$. Then there exist continuous functions $\varphi_1, \ldots, \varphi_N : K \to [0, 1]$ such that $\overline{\{x \in K : \varphi_n(x) \neq 0\}} \subseteq G_n$ and*

$$\varphi_1(x) + \cdots + \varphi_N(x) = 1 \quad \text{for} \quad x \in K.$$

4.2 Banach Spaces and Quasi-Banach Spaces

In this section we recall basic definitions and facts regarding Banach[3] and quasi-Banach spaces. We start by introducing the corresponding quasi-norms:

Definition 4.2.1 Let E be a linear space (over $\mathbb{R}$ or $\mathbb{C}$), $c \geqslant 1$, and $0 < p \leqslant 1$.

(1) A map $\|\cdot\|_E : E \to [0, \infty)$ is called **norm** and $[E, \|\cdot\|_E]$ **normed space** provided that

   (N1) $\|x\|_E = 0$ if and only if $x = 0$,
   (N2) $\|\lambda x\|_E = |\lambda| \|x\|_E$ for all $\lambda \in \mathbb{R}$ ($\lambda \in \mathbb{C}$) and $x \in E$,
   (N3) $\|x + y\|_E \leq \|x\|_E + \|y\|_E$ for all $x, y \in E$.

(2) If condition (N3) is replaced by

   (N4) $\|x + y\|_E \leq c[\|x\|_E + \|y\|_E]$ for all $x, y \in E$,

   then $\|\cdot\|_E$ is called a **quasi-norm** and $[E, \|\cdot\|]$ a **quasi-normed space**.

(3) If condition (N3) is replaced by

   (N5) $\|x + y\|_E^p \leq \|x\|_E^p + \|y\|_E^p$ for all $x, y \in E$,

   then $\|\cdot\|_E$ is called a *p*-**norm** and $[E, \|\cdot\|]$ a *p*-**normed space**.

We have the following relations:

Theorem 4.2.2 *Let E be a linear space over $\mathbb{R}$ or $\mathbb{C}$.*

(1) *A norm is a p-norm with $p = 1$ and a quasi-norm with $c = 1$.*
(2) *A p-norm is a quasi-norm with constant $c = 2^{\frac{1}{p}-1}$.*
(3) **Aoki**[4]-**Rolewicz**[5]: *Assume a quasi-norm with constant $c > 1$ and define $p \in (0, 1)$ by $2 = (2c)^p$. Then there is a p-norm $\|\cdot\|_{E,p}$ on E such that*

$$\|x\|_{E,p} \leqslant \|x\|_E \leqslant 2^{\frac{1}{p}} \|x\|_{E,p}.$$

---

[3] Stefan Banach, 30/03/1892 (Krakow, Poland)–31/08/1945 (Lvov, now Lviv, Ukraine).
[4] Tosio Aoki, 14/08/1910 (Kasai, Japan)–18/01/1989 (Tokyo, Japan).
[5] Stefan Henryk Rolewicz, 15/03/1932 (Brest, Belarus)–09/07/2015 (Warsaw, Poland).

4.2 Banach Spaces and Quasi-Banach Spaces

(4) *If $[E, \|\cdot\|_E]$ is a p-normed space for some $0 < p \leq 1$, then*

$$d_{E,p}(x, y) := \|x - y\|_E^p$$

defines a metric on E.

Proof (1) follows directly from the definition. (2) We can assume that $0 < p < 1$. Letting $A := 1/p \in (1, \infty)$ and $B := 1/(1-p) \in (1, \infty)$ we get $1 = (1/A)+(1/B)$ and, by Hölder's inequality, that

$$\|x + y\|_E \leq [\|x\|_E^p + \|y\|_E^p]^{\frac{1}{p}} \leq [\|x\|_E^{Ap} + \|y\|_E^{Ap}]^{\frac{1}{Ap}} 2^{\frac{1}{pB}} = c[\|x\|_E + \|y\|_E].$$

(3) We define the quantity

$$\|x\|_{E,p} := \inf \left\{ \left(\sum_{i=1}^n \|x_i\|_E^p \right)^{\frac{1}{p}} : x = x_1 + \cdots + x_n, n \in \mathbb{N} \right\}.$$

Letting $n = 1$ we see that $\|x\|_{E,p} \leq \|x\|_E$. To verify the opposite inequality we first prove the following fact: If $x = \sum_{i=1}^n x_i$, then

$$\|x\|_E \leq \max \left\{ 2^{\frac{n_i}{p}} \|x_i\|_E : i = 1, \ldots, n \right\} \tag{4.1}$$

for all $n_i \in \mathbb{N}_0$ with $\sum_{i=1}^n 2^{-n_i} \leq 1$. We proceed by induction over $n \in \mathbb{N}$ and start with $n = 1$, so that $x = x_1$ and the inequality is satisfied. Assume that we did prove the inequality for $n - 1$ for some $n \geq 2$. To show the inequality for n we assume $\sum_{i=1}^n 2^{-n_i} \leq 1$ and find two non-empty disjoint sets $I_1 \cup I_2 = \{1, \ldots, n\}$ such that $\sum_{i \in I_1} 2^{-n_i} \leq 1/2$ and $\sum_{i \in I_2} 2^{-n_i} \leq 1/2$. With this we get that

$$\|x\|_E = \left\| \left(\sum_{i \in I_1} x_i \right) + \left(\sum_{i \in I_2} x_i \right) \right\|_E$$

$$\leq c \left[\left\| \sum_{i \in I_1} x_i \right\|_E + \left\| \sum_{i \in I_2} x_i \right\|_E \right]$$

$$\leq c \left[\max \left\{ 2^{\frac{n_i-1}{p}} \|x_i\|_E : i \in I_1 \right\} + \max \left\{ 2^{\frac{n_i-1}{p}} \|x_i\|_E : i \in I_2 \right\} \right]$$

$$\leq 2c \max \left\{ 2^{\frac{n_i-1}{p}} \|x_i\|_E : i = 1, \ldots, n \right\}$$

$$= (2c) 2^{-\frac{1}{p}} \max \left\{ 2^{\frac{n_i}{p}} \|x_i\|_E : i = 1, \ldots, n \right\}$$

$$= \max \left\{ 2^{\frac{n_i}{p}} \|x_i\|_E : i = 1, \ldots, n \right\}.$$

Now we use (4.1) to finish the proof of (3): For a given sum $x = \sum_{i=1}^n x_i$, where we may assume $\|x_i\|_E > 0$ for all i, we find $n_i \in \mathbb{N}_0$ with

$$\frac{1}{2^{n_i}} \leqslant \frac{\|x_i\|_E^p}{\sum_{j=1}^n \|x_j\|_E^p} \leqslant \frac{2}{2^{n_i}}.$$

This implies that $\sum_{i=1}^n 2^{-n_i} \leqslant 1$ so that

$$\|x\|_E^p \leqslant \max\{2^{n_i}\|x_i\|_E^p : i = 1, \ldots, n\} \leqslant 2\sum_{j=1}^n \|x_j\|_E^p.$$

The proof that $\|\cdot\|_{E,p}$ is a p-norm is the subject of Exercise 1.
(4) We have $d_{E,p}(x, y) = 0$ if and only if $\|x - y\|_E = 0$, and this is equivalent to $x = y$ by (N1). The property (N2) for $\lambda = -1$ gives $d_{E,p}(x, y) = \|x - y\|_E^p = \|y - x\|_E^p = d_{E,p}(y, x)$. Finally, (N5) implies for $x, y, z \in E$ that

$$d_{E,p}(x, z) = \|x - z\|_E^p = \|(x - y) + (y - z)\|_E^p$$
$$\leqslant \|x - y\|_E^p + \|y - z\|_E^p = d_{E,p}(x, y) + d_{E,p}(y, z).$$

□

By Theorem 4.2.2((3),(4)) we can equip any quasi-normed space with an equivalent p-norm for some $0 < p \leqslant 1$, and this p-norm yields to a metric. Therefore we can define the completeness by this metric:

Definition 4.2.3

(1) Let $0 < p \leqslant 1$. Then $[E, \|\cdot\|_E]$ is called a p-**normed Banach space** provided that $\|\cdot\|_E$ is a p-norm and the metric space $[E, d_{E,p}]$ is a complete. For $p = 1$ we simply say that $[E, \|\cdot\|_E]$ is a **Banach space**.
(2) A quasi-normed space $[E, \|\cdot\|_E]$ is called **quasi-Banach space** provided that $[E, \|\cdot\|_{E,p}]$ is a p-normed Banach space under a p-norm $\|\cdot\|_{E,p}$ that satisfies $\frac{1}{\kappa}\|x\|_E \leqslant \|\cdot\|_{E,p} \leqslant \kappa\|x\|_E$ for all $x \in E$ for some $\kappa \geqslant 1$.

Remark 4.2.4 One can formulate the convergence in and the completeness of a quasi-normed space $[E, \|\cdot\|_E]$ directly without referring to metric spaces:

(1) A sequence $(x_n)_{n=1}^\infty \subseteq E$ converges to $x \in E$ if and only if $\lim_{n\to\infty} \|x - x_n\|_E = 0$.
(2) A sequence $(x_n)_{n=1}^\infty \subseteq E$ is a Cauchy sequence if and only if for all $\varepsilon > 0$ there is an $n(\varepsilon) \in \mathbb{N}$ such that $\|x_n - x_m\|_E \leqslant \varepsilon$ for $m, n \geqslant n(\varepsilon)$.
(3) A quasi-normed space $[E, \|\cdot\|_E]$ is a quasi-Banach space if for any Cauchy sequence $(x_n)_{n=1}^\infty \subseteq E$ there is a limit $x \in E$.

The following criteria for the completeness of a Banach space is useful:

Lemma 4.2.5 *Let $[E, \|\cdot\|_E]$ be a normed space. Then the following assertions are equivalent:*

(1) *$[E, \|\cdot\|_E]$ is a Banach space.*
(2) *For all sequences $(x_n)_{n=1}^\infty \subseteq E$ such that $\sum_{n=1}^\infty \|x_n\|_E < \infty$ one has that there is an $x \in E$ such that $x = \lim_{N\to\infty} \sum_{n=1}^N x_n$ in E.*

Proof (1) $\Rightarrow$ (2) We let $S_N := \sum_{n=1}^N x_n$. Then, for $N > M \geq 1$ one has

$$\|S_N - S_M\|_E = \left\|\sum_{n=M+1}^N x_n\right\|_E \leq \sum_{n=M+1}^N \|x_n\|_E < \varepsilon$$

whenever $M \geq N(\varepsilon)$. Therefore $(S_N)_{N=1}^\infty$ is a Cauchy sequence that converges as E is complete.

(2) $\Rightarrow$ (1) We assume a Cauchy sequence $(S_N)_{N=1}^\infty \subseteq E$ and find $1 \leq N_1 < N_2 < N_3 < \cdots$ such that

$$\|S_N - S_M\|_E \leq \frac{1}{2^l} \quad \text{for} \quad N > M \geq N_l.$$

Therefore $\sum_{l=1}^\infty \|S_{N_{l+1}} - S_{N_l}\|_E < \infty$ and

$$S := S_{N_1} + \sum_{l=1}^\infty (S_{N_{l+1}} - S_{N_l})$$

exist in E. To show $S = \lim_{N\to\infty} S_N$ we consider

$$\|S - S_N\|_E \leq \|S - S_{N_l}\|_E + \|S_{N_l} - S_N\|_E$$

for $N, l \in \mathbb{N}$. If $N \geq N_l$, then $\|S_{N_l} - S_N\|_E \leq \frac{1}{2^l}$ and

$$\|S - S_{N_l}\|_E = \lim_{k\to\infty} \|S_{N_k} - S_{N_l}\|_E \leq \frac{1}{2^l}.$$

Hence $\|S - S_N\|_E \leq \frac{2}{2^l}$ for $N \geq N_l$. $\square$

4.3 Linear Operators Between Banach Spaces

Proposition 4.3.1 *Let E and F be Banach spaces, both over $\mathbb{K}$ with $\mathbb{K} \in \{\mathbb{R}, \mathbb{C}\}$. Assume a linear map $T : E \to F$, i.e.*

$$T(\alpha x + \beta y) = \alpha T x + \beta T y$$

for all $x, y \in E$ and $\alpha, \beta \in \mathbb{K}$. Then the following assertions are equivalent:

(1) *The map $T : E \to F$ is continuous as map between metric spaces.*
(2) *There exists a constant $c \geq 0$ such that $\|Tx\|_F \leq c\|x\|_E$ for all $x \in E$.*

Proof (2) $\Rightarrow$ (1) is obvious as we get $\|Tx_n - Tx\|_F \leq c\|x_n - x\|_E$ for all $x \in E$.
(1) $\Rightarrow$ (2) We assume $\|x_n\|_E = 1$ but $\lim_{n \to \infty} \|Tx_n\|_F = \infty$ with $\|Tx_n\|_F > 0$ for all $n \in \mathbb{N}$. Then we define $x_n^0 := x_n/\|Tx_n\|_F$ and get that $\|x_n^0\|_E \to 0$ but $\|Tx_n^0\|_F = 1$ which is a contradiction. $\square$

Definition 4.3.2 Let E and F be Banach spaces, both over $\mathbb{K}$ with $\mathbb{K} \in \{\mathbb{R}, \mathbb{C}\}$. The collection of all linear and continuous maps $T : E \to F$ is denoted by $\mathcal{L}(E, F)$. Furthermore, we let

$$\|T\|_{\mathcal{L}} := \inf\{c \geq 0 : \|Tx\|_F \leq c\|x\|_E \text{ for all } x \in E\}.$$

From the definition it is obvious that the infimum is attained and that one has

$$\|T\|_{\mathcal{L}} = \sup\{\|Tx\|_F : \|x\|_E \leq 1\}.$$

Proposition 4.3.3 *The space $[\mathcal{L}(E, F), \|\cdot\|_{\mathcal{L}}]$ is a Banach space.*

Proof We leave it to the reader to check that $[\mathcal{L}(E, F), \|\cdot\|_{\mathcal{L}}]$ is a normed space. To verify the completeness we assume a Cauchy sequence $(T_n)_{n \in \mathbb{N}} \subseteq \mathcal{L}(E, F)$. By definition this means that for all $\varepsilon > 0$ there is an $n(\varepsilon) \in \mathbb{N}$ such that

$$\|T_n - T_m\|_{\mathcal{L}} \leq \varepsilon \quad \text{for} \quad m, n \geq n(\varepsilon).$$

From this we get that

$$\|T_n\|_{\mathcal{L}} \leq \|T_{n(\varepsilon)}\|_{\mathcal{L}} + \varepsilon \quad \text{for} \quad n \geq n(\varepsilon),$$

which implies $A := \sup_{n \in \mathbb{N}} \|T_n\|_{\mathcal{L}} < \infty$, and also

$$\|T_n x - T_m x\|_F \leq \varepsilon \|x\|_E \quad \text{for} \quad m, n \geq n(\varepsilon).$$

Consequently, $(T_n x)_{n \in \mathbb{N}}$ is a Cauchy sequence in F with a limit

$$L(x) := \lim_{n \to \infty} T_n x \in F.$$

The map $x \mapsto L(x)$ is obviously linear and satisfies

$$\|L(x)\|_F = \lim_{n \to \infty} \|T_n x\|_F \leq A\|x\|_E.$$

Hence $L \in \mathcal{L}(E, F)$. Finally, for $n \geq n(\varepsilon)$ we get

$$\begin{aligned}
\|L - T_n\|_{\mathcal{L}} &= \sup\{\|L(x) - T_n x\|_F : \|x\|_E \leq 1\} \\
&= \sup\left\{\lim_{m \to \infty, m \geq n} \|T_m x - T_n x\|_F : \|x\|_E \leq 1\right\} \\
&\leq \sup_{m \geq n} \|T_m - T_n\|_{\mathcal{L}} < \varepsilon,
\end{aligned}$$

so that $\lim_{n \to \infty} \|T_n - L\|_{\mathcal{L}} = 0$. $\square$

4.4 Main Theorems

For this section we assume a Banach space E over $\mathbb{R}$ and exclude the degenerated case $E = \{0\}$.

Definition 4.4.1 The space E^* of all linear functionals $a : E \to \mathbb{R}$ such that

$$\|a\|_{E^*} := \sup\{|\langle x, a \rangle| : \|x\|_E \leq 1\} < \infty,$$

where $\langle x, a \rangle$ denotes the application of a to $x \in E$, is called the **dual space** of E.

The definition is nothing else than $E^* = \mathcal{L}(E, \mathbb{R})$. By Proposition 4.3.3 we know that $[E^*, \|\cdot\|_{E^*}]$ is a Banach space. Now, as we know that E^* is a Banach space, we can define the bidual of a Banach space E as the dual of E^*, where we use the notation

$$E^{**} := (E^*)^*.$$

We denote the closed unit ball of E by

$$B_E := \{x \in E : \|x\|_E \leq 1\}.$$

The following Theorems 4.4.2, 4.4.3, and 4.4.4 are cornerstones of the Banach space theory. Theorems 4.4.2 and 4.4.3 belong to the family of *Hahn-Banach theorems*. The *closed graph* Theorem 4.4.4. is a tool to check the continuity of a linear operator $T : E \to F$.

Theorem 4.4.2 (Hahn-Banach Theorem) *Let E be a Banach space and $x \in E$. Then there exists an $a \in B_{E^*}$ such that $\|x\|_E = \langle x, a \rangle$.*

By the Hahn-Banach theorem we have that

$$\|x\|_E = \sup\{|\langle x, a \rangle| : a \in B_{E^*}\}, \qquad (4.2)$$

in particular one has $E^* \neq \{0\}$ whenever $E \neq \{0\}$.

Theorem 4.4.3 (Separation Theorem) *Assume a Banach space E and nonempty, convex, and disjoint subsets $A, B \subseteq E$, where A is compact and B is closed. Then there exists an $a \in E^*$ and $-\infty < \gamma_1 < \gamma_2 < \infty$ such that*

$$\langle x, a \rangle < \gamma_1 < \gamma_2 < \langle y, a \rangle \quad \text{for all} \quad x \in A \text{ and } y \in B.$$

Theorem 4.4.4 (Closed Graph Theorem) *Assume a linear operator $T : E \to F$ between Banach spaces E and F such that T is closed, i.e. if $x_n \to x$ in E and $Tx_n \to y$ in F, then $Tx = y$. Then the operator T is continuous.*

The name closed graph theorem originates from the fact that the graph $\{(x, Tx) : x \in E\} \subseteq E \times F$ is supposed to be closed to deduce the continuity of the operator.

Coming back to (4.2), in order to describe the norm $\|x\|_E$ the full ball B_{E^*} might not be needed:

Definition 4.4.5 We call a subset $A \subseteq B_{E^*}$ norming if $\|x\|_E = \sup_{a \in A} |\langle x, a \rangle|$.

As a Banach space is a metric space, we have the notion of separability: A Banach space $[E, \|\cdot\|_E]$ is separable provided that there is a countable subset $A \subseteq E$ such that for all $x \in E$ and $\varepsilon > 0$ there is an element $y \in A$ such that $\|x - y\|_E < \varepsilon$.

Proposition 4.4.6 *Assume that the Banach space E is separable, then there exists a countable norming set $A \subseteq B_{E^*}$.*

Proof There is a countable dense set $D \subseteq E$, i.e. for all $\varepsilon > 0$ and $x \in E$ there is a $d \in D$ such that $\|x - d\|_E < \varepsilon$. By the Hahn-Banach theorem we choose for each $d \in D$ an $a_d \in B_{E^*}$ such that $\|d\|_E = \langle d, a_d \rangle$. Then $A := \{a_d \in B_{E^*} : d \in D\}$ is norming: Indeed, for $x \in E$ and $\varepsilon > 0$ we find a $d(x, \varepsilon) \in D$ such that $\|x - d(x, \varepsilon)\|_E < \varepsilon$ and get

$$\|x\|_E \geq \sup_{a \in A} |\langle x, a \rangle| \geq \sup_{a \in A} |\langle d(x, \varepsilon), a \rangle| - \varepsilon$$

$$\geq |\langle d(x, \varepsilon), a_{d(x,\varepsilon)} \rangle| - \varepsilon = \|d(x, \varepsilon)\|_E - \varepsilon \geq \|x\|_E - 2\varepsilon.$$

□

4.5 Comments

Section 4.1 For an introduction to metric spaces the reader can consult [47], in particular the proof of Proposition 4.1.3(5) one finds in [47, (3.16.1)]. Urysohn's Lemma (Lemma 4.1.7) was proven in [185]. A proof of Theorem 4.1.8 can be found in [165, Theorem 2.13].

Section 4.2 Item (3) of Theorem 4.2.2 is due Aoki [2] and Rolewicz [160], see also [161, Section 3.2] and [98, Theorem 1.3]. The result is more general than what we stated here. Our proof is taken from [11, Lemma 3.10.1].

Section 4.4 For an introduction to functional analysis the reader is referred to, among others, Conway [42], Rudin [166], and Werner [197]. The Hahn-Banach theorems 4.4.2 and 4.4.3 can be found [166, page 53, Theorem 3.4], the closed graph Theorem 4.4.4 in [166, Theorem 2.15].

Chapter 5
*Measures on Metric Spaces

In this chapter we discuss measures on metric spaces $[M, d]$ and their regularity with respect to closed, compact, and open sets. On a metric space we have the notion of open sets. So also the Borel σ-algebra as the smallest σ-algebra containing the open sets can be defined. For a measure given on the Borel σ-algebra of a metric space the concepts of outer and inner regularity are introduced. The main result of this chapter is Ulam's theorem about the tightness of a finite measure on a complete separable metric space. The benefit of considering measures on metric spaces, instead of using the Euclidean space $\mathbb{R}^d$, comes from the fact that many techniques known from $\mathbb{R}^d$ work for metric spaces as well and the notation might even be more simple compared to $\mathbb{R}^d$. But we will see differences as well.

5.1 The Borel σ-Algebra $\mathcal{B}(M)$

The Borel-σ-algebra on metric spaces is defined in accordance with $\mathcal{B}(\mathbb{R})$:

Definition 5.1.1 Given a metric space $[M, d]$, the Borel-σ-algebra $\mathcal{B}(M)$ is the smallest σ-algebra that contains all open sets $G \subseteq M$.

As the closed sets are the complements of the open sets, we also have

$$\mathcal{B}(M) = \sigma(F \subseteq M : F \text{ closed}). \tag{5.1}$$

But there are also some differences compared to $M = \mathbb{R}$. For example, in general the open balls do not generate $\mathcal{B}(M)$, as shown by the following example, so that Proposition 3.3.6(5) does not hold for general metric spaces:

Example 5.1.2 Consider a metric space $[M, d]$ with $d(x, y) := 1$ if $x \neq y$ and $d(x, y) := 0$ if $x = y$. With this metric we get

$$U_\varepsilon(x) = \begin{cases} \{x\} & : \varepsilon \in (0, 1) \\ M & : \varepsilon \in [1, \infty) \end{cases}.$$

Consequently,

$$\sigma\left(U_\varepsilon(x) : x \in M, \varepsilon > 0\right) = \sigma\left(\{x\} : x \in M\right)$$
$$= \{A \subseteq M : A \text{ is countable or } A^c \text{ is countable}\}.$$

On the other side, each subset $A \subseteq M$ is open in this metric space, so that

$$\mathcal{B}(M) = 2^M.$$

For example, for $M = [0, 1]$ equipped with the above discrete metric we have $[0, 1/2] \in \mathcal{B}(M)$ but $[0, 1/2] \notin \sigma\left(U_\varepsilon(x) : x \in M, \varepsilon > 0\right)$.

The problem described in Example 5.1.2 does not occur for separable metric spaces:

Proposition 5.1.3 *For a separable metric space $[M, d]$ we have that*

$$\mathcal{B}(M) = \sigma\left(U_\varepsilon(x) : x \in M, \varepsilon > 0\right) = \sigma\left(U_\varepsilon(x) : x \in D, \varepsilon \in \mathbb{Q} \cap (0, \infty) > 0\right),$$

where $D \subseteq M$ is any countable dense subset.

Proof The statement follows immediately from Proposition 4.1.4. □

5.2 Regularity of Measures on Metric Spaces

Given a metric space $[M, d]$, a Borel set $A \in \mathcal{B}(M)$, and $\varepsilon > 0$, we look for a closed or compact set F and an open set G such that

$$F \subseteq A \subseteq G \quad \text{and} \quad \mu(G \setminus F) < \varepsilon.$$

So we aim to approximate a measurable set by a closed, compact, or open set with respect to μ. Note that the above inequality implies that $\mu(G \setminus A) < \varepsilon$, called outer regularity, and $\mu(A \setminus F) < \varepsilon$, called inner regularity with respect to closed or compact sets. This regularity provides an intrinsic link between real analysis and probability.

5.2 Regularity of Measures on Metric Spaces

We start by showing that any finite measure is inner and outer regular with respect to closed and open sets, respectively:

Theorem 5.2.1 *Let $[M, d]$ be a metric space, μ be a finite measure on $\mathcal{B}(M)$, and $\varepsilon > 0$. Then for all $A \in \mathcal{B}(M)$ there exists a closed set $F \subseteq M$ and an open set $G \subseteq M$ such that*

$$F \subseteq A \subseteq G \quad \text{and} \quad \mu(G \setminus F) < \varepsilon.$$

Proof We define $\mathcal{G}$ to be the family of all $A \in \mathcal{B}(M)$ such that for all $\varepsilon > 0$ there are a closed set F and an open set G with the properties that

$$F \subseteq A \subseteq G \quad \text{and} \quad \mu(G \setminus F) < \varepsilon.$$

We verify that $\mathcal{G}$ is a sub-σ-algebra of $\mathcal{B}(M)$ containing all closed sets, whence $\mathcal{G} = \mathcal{B}(M)$ by relation (5.1).

(a) Let $A \subseteq M$ be a closed set. If $A = \emptyset$ or $A = M$, then we can choose $F = G := A$ so that $\mu(G \setminus F) = \mu(\emptyset) = 0$. Therefore we assume that $A \neq \emptyset$ and $A \neq M$. Now, for A being closed, we choose of course $F := A$ so that $\mu(A \setminus F) = \mu(\emptyset) = 0$. Regarding the open set G we consider the sequence of open sets $G_n := \bigcup_{x \in A} U_{\frac{1}{n}}(x)$. As $G_1 \supseteq G_2 \supseteq G_3 \supseteq \cdots \supseteq A$ and $\bigcap_{n=1}^{\infty} G_n = A$ it follows that $\lim_{n \to \infty} \mu(G_n \setminus A) = 0$ by the continuity from above of the finite measure μ.

(b) We show that $\mathcal{G}$ is a σ-algebra. As M is closed, we have $M \in \mathcal{G}$. Moreover, as $F \subseteq A \subseteq G$ implies that $G^c \subseteq A^c \subseteq F^c$ (where G^c is closed and F^c is open) and $G \setminus F = F^c \setminus G^c$, we get that $A \in \mathcal{G}$ implies $A^c \in \mathcal{G}$. Finally, let us assume that $A_1, A_2, \ldots \in \mathcal{G}$. Given $\varepsilon > 0$, there are $F_n \subseteq A_n \subseteq G_n$ with $\mu(G_n \setminus F_n) < \frac{\varepsilon}{2^n}$. Define the open set $G := \bigcup_{n=1}^{\infty} G_n$, the closed sets $C_N := \bigcup_{n=1}^{N} F_n$, and the measurable set $C := \bigcup_{n=1}^{\infty} F_n$. Then

$$C \subseteq \bigcup_{n=1}^{\infty} A_n \subseteq G$$

and

$$\mu(G \setminus C) \leq \mu\left(\bigcup_{n=1}^{\infty} (G_n \setminus F_n)\right) \leq \sum_{n=1}^{\infty} \mu(G_n \setminus F_n) < \sum_{n=1}^{\infty} \frac{\varepsilon}{2^n} = \varepsilon.$$

Finally, we observe that

$$\lim_{N \to \infty} \mu(G \setminus C_N) = \mu(G \setminus C) < \varepsilon \quad \text{and} \quad C_N \subseteq A \subseteq G.$$

Hence we find an $N_\varepsilon \in \mathbb{N}$ such that $\mu(G \setminus C_{N_\varepsilon}) < \varepsilon$. □

Remark 5.2.2 A set $A \subseteq M$ in a metric space $[M, d]$ is called G_δ-set provided A is the countable intersection of open sets. So we did show in part (a) of the proof of Theorem 5.2.1 that each closed set is a G_δ-set.

Example 5.2.3 The outer regularity with respect to open sets does not hold for σ-finite measures in general: We equip $M := [0, 1]$ with the Euclidean distance $d(x, y) := |x - y|$, take $\mathcal{F} := \mathcal{B}([0, 1])$, and define the σ-finite measure $\mu := \sum_{n=1}^\infty \delta_{\frac{1}{n}}$. Then $\mu(\{0\}) = 0$, but for any open set $G \supseteq \{0\}$ we have that $\mu(G) = \infty$ as this neighborhood always contains the points $1/n$ for some $n \geq n_0$. So, this measure μ is not outer-regular with respect to the open sets.

In view of Example 5.2.3 a modification of Theorem 5.2.1 regarding the outer regularity is needed for the case of σ-finite measures:

Theorem 5.2.4 *Let $[M, d]$ be a metric space, μ be σ-finite measure, assume $A \in \mathcal{F}$, and let $\varepsilon > 0$.*

(1) **Inner regularity:** *If additionally $\mu(A) < \infty$, then there exists a closed set $F \subseteq M$ with*

$$F \subseteq A \quad \text{and} \quad \mu(A \setminus F) < \varepsilon.$$

(2) **Inner and outer regularity:** *If there are open sets $\emptyset \neq O_1 \subseteq O_2 \subseteq O_3 \subseteq \cdots \subseteq M$ such that $\bigcup_{n=1}^\infty O_n = M$ and $\mu(O_n) < \infty$ for all $n \in \mathbb{N}$, then there exists a closed set $F \subseteq M$ and an open set $G \subseteq M$ with*

$$F \subseteq A \subseteq G \quad \text{and} \quad \mu(G \setminus F) < \varepsilon.$$

Proof We may assume that $\mu(M) = \infty$, otherwise Theorem 5.2.1 applies, where in (2) we can take $O_n := M$.
(1) We assume a partition $M = \bigcup_{n=1}^\infty M_n$ with $\emptyset \neq M_n \in \mathcal{B}(M)$ and $\mu(M_n) \in (0, \infty)$ and define the finite measures μ_n on $(M, \mathcal{B}(M))$ by

$$\mu_n(A) := \mu(A \cap M_n).$$

By Theorem 5.2.1 the measures μ_n are inner regular so that for each $A \in \mathcal{B}(M)$ and $A_n := A \cap M_n$ there is a closed set $F_n \subseteq A_n \subseteq M_n$ with

$$\mu_n(A_n \setminus F_n) < \frac{\varepsilon}{2^n}.$$

Now we define the closed sets $C_N := \bigcup_{n=1}^N F_n$. Then

$$\mu(A \setminus C_N) = \sum_{n=1}^N \mu_n(A_n \setminus F_n) + \sum_{n=N+1}^\infty \mu_n(A_n)$$

5.2 Regularity of Measures on Metric Spaces

$$< \sum_{n=1}^{N} \frac{\varepsilon}{2^n} + \mu\left(A \cap \left(\bigcup_{n=N+1}^{\infty} M_n\right)\right)$$

$$< \varepsilon + \mu\left(A \cap \left(\bigcup_{n=N+1}^{\infty} M_n\right)\right).$$

As $\mu(A) < \infty$ we have that

$$\lim_{N \to \infty} \mu\left(A \cap \left(\bigcup_{n=N+1}^{\infty} M_n\right)\right) = 0 \quad \text{so that} \quad \limsup_{N \to \infty} \mu(A \setminus C_N) \leq \varepsilon.$$

As $\varepsilon > 0$ was arbitrary, the proof of (1) is complete.

(2) We define $\nu_n(A) := \mu(A \cap O_n)$ for $A \in \mathcal{B}(M)$ and obtain finite measures ν_n on $(M, \mathcal{B}(M))$. These measures are outer regular so that there are open sets $G_n \supseteq O_n \cap A$ with

$$\nu_n(G_n \setminus (A \cap O_n)) < \frac{\varepsilon}{2^{n+1}}.$$

As G_n is open we can replace G_n by $G_n \cap O_n$, or equivalently, we may assume that $G_n \subseteq O_n$. This implies that

$$\mu(G_n \setminus A) = \mu(G_n \setminus (O_n \cap A)) = \nu_n(G_n \setminus (A \cap O_n)) < \frac{\varepsilon}{2^{n+1}}.$$

We define the open set $G := \bigcup_{n=1}^{\infty} G_n$ and get

$$\mu(G \setminus A) \leq \sum_{n=1}^{\infty} \mu(G_n \setminus A) < \sum_{n=1}^{\infty} \frac{\varepsilon}{2^{n+1}} = \frac{\varepsilon}{2}.$$

Now we apply the previous steps to A^c and find an open set $G' \subseteq M$ such that

$$A^c \subseteq G' \quad \text{and} \quad \mu(G' \setminus A^c) < \frac{\varepsilon}{2}.$$

Setting $F := (G')^c$ we get a closed set F with

$$\mu(A \setminus F) = \mu(F^c \setminus A^c) = \mu(G' \setminus A^c) < \frac{\varepsilon}{2}$$

so that $\mu(G \setminus F) < \varepsilon$. $\square$

Example (Example 5.2.3 Continued) We show that in Theorem 5.2.4(1) the condition $\mu(A) < \infty$ is needed: We take $A := \mathbb{R} \setminus \{0\}$ so that $\mu(A) = \infty$. Any closed set $F \subseteq A$ can only contain finitely many of the points $1/n$, $n = 1, 2, \ldots$ (otherwise we would get $0 \in A$), so that we have $\mu(F) < \infty$.

So far we considered the inner regularity with respect to closed sets. Now we aim to use compact sets (any compact set is closed, but the converse is not true in general) which takes us to Ulam's Theorem. As a preparation we start to consider pre-compact sets:

Proposition 5.2.5 *Assume that the metric space $[M, d]$ is separable and that μ is a finite measure on $\mathcal{B}(M)$. Then, given $\varepsilon > 0$, there is a pre-compact set $P_\varepsilon \subseteq M$ such that $\mu(M \setminus P_\varepsilon) < \varepsilon$.*

Proof Let $(x_l)_{l=1}^\infty \subseteq M$ be dense. Then

$$M = \bigcup_{l=1}^\infty U_{\frac{1}{n}}(x_l) \quad \text{for all} \quad n \in \mathbb{N}.$$

Given $\varepsilon > 0$ we choose an $L_{n,\varepsilon} \in \mathbb{N}$ such that

$$\mu(M \setminus G_{n,\varepsilon}) < \frac{\varepsilon}{2^n} \quad \text{for} \quad G_{n,\varepsilon} := \bigcup_{l=1}^{L_{n,\varepsilon}} U_{\frac{1}{n}}(x_l).$$

Finally, we let

$$P_\varepsilon := \bigcap_{n=1}^\infty G_{n,\varepsilon},$$

so that $\mu(M \setminus P_\varepsilon) \leq \sum_{n=1}^\infty (M \setminus G_{n,\varepsilon}) < \varepsilon$ and the set P_ε is pre-compact. □

Definition 5.2.6 Let $[M, d]$ be a metric space and μ be a finite measure on $\mathcal{B}(M)$. The measure μ is **tight** provided that for all $\varepsilon > 0$ there is a compact set $K \subseteq M$ such that

$$\mu(M \setminus K) < \varepsilon.$$

Now we are ready for Ulam's theorem:

Theorem 5.2.7 (Ulam's[1] Theorem) *Let $[M, d]$ be a complete separable metric space and μ be a σ-finite measure on $\mathcal{B}(M)$.*

(1) *For each $A \in \mathcal{B}(M)$ with $\mu(A) < \infty$ and $\varepsilon > 0$ there is a compact set $K \subseteq M$ such that $\mu(A \setminus K) < \varepsilon$.*
(2) *If μ is a finite measure, then μ is tight.*

---

[1] Stanisław Marcin Ulam, 13/04/1909 (Lemberg, Austrian Empire, now Lviv, Ukraine)–13/05/1984 (Santa Fe, New Mexico, USA).

Proof (1) By Theorem 5.2.4(1) there is a closed set $F \subseteq A$ such that $\mu(A \setminus F) < \varepsilon/2$. By Proposition 5.2.5 we find a pre-compact set $P \subseteq F$ with $\mu(F \setminus P) < \varepsilon/2$. Let K be the closure of P. Then K is compact (as the metric space is complete) and satisfies $K \subseteq F$ with $\mu(F \setminus K) < \varepsilon/2$. Consequently, $\mu(A \setminus K) < \varepsilon$.
(2) follows directly from (1) by choosing $A := M$. □

In Corollary B.4.6 we give an example that shows that Theorem 5.2.7 fails without the assumption that the metric space is complete.

5.3 The Support of a Measure

Given a separable metric space $[M, d]$ and a measure μ on $\mathcal{B}(M)$, one looks for a set $S \subseteq M$ where, in some sense, the measure lives on. This is the support which we define now:

Definition 5.3.1 (Support of a Measure) Let $[M, d]$ be a separable metric space and μ be a finite measure on $\mathcal{B}(M)$. Then we let $\mathrm{supp}(\mu)$ be the union of all $x \in M$ such that $\mu(U_\varepsilon(x)) > 0$ for all $\varepsilon > 0$. The set $\mathrm{supp}(\mu)$ is called the **support of** μ.

The following properties show that the above definition of the support yields to what one would expect:

Proposition 5.3.2 *Let $[M, d]$ be a separable metric space and μ be a finite measure on $\mathcal{B}(M)$. Then the set $\mathrm{supp}(\mu) \subseteq M$ is the smallest closed set such that $\mu(\mathrm{supp}(\mu)) = \mu(M)$.*

Proof

(a) The set $\mathrm{supp}(\mu)$ is closed: By definition we have that $x \notin \mathrm{supp}(\mu)$ provided that there is an $\varepsilon > 0$ such that $\mu(U_\varepsilon(x)) = 0$. Then for $y \in U_\varepsilon(x)$ we find an $\eta > 0$ such that

$$U_\eta(y) \subseteq U_\varepsilon(x)$$

so that $\mu(U_\eta(y)) = 0$ as well and therefore $y \notin \mathrm{supp}(\mu)$. Consequently, $U_\varepsilon(x) \subseteq \mathrm{supp}(\mu)^c$, so that $\mathrm{supp}(\mu)^c$ is open and $\mathrm{supp}(\mu)$ is closed.

(b) We prove $\mu(\mathrm{supp}(\mu)) = \mu(M)$, which is equivalent to $\mu(\mathrm{supp}(\mu)^c) = 0$: The complement can be written as

$$\mathrm{supp}(\mu)^c = \bigcup_{x \notin \mathrm{supp}(\mu)} U_{\varepsilon_x}(x)$$

where $\varepsilon_x > 0$ is chosen such that $\mu(U_{\varepsilon_x}(x)) = 0$. By Proposition 4.1.4 there is a countable collection of open balls $U_n := U_{\varepsilon_n}(x_n)$ with $x_n \in M$ and $\varepsilon_n > 0$ such that for all $x \notin \mathrm{supp}(\mu)$,

$$U_{\varepsilon_x}(x) = \bigcup_{n \in I_x} U_n.$$

For $n \in I_x$ we deduce $\mu(U_n) = 0$. Therefore,

$$\mathrm{supp}(\mu)^c = \bigcup_{n \in \bigcup_{x \notin \mathrm{supp}(\mu)} I_x} U_n.$$

But this is a countable union of open balls with measure zero, so that $\mu(\mathrm{supp}(\mu)^c) = 0$.

(c) Assume a closed set $F \subseteq M$ such that $\mu(F) = \mu(M)$, which is equivalent to $\mu(F^c) = 0$. As F^c is open, for each $x \in F^c$ we find an $\varepsilon > 0$ such that $U_\varepsilon(x) \subseteq F^c$ implying $\mu(U_\varepsilon(x)) = 0$ and $x \notin \mathrm{supp}(\mu)$. Hence $F^c \subseteq \mathrm{supp}(\mu)^c$, so that $\mathrm{supp}(\mu) \subseteq F$. □

5.4 Exercises

Ex 1: Let λ be the Lebesgue measure on $(\mathbb{R}, \mathcal{B}(\mathbb{R}))$. Show that a set $N \in \mathcal{B}(\mathbb{R})$ satisfies $\lambda(N) = 0$ if and only if for all $\varepsilon > 0$ there are open intervals (a_n, b_n) with $-\infty < a_n < b_n < \infty$ such that

$$N \subseteq \bigcup_{n=1}^\infty (a_n, b_n) \quad \text{and} \quad \sum_{n=1}^\infty |b_n - a_n| < \varepsilon.$$

Ex 2: Let μ, ν be finite measures on $\mathcal{B}(M)$, where $[M, d]$ is a separable metric space. Prove that $\mathrm{supp}(\mu) \subseteq \mathrm{supp}(\nu)$ if $\mu \ll \nu$.

Ex 3: Let μ be a finite measure on $\mathcal{B}(M)$, where $[M, d]$ is a separable metric space. Prove the following:

(a) If $\mu = \sum_{n \in I} p_n \delta_{x_n}$, where $I \neq \emptyset$ is countable with $p_n > 0$ and where $x_n \neq x_m$ for $n \neq m$, then $\mathrm{supp}(\mu)$ is the closure of the set $\{x_n : n \in I\} \subseteq M$.

(b) The support $\mathrm{supp}(\mu)$ is finite ($\mathrm{supp}(\mu)$ might be empty) if and only if μ is a finite sum of Dirac measures.

(c) Give an example of a measure μ, where $\mathrm{supp}(\mu)$ is the closure of a countable set, but μ does not have any atoms (in particular, is not a countable sum of Dirac measures).

5.5 Comments

Measures on metric space and more general topological structures have been investigated intensively in the literature, see [144, 172, 189]. The proof of Theorem 5.2.4 for finite measures is taken from [17, Theorem 1.1, page 7]. Ulam's theorem Theorem 5.2.7 was proven in [141]. Regarding the support of a measure the reader is referred to Parthasarathy [144, pp. 27], also called spectrum of a measure.

Chapter 6
Random Variables and Measurable Maps

So far we made ourselves familiar with probability spaces $(\Omega, \mathcal{F}, \mathbb{P})$. Our next step is to introduce the concept of random variables, which reveals the full strength of probability theory and enables its application in other branches of pure mathematics, to statistics, and to stochastic modelling. For example, given a probability space $(\Omega, \mathcal{F}, \mathbb{P})$, we wish to consider functions $f : \Omega \to \mathbb{R}$ that describe certain random phenomena. In order to study probabilistic properties of these phenomena, one is naturally interested in expressions of the form

$$\mathbb{P}(\{\omega \in \Omega : f(\omega) \in (a, b)\}) \quad \text{where} \quad -\infty < a < b < \infty.$$

This leads us to the condition

$$\{\omega \in \Omega : f(\omega) \in (a, b)\} \in \mathcal{F}$$

which will mean that the function f is measurable and f will be called a random variable. However the chapter starts with introducing random variables as pointwise limits of measurable simple functions and the above description will be given as a characterization. After considering important properties of random variables, generalizations of measurable maps with values in $\mathbb{R}^d$ and metric spaces are introduced. Finally the difference between discrete and continuous random variables is clarified.

Before we start, we recall the notion of the **pre-image**. Given any non-empty sets Ω and M, and a map $f : \Omega \to M$, we let

$$f^{-1}(B) := \{\omega \in \Omega : f(\omega) \in B\} \quad \text{for} \quad B \subseteq M.$$

Notice that f^{-1} maps a *set* from M to a *set* from Ω, that means $f^{-1} : 2^M \to 2^\Omega$. It is *not* a map $f^{-1} : M \to \Omega$.

6.1 Measurable Maps with Values in $\mathbb{R}$ and Random Variables

We start with the most simple random variables.

Definition 6.1.1 (Simple Function) Let $(\Omega, \mathcal{F})$ be a measurable space. A function $f : \Omega \to \mathbb{R}$ is called **simple function**, provided that there are $n \in \mathbb{N}$, $\alpha_1, \ldots, \alpha_n \in \mathbb{R}$, and $A_1, \ldots, A_n \in \mathcal{F}$ such that f can be written as

$$f(\omega) = \sum_{i=1}^{n} \alpha_i \mathbb{1}_{A_i}(\omega) \quad \text{where} \quad \mathbb{1}_{A_i}(\omega) = \begin{cases} 1 : \omega \in A_i \\ 0 : \omega \notin A_i \end{cases}.$$

If the $\alpha_1, \ldots, \alpha_n \in \mathbb{R}$ are distinct and the $A_1, \ldots, A_n \in \mathcal{F}$ are non-empty and form a partition of Ω, then the representation is called **canonical representation** of f.

Remark 6.1.2

(1) Note that each simple function has a unique canonical representation (see Exercise 1).
(2) It follows from (1) that all level sets of a simple function are measurable. In fact, for all $-\infty \leqslant a < b \leqslant \infty$ we have that

$$f^{-1}((a, b)) = \bigcup_{\alpha_i \in (a,b)} A_i \in \mathcal{F}$$

provided that f is given in canonical representation.

Definition 6.1.1 concerns only functions that take finitely many values, which will be too restrictive in the future. So we wish to extend this definition.

Definition 6.1.3 (Measurable Function, Random Variable) Let $(\Omega, \mathcal{F})$ be a measurable space. A function $f : \Omega \to \mathbb{R}$ is called **measurable** provided that there is a sequence $(f_n)_{n=1}^{\infty}$ of simple functions $f_n : \Omega \to \mathbb{R}$ such that

$$f(\omega) = \lim_{n \to \infty} f_n(\omega) \quad \text{for all} \quad \omega \in \Omega.$$

It is common to call a measurable function a **random variable** in the presence of a probability measure $\mathbb{P}$ on the measurable space $(\Omega, \mathcal{F})$, i.e. when we are given a probability space $(\Omega, \mathcal{F}, \mathbb{P})$. One can also couple the notion of a random variable on the underlying measure $\mathbb{P}$ by only requiring that there is set $\Omega_0 \in \mathcal{F}$ with $\mathbb{P}(\Omega_0) = 1$ such that $f(\omega) = \lim_{n \to \infty} f_n(\omega)$ only for $\omega \in \Omega_0$. This notion would not be independent of the measure and we will *not* follow this approach.

6.1 Measurable Maps with Values in $\mathbb{R}$ and Random Variables

Does Definition 6.1.3 give what we would like to have? Yes, as we see from the following result:

Proposition 6.1.4 *Let $(\Omega, \mathcal{F})$ be a measurable space and let $f : \Omega \to \mathbb{R}$ be a function. Then the following conditions are equivalent:*

(1) *f is measurable.*
(2) *For all $-\infty < a < b < \infty$ one has that*

$$f^{-1}((a, b)) = \{\omega \in \Omega : a < f(\omega) < b\} \in \mathcal{F}.$$

Proof (1) $\Rightarrow$ (2) Assume that

$$f(\omega) = \lim_{n \to \infty} f_n(\omega)$$

where $f_n : \Omega \to \mathbb{R}$ are simple functions. For a simple function one has that

$$f_n^{-1}((a, b)) \in \mathcal{F}$$

so that

$$f^{-1}((a, b)) = \left\{\omega \in \Omega : a < \lim_{n \to \infty} f_n(\omega) < b\right\}$$

$$= \bigcup_{m=1}^{\infty} \bigcup_{N=1}^{\infty} \bigcap_{n=N}^{\infty} \left\{\omega \in \Omega : a + \frac{1}{m} < f_n(\omega) < b - \frac{1}{m}\right\} \in \mathcal{F}$$

because $\left\{\omega \in \Omega : a + \frac{1}{m} < f_n(\omega) < b - \frac{1}{m}\right\} \in \mathcal{F}$ by Remark 6.1.2(2). To express $\{\omega \in \Omega : a < \lim_{n \to \infty} f_n(\omega) < b\}$ as above using unions and intersections one observes that $a < \lim_{n \to \infty} f_n(\omega) < b$ if and only if

$$\exists m \in \mathbb{N} \; \exists N \in \mathbb{N} \; \forall n \geq N \; \left(a + \frac{1}{m} < f_n(\omega) < b - \frac{1}{m}\right).$$

To understand why the intervals $(a + \frac{1}{m}, b - \frac{1}{m})$ are needed instead of (a, b) see Exercise 2.

(2) $\Rightarrow$ (1) First we observe that we also have that

$$f^{-1}([a, b)) = \{\omega \in \Omega : a \leq f(\omega) < b\}$$

$$= \bigcap_{m=1}^{\infty} \left\{\omega \in \Omega : a - \frac{1}{m} < f(\omega) < b\right\} \in \mathcal{F} \text{ for } -\infty < a < b < \infty.$$

Therefore we can use the simple functions

$$f_n(\omega) := \sum_{k=-4^n}^{4^n-1} \frac{k}{2^n} \mathbb{1}_{\{\frac{k}{2^n} \leq f < \frac{k+1}{2^n}\}}(\omega).$$

$\square$

Measurability is stable with respect to pointwise convergence:

Proposition 6.1.5 *Assume a measurable space $(\Omega, \mathcal{F})$ and a sequence of measurable functions $f_n : \Omega \to \mathbb{R}$ such that $f(\omega) := \lim_n f_n(\omega)$ exists for all $\omega \in \Omega$. Then $f : \Omega \to \mathbb{R}$ is measurable.*

Proof In fact, we did already verify this statement in the proof of Proposition 6.1.4 as for $-\infty < a < b < \infty$ we again have that

$$f^{-1}((a,b)) = \bigcup_{m=1}^{\infty} \bigcup_{N=1}^{\infty} \bigcap_{n=N}^{\infty} \left\{ \omega \in \Omega : a + \frac{1}{m} < f_n(\omega) < b - \frac{1}{m} \right\} \in \mathcal{F}.$$

$\square$

As a direct consequence of Proposition 6.1.4 we obtain

Proposition 6.1.6 *Let $(\Omega, \mathcal{F})$ be a measurable space, $f_1, \ldots, f_N : \Omega \to \mathbb{R}$ be measurable, and $F : \mathbb{R}^N \to \mathbb{R}$ be continuous. Then $F(f_1, \ldots, f_N) : \Omega \to \mathbb{R}$ is measurable.*

Proof For all $l \in \{1, \ldots, N\}$ assume measurable simple functions $f_n^l : \Omega \to \mathbb{R}$ such that

$$\lim_{n \to \infty} f_n^l(\omega) = f_l(\omega) \quad \text{for all} \quad \omega \in \Omega.$$

Then one can show that the function $F(f_1^l, \ldots, f_N^l) : \Omega \to \mathbb{R}$ is a simple function and, by the continuity of F, one has that

$$\lim_{n \to \infty} F(f_n^1(\omega), \ldots, f_n^N(\omega)) = F(f_1(\omega), \ldots, f_N(\omega)).$$

$\square$

Later we will see that we can weaken the assumption on F and require only Borel measurability instead of continuity.

Proposition 6.1.7 (Properties of Measurable Maps) *Let $(\Omega, \mathcal{F})$ be a measurable space, $f, g : \Omega \to \mathbb{R}$ be measurable, and $\alpha, \beta \in \mathbb{R}$. Then the following is true:*

(1) $\omega \mapsto (\alpha f + \beta g)(\omega) := \alpha f(\omega) + \beta g(\omega)$ *is measurable.*
(2) $\omega \mapsto (fg)(\omega) := f(\omega)g(\omega)$ *is measurable.*

(3) If $g(\omega) \neq 0$ for all $\omega \in \Omega$, then $\omega \mapsto \left(\frac{f}{g}\right)(\omega) := \frac{f(\omega)}{g(\omega)}$ is measurable.
(4) $\omega \mapsto |f(\omega)|$ is measurable.

Proof The proof of (1), (2), and (4) follows directly from Proposition 6.1.6: In (1) we take $F(x, y) := \alpha x + \beta y$, in (2) $F(x, y) := xy$, and in (4) $F(x) := |x|$. The proof of (3) is similar (Exercise 4). □

6.2 Measurable Maps Between Measurable Spaces

Now we extend the notion of a measurable map to the notion of an $(\mathcal{F}, \Sigma)$-measurable map. This extension enables us to consider push forward measures and, at the same time, simplifies various proofs in the sequel.

Definition 6.2.1 (Measurable Map) Let $(\Omega, \mathcal{F})$ and (M, Σ) be measurable spaces. A map $f : \Omega \to M$ is called $(\mathcal{F}, \Sigma)$-**measurable**, provided that

$$f^{-1}(B) = \{\omega \in \Omega : f(\omega) \in B\} \in \mathcal{F} \quad \text{for all} \quad B \in \Sigma.$$

The connection to the previously introduced measurable maps is given by the following equivalence:

Theorem 6.2.2 *Let $(\Omega, \mathcal{F})$ be a measurable space and $f : \Omega \to \mathbb{R}$. Then the following assertions are equivalent:*

(1) *The function f is measurable.*
(2) *The map f is $(\mathcal{F}, \mathcal{B}(\mathbb{R}))$-measurable.*

For the proof we will use the following lemma which states that one only needs to check the pre-images of sets from a generating system to show measurability. It is somehow a *Lemma for Lazy People*:

Lemma 6.2.3 *Let $(\Omega, \mathcal{F})$ and (M, Σ) be measurable spaces and let $f : \Omega \to M$. Assume that $\mathcal{G} \subseteq \Sigma$ is a collection of subsets such that $\Sigma = \sigma(\mathcal{G})$. If*

$$f^{-1}(B) \in \mathcal{F} \quad \text{for all} \quad B \in \mathcal{G},$$

then

$$f^{-1}(B) \in \mathcal{F} \quad \text{for all} \quad B \in \Sigma.$$

Proof We again use the *principle of good sets* and define

$$\mathcal{A} := \left\{ B \subseteq M : f^{-1}(B) \in \mathcal{F} \right\}.$$

(a) By assumption we have $\mathcal{G} \subseteq \mathcal{A}$.
(b) We show that $\mathcal{A}$ is a σ-algebra. Because $f^{-1}(M) = \Omega \in \mathcal{F}$, we have that $M \in \mathcal{A}$. If $B \in \mathcal{A}$, then

$$f^{-1}(B^c) = (f^{-1}(B))^c \in \mathcal{F}.$$

Finally, if $B_1, B_2, \cdots \in \mathcal{A}$, then

$$f^{-1}\left(\bigcup_{i=1}^{\infty} B_i\right) = \bigcup_{i=1}^{\infty} f^{-1}(B_i) \in \mathcal{F}.$$

(c) Our assumption, step (a), and step (b) imply that

$$\Sigma = \sigma(\mathcal{G}) \subseteq \mathcal{A},$$

which implies our lemma.

□

Proof of Theorem 6.2.2 $(2) \Rightarrow (1)$ follows from $(a, b) \in \mathcal{B}(\mathbb{R})$ for $-\infty < a < b < \infty$ which implies that $f^{-1}((a, b)) \in \mathcal{F}$ so that we may apply Proposition 6.1.4((2) $\Rightarrow$ (1)).

$(1) \Rightarrow (2)$ is a consequence of Lemma 6.2.3 since $\mathcal{B}(\mathbb{R}) = \sigma((a, b) : -\infty < a < b < \infty)$, and of Proposition 6.1.4((1) $\Rightarrow$ (2)).

□

Directly from Definition 6.2.1 one obtains:

Proposition 6.2.4 (Composition of Measurable Maps) *Let* $(\Omega_1, \mathcal{F}_1)$, $(\Omega_2, \mathcal{F}_2)$, *and* $(\Omega_3, \mathcal{F}_3)$ *be measurable spaces. Assume that* $f : \Omega_1 \to \Omega_2$ *is* $(\mathcal{F}_1, \mathcal{F}_2)$-*measurable and that* $g : \Omega_2 \to \Omega_3$ *is* $(\mathcal{F}_2, \mathcal{F}_3)$-*measurable. Then*

$$g \circ f : \Omega_1 \xrightarrow{f} \Omega_2 \xrightarrow{g} \Omega_3,$$

i.e. $(g \circ f)(\omega_1) := g(f(\omega_1))$, *is* $(\mathcal{F}_1, \mathcal{F}_3)$-*measurable.*

Proof For $A_3 \in \mathcal{F}_3$ we get $A_2 := g^{-1}(A_3) \in \mathcal{F}_2$ and $(g \circ f)^{-1}(A_3) = f^{-1}(A_2) \in \mathcal{F}_1$.

□

Proposition 6.2.5 (Vectors of Measurable Maps) *Let* $(\Omega_1, \mathcal{F}_1), \ldots, (\Omega_d, \mathcal{F}_d)$ *and* $(M_1, \Sigma_1), \ldots, (M_d, \Sigma_d)$ *be measurable spaces and* $f_i : \Omega_i \to M_i$ *be* $(\mathcal{F}_i, \Sigma_i)$-*measurable. Define the product spaces*

$$\Omega := \{\omega = (\omega_1, \ldots, \omega_d) : \omega_i \in \Omega_i\} = \bigtimes_{i=1}^{d} \Omega_i,$$

$$M := \{m = (m_1, \ldots, m_d) : m_i \in M_i\} = \bigtimes_{i=1}^{d} M_i,$$

and the measurable spaces $(\Omega, \mathcal{F})$ and (M, Σ) where

$$\mathcal{F} := \sigma(A_1 \times \cdots \times A_d : A_i \in \mathcal{F}_i),$$
$$\Sigma := \sigma(B_1 \times \cdots \times B_d : B_i \in \Sigma_i).$$

Then $F = (f_1, \ldots f_d) : \Omega \to M$ with $F(\omega) := (f_1(\omega_1), \ldots f_d(\omega_d))$ is $(\mathcal{F}, \Sigma)$-measurable.

Proof By Lemma 6.2.3 we only need to show that $F^{-1}(B_1 \times \cdots \times B_d) \in \mathcal{F}$ for $B_i \in \Sigma_i, i = 1, \ldots, d$. But this follows from

$$F^{-1}(B_1 \times \cdots \times B_d) = f_1^{-1}(B_1) \times \cdots \times f_d^{-1}(B_d)$$

and the definition of $\mathcal{F}$. □

6.3* Measurable Maps Between Metric Spaces

Given a metric space $[M, d]$ we remind the reader that the Borel σ-algebra $\mathcal{B}(M)$ is the smallest σ-algebra that contains all open sets of the metric space $[M, d]$. The definition establishes an intrinsic connection between probability and topology. The following Proposition 6.3.1 and Proposition 6.3.2 are examples of this connection:

Proposition 6.3.1 (Measurability and Continuity) *Let $[M_1, d_1]$ and $[M_2, d_2]$ be metric spaces. Then any continuous map $f : M_1 \to M_2$ is $(\mathcal{B}(M_1), \mathcal{B}(M_2))$-measurable.*

Proof Since f is continuous we know that $f^{-1}(G)$ is open for all open $G \subseteq M_2$ by Proposition 4.1.6. By the definition of the Borel σ-algebra on M_1 this implies $f^{-1}(G) \in \mathcal{B}(M_1)$ for all open $G \subseteq M_2$. So we can apply Lemma 6.2.3 to conclude the proof. □

In particular, if $f : \mathbb{R} \to \mathbb{R}$ is continuous, then f is $(\mathcal{B}(\mathbb{R}), \mathcal{B}(\mathbb{R}))$-measurable.

Proposition 6.3.2 *Let $[M, d]$ be a metric space. Then the Borel σ-algebra $\mathcal{B}(M)$ is the smallest σ-algebra such that all continuous functions $f : M \to \mathbb{R}$ are $(\mathcal{B}(M), \mathcal{B}(\mathbb{R}))$-measurable.*

Proof By Proposition 6.3.1 any continuous map is $(\mathcal{B}(M), \mathcal{B}(\mathbb{R}))$-measurable. To see that $\mathcal{B}(M)$ is the smallest σ-algebra with this property we assume a non-empty closed set $F \subseteq M$ and define the continuous function $f(x) := \inf\{d(x, y) : y \in F\}$. Then we have that $f(x) = 0$ if and only $x \in F$, so that $f^{-1}(\{0\}) = F$. Therefore any σ-algebra $\mathcal{F}$ on M such that all continuous functions are measurable contains all closed sets $F \subseteq M$. Hence $\mathcal{F} \supseteq \mathcal{B}(M)$ by relation (5.1). □

6.4 Summary

For a measurable space $(\Omega, \mathcal{F})$ and a function $f : \Omega \to \mathbb{R}$ the following relations hold true:

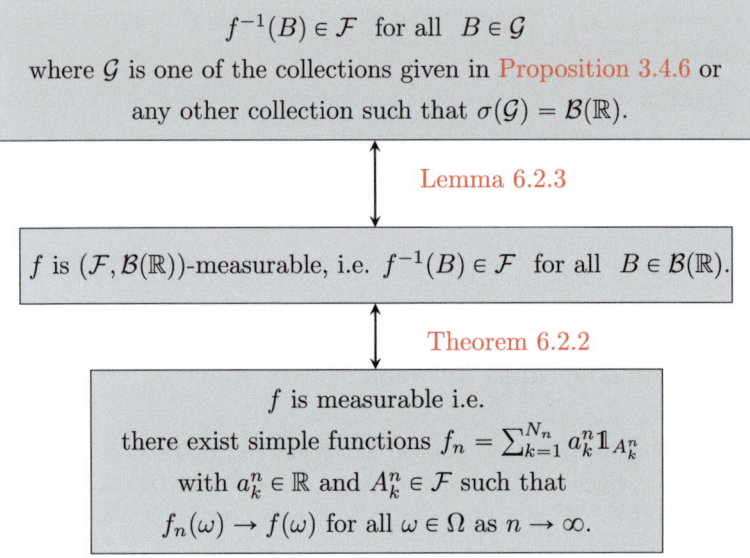

6.5 Image Measures and Distribution Functions

The notion of an *image measure* or *push forward measure* is central in probability. Our first proposition is the basis for its definition:

Proposition 6.5.1 (Image Measure) *Let $(\Omega_1, \mathcal{F}_1)$ and $(\Omega_2, \mathcal{F}_2)$ be measurable spaces, $f : \Omega_1 \to \Omega_2$ be a measurable map, and μ_1 be a measure on $\mathcal{F}_1$. Define*

$$\mu_2(A_2) := \mu_1(f^{-1}(A_2)) \quad \text{for} \quad A_2 \in \mathcal{F}_2.$$

Then $(\Omega_2, \mathcal{F}_2, \mu_2)$ is a measure space with $\mu_1(\Omega_1) = \mu_2(\Omega_2)$.

6.5 Image Measures and Distribution Functions

Proof First we observe that $\mu_2(\emptyset) = \mu_1(f^{-1}(\emptyset)) = \mu_1(\emptyset) = 0$. Moreover, let $A_1, A_2, \ldots \in \mathcal{F}_2$ with $A_i \cap A_j = \emptyset$ for $i \neq j$. Then $f^{-1}(A_i) \in \mathcal{F}_1$ with $f^{-1}(A_i) \cap f^{-1}(A_j) = \emptyset$ for $i \neq j$ and

$$\mu_2\left(\bigcup_{i=1}^{\infty} A_i\right) = \mu_1\left(f^{-1}\left(\bigcup_{i=1}^{\infty} A_i\right)\right) = \mu_1\left(\bigcup_{i=1}^{\infty} f^{-1}(A_i)\right)$$
$$= \sum_{i=1}^{\infty} \mu_1\left(f^{-1}(A_i)\right) = \sum_{i=1}^{\infty} \mu_2(A_i).$$

$\square$

Remark 6.5.2 In Proposition 6.5.1 one has that μ_2 is a finite measure (probability measure) whenever μ_1 is a finite measure (probability measure). However, there are examples that μ_1 is σ-finite, but μ_2 is not.

Definition 6.5.3 The measure μ_2 is called **image measure** or **push forward measure** of μ_1. In case μ_1 is a probability measure, the measure μ_2 is also called **law** or **distribution** of the random variable f.

We introduce the distribution function or cumulative distribution function of a finite measure and of a random variable.

Definition 6.5.4 (Distribution Function)

(1) Given a finite measure μ on $(\mathbb{R}, \mathcal{B}(\mathbb{R}))$,

$$F_\mu(x) := \mu((-\infty, x])$$

is called **distribution function** or **cumulative distribution function** of μ.

(2) Given a measurable map $f : \Omega \to \mathbb{R}$ on a probability space $(\Omega, \mathcal{F}, \mathbb{P})$,

$$F_f(x) := \mathbb{P}(\{\omega \in \Omega : f(\omega) \leqslant x\})$$

is called **distribution function** or **cumulative distribution function** of f.

The notion of a distribution function in item (1) is usually restricted to the case when μ is a probability measure. We use the same notion for finite measures μ. The interplay between (1) and (2) is as follows: If μ is the law of a measurable map $f : \Omega \to \mathbb{R}$ defined on a probability space $(\Omega, \mathcal{F}, \mathbb{P})$, then it follows from the definition that

$$F_f(x) = F_\mu(x) \quad \text{for all} \quad x \in \mathbb{R}.$$

Proposition 6.5.5 (Properties of Distribution Functions) *The distribution function $F_\mu : \mathbb{R} \to [0, \infty)$ for a finite measure μ on $(\mathbb{R}, \mathcal{B}(\mathbb{R}))$ has the following properties:*

(1) *F_μ non-decreasing.*
(2) *F_μ is right-continuous.*
(3) *One has $\lim_{x \to -\infty} F_\mu(x) = 0$ and $\lim_{x \to \infty} F_\mu(x) = \mu(\mathbb{R})$.*

Proof (2) F_μ is non-decreasing as for $x_1 < x_2$ one has that $(-\infty, x_1] \subseteq (-\infty, x_2]$ and therefore

$$F_\mu(x_1) = \mu((-\infty, x_1]) \leq \mu((-\infty, x_2]) = F_\mu(x_2).$$

(1) F_f is right-continuous: let $x \in \mathbb{R}$ and $x_n \downarrow x$. Then by Proposition 2.3.7(6)

$$F_\mu(x) = \mu((-\infty, x]) = \mu\left(\bigcap_{n=1}^{\infty}(-\infty, x_n]\right) = \lim_n \mu\left((-\infty, x_n]\right) = \lim_n F_\mu(x_n).$$

(3) This item is the subject of Exercise 5. □

The following proposition implies especially that the law of a random variable is completely characterized by its distribution function.

Theorem 6.5.6 *Assume finite measures μ_1 and μ_2 on $\mathcal{B}(\mathbb{R})$ and that F_1 and F_2 are the corresponding distribution functions. Then the following assertions are equivalent:*

(1) *$\mu_1 = \mu_2$.*
(2) *$F_1 = F_2$.*

Proof (1) $\Rightarrow$ (2) is obvious, the implication (2) $\Rightarrow$ (1) follows from Theorem 3.3.8 (4) because $F_1 = F_2$ implies, for $-\infty < a < b < \infty$, that

$$\mu_1((a, b]) = F_1(b) - F_1(a) = F_2(b) - F_2(a) = \mu_2((a, b]).$$

□

We close this section with the important converse of Proposition 6.5.5 which is an application of Theorem 3.3.8:

Theorem 6.5.7 *Assume a function $F : \mathbb{R} \to [0, \infty)$ that is a right-continuous, non-decreasing, and such that*

$$\lim_{x \to -\infty} F(x) = 0 \quad \text{and} \quad \lim_{x \to \infty} F(x) = m < \infty.$$

Then there is exists a unique measure μ on $(\mathbb{R}, \mathcal{B}(\mathbb{R}))$ such that

$$\mu((-\infty, x]) = F(x).$$

For this measure we have that $\mu(\mathbb{R}) = m$.

6.6 Discrete and Continuous Random Variables

In the literature the notion of discrete and continuous random variables frequently appears, sometimes in a heuristic manner. Here we want to give a rigorous approach which is based on the law of the random variable. For this purpose assume a random variable $f : \Omega \to \mathbb{R}$ on a probability space $(\Omega, \mathcal{F}, \mathbb{P})$.

(1) We say that f is a **discrete random variable** provided that the image measure has the form

$$\mathbb{P}_f := \mathbb{P} \circ f^{-1} = \sum_{i \in I} a_i \delta_{x_i}$$

for $a_i > 0$ with $\sum_{i \in I} a_i = 1$ and for pairwise distinct $(x_i)_{i \in I} \subseteq \mathbb{R}$, where $I = \{1, \ldots, N\}$ or $I = \mathbb{N}$. In other words, we have $\mathbb{P}(f = x_i) = a_i$ for $i \in I$.

(2) We say that f is a **continuous random variable** provided that the distribution function $F_f : \mathbb{R} \to [0, 1]$ is continuous, or equivalently, provided that $\mathbb{P}(f = x) = 0$ for all $x \in \mathbb{R}$.

If a random variable takes only countably many values, then it is necessarily a discrete random variable. Consequently, a continuous random variable must take uncountable many values, however this is not sufficient for a random variable to be continuous. For example, a discrete random variable can take uncountable many values as well: Take $\Omega := [0, 1]$, $\mathcal{F} := \mathcal{B}([0, 1])$, $\mathbb{P} := \delta_{\frac{1}{2}}$, and $f(t) := t$. So f as function takes uncountable many values, however only the value $1/2$ with probability one. Finally, we remark that there are random variables that are neither discrete nor continuous.

6.7* Distribution Functions on $\mathbb{R}^d$

In this section we will extend Theorem 6.5.7 to finite measures defined on $(\mathbb{R}^d, \mathcal{B}(\mathbb{R}^d))$. To obtain the characterizing properties of a d-dimensional distribution function, let us assume a finite measure μ on $(\mathbb{R}^d, \mathcal{B}(\mathbb{R}^d))$ with $\mu(\mathbb{R}^d) = m$. As distribution function $F : \mathbb{R}^d \to [0, m]$ we use

$$F(x) := \mu((-\infty, x_1] \times \cdots \times (-\infty, x_d]).$$

For $a, b \in \mathbb{R}^d$ with $a = (a_1, \ldots, a_d)$ and $b = (b_1, \ldots, b_d)$ and $a_j \leq b_j$ for all $j = 1, \ldots, d$ we derive that

$$\mu((a_1, b_1] \times \ldots \times (a_d, b_d]) = \Delta_{a_1, b_1} \circ \cdots \circ \Delta_{a_d, b_d} F$$

where for a function $G : \mathbb{R}^j \to \mathbb{R}$ and $-\infty < a_j \leq b_j < \infty$ we set

$$(\Delta_{a_j, b_j} G)(x_1, \ldots, x_{j-1}) := G(x_1, \ldots, x_{j-1}, b_j) - G(x_1, \ldots, x_{j-1}, a_j).$$

The distribution function F satisfies the following properties:

(DF1) $F((x_1, \ldots, x_d)) \to m$ if $\min\{x_1, \ldots, x_d\} \to \infty$.
(DF2) For all $j \in \{1, \ldots, d\}$ one has that $x_j \to -\infty$ implies

$$\sup_{x_1, \ldots, x_{j-1}, x_{j+1}, \ldots, x_d \in \mathbb{R}} F((x_1, \ldots, x_{j-1}, x_j, x_{j+1}, \ldots, x_d)) \to 0.$$

(DF3) $F((x_1, \ldots, x_d)) \leq F((y_1, \ldots, y_d))$ if $x_j \leq y_j$ for all $j = 1, \ldots, d$.
(DF4) $F((y_1, \ldots, y_d)) \to F((x_1, \ldots, x_d))$ if $y_j \downarrow x_j$ for all $j = 1, \ldots, d$.
(DF5) $\Delta_{a_1, b_1} \circ \cdots \circ \Delta_{a_d, b_d} F \geq 0$ if $a_j \leq b_j$ for all $j = 1, \ldots, d$.

The following theorem asserts that (DF1), (DF2), (DF4), and (DF5) imply the existence of a corresponding measure and that (DF3) will follow. This is the multivariate version of Theorem 6.5.7.

Theorem 6.7.1 *Assume a function $F : \mathbb{R}^d \to [0, m]$ with $m \in [0, \infty)$ such that (DF1), (DF2), (DF4), and (DF5) are satisfied. Then there is a unique measure μ on $(\mathbb{R}^d, \mathcal{B}(\mathbb{R}^d))$ such that*

$$F((x_1, \ldots, x_d)) = \mu((-\infty, x_1] \times \cdots \times (-\infty, x_d]).$$

For this measure we have that $\mu(\mathbb{R}^d) = m$.

Proof Here we use the same idea as in the proof of Theorem 3.3.8.
(a) Uniqueness: We define the π-system $\mathcal{P}$ that consists of $\mathbb{R}^d$ and of all sets $(-\infty, x_1] \times \ldots \times (-\infty, x_d]$ with $x_1, \ldots, x_d \in \mathbb{R}$. We have that $\mathcal{B}(\mathbb{R}^d) = \sigma(\mathcal{P})$. Assume two measures μ and ν on $(\mathbb{R}^d, \mathcal{B}(\mathbb{R}^d))$ satisfying

$$F((x_1, \ldots, x_d)) = \mu((-\infty, x_1] \times \cdots \times (-\infty, x_d]) = \nu((-\infty, x_1] \times \cdots \times (-\infty, x_d]).$$

By (DF1) it follows that $\mu(\mathbb{R}^d) = \nu(\mathbb{R}^d) = m$. Now we may apply Theorem 3.2.4 to deduce $\mu = \nu$.
(b) Existence on $(-L, L]^d$: For $L \in \mathbb{N}$ we define the algebra $\mathcal{A}_L$ of subsets of $\Omega_L := (-L, L]^d$ as finite unions of sets of type

$$(a_1, b_1] \times \ldots \times (a_d, b_d] \quad \text{with} \quad -L \leq a_j \leq b_j \leq L.$$

6.7* Distribution Functions on $\mathbb{R}^d$

For a disjoint union, as usual to understand as a union of pairwise disjoint sets,

$$Q = \bigcup_{n=1}^{N} (a_{1,n}, b_{1,n}] \times \ldots \times (a_{d,n}, b_{d,n}] \in \mathcal{A}_L$$

we define

$$\mu_{L,0}(Q) := \sum_{n=1}^{N} \Delta_{a_{1,n}, b_{1,n}} \circ \cdots \circ \Delta_{a_{d,n}, b_{d,n}} F \in [0, m]$$

where we use (DF5). The function $\mu_{L,0}$ is correctly defined and finitely additive on $\mathcal{A}_L$. Now we show that the function is σ-additive on $\mathcal{A}_L$. Consider a disjoint union

$$Q = \bigcup_{n=1}^{\infty} Q_n \quad \text{with} \quad Q, Q_1, Q_2, \cdots \in \mathcal{A}_L.$$

On the one side we get that

$$\sum_{n=1}^{\infty} \mu_{L,0}(Q_n) = \lim_{N \to \infty} \sum_{n=1}^{N} \mu_{L,0}(Q_n) = \lim_{N \to \infty} \mu_{L,0}\left(\bigcup_{n=1}^{N} Q_n \right) \leq \mu_{L,0}(Q).$$

For the other direction we may assume w.lo.g. that

$$Q_n = (a_{1,n}, b_{1,n}] \times \cdots \times (a_{d,n}, b_{d,n}] \quad \text{and} \quad -L \leq a_{j,n} < b_{j,n} \leq L$$

and use a compactness argument: We let $\varepsilon > 0$ and define the set

$$K := [\tilde{a}_1, b_1] \times \ldots \times [\tilde{a}_d, b_d]$$

with $\tilde{a}_j \in (a_j, b_j)$ such that

$$\Delta_{a_1, b_1} \circ \cdots \circ \Delta_{a_d, b_d} F \leq \Delta_{\tilde{a}_1, b_1} \circ \cdots \circ \Delta_{\tilde{a}_d, b_d} F + \varepsilon$$

which is possible because of assumption (DF4). Similarly, we choose $\tilde{b}_{j,n} \in (b_{j,n}, \infty)$ such that

$$\Delta_{a_{1,n}, \tilde{b}_{1,n}} \circ \cdots \circ \Delta_{a_{d,n}, \tilde{b}_{d,n}} F \leq \Delta_{a_{1,n}, b_{1,n}} \circ \cdots \circ \Delta_{a_{d,n}, b_{d,n}} F + 2^{-n}\varepsilon,$$

again by assumption (DF4). We get the open covering

$$K \subseteq \bigcup_{n=1}^{\infty} (a_{1,n}, \tilde{b}_{1,n}) \times \cdots \times (a_{d,n}, \tilde{b}_{d,n})$$

of the compact set K. Hence we can choose a finite sub-cover indexed by $I(\varepsilon)$, and conclude that

$$\Delta_{a_1,b_1} \circ \cdots \circ \Delta_{a_d,b_d} F \leq \varepsilon + \Delta_{\tilde{a}_1,b_1} \circ \cdots \circ \Delta_{\tilde{a}_d,b_d} F$$

$$\leq \varepsilon + \sum_{n \in I(\varepsilon)} \Delta_{a_{1,n},\tilde{b}_{1,n}} \circ \cdots \circ \Delta_{a_{d,n},\tilde{b}_{d,n}} F$$

$$\leq \varepsilon + \sum_{n=1}^{\infty} \left[\Delta_{a_{1,n},b_{1,n}} \circ \cdots \circ \Delta_{a_{d,n},b_{d,n}} F + \frac{\varepsilon}{2^n} \right]$$

$$= 2\varepsilon + \sum_{n=1}^{\infty} \Delta_{a_{1,n},b_{1,n}} \circ \cdots \circ \Delta_{a_{d,n},b_{d,n}} F.$$

By Theorem 3.2.1 we obtain a measure μ_L on $((-L, L]^d, \mathcal{B}((-L, L]^d))$.
(c) Extension to $\mathbb{R}^d$: For $A \in \mathcal{B}(\mathbb{R}^d)$ we define

$$\mu(A) := \lim_{L \to \infty} \mu_L(A \cap (-L, L]^d).$$

Then

$$\mu(\mathbb{R}^d) = \lim_{L \to \infty} \mu_L((-L, L]^d) = \lim_{L \to \infty} \Delta_{-L,L} \circ \cdots \circ \Delta_{-L,L} F = m$$

where we use (DF1) and (DF2). Finally, it is easy to see that μ is the measure we were looking for. $\square$

6.8 Exercises

Ex 1: Show that each simple function has a unique canonical representation.
Ex 2: Assume that $f(\omega) = \lim_{n \to \infty} f_n(\omega)$. Show that in general

$$\bigcup_{N=1}^{\infty} \bigcap_{n=N}^{\infty} \{\omega \in \Omega : a < f_n(\omega) < b\} \neq \left\{ \omega \in \Omega : a < \lim_{n \to \infty} f_n(\omega) < b \right\}.$$

Take, for example, $f_n(\omega) = b - \frac{1}{n}$ for all $\omega \in \Omega$.
Ex 3: Assume that Ω and M are non-empty sets and that $f : \Omega \to M$ is a map. Assume $(A_i)_{i \in I} \subseteq 2^M$, where I is an arbitrary index set. Show that

$$f^{-1}\left(\bigcap_{i \in I} A_i\right) = \bigcap_{i \in I} f^{-1}(A_i) \quad \text{and} \quad f^{-1}\left(\bigcup_{i \in I} A_i\right) = \bigcup_{i \in I} f^{-1}(A_i).$$

Ex 4: Verify Proposition 6.1.7(3): Let $(\Omega, \mathcal{F})$ be a measurable space, $f, g : \Omega \to \mathbb{R}$ random variables such that $g(\omega) \neq 0$ for all $\omega \in \Omega$. Show that $\left(\frac{f}{g}\right)(\omega) := \frac{f(\omega)}{g(\omega)}$ is measurable.

Ex 5: Verify Proposition 6.5.5(3): Let $F_\mu : \mathbb{R} \to [0, \infty)$ be the distribution function of a finite measure μ on $(\mathbb{R}, \mathcal{B}(\mathbb{R}))$. Show that $\lim_{x \to -\infty} F_\mu(x) = 0$ and $\lim_{x \to \infty} F_\mu(x) = \mu(\mathbb{R})$.

Ex 6: (a) Find the distribution function of $\mathcal{U}_{[0,2]}$, the uniform distribution on $[0, 2]$.
(b) Find the distribution function of a binomially distributed random variable f with parameters $p = \frac{1}{3}$ and $n = 2$.
Here both measures are considered on $(\mathbb{R}, \mathcal{B}(\mathbb{R}))$.

Ex 7: Assume two metric spaces (M_1, d_1) and (M_2, d_2) and a continuous map $f : M_1 \to M_2$ in the sense that for all $x_1 \in M_1$ and all $\varepsilon > 0$ there is a $\delta > 0$ such that $d_2(f(x_1), f(y_1)) < \varepsilon$ if $d_1(x_1, y_1) < \delta$. Show that f is continuous if and only if $f^{-1}(G)$ is open in M_1 for all open $G \subseteq M_2$.

6.9 Comments

Section 6.3: Proposition 6.3.2 states that the Borel σ-algebra coincides with the Baire σ-algebra as introduced in [50, Section 7.1] on metric spaces. This section is also recommended for further reading on the topic.

Section 6.7: The statement about multidimensional distribution functions Theorem 6.7.1 is standard in the literature, our proof is adapted from [173, Section II.3.3].

Chapter 7
Independence

In real life *independence* is an intuitive notion, which might have different meanings depending on the context. Even in mathematics *independence* is used with different meanings. In vector spaces vectors might be *linearly* independent, in mathematical logic *propositions* might be independent. Here in probability we speak about *stochastic independence*. In Definition 2.5.5 we did already introduce the (stochastic) *independence of events*, which serves as a mathematical model for the intuitive notion of independence in real life. In this chapter we extend this stochastic concept further. We provide an overview on and the connection between three levels of independence: independence of a family of sets, of a family of random variables, and independence of σ-algebras. The existence of a sequence of independent random variables is shown by constructing the probability space carrying such a sequence. This involves the product of probability spaces and the proof of the existence of a product measure. As applications Kolmogorov's zero-one law and Hewitt-Savage's zero-one law are shown.

7.1 The Basic Concept

We introduce two more concepts of independence so that we have:

(A) Independence of events (Definition 2.5.5)
(B) Independence of random variables (Definition 7.1.2)
(C) Independence of σ-algebras (Definition 7.1.4)

The concepts build on each other, where (C) is the most general concept. Each concept has its own place in the theory and in applications. So it is good to know all three concepts. This enables to simplify arguments and to understand phenomena in the correct context. The concept (A) was already introduced in Definition 2.5.5. So we turn to the next more general level, the concept (B)—the independence of random variables. We start with the notion of a finite family of independent random variables.

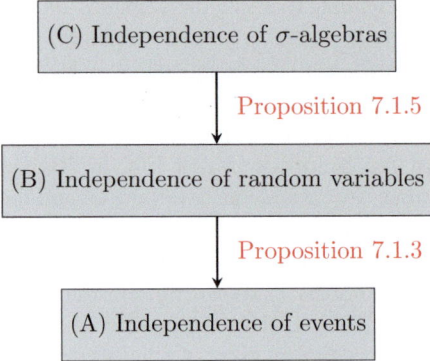

Definition 7.1.1 (Independence of a Finite Family of Random Variables) Let $(\Omega, \mathcal{F}, \mathbb{P})$ be a probability space and $f_i : \Omega \to \mathbb{R}, i = 1, \ldots, n$, random variables. The random variables $f_1, \ldots, f_n$ are called **independent** provided that for all $B_1, \ldots, B_n \in \mathcal{B}(\mathbb{R})$ one has that

$$\mathbb{P}(f_1 \in B_1, \ldots, f_n \in B_n) = \mathbb{P}(f_1 \in B_1) \cdots \mathbb{P}(f_n \in B_n).$$

Recall that $\mathbb{P}(f_k \in B_k)$ is the short notation for $\mathbb{P}(\{\omega \in \Omega : f_k(\omega) \in B_k\}) = \mathbb{P}(f^{-1}(B_k))$. Moreover, note that

$$\{f_1 \in B_1, \ldots, f_n \in B_n\} = f_1^{-1}(B_1) \cap \ldots \cap f_n^{-1}(B_n).$$

The following definition of independence suits for any family of random variables:

Definition 7.1.2 (Independence of a Family of Random Variables) Let $(\Omega, \mathcal{F}, \mathbb{P})$ be a probability space, I be a non-empty index-set, and let $f_i : \Omega \to \mathbb{R}$,

7.1 The Basic Concept

$i \in I$, be random variables. The family $(f_i)_{i \in I}$ is called **independent** provided that for all $n \in \mathbb{N}$, distinct $i_1, \ldots, i_n \in I$, and $B_1, \ldots, B_n \in \mathcal{B}(\mathbb{R})$ one has that

$$\mathbb{P}\left(f_{i_1} \in B_1, \ldots, f_{i_n} \in B_n\right) = \mathbb{P}\left(f_{i_1} \in B_1\right) \cdots \mathbb{P}\left(f_{i_n} \in B_n\right).$$

In case we have a finite index set I both definitions coincide as in Definition 7.1.1 one can chose $B_i = \mathbb{R}$ which corresponds to leaving out the particular index i. Definition 7.1.2 directly extends Definition 2.5.5:

Proposition 7.1.3 *Let $(\Omega, \mathcal{F}, \mathbb{P})$ be a probability spaces and $A_i \in \mathcal{F}$, where I is a non-empty index-set. Then the following assertions are equivalent:*

(1) *The events $(A_i)_{i \in I}$ are independent.*
(2) *The random variables $(\mathbb{1}_{A_i})_{i \in I}$ are independent.*

The proof is the subject of Exercise 1. Now we turn to the most general level, the independence of σ-algebras:

Definition 7.1.4 (Independence of a Family of σ-Algebras) Let $(\Omega, \mathcal{F}, \mathbb{P})$ be a probability space, I be a non-empty index-set, and let $\mathcal{G}_i \subseteq \mathcal{F}$, $i \in I$, be σ-algebras. The family $(\mathcal{G}_i)_{i \in I}$ is called **independent** provided that for all $n \in I$, distinct $i_1, \ldots, i_n \in I$, and $A_{i_1} \in \mathcal{G}_{i_1}, \ldots, A_{i_n} \in \mathcal{G}_{i_n}$ one has that

$$\mathbb{P}(A_{i_1} \cap \cdots \cap A_{i_n}) = \mathbb{P}(A_{i_1}) \cdots \mathbb{P}(A_{i_n}).$$

Now we have the following connection:

Proposition 7.1.5 *Let $(\Omega, \mathcal{F}, \mathbb{P})$ be a probability space and $f_i : \Omega \to \mathbb{R}$, $i \in I$, be random variables, where I is a non-empty index-set. Then the following assertions are equivalent.*

(1) *The σ-algebras $\sigma(f_i) := \{f_i^{-1}(B) : B \in \mathcal{B}(\mathbb{R})\}$ are independent.*
(2) *The family $(f_i)_{i \in I}$ is independent.*
(3) *For all families $(B_i)_{i \in I}$ of Borel sets $B_i \in \mathcal{B}(\mathbb{R})$ one has that the events $(\{\omega \in \Omega : f_i(\omega) \in B_i\})_{i \in I}$ are independent.*

Proof (1) $\Leftrightarrow$ (2) and (2) $\Leftrightarrow$ (3) follow directly from the definition. □

Definition 7.1.6 Let $(\Omega, \mathcal{F}, \mathbb{P})$ be a probability space and $f_i : \Omega \to \mathbb{R}$, $i \in I$, be a family of random variables. Then $\sigma(f_i : i \in I)$ is the smallest σ-algebra which contains all sets of the form

$$\{\omega \in \Omega : f_i(\omega) \in B\} \quad \text{where } i \in I \text{ and } B \in \mathcal{B}(\mathbb{R}).$$

Obviously, $\mathcal{G} := \sigma(f_i : i \in I)$ is the smallest σ-algebra on Ω such that all $f_i : \Omega \to \mathbb{R}$ are $(\mathcal{G}, \mathcal{B}(\mathbb{R}))$ measurable.

We close this section with an intuitive statement that is often used, sometimes without saying, and indicates the strength of the concept of independence of σ-algebras,—but that requires a serious proof. In Exercise 4 we guide the reader step by step through its proof.

Proposition 7.1.7 *Let* $(\Omega, \mathcal{F}, \mathbb{P})$ *be a probability space and* $f_i : \Omega \to \mathbb{R}$, $i \in I$, *be independent random variables, where* I *is a non-empty index-set. Assume* $I = J \cup K$, *where* $J \cap K = \emptyset$ *and both*, J *and* K, *are non-empty. If*

$$\mathcal{G} := \sigma(f_j : j \in J) \quad \text{and} \quad \mathcal{H} := \sigma(f_k : k \in K),$$

then $\mathcal{G}$ *and* $\mathcal{H}$ *are independent* σ-*algebras*.

7.2 Kolmogorov's Zero-One Law

Kolmogorov's zero-one law is a first example of what independence might imply. An example to start with is the following: Assume a probability space $(\Omega, \mathcal{F}, \mathbb{P})$ and independent and identically distributed random variables $f_1, f_2, \ldots : \Omega \to \mathbb{R}$ such that $|f_n(\omega)| \leq 1$ for all $n \in \mathbb{N}$ and $\omega \in \Omega$. We consider the random walk $(S_n)_{n=0}^\infty$, $S_n : \Omega \to \mathbb{R}$, given by $S_0 \equiv 0$ and

$$S_n(\omega) := f_1(\omega) + \cdots + f_n(\omega) \quad \text{for} \quad n \geq 1.$$

We are interested in the asymptotic behaviour as $n \to \infty$ in the following way: we assume a non-decreasing function $\psi : \mathbb{N} \to (0, \infty)$ with $\lim_{n \to \infty} \psi(n) = \infty$ and ask for $\mathbb{P}(A)$ where

$$A := \left\{ \omega \in \Omega : \limsup_{n \to \infty} \frac{S_n(\omega)}{\psi(n)} \leq 1 \right\}. \tag{7.1}$$

Possible choices for $\psi(n)$ could be n, $\sqrt{n}$ or $\sqrt{2n \log \log n}$ (see Theorem 13.1.1, Theorem 15.1.1, and Theorem 21.1.1). The measurability of A is the subject of Exercise (3a). What is the typical property of the set A? The set A does not depend on $f_1, \ldots, f_k$ for all $k \in \mathbb{N}$ in the sense that $\limsup_{n \to \infty} \frac{f_1(\omega) + \cdots + f_n(\omega)}{\psi(n)} = \limsup_{n \to \infty, n > k} \frac{f_{k+1}(\omega) + \cdots + f_n(\omega)}{\psi(n)}$. This is already sufficient to deduce that $\mathbb{P}(A) \in \{0, 1\}$ in Corollary 7.2.3 below. The abstract result about this is the 0-1 law of Kolmogorov:

Theorem 7.2.1 (Zero-One Law of Kolmogorov) *Assume independent random variables* $f_1, f_2, \ldots : \Omega \to \mathbb{R}$ *on some probability space* $(\Omega, \mathcal{F}, \mathbb{P})$ *and let*

$$\mathcal{F}_N^\infty := \sigma(f_N, f_{N+1}, \ldots) \quad \text{and} \quad \mathcal{F}^\infty := \bigcap_{N=1}^\infty \mathcal{F}_N^\infty.$$

Then $\mathbb{P}(A) \in \{0, 1\}$ *for all* $A \in \mathcal{F}^\infty$.

7.2 Kolmogorov's Zero-One Law

The proof will use the following lemma which states that for any set of a σ-algebra we can find a set in a generating algebra such that the probability of their symmetric difference is arbitrarily small.

Lemma 7.2.2 *Let $\mathcal{A} \subseteq \mathcal{F}$ be an algebra and let $\sigma(\mathcal{A}) = \mathcal{F}$. Then, for all $\varepsilon > 0$ and $B \in \mathcal{F}$, there is an $A \in \mathcal{A}$ such that $\mathbb{P}(A \triangle B) < \varepsilon$.*

Proof We use Theorem 3.2.2 and find a cover $A \subseteq \bigcup_{n=1}^{\infty} A_n$, $A_n \in \mathcal{A}$ with $\mathbb{P}\left(\bigcup_{n=1}^{\infty} A_n\right) - \mathbb{P}(A) < \frac{\varepsilon}{2}$. Then we choose $N \geq 1$ such that $\sum_{n=N+1}^{\infty} \mathbb{P}(A_n) < \frac{\varepsilon}{2}$. This implies $\mathbb{P}\left(A \triangle \left(\bigcup_{n=1}^{N} A_n\right)\right) < \varepsilon$. Moreover, we have that $\bigcup_{n=1}^{N} A_n \in \mathcal{A}$. □

Proof of Theorem 7.2.1 The idea of the proof is to show that $\mathbb{P}(A) = \mathbb{P}(A)^2$. Define the algebra

$$\mathcal{A} := \bigcup_{n=1}^{\infty} \sigma(f_1, \ldots, f_n).$$

We have that $\mathcal{F}^{\infty} \subseteq \sigma(\mathcal{A})$. Hence Lemma 7.2.2 implies that for $A \in \mathcal{F}^{\infty}$ there are $A_n \in \sigma(f_1, \ldots, f_{N_n})$ such that

$$\mathbb{P}(A \triangle A_n) \to_{n \to \infty} 0.$$

We also get that

$$\mathbb{P}(A_n \cap A) \to_{n \to \infty} \mathbb{P}(A) \quad \text{and} \quad \mathbb{P}(A_n) \to_{n \to \infty} \mathbb{P}(A).$$

The first relation can be seen as follows: Since

$$\mathbb{P}(A_n \cap A) + \mathbb{P}(A_n \triangle A) = \mathbb{P}(A_n \cup A) \geq \mathbb{P}(A)$$

we get that

$$\liminf_{n \to \infty} \mathbb{P}(A_n \cap A) \geq \mathbb{P}(A) \geq \limsup_{n \to \infty} \mathbb{P}(A_n \cap A).$$

Also the second relation can be checked easily. Since $\mathcal{F}^{\infty} := \bigcap_{N=N_n}^{\infty} \mathcal{F}_N^{\infty}$ we have that $\mathcal{F}^{\infty}$ and $\sigma(f_1, \ldots, f_{N_n})$ are independent (see Proposition 7.1.7). Therefore we conclude that

$$\mathbb{P}(A) = \lim_{n \to \infty} \mathbb{P}(A \cap A_n) = \lim_{n \to \infty} \mathbb{P}(A)\mathbb{P}(A_n) = \mathbb{P}(A)^2$$

so that $\mathbb{P}(A) \in \{0, 1\}$. □

Corollary 7.2.3 *For the set A defined in (7.1) we have that $\mathbb{P}(A) \in \{0, 1\}$.*

Another corollary is:

Corollary 7.2.4 *Let $f_1, f_2, \ldots : \Omega \to \mathbb{R}$ be independent random variables on $(\Omega, \mathcal{F}, \mathbb{P})$.*

(1) *One has $\{\omega \in \Omega : \lim_{n \to \infty} f_n(\omega) \text{ exists}\} \in \mathcal{F}^\infty$, so that the sequence $(f_n)_{n=1}^\infty$ a.s. either does or does not converge.*
(2) *Any $\mathcal{F}^\infty$ measurable random variable is a.s. constant. Especially this holds for $\limsup_{n \to \infty} f_n$ and $\liminf_{n \to \infty} f_n$.*

The proof is the subject of Exercise 3.

7.3 Products of Measure Spaces

Products of measure spaces might look slightly technical at first glance. This is mainly due to the notation, otherwise the concept is natural. Products of measure spaces are used in various places. We will use them to show the existence of independent random variables, and they are the basis for Fubini's theorem, one of the main theorems in integration theory.

In this section we construct in Theorem 7.3.3 countable products of probability spaces by an application of the extension Theorem 3.2.1. We begin with the definition of $\bigotimes_{i \in I}(\Omega_i, \mathcal{F}_i)$:

Definition 7.3.1 Assume measurable spaces $(\Omega_i, \mathcal{F}_i)$ for $i \in I$, where I is a non-empty index-set, and let

$$\Omega := \bigtimes_{i \in I} \Omega_i = \{(\omega_i)_{i \in I} : \omega_i \in \Omega_i \text{ for } i \in I\}.$$

Then the **product σ-algebra** $\bigotimes_{i \in I} \mathcal{F}_i$ on Ω is generated by all sets

$$\bigtimes_{i \in I} E_i := \{(\omega_i)_{i \in I} \in \Omega : \omega_i \in E_i\},$$

where $E_i \in \mathcal{F}_i$, and only finitely many E_i do not coincide with Ω_i.

A set $\bigtimes_{i \in I} E_i$ with $E_i \in \mathcal{F}_i$ is also called *cylinder* set. Or, one can think of $\bigtimes_{i \in I} E_i$ as some sort of *cuboid* with the *edges* E_i.

The following relation turns out to be useful in various situations:

Proposition 7.3.2 *Assume measurable spaces $(\Omega_i, \mathcal{F}_i)$, $i \in I \ne \emptyset$, and $\mathcal{F}_i = \sigma(\mathcal{G}_i)$ for non-empty collections $\mathcal{G}_i \subseteq 2^{\Omega_i}$. Then*

$$\bigotimes_{i \in I} \mathcal{F}_i = \sigma \left(\bigtimes_{i \in I} E_i : E_i \in \mathcal{G}_i \text{ for finitely many } i \in I, \text{ otherwise } E_i = \Omega_i \right).$$

7.3 Products of Measure Spaces

Proof Let us denote the right-hand side of the assertion by $\mathcal{H}$. The inclusion $\mathcal{H} \subseteq \bigotimes_{i \in I} \mathcal{F}_i$ follows from the definition of $\bigotimes_{i \in I} \mathcal{F}_i$. For the inclusion $\bigotimes_{i \in I} \mathcal{F}_i \subseteq \mathcal{H}$ we first notice that $\bigotimes_{i \in I} \mathcal{F}_i$ can be generated by sets $\bigtimes_{i \in I} E_i$ where $E_j \in \mathcal{F}_j$ and $E_j \neq \Omega_j$ for *one* $j \in I$ and $E_i = \Omega_i$ for all other $i \in I$. Indeed, the assumption of finitely many $E_i \neq \Omega_i$ in Theorem 7.3.3 we realize by taking a finite intersection of those sets with *one* $E_i \neq \Omega_i$. We define the projection

$$\pi_j : \Omega = \bigtimes_{i \in I} \Omega_i \to \Omega_j \quad \text{whith} \quad \pi_j((\omega_i)_{i \in I}) := \omega_j.$$

Since $\pi_j^{-1}(E_j) \in \mathcal{H}$ for all $E_j \in \mathcal{G}_i$, we conclude by Lemma 6.2.3 that $\pi_j^{-1}(E_j) \in \mathcal{H}$ for all $E_j \in \sigma(\mathcal{G}_j) = \mathcal{F}_j$. Hence a generating system for $\bigotimes_{i \in I} \mathcal{F}_i$ is contained in $\mathcal{H}$, so that $\bigotimes_{i \in I} \mathcal{F}_i \subseteq \mathcal{H}$. □

Now we prove the existence of countable products of probability spaces:

Theorem 7.3.3 (Countable Products of Probability Spaces) *Assume probability spaces $(\Omega_i, \mathcal{F}_i, \mathbb{P}_i)$, $i \in I$, where either $I := \{1, \ldots, d\}$ or $I := \mathbb{N}$. Then there is a unique probability measure $\mathbb{P} = \bigotimes_{i \in I} \mathbb{P}_i$ on $(\Omega, \mathcal{F}) := (\bigtimes_{i \in I} \Omega_i, \bigotimes_{i \in I} \mathcal{F}_i)$ such that*

$$\mathbb{P}\left(\bigtimes_{i \in I} E_i\right) = \prod_{i \in I} \mathbb{P}_i(E_i), \tag{7.2}$$

where only finitely many of the $E_i \in \mathcal{F}_i$ differ from Ω_i in the case of $I = \mathbb{N}$.

The product space obtained by Theorem 7.3.3 may also be denoted by

$$(\Omega, \mathcal{F}, \mathbb{P}) = \bigotimes_{i \in I} (\Omega_i, \mathcal{F}_i, \mathbb{P}_i).$$

Proof of Theorem 7.3.3 We prove the statement for the more difficult case $I = \mathbb{N}$. The case $I := \{1, \ldots, d\}$ can be proven either in the same way, where the argument is easier, or can be deduced from the case $I = \mathbb{N}$ by choosing $(\Omega_i, \mathcal{F}_i, \mathbb{P}_i) := (M, \Sigma, \mu)$ for $i > d$, where (M, Σ, μ) is a trivial probability space with $\#M = 1$.

Uniqueness of $\mathbb{P}$ Assume that we have constructed a measure as in (7.2). Then $\mathbb{P}(\Omega) = \prod_{i \in \mathbb{N}} \mathbb{P}_i(\Omega_i) = 1$, so that we have a probability measure. The uniqueness follows from Theorem 3.2.4 as the sets $\bigtimes_{i \in \mathbb{N}} E_i$, where $E_i \in \mathcal{F}_i$ and only finitely many E_i do not coincide with Ω_i, form a generating π-system for $\bigotimes_{i \in I} \mathcal{F}_i$.

Existence of $\mathbb{P}$ **(a) Construction of the algebra:** We define the algebra $\mathcal{A}$ as the system of all finite unions of cuboids

$$C = \bigtimes_{i=1}^{\infty} E_i := \{(\omega_i)_{i \in I} : \omega_i \in E_i\}$$

with $E_i \in \mathcal{F}_i$ are the edges of the cuboids, and where for only finitely many of the edges E_i it holds that $E_i \neq \Omega_i$.

(b) Construction of the pre-measure $\mathbb{P}_0$: Let $A \in \mathcal{A}$ have the representation

$$A = C^1 \cup \cdots \cup C^L,$$

where the cuboids C^l are pairwise disjoint. We define $\mathbb{P}_0 : \mathcal{A} \to [0, 1]$ by

$$\mathbb{P}_0(A) := \sum_{l=1}^{L} \left(\prod_{i \in \mathbb{N}} \mathbb{P}_i(E_i^l) \right) \quad \text{if} \quad C^l = \underset{i \in \mathbb{N}}{\times} E_i^l.$$

By construction,

- $\mathbb{P}_0(A)$ does not depend on the representation of A,
- $\mathbb{P}_0(A_1 \cup \cdots \cup A_n) = \mathbb{P}_0(A_1) + \cdots + \mathbb{P}_0(A_n)$ if $A_1, \ldots, A_n \in \mathcal{A}$ are pairwise disjoint, which implies
- $\mathbb{P}_0(A_1) \leq \mathbb{P}_0(A_2)$ if $A_1 \subseteq A_2$, i.e. $\mathbb{P}_0$ is monotone.

(c) Proof of the σ-additivity of $\mathbb{P}_0$ on $\mathcal{A}$: For $A_1, A_2, \ldots \in \mathcal{A}$ with $A_k \cap A_l = \emptyset$ if $k \neq l$ we prove that

$$\mathbb{P}_0 \left(\bigcup_{n=1}^{\infty} A_n \right) = \sum_{n=1}^{\infty} \mathbb{P}_0(A_n) \quad \text{whenever} \quad \bigcup_{n=1}^{\infty} A_n \in \mathcal{A}.$$

One inequality is immediate by monotonicity and finite additivity, because

$$\mathbb{P}_0 \left(\bigcup_{n=1}^{\infty} A_n \right) \geq \mathbb{P}_0 \left(\bigcup_{n=1}^{N} A_n \right) = \sum_{n=1}^{N} \mathbb{P}_0(A_n),$$

so that

$$\mathbb{P}_0 \left(\bigcup_{n=1}^{\infty} A_n \right) \geq \sum_{n=1}^{\infty} \mathbb{P}_0(A_n). \tag{7.3}$$

To prove the equality we assume, by contradiction, an $\varepsilon > 0$ with

$$\varepsilon + \sum_{n=1}^{\infty} \mathbb{P}_0(A_n) \leq \mathbb{P}_0 \left(\bigcup_{n=1}^{\infty} A_n \right) = \sum_{n=1}^{N} \mathbb{P}_0(A_n) + \mathbb{P}_0 \left(\bigcup_{n=N+1}^{\infty} A_n \right)$$

for $N \in \mathbb{N}$. If we set $B_N := \bigcup_{n=N+1}^{\infty} A_n$, then this implies

- $B_N \in \mathcal{A}$ for $N \in \mathbb{N}$,
- $B_1 \supseteq B_2 \supseteq B_3 \supseteq \cdots$,

7.3 Products of Measure Spaces

- $\bigcap_{N=1}^{\infty} B_N = \emptyset$,
- $\mathbb{P}_0(B_N) \geq \varepsilon$ for $N \in \mathbb{N}$.

Our aim is to prove that there is an $\omega \in \bigcap_{N=1}^{\infty} B_N$ which contradicts $\bigcap_{N=1}^{\infty} B_N = \emptyset$ and means that in (7.3) the equality must hold. To show this, we define

$$(\Omega^{(1)}, \mathcal{A}^{(1)}, \mathbb{P}_0^{(1)})$$

just as we have done this with $(\Omega, \mathcal{A}, \mathbb{P}_0)$, but without the first component $(\Omega_1, \mathcal{F}_1, \mathbb{P}_1)$. That means we will now consider $\Omega^{(1)} = \bigtimes_{i=2}^{\infty} \Omega_i$, and the algebra $\mathcal{A}^{(1)}$ is defined like in (a) but now using finite unions of $\bigtimes_{i=2}^{\infty} E_i$, again with only finitely many $E_i \neq \Omega_i$. On $\mathcal{A}^{(1)}$ we define $\mathbb{P}_0^{(1)}$ like we did in (b) for $\mathbb{P}_0$ on $\mathcal{A}$.

Moreover, we define for each $\omega_1 \in \Omega_1$ the sections

$$B_N(\omega_1) := \{(\omega_i)_{i \in \mathbb{N} \setminus \{1\}} : (\omega_1, \omega_2, \omega_3, \ldots) \in B_N\}.$$

As $B_N \in \mathcal{A}$, B_N can be represented as some finite union of disjoint cuboids $B_N = \bigcup_{k=1}^{m_N} Q_{N,k}$ for some $m_N \in \mathbb{N}$. Since $B_N(\omega_1) = \bigcup_{k=1}^{m_N} Q_{N,k}(\omega_1)$ we have that $B_N(\omega_1) \in \mathcal{A}^{(1)}$ and

$$\Omega_1 \ni \omega_1 \mapsto \mathbb{P}_0^{(1)}(B_N(\omega_1)) = \sum_{k=1}^{m_N} \mathbb{P}_0^{(1)}(Q_{N,k}(\omega_1))$$

is a simple function. Hence

$$C_{1,N} := \left\{\omega_1 \in \Omega_1 : \mathbb{P}_0^{(1)}(B_N(\omega_1)) \geq \frac{\varepsilon}{2}\right\} \in \mathcal{F}_1.$$

Then we use the estimate

$$\mathbb{P}_0^{(1)}(B_N(\omega_1)) \leq \begin{cases} 1 & \text{if } \omega_1 \in C_{1,N} \\ \frac{\varepsilon}{2} & \text{if } \omega_1 \in C_{1,N}^c \end{cases}$$

and get (see Exercise 2)

$$\mathbb{P}_0(B_N) \leq \mathbb{P}_1(C_{1,N}) + \frac{\varepsilon}{2}\mathbb{P}_1(C_{1,N}^c) \leq \mathbb{P}_1(C_{1,N}) + \frac{\varepsilon}{2}.$$

Using $\mathbb{P}_0(B_N) \geq \varepsilon$ we conclude that

$$\mathbb{P}_1(C_{1,N}) \geq \frac{\varepsilon}{2}.$$

We derive from $B_1 \supseteq B_2 \supseteq B_3 \supseteq \cdots$ that $C_{1,1} \supseteq C_{1,2} \supseteq \cdots$. By continuity from above, we conclude that

$$\mathbb{P}_1\left(\bigcap_{N=1}^{\infty} C_{1,N}\right) = \lim_{N \to \infty} \mathbb{P}_1(C_{1,N}) \geq \frac{\varepsilon}{2}.$$

This implies that there is an $\omega_1^0 \in \Omega_1$ such that $\omega_1^0 \in \bigcap_{N=1}^{\infty} C_{1,N}$ and therefore

- $B_N(\omega_1^0) \in \mathcal{A}^{(1)}$ for $N \in \mathbb{N}$,
- $B_1(\omega_1^0) \supseteq B_2(\omega_1^0) \supseteq B_3(\omega_1^0) \supseteq \cdots$,
- $\mathbb{P}_0^{(1)}(B_N(\omega_1^0)) \geq \frac{\varepsilon}{2}$ for $N \in \mathbb{N}$.

Now we repeat this construction using $(\Omega^{(1)}, \mathcal{A}^{(1)}, \mathbb{P}_0^{(1)})$, the family $(B_N(\omega_1^0))_{l=1}^{\infty}$, and $\varepsilon/2$ instead of ε. We get in the next step

- $B_N(\omega_1^0, \omega_2^0) \in \mathcal{A}^{(2)}$ for $N \in \mathbb{N}$,
- $B_1(\omega_1^0, \omega_2^0) \supseteq B_2(\omega_1^0, \omega_2^0) \supseteq B_3(\omega_1^0, \omega_2^0) \supseteq \cdots$,
- $\mathbb{P}_0^{(2)}(B_N(\omega_1^0, \omega_2^0)) \geq \frac{\varepsilon}{2^2}$ for $N \in \mathbb{N}$.

Continuing in this way, we obtain a sequence $\omega^0 = (\omega_1^0, \omega_2^0, \omega_3^0, \ldots)$. It remains to show that

$$(\omega_i^0)_{i \in I} \in \bigcap_{N=1}^{\infty} B_N.$$

Let us fix $M \in \mathbb{N}$. Since $B_M \in \mathcal{A}$ there exists a $n_M \in \mathbb{N}$ such that

$$B_M = D_{n_M} \times \Omega_{n_M+1} \times \Omega_{n_M+2} \times \cdots \tag{7.4}$$

where $D_{n_M} \subseteq \Omega_1 \times \cdots \times \Omega_{n_M}$ is a finite union of cuboids. By construction we have that

$$\bigcap_{N=1}^{\infty} B_N(\omega_1^0, \ldots, \omega_{n_M}^0) \neq \emptyset \quad \text{and therefore} \quad B_M(\omega_1^0, \ldots, \omega_{n_M}^0) \neq \emptyset.$$

By (7.4) we conclude that $\omega^0 \in B_M$.

(d) We conclude by Theorem 3.2.1 to obtain an extension of $\mathbb{P}_0 : \mathcal{A} \to [0,1]$ to $\mathbb{P} : \mathcal{F} = \sigma(\mathcal{A}) \to [0,1]$. □

7.3 Products of Measure Spaces

We deduce the existence of finite products of σ-finite measure spaces:

Corollary 7.3.4 (Finite Products of σ-Finite Measure Spaces) *For $d \geqslant 2$ and σ-finite measure spaces $(\Omega_i, \mathcal{F}_i, \mu_i)$, $i = 1, \ldots, d$, and*

$$(\Omega, \mathcal{F}) := \left(\bigtimes_{i=1}^d \Omega_i, \bigotimes_{i=1}^d \mathcal{F}_i \right)$$

there exists a unique measure μ on $(\Omega, \mathcal{F})$, denoted by $\otimes_{i=1}^d \mu_i$, such that

$$\mu\left(\bigtimes_{i=1}^d E_i \right) = \prod_{i=1}^d \mu_i(E_i) \quad \text{for} \quad E_i \in \mathcal{F}_i \text{ with } \mu_i(E_i) < \infty \tag{7.5}$$

for $i = 1, \ldots, d$. The measure μ is σ-finite.

Again we may write

$$(\Omega, \mathcal{F}, \mu) = \bigotimes_{i=1}^d (\Omega_i, \mathcal{F}_i, \mu_i).$$

Proof of Corollary 7.3.4 By assumption there are partitions $\Omega_i = \bigcup_{j=1}^\infty \Omega_{i,j}$ with $\Omega_{i,j} \in \mathcal{F}_i$ and $\mu_i(\Omega_{i,j}) < \infty$. So we let $\mu_{i,j}(E_i) := \mu_i(E_i \cap \Omega_{i,j})$ and define with Theorem 7.3.3 (which obviously holds for finite products of finite measures by a re-normalization) the product measures

$$\mu_{j_1,\ldots,j_d}(A) := \mu_{1,j_1} \otimes \cdots \otimes \mu_{d,j_d}(A) \quad \text{for} \quad A \in \mathcal{F}.$$

The measure, we are looking for, is

$$\mu(A) := \sum_{j_1,\ldots,j_d=1}^\infty \mu_{j_1,\ldots,j_d}(A).$$

As $\mu(\Omega_{1,j_1} \times \cdots \times \Omega_{d,j_d}) = \mu_1(\Omega_{1,j_1}) \cdots \mu_d(\Omega_{d,j_d}) < \infty$ and where $(\Omega_{1,j_1} \times \cdots \times \Omega_{d,j_d})_{j_1,\ldots,j_d=1}^\infty$ is a partition, the measure μ is σ-finite. Moreover, for $A \in \mathcal{F}$ with $A \subseteq \Omega_{1,j_1} \times \cdots \times \Omega_{d,j_d}$ the measure μ is unique, so the measure μ is unique on $\mathcal{F}$. Finally, we have

$$\mu\left(\bigtimes_{i=1}^d E_i \right) = \sum_{j_1,\ldots,j_d=1}^\infty \mu_{j_1,\ldots,j_d}\left(\bigtimes_{i=1}^d E_i \right)$$

$$= \sum_{j_1,\ldots,j_d=1}^\infty \mu_1(E_1 \cap \Omega_{1,j_1}) \cdots \mu_d(E_d \cap \Omega_{d,j_d})$$

$$= \prod_{i=1}^{d} \left[\sum_{j=1}^{\infty} \mu_i(E_i \cap \Omega_{i,j}) \right]$$

$$= \prod_{i=1}^{d} \mu_i(E_i).$$

$\square$

Now we extend the Lebesgue measure from $\mathbb{R}$ to $\mathbb{R}^d$. To do so, we equip $\mathbb{R}^d$ with the metric obtained from the Euclidean distance $|x - y| = \sqrt{|x_1 - y_1|^2 + \cdots + |x_d - y_d|^2}$ and obtain a complete separable metric space. By Definition 5.1.1 the Borel σ-algebra $\mathcal{B}(\mathbb{R}^d)$ is the smallest σ-algebra containing all open sets from $\mathbb{R}^d$. Now we connect this with our previous results in this section:

Corollary 7.3.5 *For $d \in \mathbb{N}$ we have the following:*

(1) $\mathcal{B}(\mathbb{R}^d) = \mathcal{B}(\mathbb{R}) \otimes \cdots \otimes \mathcal{B}(\mathbb{R})$.
(2) *There is a unique measure λ_d on $(\mathbb{R}^d, \mathcal{B}(\mathbb{R}^d))$ with*

$$\lambda_d(E_1 \times \ldots \times E_d) = \prod_{i=1}^{d} \lambda(E_i)$$

for $E_1, \ldots, E_d \in \mathcal{B}(\mathbb{R})$ with $\lambda(E_i) < \infty$ for $i = 1, \ldots, d$, where λ is the Lebesgue measure on $(\mathbb{R}, \mathcal{B}(\mathbb{R}))$. For this measure one has $\lambda_d = \otimes_{i=1}^{d} \lambda$.

Proof (1) By Proposition 3.3.6 and Proposition 7.3.2 the collection of open sets

$$\{(a_1, b_1) \times \cdots \times (a_d, b_d) : -\infty < a_i < b_i < \infty\}$$

generates $\mathcal{B}(\mathbb{R}) \otimes \cdots \otimes \mathcal{B}(\mathbb{R})$. As each open set $G \subseteq \mathbb{R}^d$ is a countable union of such cuboids $(a_1, b_1) \times \cdots \times (a_d, b_d)$, we have $\mathcal{B}(\mathbb{R}^d) \subseteq \mathcal{B}(\mathbb{R}) \otimes \cdots \otimes \mathcal{B}(\mathbb{R})$. The opposite inclusion is evident as each such cuboid $(a_1, b_1) \times \cdots \times (a_d, b_d)$ is open.
(2) follows from Corollary 7.3.4. $\square$

Definition 7.3.6 The measure λ_d in Corollary 7.3.5 is called d-dimensional Lebesgue measure.

7.4* A Remark on Non σ-Finite Measure Spaces

For products of measure spaces, where at least one component fails to be σ-finite, (7.5) does not imply the uniqueness of the measure μ as we demonstrate now:
Let $(\Omega_1, \mathcal{F}_1, \mu_1) = (\mathbb{R}, \mathcal{B}(\mathbb{R}), \lambda)$ and let
(a) $\Omega_2 := \mathbb{R}$,
(b) $\mathcal{F}_2 := \sigma(B : B \in \mathcal{B}((-\infty, 0))$ or $B = [0, \infty))$,
(c) $\mu_2(B) := \lambda(B)$ if $B \in \mathcal{B}((-\infty, 0))$, otherwise $\mu_2(B) := \infty$.

The measure space $(\Omega_2, \mathcal{F}_2, \mu_2)$ is not σ-finite as $[0, \infty) \in \mathcal{F}_2$ cannot be divided. Now we define two measures on $(\Omega_1, \mathcal{F}_1) \otimes (\Omega_2, \mathcal{F}_2)$: The measure μ is the restriction of $\lambda \otimes \lambda$ to $\mathcal{F}_1 \otimes \mathcal{F}_2$. The measure ν is defined as

$$\nu(B) := \begin{cases} (\lambda \otimes \lambda)(B) & : B \in \mathcal{B}(\mathbb{R} \times (-\infty, 0)) \\ \infty & : \text{else} \end{cases}.$$

Both measures satisfy relation (7.5), but do not coincide as

$$(\lambda \otimes \lambda)(\{0\} \times \mathbb{R}) = 0 \quad \text{but} \quad \nu(\{0\} \times \mathbb{R}) = \infty.$$

Here we used that for $\lambda \otimes \lambda$ defined on $(\mathbb{R}^2, \mathcal{B}(\mathbb{R}^2))$ we have

$$(\lambda \otimes \lambda)(\{0\} \times \mathbb{R}) = \lim_{n \to \infty} (\lambda \otimes \lambda)(\{0\} \times (-n, n)) = \lim_{n \to \infty} \lambda(\{0\})\lambda((-n, n)) = 0.$$

7.5 Construction of Independent Random Variables

When dealing with families of independent random variables the question arises how to construct probability spaces where such families exist. In this section we deduce the existence straight from Theorem 7.3.3:

Corollary 7.5.1 (Existence of Independent Random Variables) *Let $I = \{1, 2, ..., n\}$ or $I = \mathbb{N}$. Given probability measures μ_i on $\mathcal{B}(\mathbb{R})$, $i \in I$, there exists a probability space $(\Omega, \mathcal{F}, \mathbb{P})$ and independent random variables $(f_i)_{i \in I}$, $f_i : \Omega \to \mathbb{R}$ such that $\mathbb{P}_{f_i} = \mu_i$, where $\mathbb{P}_{f_i}$ is the law of f_i.*

Proof Using Theorem 7.3.3 we set

$$(\Omega, \mathcal{F}, \mathbb{P}) := \bigotimes_{i \in I} (\mathbb{R}, \mathcal{B}(\mathbb{R}), \mu_i)$$

and define the maps

$$f_i : \Omega \to \mathbb{R} \quad \text{with} \quad f_i(x) := x_i \quad \text{if} \quad x = (x_j)_{j \in I}.$$

The maps f_i are random variables as for $B \in \mathcal{B}(\mathbb{R})$ we have that

$$f_i^{-1}(B) = \bigtimes_{j \in I} E_j \in \mathcal{A} \subseteq \mathcal{F} \quad \text{with} \quad E_j := \begin{cases} B & j = i \\ \mathbb{R} & j \neq i \end{cases},$$

where the algebra $\mathcal{A}$ was introduced in step (a) of the proof of Theorem 7.3.3. Finally, for any $n \in I$ and $B_1, \ldots, B_n \in \mathcal{B}(\mathbb{R})$,

$$\mathbb{P}(f_1 \in B_1, \ldots, f_n \in B_n) = \mathbb{P}(B_1 \times \cdots \times B_n \times \mathbb{R} \times \mathbb{R} \times \cdots) = \prod_{i=1}^{n} \mu_i(B_i)$$

and, by fixing i and letting $B_j = \mathbb{R}$ for $j \neq i$, we have

$$\mathbb{P}(f_i \in B_i) = \mathbb{P}(f_1 \in \mathbb{R}, \ldots, f_{i-1} \in \mathbb{R}, f_i \in B_i, f_{i+1} \in \mathbb{R}, \ldots, f_n \in \mathbb{R})$$
$$= \mu_1(\mathbb{R}) \cdots \mu_{i-1}(\mathbb{R}) \mu_i(B_i) \mu_{i+1}(\mathbb{R}) \cdots \mu_n(\mathbb{R})$$
$$= \mu_i(B_i)$$

which proves that $\mathbb{P}_{f_i} = \mu_i$ and, by the computation before, that

$$\mathbb{P}(f_1 \in B_1, \ldots, f_n \in B_n) = \prod_{i=1}^{n} \mathbb{P}(f_i \in B_i).$$

□

7.6 Product Spaces and Independence

We continue to exploit the concept of product spaces to learn more about independence. We start by showing that the independence of random variables $f_1, \ldots, f_d$ is only a property of the distribution (or in other words the law) of the random vector

$$(f_1, \ldots, f_d) : \Omega \to \mathbb{R}^d.$$

In particular, this implies the following: If $f_1, \ldots, f_d : \Omega \to \mathbb{R}$ are independent and if $f_1', \ldots, f_d' : \Omega' \to \mathbb{R}$ are random variables such that the laws of $(f_1, \ldots, f_d)$ and $(f_1', \ldots, f_d')$ on $(\mathbb{R}^d, \mathcal{B}(\mathbb{R}^d))$ coincide, then $f_1', \ldots, f_d' : \Omega' \to \mathbb{R}$ are independent as well.

7.6 Product Spaces and Independence

Proposition 7.6.1 (Independence and Product of Laws) *Assume that $(\Omega, \mathcal{F}, \mathbb{P})$ is a probability space and that $f_1, \ldots, f_d : \Omega \to \mathbb{R}$ are random variables with laws $\mathbb{P}_{f_1}, \ldots, \mathbb{P}_{f_d}$, and distribution-functions $F_1, \ldots, F_d$. Then the following assertions are equivalent:*

(1) $f_1, \ldots, f_d$ *are independent.*
(2) $\mathbb{P}((f_1, \ldots, f_d) \in B) = (\otimes_{i=1}^d \mathbb{P}_{f_i})(B)$ *for all $B \in \mathcal{B}(\mathbb{R}^d)$.*
(3) $\mathbb{P}(f_1 \leqslant x_1, \ldots, f_d \leqslant x_d) = F_1(x_1) \cdots F_d(x_d)$ *for all $x_1, \ldots, x_d \in \mathbb{R}$.*

Proof We define the map

$$f : \Omega \to \mathbb{R}^d \quad \text{by} \quad f(\omega) := (f_1(\omega), \ldots, f_n(\omega)).$$

The map f is $(\mathcal{F}, \mathcal{B}(\mathbb{R}^d))$-measurable: We take a cuboid $B = E_1 \times \cdots \times E_d$ with $E_i \in \mathcal{B}(\mathbb{R})$ and get

$$f^{-1}(B) = \{\omega \in \Omega : f_1(\omega) \in E_1, \ldots, f_d(\omega) \in E_d\}$$
$$= f_1^{-1}(E_1) \cap \cdots \cap f_d^{-1}(E_d) \in \mathcal{F}.$$

As all such cuboids B generate $\mathcal{B}(\mathbb{R}^d)$, we apply Lemma 6.2.3 and obtain that $f^{-1}(B) \in \mathcal{F}$ for all $B \in \mathcal{B}(\mathbb{R}^d)$. Let $\mathbb{P}_f$ be the image measure of f on $(\mathbb{R}^d, \mathcal{B}(\mathbb{R}^d))$, i.e.

$$\mathbb{P}_f(B) := \mathbb{P}(\{\omega \in \Omega : (f_1(\omega), \ldots, f_d(\omega)) \in B\}) \quad \text{for} \quad B \in \mathcal{B}(\mathbb{R}^d).$$

(1) $\Leftrightarrow$ (2) The random variables $f_1, \ldots, f_d$ are independent if and only if for all $E_i \in \mathcal{B}(\mathbb{R})$ one has for $B = E_1 \times \cdots \times E_d$ that

$$\mathbb{P}((f_1, \ldots, f_d) \in B) = \mathbb{P}(f_1 \in E_1, \ldots, f_d \in E_d) = \mathbb{P}(f_1 \in E_1) \cdots \mathbb{P}(f_d \in E_d).$$

Reformulating this equality gives

$$\mathbb{P}_f(B) = (\mathbb{P}_{f_1} \otimes \cdots \otimes \mathbb{P}_{f_d})(B) \quad \text{for all} \quad B = E_1 \times \cdots \times E_d.$$

Because these cuboids B form a π-system and generate $\mathcal{B}(\mathbb{R}^d)$, this is equivalent to $\mathbb{P}_f = \mathbb{P}_{f_1} \otimes \cdots \otimes \mathbb{P}_{f_d}$, which is exactly item (2).

(2) $\Leftrightarrow$ (3) Item (3) is equivalent to the statement that for all $E_i = (-\infty, x_i] \in \mathcal{B}(\mathbb{R})$ one has

$$\mathbb{P}((f_1, \ldots, f_d) \in B) = \mathbb{P}(f_1 \in E_1, \ldots, f_d \in E_d) = \mathbb{P}(f_1 \in E_1) \cdots \mathbb{P}(f_d \in E_d).$$

Again the cuboids $B := (-\infty, x_1] \times \cdots \times (-\infty, x_d]$ form a π-system and generate $\mathcal{B}(\mathbb{R}^d)$, this is equivalent to $\mathbb{P}_f = \mathbb{P}_{f_1} \otimes \cdots \otimes \mathbb{P}_{f_d}$ exactly as in (1) $\Leftrightarrow$ (2). $\square$

The next statement builds on Proposition 7.1.7 and is about grouping independent random variables:

Proposition 7.6.2 (Groups of Independent Random Variables) *Let $(\Omega, \mathcal{F}, \mathbb{P})$ be a probability space and $f_i : \Omega \to \mathbb{R}$, $i = 1, 2, 3, \ldots$ be independent random variables. Assume Borel functions $g_k : \mathbb{R}^{n_k} \to \mathbb{R}$ for $k \in \mathbb{N}$ and $n_k \in \mathbb{N}$. Then the random variables $g_1(f_1, \ldots, f_{n_1})$, $g_2(f_{n_1+1}, \ldots, f_{n_1+n_2})$, $g_3(f_{n_1+n_2+1}, \ldots, f_{n_1+n_2+n_3})$, $\ldots$ are independent as well.*

Proof We fix $K \in \mathbb{N}$, let $N_0 := 0$, and $N_k := n_1 + \cdots + n_k$ for $k = 1, \ldots, K$. We define $G_k : \Omega \to \mathbb{R}$ by

$$G_k(\omega) := g_k(f_{N_{k-1}+1}(\omega), \ldots, f_{N_k}(\omega)).$$

Propositions 6.2.5 and 6.2.4 imply that

$$G_k : \Omega \to \mathbb{R}$$

is $(\sigma(f_{N_{K-1}+1}, \ldots, f_{N_K}), \mathcal{B}(\mathbb{R}))$-measurable, where we use Definition 7.1.6. If $A_1, \ldots, A_K \in \mathcal{B}(\mathbb{R})$, then $\{G_1 \in A_1\} \in \sigma(f_1, \ldots, f_{N_1})$ and $\{G_2 \in A_2, \ldots, G_K \in A_K\} \in \sigma(f_{N_1+1}, \ldots, f_{N_K})$ so that Proposition 7.1.7 yields to

$$\mathbb{P}(G_1 \in A_1, \ldots, G_K \in A_K) = \mathbb{P}(G_1 \in A_1)\mathbb{P}(G_2 \in A_2, \ldots, G_K \in A_K).$$

Now we proceed with the term $\mathbb{P}(G_2 \in A_2, \ldots, G_K \in A_K)$ in the same way and obtain the desired result by induction. □

7.7* Hewitt-Savage's 0-1 Law

We already know Kolomogorov's 0-1 law. In this section we prove another 0-1 law. To explain its usage we start with an example based on random walks.

Definition 7.7.1 For a probability space $(\Omega, \mathcal{F}, \mathbb{P})$ and $p \in (0, 1)$ we denote by $\varepsilon_1^{(p)}, \varepsilon_2^{(p)}, \ldots : \Omega \to \mathbb{R}$ a sequence of independent random variables such that

$$\mathbb{P}(\varepsilon_n^{(p)} = -1) = p \quad \text{and} \quad \mathbb{P}(\varepsilon_n^{(p)} = 1) = 1 - p.$$

Moreover, we let $S_n^{(p)} := \sum_{i=1}^n \varepsilon_i^{(p)}$ for $n \in \mathbb{N}$.

Consider the event

$$A^{(p)} := \left\{\omega \in \Omega : \#\left\{n \in \mathbb{N} : S_n^{(p)}(\omega) = 0\right\} = \infty\right\}, \tag{7.6}$$

i.e. $A^{(p)}$ is the event, that the path of the random walk $(S_n^{(p)})_{n=1}^\infty$ visits 0 infinitely often. We would like to have a 0-1 law for this type of event. What is the typical property of $A^{(p)}$? We can rearrange *finitely* many $\varepsilon_i^{(p)}$ and will get the same event since from some $N \in \mathbb{N}$ on, this rearrangement does not affect the random walk any more. An abstract framework for this is the following:

Definition 7.7.2 A map $\pi : \mathbb{N} \to \mathbb{N}$ is called **finite permutation** if

(1) the map π is a bijection,
(2) there is some $N \in \mathbb{N}$ such that $\pi(n) = n$ for all $n \geqslant N$.

A non-empty set $B \subseteq \mathbb{R}^\mathbb{N}$ is called **symmetric** provided that for all finite permutations $\pi : \mathbb{N} \to \mathbb{N}$ one has that

$$B = \pi(B) := \{\pi(x) : x \in B\}.$$

For example, to treat our introducing example, we choose the following set:

Example 7.7.3 We let $B \subseteq \mathbb{R}^\mathbb{N}$ be the set of all sequences $(x_n)_{n\in\mathbb{N}}$, $x_n \in \{-1, 1\}$, such that

$$\#\{n \in \mathbb{N} : x_1 + \cdots + x_n = 0\} = \infty.$$

The main result of this section is the following theorem:

Theorem 7.7.4 (Zero-One Law of Hewitt[1] and Savage[2]) *Let* $(\mathbb{R}, \mathcal{B}(\mathbb{R}), \mu)$ *be a probability space and* $(\mathbb{R}^\mathbb{N}, \mathcal{B}(\mathbb{R}^\mathbb{N}), Q) := \bigotimes_{i=1}^\infty (\mathbb{R}, \mathcal{B}(\mathbb{R}), \mu)$. *If* $B \in \mathcal{B}(\mathbb{R}^\mathbb{N})$ *is symmetric, then* $Q(B) \in \{0, 1\}$.

Proof We use the permutations

$$\pi_n(k) := \begin{cases} n+k : k \in \{1, \ldots, n\} \\ k-n : k \in \{n+1, \ldots, 2n\} \\ k : k > 2n \end{cases}.$$

Let $\mathcal{A}$ be the algebra that consists of all $C \times \mathbb{R}^\mathbb{N}$ with $C \in \mathcal{B}(\mathbb{R}^n)$ and $n \in \mathbb{N}$, so that we have $\sigma(\mathcal{A}) = \mathcal{B}(\mathbb{R}^\mathbb{N})$. Let $B \in \mathcal{B}(\mathbb{R}^\mathbb{N})$ be symmetric. By Lemma 7.2.2 we find $B_n^0 \in \mathcal{B}(\mathbb{R}^n)$ such that

$$\lim_{n\to\infty} Q(B \triangle B_n) = 0 \quad \text{for} \quad B_n := B_n^0 \times \mathbb{R}^\mathbb{N}.$$

---

[1] Edwin Hewitt 20/01/1920 (Everett, USA)–21/06/1999 (Seattle, USA).
[2] Leonard Jimmie Savage 20/11/1917 (Detroit, USA)–01/11/1971 (New Haven, USA).

This implies
$$\lim_{n\to\infty} Q(B_n) = Q(B)$$
so that
$$\lim_{n\to\infty} Q(\pi_n(B_n)) = Q(B)$$
as well since $Q(\pi_n(B_n)) = Q(B_n)$ by assumption. Moreover,
$$Q(B\Delta B_n) = Q(\pi_n(B\Delta B_n)) = Q(\pi_n(B)\Delta\pi_n(B_n)) = Q(B\Delta\pi_n(B_n)).$$
Hence from $\lim_{n\to\infty} Q(B\Delta B_n) = 0$ we conclude $\lim_{n\to\infty} Q(B\Delta\pi_n(B_n)) = 0$, which implies
$$Q(B\Delta(B_n\cap\pi_n(B_n))) \to_n 0 \quad \text{and} \quad Q(B_n\cap\pi_n(B_n)) \to_n Q(B)$$
as $n\to\infty$. Since B_n and $\pi_n(B_n)$ are independent, we have
$$Q(B_n\cap\pi_n(B_n)) = Q(B_n)Q(\pi_n(B_n)) = Q(B_n)^2 \to_n Q(B)^2$$
with $n\to\infty$. Hence $Q(B) = Q(B)^2$ so that $Q(B) \in \{0,1\}$. □

As an application we obtain for the symmetric random walk:

Corollary 7.7.5 (Symmetric Random Walk) *For the event from (7.6) we have* $\mathbb{P}(A^{(1/2)}) = 1$.

Proof Consider the sets
$$A_+ := \{\omega\in\Omega : \#\{n\in\mathbb{N} : S_n(\omega) = 0\} < \infty\} \cap \left\{\omega\in\Omega : \liminf_{n\to\infty} S_n(\omega) > 0\right\},$$
$$A := \{\omega\in\Omega : \#\{n\in\mathbb{N} : S_n(\omega) = 0\} = \infty\},$$
$$A_- := \{\omega\in\Omega : \#\{n\in\mathbb{N} : S_n(\omega) = 0\} < \infty\} \cap \left\{\omega\in\Omega : \limsup_{n\to\infty} S_n(\omega) < 0\right\}.$$

We define the measurable map $f : \Omega \to \mathbb{R}^\mathbb{N}$ by $f(\omega) := (\varepsilon_n^{(1/2)}(\omega))_{n\in\mathbb{N}}$ and let Q be the image measure of $\mathbb{P}$ with respect to f. Furthermore, we let
$$B_+ := \left\{(x_n)_{n\in\mathbb{N}} \in \{-1,1\}^\mathbb{N} : \#\{n\in\mathbb{N} : s_n := x_1 + \cdots + x_n = 0\} < \infty,\right.$$
$$\left.\liminf_{n\to\infty} s_n > 0\right\},$$
$$B := \left\{(x_n)_{n\in\mathbb{N}} \in \{-1,1\}^\mathbb{N} : \#\{n\in\mathbb{N} : x_1 + \cdots + x_n = 0\} = \infty\right\},$$

7.7* Hewitt-Savage's 0-1 Law

$$B_- := \left\{(x_n)_{n\in\mathbb{N}} \in \{-1, 1\}^\mathbb{N} : \#\{n \in \mathbb{N} : s_n := x_1 + \cdots + x_n = 0\} < \infty,\right.$$
$$\left.\limsup_{n\to\infty} s_n < 0\right\}.$$

The sets B_+, B, B_- are symmetric, and we have $A_\pm = f^{-1}(B_\pm)$ and $A = f^{-1}(B)$. The zero-one law of Hewitt-Savage applies to the sets $B_\pm$ and B, which finally gives $\mathbb{P}(A_+), \mathbb{P}(A), \mathbb{P}(A_-) \in \{0, 1\}$. Since the random walk is symmetric we also have $\mathbb{P}(A_+) = \mathbb{P}(A_-)$. As the only solution to that we obtain

$$\mathbb{P}(A) = 1 \quad \text{and} \quad \mathbb{P}(A_+) = \mathbb{P}(A_-) = 0.$$

□

The non-symmetric random walk behaves differently as the symmetric one:

Proposition 7.7.6 (Non-symmetric Random Walk) *For $p \in (0, 1) \setminus \{\frac{1}{2}\}$ and for the event from (7.6) we have $\mathbb{P}(A^{(p)}) = 0$.*

Proof Let $q := 1 - p$. First we recall Stirling's[3] formula

$$n! = \sqrt{2\pi n} \left(\frac{n}{e}\right)^n e^{\frac{\theta}{12n}}$$

for $n = 1, 2, \ldots$ and some $\theta \in (0, 1)$ depending on n. This gives

$$\binom{2n}{n} = \frac{(2n)!}{(n!)^2} = \frac{\sqrt{2\pi(2n)} \left(\frac{2n}{e}\right)^{2n} e^{\frac{\theta_1}{12\cdot 2n}}}{(\sqrt{2\pi n})^2 \left(\frac{n}{e}\right)^{2n} \left(e^{\frac{\theta_2}{12n}}\right)^2} \sim \frac{4^n}{\sqrt{\pi n}}.$$

Letting $B_n := \{S_{2n} = 0\}$ (for odd n the random walk S_n cannot reach zero) we obtain

$$\mathbb{P}(B_n) = \binom{2n}{n}(pq)^n \sim \frac{(4pq)^n}{\sqrt{\pi n}}$$

by Stirling's formula. Since $p \neq q$ gives $4pq < 1$, we have $\sum_{n=1}^\infty \mathbb{P}(B_n) < \infty$ so that the Lemma of Borel-Cantelli Theorem 2.5.7 implies that

$$\mathbb{P}(\{\omega \in \Omega : \omega \in B_n \text{ infinitely often }\}) = 0.$$

□

---

[3] James Stirling, May 1692 (Garden, Scotland)–05/12/1770 (Edinburgh, Scotland).

7.8 Exercises

Ex 1: Let $(\Omega, \mathcal{F}, \mathbb{P})$ be a probability space and let $A_1, A_2, \ldots \in \mathcal{F}$. Prove that $\mathbb{1}_{A_1}, \mathbb{1}_{A_2}, \ldots$ are independent random variables if and only if $A_1, A_2, A_3, \ldots$ are independent events.

Ex 2: Assume the probability spaces $(\Omega_i, \mathcal{F}_i, \mathbb{P}_i), i \in I$. Let $\mathcal{A}$ consist of all finite unions of cuboids

$$C = \bigtimes_{i \in I} E_i \quad \text{with} \quad E_i \in \mathcal{F}_i, \quad \#\{i \in I : E_i \neq \Omega_i\} < \infty.$$

Define a pre-measure $\mathbb{P}_0$ on $\mathcal{A}$ setting

$$\mathbb{P}_0(C_1 \cup \cdots \cup C_L) := \sum_{l=1}^{L} \left(\prod_{i \in I} \mathbb{P}_i(E_i^l) \right)$$

for any pairwise disjoint $C_l = \bigtimes_{i \in I} E_i^l$ where $l = 1, \ldots, L$ (with $\#\{i : E_i^l \neq \Omega_i\} < \infty$)). Show that for any $\varepsilon > 0$ and $B \in \mathcal{A}$ the estimate

$$\mathbb{P}_0(B) \leqslant \mathbb{P}_1\left(\omega_1 \in \Omega_1 : \mathbb{P}_0^{(1)}(B(\omega_1)) > \varepsilon\right) + \varepsilon \mathbb{P}_1\left(\omega_1 \in \Omega_1 : \mathbb{P}_0^{(1)}(B(\omega_1)) \leqslant \varepsilon\right)$$

holds. Here $B(\omega_1) = \{(\omega_i)_{i \in I \setminus \{1\}} : (\omega_1, \omega_2, \omega_3, \ldots) \in B\}$ and $\mathbb{P}_0^{(1)}$ stands for the pre-measure which is defined like $\mathbb{P}_0$ on $\mathcal{A}$ but now with the index set $\#I \setminus \{1\}$ and appropriate changes for the algebra $\mathcal{A}^{(1)}$.

Ex 3: Let $(\Omega, \mathcal{F}, \mathbb{P})$ be a probability space and let $f_1, f_2, \ldots : \Omega \to \mathbb{R}$ be random variables.

(a) Prove that $\{\limsup_{n \to \infty} f_n \leqslant x\} \in \mathcal{F}$ for all $x \in \mathbb{R}$.
(b) For $A_1, A_2, \ldots \in \mathcal{F}$ verify $\limsup_{n \to \infty} \mathbb{1}_{A_n} = \mathbb{1}_{\{\limsup_{n \to \infty} A_n\}}$.
(c) Assume that $f_1, f_2, \ldots$ are independent. Recall the notation

$$\mathcal{F}^\infty = \bigcap_{n=1}^{\infty} \sigma(f_n, f_{n+1}, \ldots)$$

and show that $\{\limsup_{n \to \infty} f_n \in B\} \in \mathcal{F}^\infty$ for $B \in \mathcal{B}(\mathbb{R})$. Using the 0-1 law of Kolmogorov deduce that for independent $A_1, A_2, \ldots \in \mathcal{F}$ one has $\mathbb{P}(\limsup_{n \to \infty} A_n) \in \{0, 1\}$. Compare this with the lemma of Borel-Cantelli. Conclude that any $\mathcal{F}^\infty$-measurable random variable is a.s. constant.
(d) Let $S_n := f_1 + \cdots + f_n$ and assume again that $f_1, f_2, \ldots$ are independent. Explain why $\{\limsup_{n \to \infty} S_n \in B\} \notin \mathcal{F}^\infty$ in general for $B \in \mathcal{B}(\mathbb{R})$.

Ex 4: Let $(\Omega, \mathcal{F}, \mathbb{P})$ be a probability space and $f_i : \Omega \to \mathbb{R}, i \in I$, be independent random variables, where I is a non-empty index-set. Assume $I = J \cup K$,

where $J \cap K = \emptyset$, and both, J and K, are non-empty. Define
$$\mathcal{G} := \sigma(f_j : j \in J) \quad \text{and} \quad \mathcal{H} := \sigma(f_k : k \in K).$$

Show that $\mathcal{G}$ and $\mathcal{H}$ are independent σ-algebras as follows:

(a) Choose $C_1, \ldots, C_m \in \mathcal{B}(\mathbb{R})$ and distinct $j_1, \ldots, j_m \in J$ such that
$$\mathbb{P}(\{\omega \in \Omega : f_{j_1}(\omega) \in C_1, \ldots, f_{j_m}(\omega) \in C_m\}) > 0.$$
Then $E := \{\omega \in \Omega : f_{j_1}(\omega) \in C_1, \ldots, f_{j_m}(\omega) \in C_m\} \in \mathcal{G}$. Define the collection Π to be all sets of the form
$$B := \{\omega \in \Omega : f_{k_1}(\omega) \in D_1, \ldots, f_{k_n}(\omega) \in D_n\} \in \mathcal{H}$$
where $n \geq 1$, $k_1, \ldots, k_n \in K$ are distinct, and $D_1, \ldots, D_n \in \mathcal{B}(\mathbb{R})$. Observe that Π is a π-system containing Ω and prove that $\mathbb{P}(B|E) = \mathbb{P}(B)$ for all $B \in \Pi$.

(b) Which theorem can one apply to get that $\mathbb{P}(B|E) = \mathbb{P}(B)$ for all $B \in \mathcal{H}$?

(c) Observe that this implies that $\mathbb{P}(E \cap B) = \mathbb{P}(E)\mathbb{P}(B)$ for all $B \in \mathcal{H}$.

(d) Now repeat the argument above (by interchanging the roles) and prove that $\mathbb{P}(E \cap F) = \mathbb{P}(E)\mathbb{P}(F)$ for all $E \in \mathcal{G}$ and $F \in \mathcal{H}$.

Ex 5: On the probability space $([0, 1), \mathcal{B}([0, 1)), \lambda)$ one can construct an independent and identically distributed sequence of random variables $(f_n)_{n=1}^\infty$ with $\mathbb{P}(f_1 = 0) = \mathbb{P}(f_1 = 1) = \mathbb{P}(f_1 = 2) = \frac{1}{3}$ as follows: For $n = 1$ we let $f_1(t) := 1_{[\frac{1}{3}, \frac{2}{3})}(t) + 2 \cdot 1_{[\frac{2}{3}, 1)}(t)$. For $n \geq 2$ and $t \in [0, 1)$ we find the $k \in \{1, \ldots, 3^{n-1}\}$ such that $t \in [\frac{k-1}{3^{n-1}}, \frac{k}{3^{n-1}})$ and set $f_n(t) := 0$ if $t \in [\frac{3k-3}{3^n}, \frac{3k-2}{3^n})$, $f_n(t) := 1$ if $t \in [\frac{3k-2}{3^n}, \frac{3k-1}{3^n})$, and $f_n(t) := 2$ if $t \in [\frac{3k-1}{3^n}, \frac{3k}{3^n})$. Show that $(f_n)_{n=1}^\infty$ is a sequence of independent random variables with $\mathbb{P}(f_n = 0) = \mathbb{P}(f_n = 1) = \mathbb{P}(f_n = 2) = \frac{1}{3}$.

7.9 Comments

Section 7.2: Kolomogorov's zero-one law Theorem 7.2.1 can be found in [105, Anhang].

Section 7.3: To prove Theorem 7.3.3 we use the same argument as in [22, Section 5.3]. The methodology can also be applied to verify the Daniell-Kolmogorov consistency theorem, which can be seen as a generalization of Theorem 7.3.3, see [100, Section 2.2].

Section 7.7: Theorem 7.7.4 is due to Hewitt and Savage [85, Theorem 11.3].

Chapter 8
Integration

The introduction of the Riemann integral on the real line in calculus is based on the concept of upper and lower sums. We will recall this concept in Sect. 8.6.2. The Riemann integration is coming up against its limits already in simple cases: Take the Dirichlet[1] function

$$f : [0, 1] \to \mathbb{R} \quad \text{given by} \quad f(x) := \mathbb{1}_{\mathbb{Q} \cap [0,1]}(x).$$

This function is a prototype of a function that is not Riemann integrable, but integrable with respect to the Lebesgue measure. In this chapter we introduce the Lebesgue integral denoted by

$$\int_\Omega f \mathrm{d}\mu = \int_\Omega f(\omega) \mathrm{d}\mu(\omega)$$

for a measure space $(\Omega, \mathcal{F}, \mu)$ and a measurable map $f : \Omega \to \mathbb{R}$. In this way we obtain two generalizations. Firstly, if $(\Omega, \mathcal{F}, \mu) = (\mathbb{R}, \mathcal{B}(\mathbb{R}), \lambda)$, then the integration extends to Lebesgue integrable functions and, for example, the Dirichlet function can be integrated. Secondly, introducing the Lebesgue integral using any measure space $(\Omega, \mathcal{F}, \mu)$ means the generalization to integration on a general set Ω with respect to a measure μ.

The definition of the Lebesgue integral is done in three steps, first for integrands which are simple nonnegative functions, then for nonnegative measurable functions, and eventually for all measurable functions for which the integral exists. Basic properties of the Lebesgue integral are shown. The lemma of Fatou for functions, the theorem about monotone convergence, and Lebesgue's theorem on dominated convergence are proven which provide conditions when limits and the Lebesgue

---

[1] Johann Peter Gustav Lejeune Dirichlet, 13/02/1805 (Düren, French Empire, now Germany)–05/05/1859 (Göttingen, Hanover, now Germany).

integral may be interchanged. Lebesgue's criterion for Riemann integrability is included and provides a practical criterion under which conditions the integral with respect to the Lebesgue measure coincides with the Riemann integral. The chapter continues with the change of variable formula, theorems of Tonelli and Fubini, and the inequalities of Markov, Jensen, Hölder, and Minkowski. We conclude with the proof of Hoeffding's inequality, Wald's identity, and a short excursion to atomless probability spaces.

From now on we use the following convention:

Convention Given a measure space $(\Omega, \mathcal{F}, \mu)$, we say that a property $\mathcal{P}(\omega)$, depending on ω, holds μ-**almost everywhere** (a.e.) (or **almost surely** (a.s.) in case μ is a probability measure) if $\{\omega \in \Omega : \mathcal{P}(\omega) \text{ holds}\} \in \mathcal{F}$ and that

$$\mu(\{\omega \in \Omega : \mathcal{P}(\omega) \text{ does not hold}\}) = 0.$$

8.1 Definition of the Lebesgue Integral

In the following we assume a measure space $(\Omega, \mathcal{F}, \mu)$. The definition of the Lebesgue integral $\int_\Omega f d\mu$ (also called expected value if μ is a probability measure) is done in three steps.

Definition 8.1.1 (Step One: Non-negative Simple Functions) Given a measurable $g : \Omega \to \mathbb{R}$ with representation

$$g = \sum_{i=1}^n \alpha_i \mathbb{1}_{A_i} \tag{8.1}$$

where $\alpha_i \in [0, \infty)$ and $A_i \in \mathcal{F}$, we let

$$\int_\Omega g d\mu = \int_\Omega g(\omega) d\mu(\omega) := \sum_{i=1}^n \alpha_i \mu(A_i).$$

Before we proceed, we need to explain how we interpret $\alpha_i \mu(A_i)$ if $\alpha_i = 0$, but $\mu(A_i) = \infty$. In this case we set $\alpha_i \mu(A_i) := 0$. This convention will be used in the sequel.

Now we have to check that $\int_\Omega g d\mu$ is well-defined, since it might be that different representations of g give different integrals $\int_\Omega g d\mu$. However, this is not the case:

8.1 Definition of the Lebesgue Integral

Lemma 8.1.2 *If one has two representations[2]*

$$g = \sum_{i=1}^{n} \alpha_i \mathbb{1}_{A_i} = \sum_{j=1}^{m} \beta_j \mathbb{1}_{B_j} \tag{8.2}$$

where $\alpha_i, \beta_j \geq 0$ and $A_i, B_j \in \mathcal{F}$, then $\sum_{i=1}^{n} \alpha_i \mu(A_i) = \sum_{j=1}^{m} \beta_j \mu(B_j)$.

Proof We define a collection of sets $C_1, ..., C_N \in \mathcal{F}$ given by

$$\{C_1, ..., C_N\} = \{D_1 \cap ... \cap D_{n+m} : D_i \in \{A_i, A_i^c\} \text{ for } i = 1, ..., n,$$
$$\text{and } D_{n+j} \in \{B_j, B_j^c\} \text{ for } j = 1, ..., m\}.$$

It is easy to see that $N = 2^{n+m}$. Moreover, it holds

(a) $C_k \cap C_l = \emptyset$ if $k \neq l$,
(b) $\bigcup_{k=1}^{N} C_k = \Omega$,
(c) for all A_i there is a set $I_i \subseteq \{1, ..., N\}$ such that $A_i = \bigcup_{k \in I_i} C_k$,
(d) for all B_j there is a set $J_j \subseteq \{1, ..., N\}$ such that $B_j = \bigcup_{k \in J_j} C_k$.

From (8.2) we conclude that for $C_k \neq \emptyset$ and $\omega \in C_k$ we have $\sum_{i=1}^{n} \alpha_i \mathbb{1}_{A_i}(\omega) = \sum_{j=1}^{m} \beta_j \mathbb{1}_{B_j}(\omega)$, so that (c) and (d) imply

$$\sum_{i: k \in I_i} \alpha_i = \sum_{j: k \in J_j} \beta_j \quad \text{whenever} \quad C_k \neq \emptyset$$

and

$$\sum_{i=1}^{n} \alpha_i \mu(A_i) = \sum_{i=1}^{n} \sum_{k \in I_i} \alpha_i \mu(C_k) = \sum_{k=1}^{N} \left(\sum_{i: k \in I_i} \alpha_i \right) \mu(C_k)$$
$$= \sum_{k=1}^{N} \left(\sum_{j: k \in J_j} \beta_j \right) \mu(C_k) = \sum_{j=1}^{n} \beta_j \mu(B_j).$$

□

Remark 8.1.3 One might wonder why we assumed that the simple functions are non-negative. If we do not assume this, then if the measure μ is not finite the definition $\int_\Omega g d\mu$ might fail as we could get in the sum $\sum_{i=1}^{n} \alpha_i \mu(A_i)$ a situation like $\infty - \infty$.

---

[2] We do not assume that the $(A_i)_{i=1}^{n}$ and $(B_j)_{j=1}^{m}$ are pairwise disjoint nor cover Ω.

Definition 8.1.4 (Step Two: Non-negative Functions) Given a measurable map $f : \Omega \to \mathbb{R}$ with $f(\omega) \geq 0$ for all $\omega \in \Omega$, we let

$$\int_\Omega f \, d\mu = \int_\Omega f(\omega) d\mu(\omega)$$
$$:= \sup \left\{ \int_\Omega g \, d\mu : 0 \leq g(\omega) \leq f(\omega), g \text{ is a simple function} \right\}.$$

Remark 8.1.5 If f and g are measurable functions such that $0 \leq f \leq g$, then we see from the above definition the monotonicity $0 \leq \int_\Omega f d\mu \leq \int_\Omega g d\mu$.

In the last step we define the Lebesgue integral for a general measurable map. To this end we decompose $f : \Omega \to \mathbb{R}$ into its positive and negative part

$$f(\omega) = f^+(\omega) - f^-(\omega)$$

with

$$f^+(\omega) := \max\{f(\omega), 0\} \geq 0 \quad \text{and} \quad f^-(\omega) := \max\{-f(\omega), 0\} \geq 0.$$

Definition 8.1.6 (Step Three: The General Case) Let $f : \Omega \to \mathbb{R}$ be a measurable function.

(1) If $\int_\Omega f^+ d\mu < \infty$ or $\int_\Omega f^- d\mu < \infty$, then we say that the Lebesgue integral $\int_\Omega f d\mu$ **exists** and set

$$\int_\Omega f d\mu := \int_\Omega f^+ d\mu - \int_\Omega f^- d\mu \in [-\infty, \infty].$$

(2) The map f is called **integrable** provided that

$$\int_\Omega f^+ d\mu < \infty \quad \text{and} \quad \int_\Omega f^- d\mu < \infty.$$

(3) If the Lebesgue integral $\int_\Omega f d\mu$ exists and $A \in \mathcal{F}$, then

$$\int_A f d\mu = \int_A f(\omega) d\mu(\omega) := \int_\Omega f(\omega) \mathbb{1}_A(\omega) d\mu(\omega).$$

The Lebesgue integral $\int_\Omega f d\mu$ is called **expectation** or **expected value** of the random variable f in case μ is a probability measure, and then it is also denoted by $\mathbb{E} f$ or $\mathbb{E}_\mu f$.

Remark 8.1.7 (Exercise 1) In Definition 8.1.6(3) we used the fact that if $\int_\Omega f d\mu$ exists, then $\int_\Omega f \mathbb{1}_A d\mu$ exists as well.

8.2 Theorem About Monotone Convergence

The following lemma is intuitive and clarifies that null sets are negligible while integrating. The proof is the subject of Exercise 1. This lemma will be needed later at various places to obtain the natural assumptions for our statements:

Lemma 8.1.8 *If $f, g : \Omega \to \mathbb{R}$ are measurable functions with*

$$\mu(\{\omega \in \Omega : f(\omega) \neq g(\omega)\}) = 0$$

and if $\int_\Omega f \mathrm{d}\mu \in [-\infty, \infty]$ exists, then $\int_\Omega g \mathrm{d}\mu$ exists and one has

$$\int_\Omega g \mathrm{d}\mu = \int_\Omega f \mathrm{d}\mu.$$

8.2 Theorem About Monotone Convergence

The first basic theorem about integration is Theorem 8.2.2 which concerns monotone convergence. It is typically used where one has to interchange a limit with an integral and also opens the way to compute various concrete integrals.

Let us start with an example: Assume $(\Omega, \mathcal{F}, \mu) = ([0, 1], \mathcal{B}([0, 1]), \mathcal{U}_{[0,1]})$ and simple functions $f_n := n \mathbf{1}_{(0, 1/n)}$. Then $\lim_{n \to \infty} f_n(t) = 0$ for all $t \in [0, 1]$ and

$$\int_\Omega f_n \mathrm{d}\mu = 1, \quad \text{but} \quad \int_\Omega \lim_{n \to \infty} f_n \mathrm{d}\mu = \int_\Omega 0 \mathrm{d}\mu = 0.$$

Therefore we do *not* have $\lim_{n \to \infty} \int_\Omega f_n \mathrm{d}\mu = \int_\Omega \lim_{n \to \infty} f_n \mathrm{d}\mu$ in this situation. What is the reason for this failure? One reason is that the sequence f_n is neither monotone nor it is uniformly integrable (a concept we will introduce later).

We start with an elementary version of monotone convergence. It implies that instead using Definition 8.1.4 we can approximate the Lebesgue integral of a non-negative f using any sequence of simple functions which approximate f from below.

Lemma 8.2.1 *Let $(\Omega, \mathcal{F}, \mu)$ be a measure space and let $f_n, f : \Omega \to \mathbb{R}$ be non-negative measurable functions such that the f_n are simple functions and $0 \leq f_n(\omega) \uparrow f(\omega)$ for all $\omega \in \Omega$ as $n \to \infty$. Then*

$$\int_\Omega f \mathrm{d}\mu = \lim_{n \to \infty} \int_\Omega f_n \mathrm{d}\mu.$$

Proof

(a) First we assume that $f = \mathbf{1}_A$ for some $A \in \mathcal{F}$. Let $\varepsilon \in (0, 1)$ and

$$B_n^\varepsilon := \{\omega \in A : 1 - \varepsilon \leq f_n(\omega)\} \subseteq A.$$

Then
$$(1-\varepsilon)\mathbb{1}_{B_n^\varepsilon}(\omega) \leqslant f_n(\omega) \leqslant \mathbb{1}_A(\omega).$$

Since $B_n^\varepsilon \subseteq B_{n+1}^\varepsilon$ and $\bigcup_{n=1}^\infty B_n^\varepsilon = A$ we get, by the continuity of μ from below that $\lim_n \mu(B_n^\varepsilon) = \mu(A) \in [0, \infty]$. Hence the monotonicity of the integral (see Remark 8.1.5) implies

$$(1-\varepsilon)\mu(A) \leqslant \lim_{n\to\infty} \int_\Omega f_n \mathrm{d}\mu \leqslant \int_\Omega \mathbb{1}_A \mathrm{d}\mu = \mu(A).$$

Since this is true for all $\varepsilon > 0$ we finally derive

$$\int_\Omega f \mathrm{d}\mu = \mu(A) = \lim_{n\to\infty} \int_\Omega f_n \mathrm{d}\mu.$$

(b) From (a) we deduce the statement when f is a simple function.

(c) Now let us assume that f is general. We construct simple functions $0 \leqslant h_n \uparrow f$ and show $\int_\Omega f \mathrm{d}\mu = \lim_{n\to\infty} \int_\Omega h_n \mathrm{d}\mu$. For this let

$$f_n^0(\omega) := \sum_{k=0}^{4^n-1} \frac{k}{2^n} \mathbb{1}_{\{\frac{k}{2^n} \leqslant f(\omega) < \frac{k+1}{2^n}\}}$$

so that $0 \leqslant f_n^0(\omega) \uparrow f(\omega)$ for all $\omega \in \Omega$. By Definition 8.1.4 of the Lebesgue integral there exists a sequence $0 \leqslant g_n(\omega) \leqslant f(\omega)$ of simple functions such that $\int_\Omega g_n \mathrm{d}\mu \uparrow \int_\Omega f \mathrm{d}\mu$. Hence

$$h_n := \max\left\{f_n^0, g_1, \ldots, g_n\right\}$$

is a simple function with $0 \leqslant g_n(\omega) \leqslant h_n(\omega) \uparrow f(\omega)$, so that

$$\int_\Omega g_n \mathrm{d}\mu \leqslant \int_\Omega h_n \mathrm{d}\mu \leqslant \int_\Omega f \mathrm{d}\mu$$

and

$$\lim_{n\to\infty} \int_\Omega g_n \mathrm{d}\mu = \lim_{n\to\infty} \int_\Omega h_n \mathrm{d}\mu = \int_\Omega f \mathrm{d}\mu.$$

(d) Finally we show the assertion. Consider

$$d_{k,n} := f_k \wedge h_n.$$

8.2 Theorem About Monotone Convergence

Clearly, $d_{k,n} \uparrow f_k$ as $n \to \infty$ and $d_{k,n} \uparrow h_n$ as $k \to \infty$. Therefore we get

$$\int_\Omega f \, d\mu = \lim_{n\to\infty} \int_\Omega h_n \, d\mu = \lim_{n\to\infty} \lim_{k\to\infty} \int_\Omega d_{k,n} \, d\mu = \lim_{k\to\infty} \lim_{n\to\infty} \int_\Omega d_{k,n} \, d\mu$$

$$= \lim_{k\to\infty} \int f_k \, d\mu$$

where we use $\lim_{k\to\infty} \lim_{n\to\infty} z_{k,n} = \lim_{n\to\infty} \lim_{k\to\infty} z_{k,n}$ for non-negative $z_{k,n} \in \mathbb{R}$ that increase with respect to k and n (where the remaining k or n is fixed). □

Theorem 8.2.2 (Monotone Convergence) *Let $(\Omega, \mathcal{F}, \mu)$ be a measure space and $f, f_1, f_2, \ldots : \Omega \to \mathbb{R}$ be measurable.*

(1) If $0 \leq f_n \uparrow f$ a.e., then $\lim_{n\to\infty} \int_\Omega f_n \, d\mu = \int_\Omega f \, d\mu$.
(2) If $0 \geq f_n \downarrow f$ a.e., then $\lim_{n\to\infty} \int_\Omega f_n \, d\mu = \int_\Omega f \, d\mu$.

Proof (1) Let us first assume that $0 \leq f_n(\omega) \uparrow f(\omega)$ for all $\omega \in \Omega$. Because $\int_\Omega f_n \, d\mu \leq \int_\Omega f \, d\mu$ (see Remark 8.1.5) we only need to show

$$\lim_{n\to\infty} \int_\Omega f_n \, d\mu \geq \int_\Omega f \, d\mu.$$

For each f_n we take a sequence of step functions $(f_{n,k})_{k=1}^\infty$ such that $0 \leq f_{n,k} \uparrow f_n$ as $k \to \infty$ and set

$$h_L := \max_{\substack{1 \leq k \leq L \\ 1 \leq n \leq L}} f_{n,k} \leq f_L.$$

For $1 \leq n \leq L$ it holds that

$$f_{n,L} \leq h_L \leq f_L \quad \text{so that} \quad f_n \leq \lim_{L\to\infty} h_L \leq f.$$

Now $f_n \uparrow f$ yields $h_L \uparrow f$ as $L \to \infty$. Finally, Lemma 8.2.1 implies

$$\lim_{L\to\infty} \int f_L \, d\mu \geq \lim_{L\to\infty} \int h_L \, d\mu = \int f \, d\mu.$$

The general case is treated as follows: We let $A := \{\omega \in \Omega : f_n(\omega) \uparrow f(\omega)\}$ so that $\mu(A^c) = 0$ by assumption. Then we let $g := f \mathbb{1}_A$ and $g_n := \mathbb{1}_A f_n$ and use that Lemma 8.1.8 implies $\int_\Omega g_n \, d\mu = \int_\Omega f_n \, d\mu$ and $\int_\Omega g \, d\mu = \int_\Omega f \, d\mu$.
(2) is proved exactly in the same way. □

8.3 Basic Properties of the Lebesgue Integral

Now we state a first set of basic properties of the Lebesgue integral:

Proposition 8.3.1 (Properties of the Lebesgue Integral) *For a measure space* $(\Omega, \mathcal{F}, \mu)$ *and measurable* $f, g : \Omega \to \mathbb{R}$, *such that* $\int_\Omega f d\mu$ *and* $\int_\Omega g d\mu$ *exist, the following assertions hold:*

(1) **Additivity:** *If* $\int_\Omega f^+ d\mu + \int_\Omega g^+ d\mu < \infty$ *or* $\int_\Omega f^- d\mu + \int_\Omega g^- d\mu < \infty$, *then*

$$\int_\Omega (f+g)^+ d\mu < \infty \quad \text{or} \quad \int_\Omega (f+g)^- d\mu < \infty$$

and $\int_\Omega (f+g) d\mu = \int_\Omega f d\mu + \int_\Omega g d\mu$.

(2) **Homogeneity:** *If* $a \in \mathbb{R}$, *then* $\int_\Omega (af) d\mu$ *exists and, for* $a \neq 0$,

$$\int_\Omega (af) d\mu = a \int_\Omega f d\mu.$$

(3) **Monotonicity:** *If* $f \leq g$ *a.e., then* $\int_\Omega f d\mu \leq \int_\Omega g d\mu$.
(4) *If* $f \geq 0$ *a.e. and* $\int_\Omega f d\mu = 0$, *then* $f = 0$ *a.e.*
(5) *One has* $\left|\int_\Omega f d\mu\right| \leq \int_\Omega |f| d\mu = \int_\Omega f^+ d\mu + \int_\Omega f^- d\mu$.
(6) *The map* f *is integrable if and only if* $|f|$ *is integrable.*
(7) *If* f *and* g *are integrable and* $a, b \in \mathbb{R}$, *then* $af + bg$ *is integrable and*

$$\int_\Omega (af+bg) d\mu = a \int_\Omega f d\mu + b \int_\Omega g d\mu.$$

Proof (1) First we assume $f(\omega) \geq 0$ and $g(\omega) \geq 0$ for all $\omega \in \Omega$. Using the approximations f_n^0 and g_n^0 from the proof of Lemma 8.2.1 we see that

$$0 \leq f_n^0 \leq f, \quad 0 \leq g_n^0 \leq g, \quad \lim_{n \to \infty} f_n^0 = f, \quad \lim_{n \to \infty} g_n^0 = g.$$

Hence $\lim_n (f_n^0 + g_n^0) = f + g$ as well and $\int_\Omega (f_n^0 + g_n^0) d\mu = \int_\Omega f_n^0 d\mu + \int_\Omega g_n^0 d\mu$ (because we have simple functions) and Lemma 8.2.1 gives therefore that

$$\int_\Omega (f+g) d\mu = \int_\Omega f d\mu + \int_\Omega g d\mu. \tag{8.3}$$

To treat the general case we observe that

$$(f+g)^+ + f^- + g^- = f^+ + g^+ + (f+g)^-$$

8.3 Basic Properties of the Lebesgue Integral

and

$$\int_\Omega (f+g)^+ d\mu + \int_\Omega f^- d\mu + \int_\Omega g^- d\mu = \int_\Omega f^+ d\mu + \int_\Omega g^+ d\mu + \int_\Omega (f+g)^- d\mu.$$

Now we use that $(f+g)^\pm \leq f^\pm + g^\pm$. From Definition 8.1.4 one can derive that

$$\int_\Omega (f+g)^\pm d\mu \leq \int_\Omega f^\pm + g^\pm d\mu = \int_\Omega f^\pm d\mu + \int_\Omega g^\pm d\mu,$$

where the equation is justified by (8.3). Hence under the assumptions of (1) we can rearrange the above equation to

$$\int_\Omega (f+g)^+ d\mu - \int_\Omega (f+g)^- d\mu = \int_\Omega f^+ d\mu - \int_\Omega f^- d\mu + \int_\Omega g^+ d\mu - \int_\Omega g^- d\mu.$$

(2) is the subject of Exercise 1.

(3) According to Lemma 8.1.8 we only have to check the case $f(\omega) \leq g(\omega)$ for all $\omega \in \Omega$ by setting f and g to zero when $f(\omega) > g(\omega)$. Moreover, $\int_\Omega f^- d\mu = \infty$ gives $\int_\Omega f d\mu = -\infty$ and there is nothing to prove. So let us assume that $\int_\Omega f^- d\mu < \infty$ and define $h(\omega) := g(\omega) - f(\omega) \geq 0$ so that $f + h = g$ and $\int_\Omega h^- d\mu < \infty$ as well. By (1) we deduce that $\int_\Omega g d\mu = \int_\Omega f d\mu + \int_\Omega h d\mu \geq \int_\Omega f d\mu$.

(4) Again by Lemma 8.1.8 we can assume that $f(\omega) \geq 0$ for all $\omega \in \Omega$. Let $\mu(f > 0) > 0$. By considering $\{f > 0\} = \bigcup_{n=1}^\infty \{f \geq 1/n\}$ and using the monotonicity of μ from below we get $\mu(f > 0) = \lim_{n \to \infty} \mu(f \geq 1/n)$ and some $n_0 \geq 1$ with $\mu(f \geq 1/n_0) > 0$. Now

$$0 \leq \frac{1}{n_0} \mathbb{1}_{\{f \geq 1/n_0\}} \leq f$$

so that by (3) we have $\int_\Omega f d\mu \geq \frac{1}{n_0} \mu(f \geq 1/n_0) > 0$ which is a contradiction.

(5) Because of $|f| = f^+ + f^-$, item (1) implies

$$\int_\Omega |f| d\mu = \int_\Omega f^+ d\mu + \int_\Omega f^- d\mu. \tag{8.4}$$

Because $|\int_\Omega f d\mu| = |\int_\Omega f^+ d\mu - \int_\Omega f^- d\mu|$ item (5) follows.

(6) follows from equation (8.4).

(7) Since $(af+bg)^+ \leq |a||f| + |b||g|$ and $(af+bg)^- \leq |a||f| + |b||g|$ we get that $af+bg$ is integrable. The equation follows from combining (1) and (2). □

8.4 Extended Measurable Maps

In the sequel it will be necessary to work with extended measurable maps at some places. Their definition is as follows:

Definition 8.4.1 (Extended Measurable Map) Let $(\Omega, \mathcal{F})$ be a measurable space. A function $f : \Omega \to [-\infty, \infty]$ is called an **extended measurable map** if

(1) $f^{-1}(B) \in \mathcal{F}$ for all $B \in \mathcal{B}(\mathbb{R})$,
(2) $f^{-1}(\{-\infty\}) \in \mathcal{F}$ and $f^{-1}(\{\infty\}) \in \mathcal{F}$.

Given an extended map f, the positive and negative part are defined as before with the convention that ∞ is a positive number and $-\infty$ is a negative number, in particular we distinguish between these two values. There is an equivalent formulation: a map $f : \Omega \to [-\infty, \infty]$ is measurable if and only if the map

$$\tilde{f} := \arctan \circ f : \Omega \to [-1, 1] \subseteq \mathbb{R} \tag{8.5}$$

is $(\mathcal{F}, \mathcal{B}(\mathbb{R}))$-measurable.

In the following it is convenient to have the Lebesgue integral for extended measurable maps. We do not need to re-define the Lebesgue integral, we use the following definition instead:

Definition 8.4.2 Let $(\Omega, \mathcal{F}, \mu)$ be a measure space and $f : \Omega \to [-\infty, \infty]$ be an extended measurable map.

(1) If $f(\omega) \in [0, \infty]$ for all $\omega \in \Omega$, then

$$\int_\Omega f \, d\mu := \begin{cases} \int_\Omega \mathbb{1}_{\{f < \infty\}} f \, d\mu & : \mu(f = \infty) = 0 \\ +\infty & : \mu(f = \infty) > 0 \end{cases}.$$

(2) If $\int_\Omega f^+ d\mu < \infty$ or $\int_\Omega f^- d\mu < \infty$, then

$$\int_\Omega f \, d\mu := \int_\Omega f^+ d\mu - \int_\Omega f^- d\mu.$$

If $f(\omega) \in [0, \infty]$ for all $\omega \in \Omega$, then an equivalent definition for the Lebesgue integral is

$$\int_\Omega f \, d\mu = \lim_{\substack{N \to \infty \\ N \in \mathbb{N}}} \int_\Omega (f \wedge N) \, d\mu. \tag{8.6}$$

8.4 Extended Measurable Maps

Moreover, we need to extend Proposition 6.1.5. To do so we first remark that

$$\liminf_{n\to\infty} a_n := \lim_{n\to\infty} \inf_{k\geq n} a_k \quad \text{and} \quad \limsup_{n\to\infty} a_n := \lim_{n\to\infty} \sup_{k\geq n} a_k$$

are defined for $(a_n)_{n=1}^\infty \subseteq [-\infty, \infty]$ as for $(a_n)_{n=1}^\infty \subseteq \mathbb{R}$.

Proposition 8.4.3 *For a measurable space $(\Omega, \mathcal{F})$ and measurable maps $f_n : \Omega \to [-\infty, \infty]$ the pointwise limits $\liminf_{n\to\infty} f_n : \Omega \to [-\infty, \infty]$ and $\limsup_{n\to\infty} f_n : \Omega \to [-\infty, \infty]$ are measurable.*

Proof By (8.5) we may restrict the problem to maps $f_n : \Omega \to [-1, 1]$. Moreover, the problem can be further reduced to the following: If $g_n : \Omega \to [-1, 1]$, $n \in \mathbb{N}$, are measurable and nondecreasing on Ω in n (or nonincreasing on Ω in n), then $g := \lim_{n\to\infty} g_n$ is measurable. And this follows from Proposition 6.1.5. □

Now we extend our previous statement about monotone convergence Theorem 8.2.2:

Theorem 8.4.4 (Monotone Convergence II) *Let $(\Omega, \mathcal{F}, \mu)$ be a measure space and assume measurable maps $g, f, f_1, f_2, \ldots : \Omega \to [-\infty, \infty]$. If*

(1) $g \leq f_n \uparrow f$ *a.e. and* $\int_\Omega g^- d\mu < \infty$ *or*
(2) $g \geq f_n \downarrow f$ *a.e. and* $\int_\Omega g^+ d\mu < \infty$,

then $\lim_{n\to\infty} \int_\Omega f_n d\mu = \int_\Omega f d\mu$.

Proof We only verify (1), since (2) follows from (1) by multiplying (g, f, f_n) by -1. Furthermore, the integrals do not change by putting $f_n(\omega)$ and $f(\omega)$ equal to $g(\omega)$ where the a.s. assumption is not satisfied. Hence we may assume $g \leq f_n \uparrow f$ on Ω.

(a) If $\int_\Omega g d\mu = \infty$ or if there is an $n_0 \in \mathbb{N}$ such that $\int_\Omega f_{n_0} d\mu = \infty$, then $\int_\Omega f_n d\mu = \int_\Omega f d\mu = \infty$ for $n \geq n_0$.
(b) From $\int_\Omega g^- d\mu < \infty$ we know that $\mu(g = -\infty) = 0$. Since in (a) we discussed already the case $\int_\Omega g d\mu + \int_\Omega f_{n_0} d\mu = \infty$ we may assume that $g, f_n : \Omega \to \mathbb{R}$ (if not, we put f, f_n, and g equal to zero where this does not hold), and moreover, that we have $\int_\Omega |g| d\mu < \infty$ and $\int_\Omega |f_n| d\mu < \infty$ for all $n \in \mathbb{N}$.
(b1) If $\int_\Omega f^+ d\mu < \infty$, then we put $f(\omega)$ and $f_n(\omega)$ equal to $g(\omega)$ when $f(\omega) = \infty$. Then $0 \leq f_n - g \uparrow f - g$. Theorem 8.2.2 and Proposition 8.3.1 imply

$$\lim_{n\to\infty} \int_\Omega (f_n - g) d\mu = \int_\Omega (f - g) d\mu \quad \text{and} \quad \lim_{n\to\infty} \int_\Omega f_n d\mu = \int_\Omega f d\mu.$$

(b2) On the other hand, if $\int_\Omega f^+ d\mu = \infty$, then we get from $f_n^+ \uparrow f^+$ that

$$\lim_{n\to\infty} \int_\Omega f_n^+ d\mu = \lim_{n\to\infty} \lim_{N\to\infty} \int_\Omega (f_n^+ \wedge N) d\mu = \lim_{N\to\infty} \lim_{n\to\infty} \int_\Omega (f_n^+ \wedge N) d\mu$$
$$= \lim_{N\to\infty} \int_\Omega (f^+ \wedge N) d\mu = \int_\Omega f^+ d\mu = \infty$$

by (8.6). Because of $\int_\Omega f_n^- d\mu \leq \int_\Omega g^- d\mu < \infty$ we get $\lim_{n\to\infty} \int_\Omega f_n d\mu = \infty$.

$\square$

8.5 Fatou's Lemma and Lebesgue's Theorem

The following lemma is a useful tool to relate the limit inferior or superior of functions to the Lebesgue integral.

Theorem 8.5.1 (Lemma of Fatou) *Let $(\Omega, \mathcal{F}, \mu)$ be a measure space and $g, f_1, f_2, \ldots : \Omega \to \mathbb{R}$ be measurable.*

(1) *If $g \leq f_n$ a.e. and $\int_\Omega g^- d\mu < \infty$, then $\int_\Omega (\liminf_{n\to\infty} f_n)^- d\mu < \infty$ and*

$$\int_\Omega \liminf_{n\to\infty} f_n d\mu \leq \liminf_{n\to\infty} \int_\Omega f_n d\mu.$$

(2) *If $f_n \leq g$ a.e. and $\int_\Omega g^+ d\mu < \infty$, then $\int_\Omega (\limsup_{n\to\infty} f_n)^+ d\mu < \infty$ and*

$$\limsup_{n\to\infty} \int_\Omega f_n d\mu \leq \int_\Omega \limsup_{n\to\infty} f_n d\mu.$$

Proof We only prove (1) as (2) can be deduced from (1) by multiplying (g, f_n) by -1. We let $h_k := \inf_{n \geq k} f_n$ so that $h_k \uparrow \liminf_{n\to\infty} f_n$, $g \leq h_k$ a.e., and $g \leq \liminf_n f_n$ a.e. This explains $\int_\Omega (\liminf_{n\to\infty} f_n)^- d\mu \leq \int_\Omega g^- d\mu < \infty$. Furthermore, applying Theorem 8.4.4 gives that

$$\int_\Omega \liminf_{n\to\infty} f_n d\mu = \lim_{k\to\infty} \int_\Omega h_k d\mu = \lim_{k\to\infty} \left(\int_\Omega \inf_{n \geq k} f_n d\mu \right)$$
$$\leq \lim_{k\to\infty} \left(\inf_{n \geq k} \int_\Omega f_n d\mu \right) = \liminf_{k\to\infty} \int_\Omega f_k d\mu.$$

$\square$

Lebesgue's theorem states sufficient conditions under which an a.e. limit and Lebesgue integration commute.

Theorem 8.5.2 (Lebesgue's Theorem, Dominated Convergence I) Let $(\Omega, \mathcal{F}, \mu)$ be a measure space and $g, f, f_1, f_2, \ldots : \Omega \to \mathbb{R}$ be measurable and such that $|f_n| \leq g$ a.e. Assume that g is integrable and that $f = \lim_{n \to \infty} f_n$ a.e. Then f is integrable and one has that

$$\int_\Omega f \, d\mu = \lim_{n \to \infty} \int_\Omega f_n \, d\mu.$$

Proof First we observe that $-g \leq f_n \leq g$ a.e. and that $\int_\Omega (-g)^- d\mu = \int_\Omega g^+ d\mu = \int_\Omega g \, d\mu \in [0, \infty)$. Applying Fatou's lemma gives

$$\int_\Omega f \, d\mu = \int_\Omega \liminf_{n \to \infty} f_n \, d\mu \leq \liminf_{n \to \infty} \int_\Omega f_n \, d\mu$$

$$\leq \limsup_{n \to \infty} \int_\Omega f_n \, d\mu \leq \int_\Omega \limsup_{n \to \infty} f_n \, d\mu = \int_\Omega f \, d\mu.$$

□

8.6 Examples and Relations to Other Integrals

One feature of the Lebesgue integral is that it provides a unified framework to different mathematical objects, for example to sums and to the Riemann integral.

8.6.1 Countable Measure Spaces

Let Ω be finite or countably infinite, $\mathcal{F} := 2^\Omega$, assume $p_\omega \in [0, \infty]$ for $\omega \in \Omega$ and define the measure μ by

$$\mu(A) := \sum_{\omega \in A} p_\omega \quad \text{with} \quad \sum_{\omega \in \emptyset} p_\omega := 0.$$

Since $\mathcal{F} = 2^\Omega$ any function $f : \Omega \to \mathbb{R}$ is measurable, and for a nonnegative f we have

$$\int_\Omega f \, d\mu = \sum_{\omega \in \Omega} f(\omega) p_\omega.$$

The Lebesgue integral of the map $f : \Omega \to \mathbb{R}$ exists if and only if

$$\sum_{\{\omega : f(\omega) \geq 0\}} f(\omega) p_\omega < \infty \quad \text{or} \quad \sum_{\{\omega : f(\omega) \leq 0\}} f(\omega) p_\omega > -\infty.$$

If the Lebesgue integral exists, then it computes to

$$\int_\Omega f \, \mathrm{d}\mu = \sum_{\{\omega : f(\omega) \geq 0\}} f(\omega) p_\omega + \sum_{\{\omega : f(\omega) \leq 0\}} f(\omega) p_\omega.$$

The map f is integrable if and only if $\sum_{\omega \in \Omega} |f(\omega)| p_\omega < \infty$.

8.6.2* Lebesgue's Criterion for Riemann Integrability

For the following let $-\infty < a < b < \infty$. For $n \in \mathbb{N}$ we call $\mathcal{P} = (t_i)_{i=0}^n$ a partition of $[a, b]$ provided that $a = t_0 < \cdots < t_n = b$, the points $t_0, \ldots, t_n$ are called grid-points of the partition. For a bounded function $f : [a, b] \to \mathbb{R}$ we define the upper and lower sums

$$U_\mathcal{P}(f) := \sum_{i=1}^n \left[\sup_{t \in [t_{i-1}, t_i]} f(t) \right] (t_i - t_{i-1}),$$

$$L_\mathcal{P}(f) := \sum_{i=1}^n \left[\inf_{t \in [t_{i-1}, t_i]} f(t) \right] (t_i - t_{i-1}),$$

and the corresponding upper and lower Darboux[3] integrals

$$U(f) := \inf_\mathcal{P} U_\mathcal{P}(f) \quad \text{and} \quad L(f) := \sup_\mathcal{P} L_\mathcal{P}(f).$$

The function f is called Riemann integrable provided that $U(f) = L(f)$. In this case we let

$$\int_a^b f(t) \, \mathrm{d}t := U(f) = L(f)$$

be the Riemann[4] integral.

---

[3] Jean Gaston Darboux, 13/08/1842 (Nimes, Gard, Languedoc, France)–23/02/1917 (Paris, France).

[4] Georg Friedrich Bernhard Riemann, 17/09/1826 (Hanover, now Germany)–20/07/1866 (Selasca, Italy).

8.6 Examples and Relations to Other Integrals

Theorem 8.6.1 (Lebesgue criterion for Riemann integrability) *Assume that $-\infty < a < b < \infty$ and that $f : [a, b] \to \mathbb{R}$ is bounded. Then the following assertions are equivalent:*

(1) *The function f is Riemann integrable.*
(2) *There exists a Borel set $N \in \mathcal{B}(\mathbb{R})$ with $N \subseteq [a, b]$ and $\lambda(N) = 0$ such that f is continuous in all $x \in [a, b] \setminus N$.*

Proof (1) $\Rightarrow$ (2) By assumption there are partitions $\mathcal{P}_m$ and $\mathcal{Q}_m$ such that

$$\int_a^b f(t)dt = \lim_{m \to \infty} U_{\mathcal{P}_m}(f) = \lim_{m \to \infty} L_{\mathcal{Q}_m}(f).$$

Replacing $\mathcal{P}_m$ and $\mathcal{Q}_m$ by the union $\mathcal{P}_1 \cup \cdots \cup \mathcal{P}_m \cup \mathcal{Q}_1 \cup \cdots \cup \mathcal{Q}_m$, we may assume that $\mathcal{P}_m = \mathcal{Q}_m$. Furthermore, we may assume that $\mathcal{P}_m$ is a sub-partition of $\mathcal{P}_{m+1}$. So we can choose $\mathcal{P}_m$ with

$$a = t_0^m < t_1^m < \cdots < t_{n_m}^m = b$$

such that

$$L_{\mathcal{P}_1}(f) \leqslant L_{\mathcal{P}_2}(f) \leqslant \cdots \leqslant \int_a^b f(t)dt \leqslant \cdots \leqslant U_{\mathcal{P}_2}(f) \leqslant U_{\mathcal{P}_1}(f). \qquad (8.7)$$

For a subset $I \subseteq [a, b]$ define the oscillation

$$\operatorname{osc}(f, I) := \sup_{s,t \in I} |f(s) - f(t)| = \sup_{t \in I} f(t) - \inf_{t \in I} f(s)$$

so that by (8.7) we have

$$\lim_{m \to \infty} \left[\sum_{i=1}^{n_m} \operatorname{osc}(f, [t_{i-1}^m, t_i^m])(t_i^m - t_{i-1}^m) \right] = 0.$$

We define the simple functions $g_m : [a, b] \to \mathbb{R}$ by

$$g_m(t) := \sum_{i=1}^{n_m} \operatorname{osc}(f, [t_{i-1}^m, t_i^m]) \mathbb{1}_{(t_{i-1}^m, t_i^m]}(t)$$

so that $g_1(t) \geqslant g_2(t) \geqslant \cdots \geqslant 0$ for all $t \in [a, b]$ and

$$0 = \lim_{m \to \infty} \int_{[a,b]} g_m(t)d\lambda(t) = \int_{[a,b]} \lim_{m \to \infty} g_m(t)d\lambda(t)$$

by dominated convergence. Hence

$$\lambda\left(\left\{t \in [a,b] : \lim_{m\to\infty} g_m(t) \neq 0\right\}\right) = 0.$$

Now let $t \in [a,b]$ be such that $\lim_{m\to\infty} g_m(t) = 0$ and that t is not a grid-point of one of the partitions $\mathcal{P}_m$. Assume that f is not continuous in t. Then

$$\lim_{m\to\infty} g_m(t) \geq \mathrm{osc}(f,t) > 0$$

with

$$\mathrm{osc}(f,t) := \lim_{\delta\to 0}\left[\sup\left\{|f(s_0) - f(s_1)| : s_0, s_1 \in [a,b] \cap [t-\delta, t+\delta]\right\}\right],$$

which is a contradiction.

(2) $\Rightarrow$ (1) We fix arbitrary $\varepsilon, \eta > 0$. By definition we have that $\mathrm{osc}(f,t) = 0$ if and only if f is continuous at t. Moreover, $G_\varepsilon := \{t \in [a,b] : \mathrm{osc}(f,t) < \varepsilon\}$ is open in $[a,b]$ for $\varepsilon > 0$, so that $F_\varepsilon := \{t \in [a,b] : \mathrm{osc}(f,t) \geq \varepsilon\}$ is closed in $[a,b]$ and in $\mathbb{R}$. By assumption $\lambda(F_\varepsilon) = 0$, so that the compactness of F_ε and Exercise 1 from Chap. 3 imply that we find finitely many open (in $[a,b]$) intervals $O_1, \ldots, O_K$ such that

$$F_\varepsilon \subseteq \bigcup_{k=1}^{K} O_k \quad \text{and} \quad \sum_{k=1}^{K} \lambda(O_k) < \eta.$$

Without loss of generality we can arrange that the closures of the O_k are pair-wise disjoint. The complement $[a,b] \setminus \bigcup_{k=1}^{K} O_k \subseteq \{t \in [a,b] : \mathrm{osc}(f,t) < \varepsilon\}$ is a finite union of closed intervals. For each $t \in [a,b] \setminus \bigcup_{k=1}^{K} O_k$ we find an (in $[a,b]$) open interval $I_t \ni t$ such that $\mathrm{osc}(f, I_t) < \varepsilon$. Therefore we can cover $[a,b] \setminus \bigcup_{k=1}^{K} O_k$ by a finite union of $I_{t_1}, \ldots, I_{t_M}$. As $[a,b] \setminus \bigcup_{k=1}^{K} O_k$ is a finite union of closed intervals we can construct from the $I_{t_1}, \ldots, I_{t_M}$ a finite cover of closed intervals $J_1, \ldots, J_L$ with pairwise disjoint interiors and such that $\mathrm{osc}(f, J_l) < \varepsilon$ for $l = 1, \ldots, L$. Consequently,

$$U_\mathcal{P}(f) - L_\mathcal{P}(f) \leq \varepsilon(b-a) + 2\sup_{t\in[a,b]} |f(t)|\eta.$$

$\square$

Theorem 8.6.2 *Assume that* $-\infty < a < b < \infty$ *and that* $f : [a,b] \to \mathbb{R}$ *is bounded, $(\mathcal{B}([a,b]), \mathcal{B}(\mathbb{R}))$-measurable, and Riemann integrable. Then*

$$\int_a^b f(t)\mathrm{d}t = \int_{[a,b]} f\mathrm{d}\lambda.$$

8.6 Examples and Relations to Other Integrals

Proof Inspecting the proof of Theorem 8.6.1, the equality of the integrals follows from (8.7) as $L_{\mathcal{P}_m}(f) \leq \int_{[a,b]} f \, d\lambda \leq U_{\mathcal{P}_m}$. □

Remark 8.6.3

(1) There are bounded Riemann integrable functions $f : [a, b] \to \mathbb{R}$, that are not $(\mathcal{B}([a, b]), \mathcal{B}(\mathbb{R}))$-measurable: We choose sets $A \subseteq N \subseteq [a, b]$ such that $N \in \mathcal{B}([a, b])$ and $\lambda(N) = 0$, but $A \notin \mathcal{B}([a, b])$. To see the existence of such a set one can use the proof of Theorem B.4.2(4) for $[a, b] = [0, 1]$ to find a non Borel-measurable subset of the Cantor set. Then $\mathbb{1}_A$ is Riemann integrable, but not $(\mathcal{B}([a, b]), \mathcal{B}(\mathbb{R}))$-measurable.

(2) Each bounded Riemann integrable function $f : [a, b] \to \mathbb{R}$ is $(\mathcal{L}([a, b]), \mathcal{B}([a, b]))$-measurable, where $([a, b], \mathcal{L}([a, b]), \lambda)$ is the completion of $([a, b], \mathcal{B}([a, b]), \lambda)$. Indeed, by Theorem 8.6.1 there is a Borel set N with $\lambda(N) = 0$ such that f is continuous on $[a, b] \setminus N$. If $g : [a, b] \setminus N \to$ is the restriction of f and therefore continuous, we have $g^{-1}(G) \in \mathcal{B}([a, b] \setminus N)$ for each open set $G \subseteq \mathbb{R}$ where $\mathcal{B}([a, b] \setminus N)$ is the trace σ-algebra of $\mathcal{B}([a, b])$.

Let us return to an example considered before which is the prototype of example to illustrate the difference between the Lebesgue integral and the Riemann integral:

Example 8.6.4 The Dirichlet function

$$f(t) := \begin{cases} 1, & t \in [0, 1] \text{ irrational} \\ 0, & t \in [0, 1] \text{ rational} \end{cases}$$

fails to be Riemann integrable, which can be seen directly from

$$L(f) = 0 \quad \text{and} \quad U(f) = 1$$

or it can be seen from Theorem 8.6.1 because f is discontinuous in all $t \in [0, 1]$. On the other hand, the function

$$g : [0, 1] \to \mathbb{R} \quad \text{with} \quad g(x) := \begin{cases} \frac{1}{q} & : x = \frac{p}{q} > 0, \text{ where } p, q \in \mathbb{N} \\ & \quad \text{cannot be reduced} \\ 0 & : x \notin \mathbb{Q} \text{ or } x = 0, \end{cases}$$

is continuous at all $[0, 1] \setminus \mathbb{Q}$, so that this function is Riemann integrable since $\lambda(\mathbb{Q}) = 0$.

8.6.3 The Lebesgue-Stieltjes Integral

In our setup the Lebesgue-Stieltjes[5] integral is only a notation for an integral we already defined. Assume a nondecreasing and right-continuous function $F : \mathbb{R} \to \mathbb{R}$. Consider the unique measure μ on $\mathcal{B}(\mathbb{R})$ with

$$\mu((a,b]) := F(b) - F(a)$$

constructed in Theorem 3.3.8. Then we use for the Lebesgue integral the notation

$$\int_{\mathbb{R}} f(t) \mathrm{d}\mu(t) = \int_{\mathbb{R}} f(t) \mathrm{d}F(t).$$

The right-hand side is called Lebesgue-Stieltjes integral.

8.7 Change of Variable Formula

We want to prove a *change of variable formula* for the integrals $\int_\Omega g \mathrm{d}\mu$. In many cases, only by this formula it is possible to compute expected values explicitly.

Proposition 8.7.1 (Change of Variable) *Let $(\Omega, \mathcal{F}, \mu)$ be a measure space, $(E, \mathcal{E})$ be a measurable space, $f : \Omega \to E$ be $(\mathcal{F}, \mathcal{E})$-measurable, and $g : E \to \mathbb{R}$ be $(\mathcal{E}, \mathcal{B}(\mathbb{R}))$-measurable. Let us denote by μ_f the image measure of μ with respect to f, that means*

$$\mu_f(B) = \mu(\{\omega \in \Omega : f(\omega) \in B\}) \quad \text{for all} \quad B \in \mathcal{E}.$$

Then one has

$$\int_B g \mathrm{d}\mu_f = \int_{f^{-1}(B)} g(f) \mathrm{d}\mu$$

for all $B \in \mathcal{E}$ in the sense that if one integral exists, the other exists as well, and their values are equal.

Proof By replacing g by g^+ and g^- we can restrict ourselves to nonnegative $g : E \to \mathbb{R}$. Moreover, letting $\widetilde{g}(x) := \mathbb{1}_B(x) g(x)$ we have

$$\widetilde{g}(f(\omega)) = \mathbb{1}_B(f(\omega)) g(f(\omega)) = \mathbb{1}_{f^{-1}(B)}(\omega) g(f(\omega))$$

---

[5] Thomas Jan Stieltjes 29/12/1856 (Zwolle, Overijssel, The Netherlands)–31/12/1894 (Toulouse, France).

8.8 Affine Images of the Lebesgue Measure λ_d

so that it is sufficient to consider the case $B = E$. Hence we have to show that

$$\int_E g(x) \mathrm{d}\mu_f(x) = \int_\Omega g(f(\omega)) \mathrm{d}\mu(\omega).$$

Assume now a sequence of simple functions $0 \leq g_n(x) \uparrow g(x)$ for all $x \in E$ so that $g_n(f(\omega)) \uparrow g(f(\omega))$ for all $\omega \in \Omega$ as well. If we can show that

$$\int_E g_n \mathrm{d}\mu_f = \int_\Omega g_n(f) \mathrm{d}\mu,$$

then the proof is complete. By linearity it is enough to show the equality for $g_n = \mathbb{1}_B$ for some $B \in \mathcal{E}$. But here we get

$$\int_E g_n \mathrm{d}\mu_f = \mu(f^{-1}(B)) = \int_\Omega \mathbb{1}_{f^{-1}(B)} \mathrm{d}\mu = \int_\Omega \mathbb{1}_B(f) \mathrm{d}\mu = \int_\Omega g_n(f) \mathrm{d}\mu.$$

$\square$

Let us give an example for the change of variable formula.

Definition 8.7.2 (Moments) Let $n \in \mathbb{N}$ and assume that $(\Omega, \mathcal{F}, \mu)$ is a measure space.

(1) For a measurable map $f : \Omega \to \mathbb{R}$ the integral $\int_\Omega |f|^n \mathrm{d}\mu$ is called *n*-**th absolute moment** of f. If $\int_\Omega f^n \mathrm{d}\mu$ exists, then $\int_\Omega f^n \mathrm{d}\mu$ is called *n*-**th moment** of f.
(2) For a measure ν on $(\mathbb{R}, \mathcal{B}(\mathbb{R}))$ the integral $\int_\mathbb{R} |x|^n \mathrm{d}\nu(x)$ is called *n*-**th absolute moment** of ν. If $\int_\mathbb{R} x^n \mathrm{d}\nu(x)$ exists, then it is called *n*-**th moment** of ν.

Corollary 8.7.3 *Let* $(\Omega, \mathcal{F}, \mu)$ *be a measure space and* $f : \Omega \to \mathbb{R}$ *be measurable with image measure* μ_f. *Then, for all* $n \in \mathbb{N}$,

$$\int_\Omega |f|^n \mathrm{d}\mu = \int_\mathbb{R} |x|^n \mathrm{d}\mu_f(x) \quad \text{and} \quad \int_\Omega f^n \mathrm{d}\mu = \int_\mathbb{R} x^n \mathrm{d}\mu_f(x),$$

where the latter equality has to be understood as follows: if one side exists, then the other exists as well and they coincide.

8.8 Affine Images of the Lebesgue Measure λ_d

In this section we give a characterization of the Lebesgue measure λ_d (Theorem 8.8.1) and specialize the change of variable formula Proposition 8.7.1 to affine transforms of the Lebesgue measure λ_d. In particular we verify the rotation invariance of λ_d. This proof will be based on the fact that affine images of λ_d are translation invariant and that translation invariance guaranties that the affine images are multiples of λ_d. So we start with Theorem 8.8.1 and will use later that translation invariance implies condition Theorem 8.8.1(2). For its formulation we define, given $n \in \mathbb{N}$, the set of dyadic cuboids

$$\mathcal{D}(n,d) := \left\{ \left(\frac{k_1 - 1}{2^n}, \frac{k_1}{2^n} \right] \times \cdots \times \left(\frac{k_d - 1}{2^n}, \frac{k_d}{2^n} \right] : k_1, \ldots, k_d \in \mathbb{Z} \right\}.$$

Theorem 8.8.1 *For a measure μ on $(\mathbb{R}, \mathcal{B}(\mathbb{R}^n))$ with $\mu((0,1]^d) \in (0, \infty)$ the following assertions are equivalent:*

(1) *There is a constant $\kappa > 0$ such that $\mu = \kappa \lambda_d$.*
(2) $\mu(Q) = \mu(Q')$ *for all $Q, Q' \in \mathcal{D}(n,d)$ and $n \in \mathbb{N}$.*

Proof As (1) $\Rightarrow$ (2) is obvious, we turn to (2) $\Rightarrow$ (1). Condition (2) implies that $\mu(D) = 2^{-nd}\kappa$ for all $D \in \mathcal{D}(n,d)$ where $\kappa := \mu((0,1]^d)$. For $-\infty < a_i < b_i = \frac{k_i}{2^n} < \infty$ with $k_i \in \mathbb{Z}$ and $n \in \mathbb{N}$ this implies by continuity from below of a measure that

$$\mu((a_1, b_1] \times \cdots \times (a_d, b_d]) = \kappa (b_1 - a_1) \cdots (b_d - a_d). \tag{8.8}$$

Indeed, for $A := (a_1, b_1] \times \cdots \times (a_d, b_d]$ it holds $\mu \left(\bigcup_{D \subseteq A, D \in \mathcal{D}(N,d)} D \right) \to \kappa (b_1 - a_1) \cdots (b_d - a_d)$ as $N \to \infty$. Using continuity from above and that $\mu([-N, N]^d) < \infty$ for all $N \in \mathbb{N}$, we can remove the condition that the b_i are dyadic. Finally, we can deduce from (8.8) that

$$\mu(E_1 \times \cdots \times E_d) = \kappa \lambda(E_1) \cdots \lambda(E_d) \quad \text{for} \quad E_i \in \mathcal{B}(\mathbb{R}) \text{ with } \lambda(E_i) < \infty.$$

Here we first replace $(a_1, b_1]$ by E_1, while keeping the other intervals fixed and use that λ is uniquely determined by Proposition 3.3.10. We then proceed by fixing $E_1, (a_3, b_3], \ldots, (a_d, b_d]$ and replacing $(a_2, b_2]$ by E_2 on so on. By Corollary 7.3.4 and Corollary 7.3.5 we deduce $\mu = \kappa \lambda_d$. □

Corollary 8.8.2 (Affine Change of Variables) *Assume $f : \mathbb{R}^d \to \mathbb{R}^d$ with $f(x) := a + Ax$, where $x \in \mathbb{R}^d$ and A is an invertible $\mathbb{R}^{d \times d}$-matrix. Then, for a nonnegative measurable function $g : \mathbb{R}^d \to \mathbb{R}$ one has that*

$$\int_{\mathbb{R}^d} g(f) \mathrm{d}\lambda_d = |\det(A)|^{-1} \int_{\mathbb{R}^d} g \mathrm{d}\lambda_d,$$

where the equality means: if one side exists, then the other exists as well and they coincide.

Proof We let μ be the image measure of λ_d with respect to f.

(a) The measure μ satisfies Theorem 8.8.1(2) as it is translation invariant: For $B \in \mathcal{B}(\mathbb{R}^d)$ and $y \in \mathbb{R}^d$ one has

$$\mu(y+B) = \lambda_d(f^{-1}(y+B)) = \lambda_d(A^{-1}(y-a) + A^{-1}(B))$$
$$= \lambda_d(A^{-1}(B)) = \mu(B),$$

where in $A^{-1}(y-a)$ we mean the inverse matrix of A.

(b) If U is an orthogonal matrix, then we get by (a) that $\lambda_d(U^{-1}(B)) = \kappa_U \lambda_d(B)$ for all $B \in \mathcal{B}(\mathbb{R}^d)$ and some $\kappa_U > 0$. As for $B(0,1) := \{x \in \mathbb{R}^d : |x| \leq 1\}$ we have that $\lambda_d(U^{-1}(B(0,1))) = \lambda_d(B(0,1)) > 0$, we get $\kappa_U = 1$.

(c) Using the polar decomposition of a matrix, we can write $A = U D_a V$, where U, V are orthogonal matrices and $D_a(x_1, \ldots, x_d) := (\alpha_1 x_1, \ldots, \alpha_d x_d)$ with $\alpha_i > 0$. By (b) we get

$$\mu(B) = \lambda_d(V^{-1} D_a^{-1} U^{-1}(B)) = \lambda_d(D_a^{-1} U^{-1}(B))$$
$$= (a_1 \cdots a_n)^{-1} \lambda_d(U^{-1}(B))$$
$$= (a_1 \cdots a_n)^{-1} \lambda_d(B) = |\det(A)|^{-1} \lambda_d(B),$$

where the third equality follows from $\lambda_d = \otimes_{i=1}^d \lambda$ and Corollary 7.3.5. Hence $\mu = \lambda_d / \det(A)$ and the assertion follows. $\square$

8.9 The Theorems of Tonelli and Fubini

In this section we consider iterated integrals, as they appear often in applications. Tonelli's[6] and Fubini's[7] theorem state that integrals with respect to product measures (see Corollary 7.3.4) can be written as iterated integrals and that the order of integration can be changed in these iterated integrals. In many cases this provides an appropriate tool for the computation of integrals.

Theorem 8.9.1 (Tonelli's Theorem) *Let* $(\Omega_i, \mathcal{F}_i, \mu_i)$, $i = 1, 2$ *be σ-finite measure spaces and* $f : \Omega_1 \times \Omega_2 \to \mathbb{R}$ *a nonnegative* $(\mathcal{F}_1 \otimes \mathcal{F}_2, \mathcal{B}(\mathbb{R}))$-*measurable function. Then one has the following:*

(1) *The functions* $\omega_1 \mapsto f(\omega_1, \omega_2^0)$ *and* $\omega_2 \mapsto f(\omega_1^0, \omega_2)$ *are $\mathcal{F}_1$-measurable and $\mathcal{F}_2$-measurable, respectively, for all fixed* $\omega_i^0 \in \Omega_i$.

---
[6] Leonida Tonelli, 19/04/1885 (Gallipoli, Italy)–12/03/1946 (Pisa, Italy).
[7] Guido Fubini, 19/01/1879 (Venice, Italy)–06/06/1943 (New York, USA).

(2) *The functions*

$$\omega_1 \mapsto \int_{\Omega_2} f(\omega_1, \omega_2) d\mu_2(\omega_2) \quad \text{and} \quad \omega_2 \mapsto \int_{\Omega_1} f(\omega_1, \omega_2) d\mu_1(\omega_1)$$

are extended $\mathcal{F}_1$-measurable and $\mathcal{F}_2$-measurable maps, respectively.
(3) *One has that*

$$\int_{\Omega_1 \times \Omega_2} f(\omega_1, \omega_2) d(\mu_1 \otimes \mu_2)(\omega_1, \omega_2)$$
$$= \int_{\Omega_1} \left[\int_{\Omega_2} f(\omega_1, \omega_2) d\mu_2(\omega_2) \right] d\mu_1(\omega_1)$$
$$= \int_{\Omega_2} \left[\int_{\Omega_1} f(\omega_1, \omega_2) d\mu_1(\omega_1) \right] d\mu_2(\omega_2).$$

Remark 8.9.2 Under the condition

$$\int_{\Omega_1 \times \Omega_2} f(\omega_1, \omega_2) d(\mu_1 \otimes \mu_2)(\omega_1, \omega_2) < \infty$$

part (3) implies that

$$\mu_2\left(\left\{\omega_2 \in \Omega_2 : \int_{\Omega_1} f(\omega_1, \omega_2) d\mu_1(\omega_1) = \infty\right\}\right) = 0$$

and

$$\mu_1\left(\left\{\omega_1 \in \Omega_1 : \int_{\Omega_2} f(\omega_1, \omega_2) d\mu_2(\omega_2) = \infty\right\}\right) = 0.$$

Proof of Theorem 8.9.1 First we observe that we may assume that $\mu_1(\Omega_1) = \mu_2(\Omega_2) = 1$: We choose measurable partitions

$$\Omega_1 = \bigcup_{j \in I_1} \Omega_{1,j} \quad \text{and} \quad \Omega_2 = \bigcup_{j \in I_2} \Omega_{2,j}$$

with $\mu_i(\Omega_{i,j}) \in (0, \infty)$ and $I_i \subseteq \mathbb{N}$. We find $\alpha_{i,j} > 0$ such that

$$\sum_{j \in I_i} \alpha_{i,j} \mu_i(\Omega_{i,j}) = 1$$

8.9 The Theorems of Tonelli and Fubini

and define the measurable maps

$$h_i : \Omega_i \to \mathbb{R} \quad \text{by} \quad h_i(\omega_i) := \sum_{j \in I_i} \alpha_{i,j} \mathbf{1}_{\Omega_{i,j}}(\omega_i).$$

Proving (1), (2), and (3) for f or for

$$(\omega_1, \omega_2) \mapsto f(\omega_1, \omega_2) h_1(\omega_1) h_2(\omega_2)$$

does not make a difference as $h_i(\omega_i) \in (0, \infty)$. In other words, proving (1), (2), and (3) for $f h_1 h_2$ and (μ_1, μ_2) is the same as proving these statements for f and $(\mathbb{P}_1, \mathbb{P}_2)$ with

$$\mathbb{P}_i(A_i) := \sum_{j \in I_i} \alpha_{i,j} \mu_i(A_i \cap \Omega_{i,j}) \quad \text{with} \quad \mathbb{P}_i(\Omega_i) = 1.$$

For this reason we assume for the following that $\mu_1(\Omega_1) = \mu_2(\Omega_2) = 1$.

(a) We want to apply the *Monotone Class Theorem* Theorem B.1.7. Let $\mathcal{H}$ be the class of bounded $\mathcal{F}_1 \otimes \mathcal{F}_2$-measurable functions $f : \Omega_1 \times \Omega_2 \to \mathbb{R}$ such that the following holds:

(i) The functions $\omega_1 \mapsto f(\omega_1, \omega_2^0)$ and $\omega_2 \mapsto f(\omega_1^0, \omega_2)$ are $\mathcal{F}_1$-measurable and $\mathcal{F}_2$-measurable, respectively, for all $\omega_i^0 \in \Omega_i$.

(ii) The functions

$$\Omega_1 \ni \omega_1 \mapsto \int_{\Omega_2} f(\omega_1, \omega_2) d\mu_2(\omega_2),$$

$$\Omega_2 \ni \omega_2 \mapsto \int_{\Omega_1} f(\omega_1, \omega_2) d\mu_1(\omega_1)$$

are $\mathcal{F}_1$-measurable and $\mathcal{F}_2$-measurable, respectively.

(iii) One has that

$$\int_{\Omega_1 \times \Omega_2} f(\omega_1, \omega_2) d(\mu_1 \otimes \mu_2) = \int_{\Omega_1} \left[\int_{\Omega_2} f(\omega_1, \omega_2) d\mu_2(\omega_2) \right] d\mu_1(\omega_1)$$

$$= \int_{\Omega_2} \left[\int_{\Omega_1} f(\omega_1, \omega_2) d\mu_1(\omega_1) \right] d\mu_2(\omega_2).$$

As π-system we choose the collection of all $A_1 \times A_2$ with $A_i \in \mathcal{F}_i$. Now the conditions ((1)), ((2)), and ((3)) of Theorem B.1.7 are satisfied:

(1) We have $\mathbb{1}_{A_1 \times A_2} \in \mathcal{H}$ because, for example, (iii) follows from

$$\int_{\Omega_1 \times \Omega_2} f(\omega_1, \omega_2) \mathrm{d}(\mu_1 \otimes \mu_2) = (\mu_1 \otimes \mu_2)(A_1 \times A_2)$$

$$= \mu_1(A_1)\mu_2(A_2)$$

$$= \int_{\Omega_1} \mathbb{1}_{A_1}(\omega_1) \mu_2(A_2) \mathrm{d}\mu_1(\omega_1)$$

$$= \int_{\Omega_1} \left[\int_{\Omega_2} f(\omega_1, \omega_2) \mathrm{d}\mu_2(\omega_2) \right] \mathrm{d}\mu_1(\omega_1).$$

Assumption (2) of Theorem B.1.7 is obvious, and (3) follows from Proposition 8.4.3 and Theorem 8.4.4. Applying Theorem B.1.7 gives that $\mathcal{H}$ contains all bounded functions $f : \Omega_1 \times \Omega_2 \to \mathbb{R}$ measurable with respect to $\mathcal{F}_1 \otimes \mathcal{F}_2$, so that our statement holds for bounded functions f.

(b) To treat the general case, we let $N \in \mathbb{N}$ and

$$f_N(\omega_1, \omega_2) := \min\{f(\omega_1, \omega_2), N\}$$

which is bounded. As (1), (2), and (3) hold for f_N, using Proposition 8.4.3 and Theorem 8.4.4 we deduce the statement for f.

$\square$

Fubini's theorem can be seen as a generalization of Tonelli's theorem to general measurable maps $f : \Omega_1 \times \Omega_2 \to \mathbb{R}$:

Theorem 8.9.3 (Fubini's Theorem) *Let* $(\Omega_i, \mathcal{F}_i, \mu_i)$, $i = 1, 2$, *be σ-finite measure spaces and* $f : \Omega_1 \times \Omega_2 \to \mathbb{R}$ *be an $\mathcal{F}_1 \otimes \mathcal{F}_2$-measurable function such that*

$$\int_{\Omega_1 \times \Omega_2} |f(\omega_1, \omega_2)| \mathrm{d}(\mu_1 \otimes \mu_2)(\omega_1, \omega_2) < \infty. \tag{8.9}$$

Then the following holds:

(1) *The functions* $\omega_1 \mapsto f(\omega_1, \omega_2^0)$ *and* $\omega_2 \mapsto f(\omega_1^0, \omega_2)$ *are $\mathcal{F}_1$-measurable and $\mathcal{F}_2$-measurable, respectively, for all $\omega_i^0 \in \Omega_i$.*
(2) *There are $M_i \in \mathcal{F}_i$ with $\mu_i(M_i^c) = 0$ such that the integrals*

$$\int_{\Omega_1} f(\omega_1, \omega_2^0) \mathrm{d}\mu_1(\omega_1) \quad \text{and} \quad \int_{\Omega_2} f(\omega_1^0, \omega_2) \mathrm{d}\mu_2(\omega_2)$$

8.9 The Theorems of Tonelli and Fubini

exist and are finite for all $\omega_i^0 \in M_i$ and the functions

$$\omega_1 \mapsto \mathbb{1}_{M_1}(\omega_1) \int_{\Omega_2} f(\omega_1, \omega_2) d\mu_2(\omega_2),$$

$$\omega_2 \mapsto \mathbb{1}_{M_2}(\omega_2) \int_{\Omega_1} f(\omega_1, \omega_2) d\mu_1(\omega_1)$$

are $\mathcal{F}_1$-measurable and $\mathcal{F}_2$-measurable maps, respectively.
(3) One has that

$$\int_{\Omega_1 \times \Omega_2} f(\omega_1, \omega_2) d(\mu_1 \otimes \mu_2)$$

$$= \int_{\Omega_1} \left[\mathbb{1}_{M_1}(\omega_1) \int_{\Omega_2} f(\omega_1, \omega_2) d\mu_2(\omega_2) \right] d\mu_1(\omega_1)$$

$$= \int_{\Omega_2} \left[\mathbb{1}_{M_2}(\omega_2) \int_{\Omega_1} f(\omega_1, \omega_2) d\mu_1(\omega_1) \right] d\mu_2(\omega_2).$$

Remark 8.9.4 Our understanding is that writing, for example, an expression like

$$\mathbb{1}_{M_2}(\omega_2) \int_{\Omega_1} f(\omega_1, \omega_2) d\mu_1(\omega_1)$$

we only consider and compute the integral for $\omega_2 \in M_2$.

Proof of Theorem 8.9.3 We let $f = f^+ - f^-$, where $f^+ := f \vee 0$ and $f^- := (-f) \vee 0$, apply Theorem 8.9.1 to f^+ and f^- separately, and define the sets

$$M_2^{\pm} = \left\{ \omega_2 \in \Omega_2 : \int_{\Omega_1} f^{\pm}(\omega_1, \omega_2) d\mu_1(\omega_1) = \infty \right\}^c,$$

$$M_1^{\pm} = \left\{ \omega_1 \in \Omega_1 : \int_{\Omega_2} f^{\pm}(\omega_1, \omega_2) d\mu_2(\omega_2) = \infty \right\}^c.$$

By Remark 8.9.2 we have that $\mu_i((M_i^{\pm})^c) = 0$. So we define the sets

$$M_1 := M_1^+ \cap M_1^- \quad \text{and} \quad M_2 := M_2^+ \cap M_2^-.$$

Now we can combine the statements for f^+ and f^- to the statement for f. □

We close this section with an example that should remind us to check the assumptions of Fubini's Theorem in applications carefully.

Example 8.9.5 Let $\Omega := [-1, 1] \setminus \{0\}$ and λ be the Lebesgue measure on Ω. The function $f : \Omega \times \Omega \to \mathbb{R}$ with

$$f(x, y) := \frac{xy}{(x^2 + y^2)^2}$$

is not integrable on $\Omega \times \Omega$, even though the iterated integrals exist end are equal. In fact we have for all $x^0, y^0 \in \Omega$ that

$$\int_\Omega |f(x, y^0)| d\lambda(x) < \infty \quad \text{and} \quad \int_\Omega |f(x^0, y)| d\lambda(y) < \infty$$

and by antisymmetry

$$\int_\Omega f(x, y^0) d\lambda(x) = 0 \quad \text{and} \quad \int_\Omega f(x^0, y) d\lambda(y) = 0.$$

Hence

$$\int_\Omega \left(\int_\Omega f(x, y) d\lambda(x) \right) d\lambda(y) = \int_\Omega \left(\int_\Omega f(x, y) d\lambda(y) \right) d\lambda(x) = 0.$$

On the other hand, for $Q_n := \left(\frac{1}{2^n}, \frac{1}{2^{n-1}} \right] \times \left(\frac{1}{2^n}, \frac{1}{2^{n-1}} \right]$, $n \in \mathbb{N}$,

$$\int_{\Omega \times \Omega} |f(x, y)| d(\lambda \otimes \lambda)(x, y) \geq \int_{\bigcup_{n=1}^\infty Q_n} |f(x, y)| d(\lambda \otimes \lambda)(x, y)$$

$$= \sum_{n=1}^\infty \int_{Q_n} |f(x, y)| d(\lambda \otimes \lambda)(x, y)$$

$$\geq \sum_{n=1}^\infty \left(\frac{1}{2^n} \right)^2 \frac{\left(\frac{1}{2^n} \right)^2}{\left(2 \frac{1}{2^{2(n-1)}} \right)^2}$$

$$= \infty.$$

8.10 Applications of Tonelli's and Fubini's Theorem

In the following we consider several applications of Tonelli's and Fubini's theorem. We start with the formula how to compute $\int_{\mathbb{R}^2} f d\lambda_2$ in terms of polar coordinates. Here we assume it is known that

$$\lambda_2 \left(\left\{ (x, y) \in \mathbb{R}^2 : \sqrt{x^2 + y^2} < 1 \right\} \right) = \pi. \tag{8.10}$$

8.10 Applications of Tonelli's and Fubini's Theorem

Proposition 8.10.1 (Polar Coordinates) *For a nonnegative function $f : \mathbb{R}^2 \to \mathbb{R}$, that is $(\mathcal{B}(\mathbb{R}^2), \mathcal{B}(\mathbb{R}))$-measurable, one has that*

$$\int_{\mathbb{R}^2} f \, d\lambda_2 = \int_{[0,\infty)} \int_{[0,2\pi)} f(r(\cos(\varphi), \sin(\varphi))) \, d\lambda(\varphi) r \, d\lambda(r).$$

Proof For $\varphi \in [0, 2\pi)$ define the rotation matrix

$$D_\varphi := \begin{pmatrix} \cos(\varphi) & -\sin(\varphi) \\ \sin(\varphi) & \cos(\varphi) \end{pmatrix}.$$

Because of the invariance of λ_2 with respect to rotations (Corollary 8.8.2) we get by Tonelli's Theorem 8.9.1, where x^T denotes the transposed of x:

$$\int_{\mathbb{R}^2} f(x_1, x_2) \, d\lambda_2(x_1, x_2)$$

$$= \int_{[0,2\pi)} \int_{\mathbb{R}^2} f\left(\sqrt{x_1^2 + x_2^2} \, D_\varphi \left(\frac{x_1}{\sqrt{x_1^2 + x_2^2}}, \frac{x_2}{\sqrt{x_1^2 + x_2^2}}\right)^T\right) d\lambda_2(x_1, x_2) \frac{d\lambda(\varphi)}{2\pi}$$

$$= \int_{\mathbb{R}^2} \left[\int_{[0,2\pi)} f\left(\sqrt{x_1^2 + x_2^2} \, D_\varphi \left(\frac{x_1}{\sqrt{x_1^2 + x_2^2}}, \frac{x_2}{\sqrt{x_1^2 + x_2^2}}\right)^T\right) \frac{d\lambda(\varphi)}{2\pi} \right] d\lambda_2(x_1, x_2)$$

$$= \int_{\mathbb{R}^2} \left[\int_{[0,2\pi)} f\left(\sqrt{x_1^2 + x_2^2} \, (\cos(\varphi), \sin(\varphi))\right) \frac{d\lambda(\varphi)}{2\pi} \right] d\lambda_2(x_1, x_2)$$

The last equation we deduce from the observation that the integration w.r.t. $d\lambda(\varphi)$ has the effect that D_φ moves the point $\left(\frac{x_1}{\sqrt{x_1^2+x_2^2}}, \frac{x_2}{\sqrt{x_1^2+x_2^2}}\right)$ exactly once around the circle, which happens also for $(\cos(\varphi), \sin(\varphi))$.

Now we consider the map $h : \mathbb{R}^2 \to [0, \infty)$ with $h(x_1, x_2) := \sqrt{x_1^2 + x_2^2}$. Let $\rho := (\lambda_2)_h$ be the corresponding image measure. For $0 \leqslant a < b < \infty$ we get

$$\rho([a, b)) = \lambda_2\left(\left\{ (x_1, x_2) \in \mathbb{R}^2 : a \leqslant \sqrt{x_1^2 + x_2^2} < b \right\}\right)$$

$$= (b^2 - a^2)\pi = 2\pi \int_{[a,b)} r \, d\lambda(r),$$

where we use (8.10), Corollary 8.8.2 (to get the quadratic scaling in a and b), and Theorem 8.6.1 to compute $b^2 - a^2$ as integral. Consequently,

$$\int_{\mathbb{R}^2} f(x) \mathrm{d}\lambda_2(x)$$
$$= \int_{\mathbb{R}^2} \left[\int_{[0,2\pi)} f\left(\sqrt{x_1^2 + x_2^2} \, (\cos(\varphi), \sin(\varphi))\right) \frac{\mathrm{d}\lambda(\varphi)}{2\pi} \right] \mathrm{d}\lambda_2(x_1, x_2)$$
$$= \int_{[0,\infty)} \int_{[0,2\pi)} f(r(\cos(\varphi), \sin(\varphi))) \, r \, \mathrm{d}\lambda(\varphi) \mathrm{d}(r).$$

□

Our second application of Tonelli's theorem concerns Gaussian measures introduced before in Sect. 3.3.4.

Corollary 8.10.2 *The following assertions hold true:*

(1) *One has* $\int_{-\infty}^{\infty} e^{-x^2} \mathrm{d}\lambda(x) = \sqrt{\pi}$.
(2) *For* $\sigma > 0$ *and* $m \in \mathbb{R}$ *let*

$$p_{m,\sigma^2}(x) := \frac{1}{\sqrt{2\pi\sigma^2}} e^{-\frac{(x-m)^2}{2\sigma^2}}.$$

Then one has $\int_{\mathbb{R}} p_{m,\sigma^2}(x) \mathrm{d}\lambda(x) = 1$,

$$\int_{\mathbb{R}} x p_{m,\sigma^2}(x) \mathrm{d}\lambda(x) = m, \quad \text{and} \quad \int_{\mathbb{R}} (x-m)^2 p_{m,\sigma^2}(x) \mathrm{d}\lambda(x) = \sigma^2. \tag{8.11}$$

Consequently, if a random variable $f : \Omega \to \mathbb{R}$ *has as law the normal distribution* $\mathcal{N}_{m,\sigma^2}$, *then*

$$\mathbb{E} f = m \quad \text{and} \quad \mathbb{E}(f - \mathbb{E} f)^2 = \sigma^2. \tag{8.12}$$

Proof (1) Let $f : \mathbb{R} \times \mathbb{R} \to \mathbb{R}$ be a nonnegative continuous function. Tonelli's Theorem 8.9.1 gives that

$$\int_{\mathbb{R} \times \mathbb{R}} f(x,y) \mathrm{d}(\lambda \otimes \lambda)(x,y) = \int_{\mathbb{R}} \left[\int_{\mathbb{R}} f(x,y) \mathrm{d}\lambda(y) \right] \mathrm{d}\lambda(x).$$

8.10 Applications of Tonelli's and Fubini's Theorem

Letting $f(x, y) := e^{-(x^2+y^2)}$, the above yields that

$$\int_{\mathbb{R}\times\mathbb{R}} e^{-(x^2+y^2)} d(\lambda \otimes \lambda)(x,y) = \int_{\mathbb{R}} \left[\int_{\mathbb{R}} e^{-(x^2+y^2)} d\lambda(y)\right] d\lambda(x)$$

$$= \int_{\mathbb{R}} e^{-x^2} \left[\int_{\mathbb{R}} e^{-y^2} d\lambda(y)\right] d\lambda(x)$$

$$= \left[\int_{-\infty}^{\infty} e^{-x^2} d\lambda(x)\right]^2.$$

On the other side, by monotone convergence we get that

$$\int_{\mathbb{R}^2} e^{-(x^2+y^2)} d(\lambda \otimes \lambda)(x,y) = \lim_{R\to\infty} \int_{x^2+y^2 \leq R^2} e^{-(x^2+y^2)} d(\lambda \otimes \lambda)(x,y)$$

$$= \lim_{R\to\infty} \int_{[0,R]} \int_{[0,2\pi)} e^{-r^2} r d\lambda(r) d\lambda(\varphi)$$

$$= \pi \lim_{R\to\infty} \left(1 - e^{-R^2}\right)$$

$$= \pi$$

where we use Proposition 8.10.1 regarding the polar coordinates.

(2) Using Corollary 8.8.2 we perform the change of variables $y = (x - m)/(\sqrt{2}\sigma)$ and get that

$$\int_{\mathbb{R}} \frac{1}{\sqrt{2\pi\sigma^2}} e^{-\frac{(x-m)^2}{2\sigma^2}} d\lambda(x) = \int_{\mathbb{R}} \frac{1}{\sqrt{\pi}} e^{-y^2} d\lambda(y) = 1.$$

Secondly,

$$\int_{\mathbb{R}} \frac{1}{\sqrt{2\pi\sigma^2}} x e^{-\frac{(x-m)^2}{2\sigma^2}} d\lambda(x) = \int_{\mathbb{R}} \frac{1}{\sqrt{2\pi\sigma^2}} (x-m) e^{-\frac{(x-m)^2}{2\sigma^2}} d\lambda(x) + m$$

$$= \sqrt{\frac{2}{\pi}} \sigma \int_{\mathbb{R}} y e^{-y^2} d\lambda(y) + m$$

$$= m$$

where the integral term vanishes because the function that is integrated is odd. The remaining part is the subject of Exercise 4. □

Another corollary of Tonelli's theorem is the following formula, sometimes referred to as the *layer cake formula*, that allows to compute the Lebesgue integral of a measurable function by its distribution function:

Corollary 8.10.3 *Assume that $(\Omega, \mathcal{F}, \mu)$ is a σ-finite measure space and let $f : \Omega \to \mathbb{R}$ be nonnegative and measurable. Then,*

$$\int_\Omega f \, d\mu = \int_{[0,\infty)} \mu(f > t) \, d\lambda(t),$$

where λ is the Lebesgue measure.

Proof We let $A := \{(t, \omega) \in [0, \infty) \times \Omega : t < f(\omega)\}$ and verify in Exercise 14 that $A \in \mathcal{B}([0, \infty)) \otimes \mathcal{F}$. Then, by Tonelli's theorem,

$$\int_\Omega f \, d\mu = \int_\Omega \int_{[0,\infty)} \mathbb{1}_{\{t \in [0,\infty) : t < f(\omega)\}} \, d\lambda(t) \, d\mu(\omega)$$

$$= \int_{[0,\infty)} \int_\Omega \mathbb{1}_{\{t \in [0,\infty) : t < f(\omega)\}} \, d\mu(\omega) \, d\lambda(t)$$

$$= \int_{[0,\infty)} \mu(f > t) \, d\lambda(t).$$

$\square$

A more general version of this statement can be found in Exercise 7. Another corollary is a formula to compute the expected value of products of independent random variables:

Corollary 8.10.4 *If $(\Omega, \mathcal{F}, \mathbb{P})$ is a probability space and $f, g : \Omega \to \mathbb{R}$ are independent random variables such that $\mathbb{E}|f| < \infty$ and $\mathbb{E}|g| < \infty$. Then*

$$\mathbb{E}|fg| < \infty \quad \text{and} \quad \mathbb{E}fg = \mathbb{E}f \mathbb{E}g.$$

Proof We consider the product space $(\Omega, \mathcal{F}, \mathbb{P}) \otimes (\Omega, \mathcal{F}, \mathbb{P})$ and the random variable $h : \Omega \times \Omega \to \mathbb{R}$ with $h(\omega, \eta) := f(\omega)g(\eta)$. Tonelli's theorem gives

$$\int_{\Omega \times \Omega} |h(\omega, \eta)| \, d(\mathbb{P} \otimes \mathbb{P})(\omega, \eta) = \int_\Omega \int_\Omega |h(\omega, \eta)| \, d\mathbb{P}(\omega) \, d\mathbb{P}(\eta)$$

$$= \int_\Omega |f(\omega)| \, d\mathbb{P}(\omega) \int_\Omega |g(\eta)| \, d\mathbb{P}(\eta) = \mathbb{E}|f|\mathbb{E}|g| < \infty$$

and, in the same way, by Fubini's theorem, $\mathbb{E}h = \mathbb{E}f\mathbb{E}g$. On the other hand, by independence we know that $h : \Omega \times \Omega \to \mathbb{R}$ and $fg : \Omega \to \mathbb{R}$ have the same distribution (see Proposition 7.6.1), so that $\mathbb{E}|fg| = \int_{\Omega \times \Omega} |h(\omega, \eta)| \, d(\mathbb{P} \otimes \mathbb{P})(\omega, \eta)$ and $\mathbb{E}fg = \int_{\Omega \times \Omega} h(\omega, \eta) \, d(\mathbb{P} \otimes \mathbb{P})(\omega, \eta)$. $\square$

An alternative proof can be found in Exercise 5.

8.11 The Variance of a Random Variable

Besides the expected value for a random variable, its variance is another important quantity frequently used. It is a useful tool to describe the deviation of a random variable from its mean.

Definition 8.11.1 (Variance) Let $(\Omega, \mathcal{F}, \mathbb{P})$ be a probability space and $f : \Omega \to \mathbb{R}$ be an integrable random variable. Then

$$\operatorname{var}(f) = \mathbb{E}[f - \mathbb{E}f]^2 \in [0, \infty]$$

is called the **variance** (of f).

Now we summarize some basic properties of the variance:

Theorem 8.11.2

(1) If f is integrable and $\alpha, c \in \mathbb{R}$, then

$$\operatorname{var}(\alpha f - c) = \alpha^2 \operatorname{var}(f).$$

(2) If $\mathbb{E}f^2 < \infty$, then $\operatorname{var}(f) = \mathbb{E}f^2 - (\mathbb{E}f)^2 \leq \mathbb{E}f^2 < \infty$.

(3) If $f_1, \ldots, f_n$ are independent random variables such that $\mathbb{E}f_i^2 < \infty$ for $i = 1, \ldots, n$, then

$$\operatorname{var}(f_1 + \cdots + f_n) = \operatorname{var}(f_1) + \cdots + \operatorname{var}(f_n).$$

Proof (1) follows from

$$\operatorname{var}(\alpha f - c) = \mathbb{E}[(\alpha f - c) - \mathbb{E}(\alpha f - c)]^2 = \mathbb{E}[\alpha f - \alpha \mathbb{E}f]^2 = \alpha^2 \operatorname{var}(f).$$

(2) First we remark that $\mathbb{E}|f| \leq (\mathbb{E}f^2)^{\frac{1}{2}}$ as we shall see later by Hölder's inequality (Corollary 8.12.6), that means that any square integrable random variable is integrable. Then we simply get that

$$\operatorname{var}(f) = \mathbb{E}[f - \mathbb{E}f]^2 = \mathbb{E}f^2 - 2\mathbb{E}(f\mathbb{E}f) + (\mathbb{E}f)^2 = \mathbb{E}f^2 - 2(\mathbb{E}f)^2 + (\mathbb{E}f)^2.$$

(3) The formula follows from

$$\operatorname{var}(f_1 + \cdots + f_n) = \mathbb{E}((f_1 + \cdots + f_n) - \mathbb{E}(f_1 + \cdots + f_n))^2$$

$$= \mathbb{E}\left(\sum_{i=1}^n (f_i - \mathbb{E}f_i)\right)^2$$

$$= \mathbb{E} \sum_{i,j=1}^{n} (f_i - \mathbb{E} f_i)(f_j - \mathbb{E} f_j)$$

$$= \sum_{i,j=1}^{n} \mathbb{E}\left((f_i - \mathbb{E} f_i)(f_j - \mathbb{E} f_j)\right)$$

$$= \sum_{i=1}^{n} \mathbb{E}(f_i - \mathbb{E} f_i)^2 + \sum_{i \neq j} \mathbb{E}\left((f_i - \mathbb{E} f_i)(f_j - \mathbb{E} f_j)\right)$$

$$= \sum_{i=1}^{n} \operatorname{var}(f_i) + \sum_{i \neq j} \mathbb{E}(f_i - \mathbb{E} f_i)\mathbb{E}(f_j - \mathbb{E} f_j)$$

$$= \sum_{i=1}^{n} \operatorname{var}(f_i)$$

where we apply Corollary 8.10.4 to verify $\mathbb{E}\left((f_i - \mathbb{E} f_i)(f_j - \mathbb{E} f_j)\right) = \mathbb{E}(f_i - \mathbb{E} f_i)\mathbb{E}(f_j - \mathbb{E} f_j)$ for the second last equation, and use $\mathbb{E}(f_i - \mathbb{E} f_i) = \mathbb{E} f_i - \mathbb{E} f_i = 0$ for the last equation. □

8.12 Inequalities

In this section we prove basic inequalities related to the Lebesgue integral. These inequalities have an enormous importance not only in probability or measure theory, but also in all branches of analysis and numerics. Besides of this, they are of interest on its own including their proofs.

8.12.1 Markov's Inequality

We start with Markov's inequality that relates—in the probabilistic context—the distribution of a random variable to its expected value:

Theorem 8.12.1 (Markov's[8] Inequality) *Let $(\Omega, \mathcal{F}, \mu)$ be a measure space and $f : \Omega \to \mathbb{R}$ be nonnegative, measurable, and integrable. Then, for all $\varepsilon > 0$,*

$$\mu(f \geq \varepsilon) \leq \frac{\int_\Omega f \, d\mu}{\varepsilon}.$$

---

[8] Andrei Andreyevich Markov, 14/06/1856 (Ryazan, Russia)–20/07/1922 (St Petersburg, Russia).

8.12 Inequalities

Proof We simply have

$$\varepsilon\mu(\{\omega \in \Omega : f(\omega) \geq \varepsilon\}) = \varepsilon \int_\Omega \mathbb{1}_{\{f \geq \varepsilon\}} d\mu \leq \int_\Omega f \mathbb{1}_{\{f \geq \varepsilon\}} d\mu \leq \int_\Omega f d\mu.$$

□

8.12.2 Jensen's Inequality

The Jensen inequality allows to interchange the Lebesgue integral with the application of a convex function. Let us first recall the notion of convexity and concavity:

Definition 8.12.2 (Convexity and Concavity) A function $g : \mathbb{R} \to \mathbb{R}$ is **convex** if and only if

$$g(\theta x + (1-\theta)y) \leq \theta g(x) + (1-\theta)g(y)$$

for all $0 \leq \theta \leq 1$ and all $x, y \in \mathbb{R}$. A function $g : \mathbb{R} \to \mathbb{R}$ is **concave** if and only if $-g$ is convex.

Any convex function $g : \mathbb{R} \to \mathbb{R}$ is continuous (which requires a proof, see Exercise 9) and therefore $(\mathcal{B}(\mathbb{R}), \mathcal{B}(\mathbb{R}))$-measurable.

Theorem 8.12.3 (Jensen's[9] Inequality) If $(\Omega, \mathcal{F}, \mathbb{P})$ is a probability space, $g : \mathbb{R} \to \mathbb{R}$ convex, and if $f : \Omega \to \mathbb{R}$ is a random variable with $\mathbb{E}|f| < \infty$, then

$$g(\mathbb{E}f) \leq \mathbb{E}g(f)$$

where the expected value on the right-hand side might be infinite.

Proof Let $x_0 = \mathbb{E}f$. Since g is convex we find a *supporting line* in x_0, that means $a, b \in \mathbb{R}$ such that

$$ax_0 + b = g(x_0) \quad \text{and} \quad ax + b \leq g(x)$$

for all $x \in \mathbb{R}$. One approach to find such a supporting line is a follows: One takes the limit of the monotone sequence $(n(g(x_0 + \frac{1}{n}) - g(x_0)))_{n=1}^\infty$, say $a \in \mathbb{R}$, and computes $b \in \mathbb{R}$ such that $ax_0 + b = g(x_0)$. It follows $af(\omega) + b \leq g(f(\omega))$ for all

---

[9] Johan Ludwig William Valdemar Jensen, 08/05/1859 (Nakskov, Denmark)–05/03/1925 (Copenhagen, Denmark).

$\omega \in \Omega$ and

$$g(\mathbb{E}f) = a\mathbb{E}f + b = \mathbb{E}(af+b) \leq \mathbb{E}g(f).$$

□

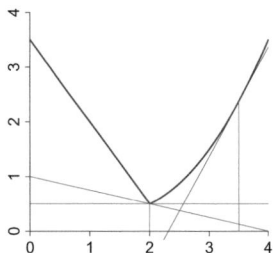

Example 8.12.4 Frequently used examples of convex functions are the following:

(1) The function $g(x) := |x|$ is convex so that, for any integrable f,

$$|\mathbb{E}f| \leq \mathbb{E}|f|.$$

(2) For $1 \leq p < \infty$ the function $g(x) := |x|^p$ is convex, so that Jensen's inequality applied to $|f|$ gives that

$$(\mathbb{E}|f|)^p \leq \mathbb{E}|f|^p.$$

8.12.3 Hölder's and Minkowski's Inequality

Hölder's and Minkowski's inequality deduce from moments of measurable functions $f, g : \Omega \to \mathbb{R}$ moments of fg and $f+g$, respectively. Besides its natural usage in measure theory these inequalities are the cornerstones in functional analysis to establish the Lebesgue spaces and their dual spaces.

Theorem 8.12.5 (Hölder's[10] Inequality) *Assume a measure space $(\Omega, \mathcal{F}, \mu)$ and measurable maps $f, g : \Omega \to \mathbb{R}$. If $1 < p, q < \infty$ with $\frac{1}{p} + \frac{1}{q} = 1$, then*

$$\int_\Omega |fg| d\mu \leq \left(\int_\Omega |f|^p d\mu \right)^{\frac{1}{p}} \left(\int_\Omega |g|^q d\mu \right)^{\frac{1}{q}}.$$

---

[10] Otto Ludwig Hölder, 22/12/1859 (Stuttgart, Germany)–29/08/1937 (Leipzig, Germany).

8.12 Inequalities

For $p = q = 2$ the inequality is also called Cauchy-Schwarz inequality.

Proof of Theorem 8.12.5 We can assume that $\int_\Omega |f|^p d\mu > 0$ and $\int_\Omega |g|^q d\mu > 0$. For example, assuming $\int_\Omega |f|^p d\mu = 0$ would imply $|f|^p = 0$ a.e. according to Proposition 8.3.1 so that $fg = 0$ a.e. and $\int_\Omega |fg| d\mu = 0$. In addition we may assume that $\int_\Omega |f|^p d\mu + \int_\Omega |g|^q d\mu < \infty$, otherwise there is nothing to prove. Hence we may set

$$\tilde{f} := \frac{f}{\left(\int_\Omega |f|^p d\mu\right)^{\frac{1}{p}}} \quad \text{and} \quad \tilde{g} := \frac{g}{\left(\int_\Omega |g|^q d\mu\right)^{\frac{1}{q}}}.$$

We notice that

$$x^a y^b \leq ax + by$$

for $x, y \geq 0$ and positive a, b with $a + b = 1$, which follows from the concavity of the logarithm (we can assume for a moment that $x, y > 0$)

$$\ln(ax + by) \geq a \ln x + b \ln y = \ln x^a + \ln y^b = \ln x^a y^b.$$

Setting $x := |\tilde{f}|^p$, $y := |\tilde{g}|^q$, $a := \frac{1}{p}$, and $b := \frac{1}{q}$, we get

$$|\tilde{f}\tilde{g}| = x^a y^b \leq ax + by = \frac{1}{p}|\tilde{f}|^p + \frac{1}{q}|\tilde{g}|^q,$$

which is known as Young's[11] inequality, and

$$\int_\Omega |\tilde{f}\tilde{g}| d\mu \leq \frac{1}{p} \int_\Omega |\tilde{f}|^p d\mu + \frac{1}{q} \int_\Omega |\tilde{g}|^q d\mu = \frac{1}{p} + \frac{1}{q} = 1.$$

On the other hand side,

$$\int_\Omega |\tilde{f}\tilde{g}| d\mu = \frac{\int_\Omega |fg| d\mu}{\left(\int_\Omega |f|^p d\mu\right)^{\frac{1}{p}} \left(\int_\Omega |g|^q d\mu\right)^{\frac{1}{q}}},$$

so that the proof is complete. □

Corollary 8.12.6 *For $0 < p < q < \infty$, a probability space $(\Omega, \mathcal{F}, \mathbb{P})$ and a random variable $f : \Omega \to \mathbb{R}$ one has that $(\mathbb{E}|f|^p)^{\frac{1}{p}} \leq (\mathbb{E}|f|^q)^{\frac{1}{q}}$.*

---

[11] William Henry Young, 20/10/1863 (London, England)–07/07/1942 (Lausanne, Switzerland).

The proof is the subject of Exercise 10. Now we combine Corollary 8.12.6 with Markov's inequality Theorem 8.12.1 to obtain a frequently used deviation inequality:

Corollary 8.12.7 (Chebyshev's[12] Inequality) *Let $(\Omega, \mathcal{F}, \mathbb{P})$ be a probability space and $f : \Omega \to \mathbb{R}$ be a random variable such that $\mathbb{E} f^2 < \infty$. Then one has, for all $\varepsilon > 0$,*

$$\mathbb{P}(|f - \mathbb{E}f| \geq \varepsilon) \leq \frac{\mathbb{E}(f - \mathbb{E}f)^2}{\varepsilon^2} \leq \frac{\mathbb{E}f^2}{\varepsilon^2}.$$

Proof From Corollary 8.12.6 we get that $\mathbb{E}|f| < \infty$ so that $\mathbb{E}f$ exists. Applying Theorem 8.12.1 to $|f - \mathbb{E}f|^2$ gives that

$$\mathbb{P}(|f - \mathbb{E}f| \geq \varepsilon) = \mathbb{P}(|f - \mathbb{E}f|^2 \geq \varepsilon^2) \leq \frac{\mathbb{E}|f - \mathbb{E}f|^2}{\varepsilon^2} \quad \text{for} \quad \varepsilon > 0.$$

Finally, we use that $\mathbb{E}(f - \mathbb{E}f)^2 = \mathbb{E}f^2 - (\mathbb{E}f)^2 \leq \mathbb{E}f^2$. □

The next corollary is Hölder's inequality in a more analytic form:

Corollary 8.12.8 (Hölder's Inequality for Sequences) *Let $(a_n)_{n=1}^\infty$ and $(b_n)_{n=1}^\infty$ be sequences of real numbers. If $1 < p, q < \infty$ with $\frac{1}{p} + \frac{1}{q} = 1$, then*

$$\sum_{n=1}^\infty |a_n b_n| \leq \left(\sum_{n=1}^\infty |a_n|^p\right)^{\frac{1}{p}} \left(\sum_{n=1}^\infty |b_n|^q\right)^{\frac{1}{q}}.$$

Proof The statement is a special case of Theorem 8.12.5 when using the measure space $\Omega := \mathbb{N}$, $\mathcal{F} := 2^\mathbb{N}$, and $\mu(A)$ is the cardinality of A. □

We close this section with the fundamental Minkowski inequality:

Theorem 8.12.9 (Minkowski's[13] Inequality) *Assume a measure space $(\Omega, \mathcal{F}, \mu)$, measurable $f, g : \Omega \to \mathbb{R}$, and $0 < p < \infty$. Then*

$$\left(\int_\Omega |f+g|^p d\mu\right)^{\frac{1}{p}} \leq c_p \left[\left(\int_\Omega |f|^p d\mu\right)^{\frac{1}{p}} + \left(\int_\Omega |g|^p d\mu\right)^{\frac{1}{p}}\right] \tag{8.13}$$

---

[12] Pafnuty Lvovich Chebyshev, 16/05/1821 (Okatovo, Russia)–08/12/1894 (St Petersburg, Russia).
[13] Hermann Minkowski, 22/06/1864 (Alexotas, Russian Empire, now Kaunas, Lithuania)–12/01/1909 (Göttingen, Germany).

8.12 Inequalities

where $c_p := 1$ for $1 \leqslant p < \infty$ and $c_p := 2^{\frac{1}{p}-1}$ for $0 < p < 1$. Moreover, for $0 < p \leqslant 1$ we have

$$\int_\Omega |f+g|^p \, d\mu \leqslant \int_\Omega |f|^p \, d\mu + \int_\Omega |g|^p \, d\mu.$$

Proof $p = 1$: This case follows from $|f + g| \leqslant |f| + |g|$.

$1 < p < \infty$: Here we note that the convexity of $x \mapsto |x|^p$ gives that

$$\left|\frac{a+b}{2}\right|^p \leqslant \frac{|a|^p + |b|^p}{2}$$

and hence

$$(a+b)^p \leqslant 2^{p-1}(a^p + b^p) \quad \text{for} \quad a, b \geqslant 0. \tag{8.14}$$

Consequently, $|f+g|^p \leqslant (|f|+|g|)^p \leqslant 2^{p-1}(|f|^p + |g|^p)$ and

$$\int_\Omega |f+g|^p \, d\mu \leqslant 2^{p-1}\left(\int_\Omega |f|^p \, d\mu + \int_\Omega |g|^p \, d\mu\right).$$

Assuming now that $\left(\int_\Omega |f|^p d\mu\right)^{\frac{1}{p}} + \left(\int_\Omega |g|^p d\mu\right)^{\frac{1}{p}} < \infty$, otherwise there is nothing to prove, we get by the above considerations that $\int_\Omega |f+g|^p d\mu < \infty$ holds as well. Taking $1 < q < \infty$ with $\frac{1}{p} + \frac{1}{q} = 1$, we continue with

$$\int_\Omega |f+g|^p \, d\mu = \int_\Omega |f+g||f+g|^{p-1} \, d\mu$$

$$\leqslant \int_\Omega (|f|+|g|)|f+g|^{p-1} \, d\mu$$

$$= \int_\Omega |f||f+g|^{p-1} \, d\mu + \int_\Omega |g||f+g|^{p-1} \, d\mu$$

$$\leqslant \left(\int_\Omega |f|^p \, d\mu\right)^{\frac{1}{p}} \left(\int_\Omega |f+g|^{(p-1)q} \, d\mu\right)^{\frac{1}{q}}$$

$$+ \left(\int_\Omega |g|^p \, d\mu\right)^{\frac{1}{p}} \left(\int_\Omega |f+g|^{(p-1)q} \, d\mu\right)^{\frac{1}{q}},$$

where we have used Hölder's inequality. Since $(p-1)q = p$, (8.13) follows by dividing the above inequality by $\left(\int_\Omega |f+g|^p d\mu\right)^{\frac{1}{q}}$ (in case this expression is positive, otherwise there is nothing to prove) and taking into account that $1 - \frac{1}{q} = \frac{1}{p}$.

$0 < p < 1$: In this case we get for $a, b \geqslant 0$ with $a + b > 0$ that

$$(a+b)^p = (a+b)^p \frac{a}{a+b} + (a+b)^p \frac{b}{a+b}$$
$$\leqslant (a+b)^p \frac{a^p}{(a+b)^p} + (a+b)^p \frac{b^p}{(a+b)^p} = a^p + b^p,$$

which implies $|a+b|^p \leqslant |a|^p + |b|^p$ for all $a, b \in \mathbb{R}$. Therefore we get

$$\left(\int_\Omega |f+g|^p \, d\mu\right)^{\frac{1}{p}} \leqslant \left(\int_\Omega |f|^p \, d\mu + \int_\Omega |g|^p \, d\mu\right)^{\frac{1}{p}}$$
$$\leqslant 2^{\frac{1}{p}-1} \left[\left(\int_\Omega |f|^p \, d\mu\right)^{\frac{1}{p}} + \left(\int_\Omega |g|^p \, d\mu\right)^{\frac{1}{p}}\right],$$

where, for $1 < q = \frac{1}{p} < \infty$, we have used $|a+b|^q \leqslant 2^{q-1}(|a|^q + |b|^q)$ which is in fact (8.14). □

For $0 < p < 1$ the constant $c_p = 2^{\frac{1}{p}-1}$ is sharp, see Exercise 13.

8.12.4* Hoeffding's Inequality

Hoeffding's inequality is a deviation inequality that provides a sub-Gaussian tail estimate. The inequality is widely used, for example in a-priori estimates for Monte-Carlo methods, see Sect. 13.4. The inequality is formulated as follows:

Theorem 8.12.10 (Hoeffding's[14] Inequality) *Assume independent random variables $f_1, \ldots, f_n : \Omega \to \mathbb{R}$ on some probability space $(\Omega, \mathcal{F}, \mathbb{P})$ such that $a_i \leqslant f_i \leqslant b_i$ on Ω with $-\infty < a_i < 0 < b_i < \infty$ and $\mathbb{E} f_i = 0$ for all $i = 1, \ldots, n$. Then one has that*

$$\mathbb{P}(|f_1 + \cdots + f_n| \geqslant \varepsilon) \leqslant 2e^{-\frac{2\varepsilon^2}{\sum_{i=1}^n (b_i - a_i)^2}} \quad \textit{for} \quad \varepsilon > 0.$$

The proof is based on Hoeffding's lemma:

Lemma 8.12.11 *Let $f : \Omega \to \mathbb{R}$ be a random variable on $(\Omega, \mathcal{F}, \mathbb{P})$ such that $a \leqslant f \leqslant b$ on Ω with $-\infty < a < 0 < b < \infty$ and $\mathbb{E} f = 0$. Then one has*

$$\mathbb{E} e^f \leqslant e^{\frac{(b-a)^2}{8}}.$$

---

[14] Wassily Hoeffding, 12/06/1914 (Mustamäki, Finland, now Mukhino, Russia)–28/02/1991 (Göttingen, Germany).

Proof For $x \in [a, b]$ we have the convex combination

$$x = \frac{b-x}{b-a}a + \frac{x-a}{b-a}b.$$

The convexity of the exponential function implies

$$e^x \leq \frac{b-x}{b-a}e^a + \frac{x-a}{b-a}e^b.$$

Integrating this equation and using $\mathbb{E}f = 0$ yields to

$$\mathbb{E}e^f \leq \frac{b}{b-a}e^a + \frac{-a}{b-a}e^b = pe^{-qh} + qe^{ph} = e^{\log(pe^{-qh}+qe^{ph})} =: e^{H(h)}$$

with $p := \frac{b}{b-a}$, $q := \frac{-a}{b-a}$, and $h := b - a$. By a computation we see that $H(0) = H'(0) = 0$ and $|H''(h)| \leq \frac{1}{4}$ for all $h \in \mathbb{R}$. As by Taylor's formula $H(h) = \frac{h^2}{2}H''(\tilde{h})$ for some $\tilde{h} \in \mathbb{R}$, we get $|H(h)| \leq h^2/8$. □

Proof of Theorem 8.12.10 For $\varepsilon, h > 0$ we get by Markov's inequality, the independence of $f_1, \ldots, f_n$, and Lemma 8.12.11 that

$$\mathbb{P}(f_1 + \cdots + f_n \geq \varepsilon) = \mathbb{P}(e^{h(f_1+\cdots+f_n)} \geq e^{h\varepsilon}) \leq e^{-h\varepsilon}\mathbb{E}e^{h(f_1+\cdots+f_n)}$$

$$= e^{-h\varepsilon}\prod_{i=1}^{n}\mathbb{E}e^{hf_i} \leq e^{-h\varepsilon}\prod_{i=1}^{n}\mathbb{E}e^{\frac{h^2(b_i-a_i)^2}{8}}.$$

Choosing $h := \frac{4\varepsilon}{\sum_{i=1}^{n}(b_i-a_i)^2}$ we arrive at

$$\mathbb{P}(f_1 + \cdots + f_n \geq \varepsilon) \leq e^{-\frac{2\varepsilon^2}{\sum_{i=1}^{n}(b_i-a_i)^2}} \quad \text{for} \quad \varepsilon \geq 0.$$

Replacing f_i by $-f_i$ we can upper bound $\mathbb{P}(f_1 + \cdots + f_n \leq -\varepsilon)$ in the same way, so that we can conclude the proof. □

8.13* Wald's Identity

Given independent and identically distributed (i.i.d.) random variables $f_1, f_2, \ldots$ on $(\Omega, \mathcal{F}, \mathbb{P})$ such that $\mathbb{E}|f_1| < \infty$ one has

$$\mathbb{E}\left(\sum_{n=1}^{N} f_n\right) = N\mathbb{E}f_1.$$

We generalize this equality to randomly stopped sums, called Wald's identity, and shall use this identity later in Chap. 22. We start with the notion of a stopping time that will replace the number of terms N used above:

Definition 8.13.1 We assume a probability space $(\Omega, \mathcal{F}, \mathbb{P})$.

(1) A sequence of σ-algebras $(\mathcal{F}_n)_{n=0}^{\infty}$ with $\mathcal{F}_0 \subseteq \mathcal{F}_1 \subseteq \cdots \subseteq \mathcal{F}$ is called **filtration**. The quadruple $(\Omega, \mathcal{F}, \mathbb{P}, (\mathcal{F}_n)_{n=0}^{\infty})$ is called a **stochastic basis** or a **filtered probability space**.
(2) A map $\tau : \Omega \to \{0, 1, \ldots\} \cup \infty$ is called **stopping time** provided that $\{\tau = n\} \in \mathcal{F}_n$ for $n \in \mathbb{N}_0$.

By definition we have that $\tau^{-1}(B) \in \mathcal{F}$ for all subsets $B \subseteq \mathbb{R}$ so that we also have $\tau^{-1}(\{\infty\}) \in \mathcal{F}$. Therefore τ can be interpreted as an extended random variable $\tau : \Omega \to [-\infty, \infty]$. Moreover, the definition implies that

$$\{\tau \leq n\} \in \mathcal{F}_n \text{ for } n \in \mathbb{N}_0 \quad \text{and} \quad \{n \leq \tau\} = \{\tau < n\}^c \in \mathcal{F}_{n-1} \text{ for } n \in \mathbb{N}. \tag{8.15}$$

The interpretation of a stopping is that it is a random time, but if this random time occurs, i.e. $\tau(\omega) = n$, then this can decided with the information of the σ-algebra $\mathcal{F}_n$, which is interpreted as the information we have to our disposal at time n.

Theorem 8.13.2 (Wald's[15] Identity) *We assume a stochastic basis $(\Omega, \mathcal{F}, \mathbb{P}, (\mathcal{F}_n)_{n=0}^{\infty})$ and i.i.d. random variables $f_1, f_2, \ldots$ on $(\Omega, \mathcal{F}, \mathbb{P})$ such that $\mathbb{E}|f_1| < \infty$ and f_n is $\mathcal{F}_n$-measurable and independent from $\mathcal{F}_{n-1}$ for all $n \in \mathbb{N}$, i.e. $\mathcal{F}_{n-1}$ and $\sigma(f_n)$ are independent. Then one has for a stopping time τ with $\mathbb{E}\tau < \infty$ that*

$$\mathbb{E}\left(\sum_{n=1}^{\tau} |f_n|\right) = \mathbb{E}|f_1|\mathbb{E}\tau < \infty \quad \text{and} \quad \mathbb{E}\left(\sum_{n=1}^{\tau} f_n\right) = \mathbb{E}\tau \mathbb{E}f_1.$$

Proof We let $f_n = f_n^+ - f_n^-$ with $f_n^+ := \max\{f_n, 0\}$ and $f_n^- := \max\{-f_n, 0\}$, and assume that the sequence $(g_n)_{n=1}^{\infty}$ is either $(f_n^+)_{n=1}^{\infty}$ or $(f_n^-)_{n=1}^{\infty}$. By assumption and the relation (8.15) we know that g_n and $\mathbb{1}_{\{n \leq \tau\}}$ are independent. Therefore monotone convergence implies that

$$\mathbb{E}\left(\sum_{n=1}^{\tau} g_n\right) = \mathbb{E}\left(\sum_{n=1}^{\infty} g_n \mathbb{1}_{\{n \leq \tau\}}\right) = \sum_{n=1}^{\infty} \mathbb{E}\left(g_n \mathbb{1}_{\{n \leq \tau\}}\right)$$

$$= \sum_{n=1}^{\infty} \mathbb{E}g_n \, \mathbb{E}\mathbb{1}_{\{n \leq \tau\}} = \mathbb{E}g_1 \left(\sum_{n=1}^{\infty} \mathbb{P}(\tau \geq n)\right)$$

---

[15] Abraham Wald, 31/12/1902 (Kolozsvár, Hungary, now Cluj, Romania)–13/12/1950 (Travancore, India).

$$= \mathbb{E} g_1 \left(\sum_{n=1}^{\infty} n \mathbb{P}(\tau = n) \right) = \mathbb{E} g_1 \mathbb{E} \tau.$$

If we consider $f_n^+ + f_n^- = |f_n|$, the first equality follows, while the second one is obtained using $f_n^+ - f_n^- = f_n$. □

8.14* Atomless Probability Spaces

In this section we investigate with Theorem 8.14.2 an intuitive property that probability spaces might share or not: to be atomless. The proof relies on the Zorn-Kuratowski Lemma A.6.4, i.e. on the axiomatic foundation of set theory. As a good news we show by Tonelli's theorem in Example 8.14.3 that each probability space can be extended to an atomless probability space.

Definition 8.14.1 A probability space $(\Omega, \mathcal{F}, \mathbb{P})$ is called **atomless** provided that for all $A \in \mathcal{F}$ with $\mathbb{P}(A) > 0$ there is some $B \subseteq A$ with $B \in \mathcal{F}$ and $0 < \mathbb{P}(B) < \mathbb{P}(A)$.

The following result of Sierpiński[16] says that the above definition implies a considerably stronger property:

Theorem 8.14.2 *Let $(\Omega, \mathcal{F}, \mathbb{P})$ be an atomless probability space, $A \in \mathcal{F}$ be of positive measure, and $\theta \in (0, 1)$. Then there is a subset $B \subseteq A$ with $B \in \mathcal{F}$ and $\mathbb{P}(B) = \theta \mathbb{P}(A)$.*

Proof

(a) We fix $\theta \in (0, 1)$ and may assume without loss of generality that $A = \Omega$. Otherwise we consider the probability space $(A, \mathcal{F}_A, \mathbb{P}_A)$ where $\mathbb{P}_A(C) := \mathbb{P}(A \cap C)/\mathbb{P}(A)$ for $C \in \mathcal{F}_A = \{F \in \mathcal{F} : F \subseteq A\}$ (see Lemma 3.3.17).

(b) We define

$$\mathcal{S} := \{A \in \mathcal{F} : \mathbb{P}(A) \leqslant \theta\}$$

and the corresponding set of equivalence classes

$$[\mathcal{S}] := \{[A] : A \in \mathcal{F} : \mathbb{P}(A) \leqslant \theta\}$$

with the equivalence relation $A \sim B$ if and only if $\mathbb{P}(A \triangle B) = 0$. We say that $[A] \leqslant [B]$ if and only if $\mathbb{P}(A \setminus B) = 0$. Assume a nonempty totally ordered set $T \subseteq [\mathcal{S}]$ and set $\varepsilon := \sup\{\mathbb{P}(A) : [A] \in T\}$. If there is an $[A] \in T$ with $\mathbb{P}(A) = \varepsilon$, then $[A] \in [\mathcal{S}]$ and $[B] \leqslant [A]$ for all $B \in T$. In particular this is

---

[16] Wacław Sierpiński, 14/03/1882 (Warsaw, Poland)–21/10/1969 (Warsaw, Poland).

true for $\varepsilon = 0$, where we can take A to be the empty set. If $\varepsilon > 0$ and if there is no such $[A]$, then we take a sequence of $[A_n]$ with $0 < \mathbb{P}(A_n) \uparrow \varepsilon$. Then for $A := \bigcup_{n=1}^{\infty} A_n$ we have $A \in \mathcal{S}$ as $\mathbb{P}(A) = \epsilon \leqslant \theta$ and $[B] \leqslant [A]$ for all $B \in T$.

(c) Because of (a) we can apply the Zorn-Kuratowski Lemma A.6.4 and find a maximal element $[A_\theta] \in [\mathcal{S}]$. If $\mathbb{P}(A_\theta) = \theta$, then our proof is complete. So we assume that $\mathbb{P}(A_\theta) < \theta$. However, we may split A_θ^c as long as we obtain a subset $D \subseteq A_\theta^c$ such that $0 < \mathbb{P}(D) < \theta - \mathbb{P}(A_\theta)$. So $A_\theta \cup D \in \mathcal{S}$ with $\mathbb{P}(A_\theta \cup D) > \mathbb{P}(A_\theta)$, which means that $[A_\theta]$ was not maximal which is a contradiction. This completes the proof. □

In case one needs an atomless probability space, but does not have information about atoms in $(\Omega, \mathcal{F}, \mathbb{P})$, then one can extend this space to an atomless space as follows:

Example 8.14.3 For any probability space the product space $(\Omega \times [0, 1], \mathcal{F} \otimes \mathcal{B}([0, 1]), \mathbb{P} \otimes \lambda)$ is atomless. This can be seen as follows: Assume $A \in \mathcal{F} \otimes \mathcal{B}([0, 1])$ and let $A(t) := A \cap (\Omega \times [0, t]) \subseteq A$. Tonelli's Theorem 8.9.1 implies that

$$(\mathbb{P} \otimes \lambda)(A(t)) = \int_{\Omega \times [0,1]} \mathbb{1}_A(\omega, s) \mathbb{1}_{[0,t]}(s) \mathrm{d}(\mathbb{P} \otimes \lambda)(\omega, s)$$

$$= \int_{[0,1]} \left[\int_\Omega \mathbb{1}_A(\omega, s) \mathrm{d}\mathbb{P}(\omega)\right] \mathbb{1}_{[0,t]}(s) \mathrm{d}\lambda(s)$$

$$= \int_{[0,t]} \left[\int_\Omega \mathbb{1}_A(\omega, s) \mathrm{d}\mathbb{P}(\omega)\right] \mathrm{d}\lambda(s).$$

Because $\lambda(\{s\}) = 0$ for all $s \in [0, 1]$, the map $t \mapsto (\mathbb{P} \otimes \lambda)(A(t))$ is continuous. Moreover, $(\mathbb{P} \otimes \lambda)(A(0)) = 0$ and $(\mathbb{P} \otimes \lambda)(A(1)) = (\mathbb{P} \otimes \lambda)(A)$, so that $t \mapsto (\mathbb{P} \otimes \lambda)(A(t))$ takes all intermediate values.

8.15 Exercises

Ex 1: The following properties of the Lebesgue integral are constantly used, but we did not formally prove them. So we recommended that the reader verifies these properties along the definition of the Lebesgue integral:

(a) Prove Remark 8.1.7: Let $(\Omega, \mathcal{F}, \mu)$ be a measure space, $f : \Omega \to \mathbb{R}$ a measurable map, and $A \in \mathcal{F}$. Show that if $\int_\Omega f \mathrm{d}\mu$ exists, then $\int_\Omega f \mathbb{1}_A \mathrm{d}\mu$ exists as well.

(b) Prove Lemma 8.1.8: Let $(\Omega, \mathcal{F}, \mu)$ be a measure space and $f, g : \Omega \to \mathbb{R}$ be measurable maps such that $\mu(\{\omega \in \Omega : f(\omega) \neq g(\omega)\}) = 0$. Show that if $\int_\Omega f \mathrm{d}\mu$ exists, then $\int_\Omega g \mathrm{d}\mu$ exists and we have $\int_\Omega f \mathrm{d}\mu = \int_\Omega g \mathrm{d}\mu$.

8.15 Exercises

(c) Prove Proposition 8.3.1(2): Let $(\Omega, \mathcal{F}, \mu)$ be a measure space, $f : \Omega \to \mathbb{R}$ be a measurable map such that $\int_\Omega f \, d\mu$ exists, and $a \in \mathbb{R}$. Prove that $\int_\Omega (af) \, d\mu$ exists and $\int_\Omega (af) \, d\mu = a \int_\Omega f \, d\mu$ if $a \neq 0$.

Ex 2: Consider the measure space $(\Omega, \mathcal{F}, \mu) := ([0, 1], 2^{[0,1]}, \mu)$, where μ is the counting measure $\mu(A) := \#A$. Then any function $f : [0, 1] \to \mathbb{R}$ is $(\mathcal{F}, \mathcal{B}(\mathbb{R}))$-measurable. Find a condition for f to be integrable and compute $\int_{[0,1]} f(t) \, d\mu(t)$.

Ex 3: Consider the probability space $([0, 1], \mathcal{B}([0, 1]), \lambda)$. Compute

(a) $\lim_{n \to \infty} \mathbb{E} f_n = \lim_{n \to \infty} \int_{[0,1]} f_n(x) \, d\lambda(x)$ for $f_n(x) := (\sin(\pi x))^n$,

(b) $\int_{[0,1]} f(x) \, d\lambda(x)$, where

$$f := \lim_{n \to \infty} f_n \quad \text{with} \quad f_n(x) := \sum_{k=1}^{n} (-1)^k 2^k \mathbb{1}_{[2^{-(2k+1)}, 2^{-2k})}(x).$$

Ex 4: Gaussian density: Prove the second relation in (8.11), i.e.

$$\int_\mathbb{R} (x - m)^2 p_{m,\sigma^2}(x) \, d\lambda(x) = \sigma^2,$$

where $\sigma > 0$, $m \in \mathbb{R}$, and $p_{m,\sigma^2}(x) := \frac{1}{\sqrt{2\pi\sigma^2}} e^{-\frac{(x-m)^2}{2\sigma^2}}$.

Ex 5: Independence and expected values I: We verify Corollary 8.10.4 in a different way:

(a) Assume two *independent simple* functions $f_0, g_0 : \Omega \to \mathbb{R}$ where f_0 takes the values $-\infty < \alpha_1 < \cdots < \alpha_m < \infty$ and g_0 takes the values $-\infty < \beta_1 < \cdots < \beta_n < \infty$, so that we have representations

$$f_0(\omega) = \sum_{k=1}^{m} \mathbb{1}_{A_k}(\omega) \alpha_k \quad \text{and} \quad g_0(\omega) = \sum_{l=1}^{n} \mathbb{1}_{B_l}(\omega) \beta_l.$$

(i) Why are $(A_k)_{k=1}^m$ and $(B_l)_{l=1}^n$ partitions of Ω?
(ii) Prove that A_k and B_l are independent events for all (k, l).
(iii) Using that $\mathbb{E} f = \sum_{k=1}^{m} \alpha_k \mathbb{P}(A_k)$ and $\mathbb{E} g = \sum_{l=1}^{n} \beta_l \mathbb{P}(B_l)$, deduce that $\mathbb{E}(f_0 g_0) = (\mathbb{E} f_0)(\mathbb{E} g_0)$.

(b) Assume that $f, g : \Omega \to \mathbb{R}$ are measurable, independent, and nonnegative. Use Proposition 7.6.2 and the Definition of f_n^0 from the proof of Lemma 8.2.1 (and define g_n^0 in the same way) to check that f_n^0 and g_n^0 are independent. From item (a) we get $\mathbb{E}(f_n^0 g_n^0) = (\mathbb{E} f_n^0)(\mathbb{E} g_n^0)$. Which statement from Chap. 8 implies that $\mathbb{E}(fg) = (\mathbb{E} f)(\mathbb{E} g)$?

Ex 6: Independence and expected values II: In Corollary 8.10.4 we found that the independence of two integrable random variables $f, g : \Omega \to \mathbb{R}$ implies that their product is integrable, and it holds $\mathbb{E}fg = \mathbb{E}f\mathbb{E}g$. Assume that $\mathbb{E}[h(f)k(g)] = \mathbb{E}h(f)\mathbb{E}k(g)$ holds for all bounded, $(\mathcal{B}(\mathbb{R}), \mathcal{B}(\mathbb{R}))$-measurable functions $h, k : \mathbb{R} \to \mathbb{R}$. Prove that f and g are independent.

Ex 7: Computation of moments: Modify the proof of Corollary 8.10.3 to show the following: If $(\Omega, \mathcal{F}, \mu)$ is a σ-finite measure space and $f : \Omega \to \mathbb{R}$ is measurable and nonnegative, then one has for $p \in (0, \infty)$ that

$$\int_\Omega f^p d\mu = \int_{[0,\infty)} \mu(f > t) p t^{p-1} d\lambda(t).$$

Ex 8: Show that for any nonnegative random variable $f : \Omega \to [0, \infty)$ it holds that:

$$\mathbb{E}f < \infty \iff \sum_{n=1}^{\infty} \mathbb{P}(f > n) < \infty$$

Ex 9: Prove that all convex functions $f : \mathbb{R} \to \mathbb{R}$ are continuous.

Ex 10: Comparison of moments (Corollary 8.12.6): For $0 < p < q < \infty$, a probability space $(\Omega, \mathcal{F}, \mathbb{P})$, and a random variable $f : \Omega \to \mathbb{R}$ prove that

$$(\mathbb{E}|f|^p)^{\frac{1}{p}} \leq (\mathbb{E}|f|^q)^{\frac{1}{q}}.$$

Ex 11: Inclusions for Lorentz sequence spaces: For $p \in (0, \infty)$ we let

$$\ell_p := \left\{ (a_n)_{n=1}^\infty \subset \mathbb{R} : \|(a_n)_{n=1}^\infty\|_{\ell_p} := \left(\sum_{n=1}^\infty |a_n|^p \right)^{\frac{1}{p}} < \infty \right\}.$$

For $0 < p < q < \infty$ prove that $\ell_p \subsetneq \ell_q$. Compare your result with Corollary 8.12.6.

Hint: You can reduce the inequality to the case $1 = p < q < \infty$. Then use $a^q + b^q \leq (a+b)^q$ for $a, b \geq 0$ and an iteration argument.

Ex 12: A reverse Hölder inequality: One deduces from the Hölder inequality the following reverse form: For a measure space $(\Omega, \mathcal{F}, \mu)$, nonnegative measurable maps $f, g : \Omega \to \mathbb{R}$, and $1 < p < \infty$ it holds that

$$\int_\Omega fg \, d\mu \geq \left(\int_\Omega f^{\frac{1}{p}} d\mu \right)^p \left(\int_\Omega g^{\frac{1}{1-p}} d\mu \right)^{1-p}$$

provided that $\int_\Omega g^{\frac{1}{1-p}} d\mu \in (0, \infty)$.

Ex 13: Find an example that shows that the constant $c_p = 2^{\frac{1}{p}-1}$ in Minkowski's inequality is sharp for $0 < p < 1$.

Ex 14: Given a measurable space $(\Omega, \mathcal{F})$ and an $(\mathcal{F}, \mathcal{B}([0, \infty)))$-measurable map $f : \Omega \to [0, \infty)$, show that

$$A := \{(t, \omega) \in [0, \infty) \times \Omega : t < f(\omega)\} \in \mathcal{B}([0, \infty)) \otimes \mathcal{F}.$$

Hint: Consider the map $g : [0, \infty) \times \Omega \to [0, \infty) \times [0, \infty)$ given by $g(t, \omega) := (t, f(\omega))$ and the set $B \subseteq [0, \infty) \times [0, \infty)$ with $B := \{(t, x) \in [0, \infty) \times [0, \infty) : t < x\}$. Check its measurability and deduce $A \in \mathcal{B}([0, \infty)) \otimes \mathcal{F}$.

8.16 Comments

Section 8.1: The Lebesgue integral goes back to Lebesgue [116].

Section 8.5: Theorem 8.5.1 was proved by Fatou [60, pages 375–376] for an interval.

Section 8.6.2: Lebesgue's criterion Theorem 8.6.1 can be found at various places in the literature. We are grateful to Roger R.-C. Chen for the permission to use his lecture notes.

Section 8.8: A sufficient condition for Theorem 8.8.1(2) is that μ is translation invariant, i.e. $\mu(B) = \mu(a + B)$ for all $a \in \mathbb{R}^d$ and $B \in \mathcal{B}(\mathbb{R}^d)$, cf. also [165, Theorem 2.20]

Section 8.9: Theorem 8.9.1 and Theorem 8.9.3 are due to Tonelli [184] and Fubini [68], see also [50, page 149].

Section 8.12: Jensen's inequality (Theorem 8.12.3) was discovered by Hölder [88, page 39] for functions with increasing differential quotients and then by Jensen [94] in a more general form. Hölder's inequality Theorem 8.12.5 for sequences (Corollary 8.12.8) was observed by Rogers [159] and Hölder [88]. Chebyshev's inequality Corollary 8.12.7 goes back to Bienaymé [14] and Chebyshev [33]. Minkowski's inequality Theorem 8.12.9 has its origin in the work of Minkowski [131, page 85, (39)] and was extended by F. Riesz [154]. Hoeffding's inequality Theorem 8.12.10 is due to Hoeffding [89, Theorem 2]. For Hoeffding's inequality and an account on other inequalities the reader is also referred to [128].

Section 8.13 Wald's identity Theorem 8.13.2 was used by Wald for hitting times, see for example [195, equation (85)].

Section 8.14: Theorem 8.14.2 is due to Sierpinski [174], where actually a more general result is proved.

Chapter 9
Convergence of Random Variables

Assume that we perform some experiment several times and get measurements denoted by $f_1, f_2, f_3, \ldots$ To get the *true* quantity (whatever this means) we naturally consider

$$S_n = \frac{1}{n}(f_1 + \cdots + f_n)$$

for large n and hope that S_n converges to this true value as $n \to \infty$. To make this precise we have at least to clarify in which sense the convergence takes place.

In probability theory, various types of convergence of random variables are considered. This chapter defines and investigates almost sure convergence, convergence in probability, and convergence in L_p. Sometimes it is desirable to switch from one type of convergence to another, hence it is important to know the implications between the various types. Besides the probabilistic point of view also the analytical formulation of the convergence of random variables is presented. For example, the Ky-Fan metric characterizes convergence in probability. It turns the space of measurable functions defined on a probability space, after taking equivalence classes, into a complete metric space. We will see that the concept of uniform integrability plays an important role in Vitali's convergence theorem which generalizes Lebesgue's theorem on dominated convergence.

For our purpose, it is convenient to introduce the space of measurable functions:

Definition 9.0.1 Given a measurable space $(\Omega, \mathcal{F})$, we let

$$\mathcal{L}_0(\Omega, \mathcal{F}) := \{f : \Omega \to \mathbb{R} \text{ is } (\mathcal{F}, \mathcal{B}(\mathbb{R}))\text{-measurable}\}.$$

In the case we have a measure space $(\Omega, \mathcal{F}, \mu)$, then we might use the notation $\mathcal{L}_0(\Omega, \mathcal{F}, \mu) = \mathcal{L}_0(\Omega, \mathcal{F})$ despite the fact that the measure μ is formally not used to define $\mathcal{L}_0(\Omega, \mathcal{F}, \mu)$.

9.1 Almost Sure Convergence

Definition 9.1.1 Let $f, f_1, f_2, \ldots \in \mathcal{L}_0(\Omega, \mathcal{F}, \mathbb{P})$. We say that $(f_n)_{n=1}^\infty$ **converges almost surely to** f if

$$\mathbb{P}(\{\omega \in \Omega : |f_n(\omega) - f(\omega)| \xrightarrow[n]{} 0\}) = 1.$$

We write $f_n \xrightarrow{a.s.} f$.

Remark 9.1.2

(1) To formulate the above definition we need that

$$\{\omega \in \Omega : |f_n(\omega) - f(\omega)| \xrightarrow[n]{} 0\} \in \mathcal{F}.$$

This follows from

$$\{\omega \in \Omega : |f_n(\omega) - f(\omega)| \xrightarrow[n]{} 0\}$$

$$= \left\{\omega \in \Omega : \forall m \geq 1\, \exists k \geq 1\, \forall n \geq k \ \left(|f_n(\omega) - f(\omega)| < \frac{1}{m}\right)\right\}$$

$$= \bigcap_{m=1}^\infty \bigcup_{k=1}^\infty \bigcap_{n=k}^\infty \left\{\omega : |f_n(\omega) - f(\omega)| < \frac{1}{m}\right\} \in \mathcal{F}.$$

(2) The above definition depends on the measure $\mathbb{P}$. In general one does not have that

$$\mathbb{P}(\{\omega \in \Omega : |f_n(\omega) - f(\omega)| \xrightarrow[n]{} 0\}) = 1$$

if and only if

$$\mathbb{Q}(\{\omega \in \Omega : |f_n(\omega) - f(\omega)| \xrightarrow[n]{} 0\}) = 1$$

if $\mathbb{Q}$ is another probability measure on $\mathcal{F}$.

(3) Only few properties of f_n are transferred to f by the almost sure convergence. Take, for example, $\Omega = [0, 1]$, $\mathcal{F} = \mathbb{B}([0, 1])$, and denote by λ the Lebesgue measure (recall $\lambda([a, b]) = b - a$). Let f_n be (Fig. 9.1)

$$f_n(\omega) := \begin{cases} n^2 2^{n+2}\omega, & \omega \in \left[0, \frac{1}{2n}\right] \\ n2^{n+2} - n^2 2^{n+2}\omega, & \omega \in \left(\frac{1}{2n}, \frac{1}{n}\right] \\ 0, & \omega \in \left(\frac{1}{n}, 1\right]. \end{cases}$$

9.1 Almost Sure Convergence

Fig. 9.1 Tent functions

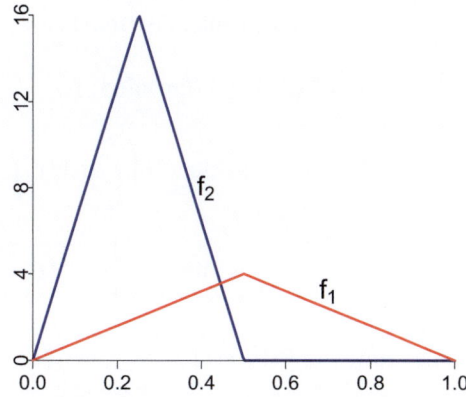

The function $f_n : [0, 1] \to \mathbb{R}$ is continuous so that f_n is a random variable. Moreover, $\lim_n f_n(\omega) = 0$ for all $\omega \in [0, 1]$. On the other side

$$\int_{[0,1]} f_n(\omega) \, d\lambda(\omega) = \int_{[0,1]} f_n(t) \, dt = 2^n \xrightarrow[n]{} \infty.$$

A useful characterization of the almost sure convergence is as follows:

Proposition 9.1.3 *Let $f, f_1, f_2, \ldots \in \mathcal{L}_0(\Omega, \mathcal{F}, \mathbb{P})$. Then the following assertions are equivalent:*

(1) $f_n \xrightarrow{a.s.} f$.
(2) $\lim_{n\to\infty} \mathbb{P}(\{\omega \in \Omega : \sup_{k \geq n} |f_k(\omega) - f(\omega)| > \varepsilon\}) = 0$ *for all $\varepsilon > 0$.*

Proof For $\varepsilon > 0$ and $k \geq 1$ define $A_k^\varepsilon := \{\omega \in \Omega : |f_k(\omega) - f(\omega)| > \varepsilon\}$. Note that

$$\left\{\omega \in \Omega : \sup_{k \geq n} |f_k(\omega) - f(\omega)| > \varepsilon \right\} = \bigcup_{k=n}^\infty A_k^\varepsilon \in \mathcal{F}$$

and $\Omega_0 := \{\omega \in \Omega : \lim_{n \to \infty} f_n(\omega) = f(\omega)\} \in \mathcal{F}$.
(1) $\Rightarrow$ (2) For all $\varepsilon > 0$ it holds

$$\Omega_0 \subseteq \liminf_{n \to \infty} (A_n^\varepsilon)^c = \bigcup_{n=1}^\infty \bigcap_{k=n}^\infty \{\omega \in \Omega : |f_k(\omega) - f(\omega)| \leq \varepsilon\}.$$

If $f_n \xrightarrow{a.s.} f$, then by continuity from below,

$$1 = \mathbb{P}(\Omega_0) \leq \mathbb{P}(\liminf_{n\to\infty}(A_n^\varepsilon)^c)$$

$$= \lim_{n\to\infty} \mathbb{P}\left(\bigcap_{k=n}^\infty (A_k^\varepsilon)^c\right)$$

$$= \lim_{n\to\infty} \mathbb{P}\left(\left\{\omega \in \Omega : \sup_{k\geq n} |f_k(\omega) - f(\omega)| \leq \varepsilon\right\}\right).$$

Hence $\lim_{n\to\infty} \mathbb{P}(\{\omega \in \Omega : \sup_{k\geq n} |f_k(\omega) - f(\omega)| > \varepsilon\}) = 0$.

(2) $\Rightarrow$ (1) For all $\varepsilon > 0$ we have that

$$0 = \lim_{n\to\infty} \mathbb{P}\left(\bigcup_{k=n}^\infty A_k^\varepsilon\right) = \mathbb{P}\left(\limsup_{n\to\infty} A_n^\varepsilon\right) = 1 - \mathbb{P}\left(\liminf_{n\to\infty}(A_n^\varepsilon)^c\right).$$

This implies

$$\mathbb{P}\left(\liminf_{n\to\infty}(A_n^\varepsilon)^c\right) = 1 \quad \text{and therefore} \quad 1 = \mathbb{P}\left(\bigcap_{N=1}^\infty \liminf_{n\to\infty}(A_n^{1/N})^c\right) = \mathbb{P}(\Omega_0).$$

Here the last equality follows from $\omega \in \bigcap_{N=1}^\infty \liminf_{n\to\infty}(A_n^{1/N})^c$ if and only if for all $N = 1, 2, \ldots$ one has $\omega \in \liminf_{n\to\infty}(A_n^{1/N})^c$, which means that for all $N = 1, 2, \ldots$ there is an $n_N(\omega)$ such that $\omega \in \bigcap_{n=n_N(\omega)}^\infty (A_n^{1/N})^c$ which is exactly the condition $\omega \in \Omega_0$. □

An analogous statement is true that describes Cauchy sequences:

Proposition 9.1.4 *Let $f_1, f_2, \ldots \in \mathcal{L}_0(\Omega, \mathcal{F}, \mathbb{P})$. Then the following conditions are equivalent:*

(1) $\mathbb{P}(\{\omega \in \Omega : (f_n(\omega))_{n=1}^\infty \text{ is a Cauchy sequence }\}) = 1$.
(2) *For all $\varepsilon > 0$ one has that*

$$\lim_{n\to\infty} \mathbb{P}\left(\sup_{k,l\geq n} |f_k - f_l| > \varepsilon\right) = 0.$$

(3) *For all $\varepsilon > 0$ one has that*

$$\lim_{n\to\infty} \mathbb{P}\left(\sup_{k\geq n} |f_k - f_n| > \varepsilon\right) = 0.$$

9.1 Almost Sure Convergence

Proof (2) $\Leftrightarrow$ (3) follows from

$$\sup_{k \geq n} |f_k - f_n| \leq \sup_{k,l \geq n} |f_k - f_l|$$
$$\leq \sup_{k \geq n} |f_k - f_n| + \sup_{l \geq n} |f_n - f_l| = 2 \sup_{k \geq n} |f_k - f_n|.$$

(1) $\Leftrightarrow$ (2) Let

$$A := \{\omega \in \Omega : (f_n(\omega))_{n=1}^\infty \text{ is a Cauchy sequence}\}$$
$$= \bigcap_{N=1,2,\ldots} \bigcup_{n=1,2,\ldots} \bigcap_{k > l \geq n} \left\{\omega \in \Omega : |f_k(\omega) - f_l(\omega)| \leq \frac{1}{N}\right\}.$$

Consequently, we have that $\mathbb{P}(A) = 1$ if and only if

$$\mathbb{P}\left(\bigcup_{n=1,2,\ldots} \bigcap_{k > l \geq n} \left\{\omega \in \Omega : |f_k(\omega) - f_l(\omega)| \leq \frac{1}{N}\right\}\right) = 1$$

for all $N = 1, 2, \ldots$, if and only if

$$\lim_{n \to \infty} \mathbb{P}\left(\bigcap_{k > l \geq n} \left\{\omega \in \Omega : |f_k(\omega) - f_l(\omega)| \leq \frac{1}{N}\right\}\right) = 1$$

for all $N = 1, 2, \ldots$, if and only if

$$\lim_{n \to \infty} \mathbb{P}\left(\bigcup_{k > l \geq n} \left\{\omega \in \Omega : |f_k(\omega) - f_l(\omega)| > \frac{1}{N}\right\}\right) = 0$$

for all $N = 1, 2, \ldots$. We conclude the proof by remarking that

$$\bigcup_{k > l \geq n} \left\{\omega \in \Omega : |f_k(\omega) - f_l(\omega)| > \frac{1}{N}\right\} = \left\{\sup_{k > l \geq n} |f_k(\omega) - f_l(\omega)| > \frac{1}{N}\right\}.$$

$\square$

An important application of almost sure convergence is the following version of the *strong law of large numbers*.

Theorem 9.1.5 (Strong Law of Large Numbers Under 4th-Moments) *Assume a sequence of independent random variables* $f_1, f_2, \ldots \in \mathcal{L}_0(\Omega, \mathcal{F}, \mathbb{P})$ *with* $\mathbb{E} f_n = 0$ *and* $c^4 := \sup_n \mathbb{E} f_n^4 < \infty$ *with* $c \geqslant 0$. *Then*

$$\frac{f_1 + \cdots + f_n}{n} \xrightarrow{a.s.} 0.$$

Proof First we note that Corollary 8.12.6 implies that

$$\mathbb{E}|f_i| \leqslant (\mathbb{E}|f_i|^2)^{\frac{1}{2}} \leqslant (\mathbb{E}|f_i|^3)^{\frac{1}{3}} \leqslant (\mathbb{E}|f_i|^4)^{\frac{1}{4}} \leqslant c.$$

For $S_n := \sum_{k=1}^n f_k$ it holds that

$$\mathbb{E} S_n^4 = \mathbb{E}\left(\sum_{k=1}^n f_k\right)^4 = \mathbb{E} \sum_{i,j,k,l=1}^n f_i f_j f_k f_l$$

$$= \sum_{k=1}^n \mathbb{E} f_k^4 + 3 \sum_{\substack{k,l=1,\ldots,n \\ k \neq l}} \mathbb{E} f_k^2 \mathbb{E} f_l^2,$$

because for distinct $\{i, j, k, l\}$ we have

$$\mathbb{E} f_i f_j^3 = \mathbb{E} f_i f_j^2 f_k = \mathbb{E} f_i f_j f_k f_l = 0$$

by independence (for example, $\mathbb{E} f_i f_j^3 = \mathbb{E} f_i \mathbb{E} f_j^3 = 0 \cdot \mathbb{E} f_j^3 = 0$). We conclude

$$\mathbb{E} S_n^4 = \sum_{k=1}^n \mathbb{E} f_k^4 + 3 \sum_{\substack{k,l=1,\ldots,n \\ k \neq l}} \mathbb{E} f_k^2 \mathbb{E} f_l^2 \leqslant nc^4 + 3n(n-1)c^2 c^2 \leqslant 3c^4 n^2,$$

so that

$$\mathbb{E} \sum_{n=1}^\infty \frac{S_n^4}{n^4} = \sum_{n=1}^\infty \mathbb{E} \frac{S_n^4}{n^4} \leqslant \sum_{n=1}^\infty \frac{3c^4}{n^2} < \infty,$$

where we use monotone convergence for the equality. This implies

$$\mathbb{P}\left(\left\{\omega \in \Omega : \sum_{n=1}^\infty \frac{S_n^4(\omega)}{n^4} < \infty\right\}\right) = 1,$$

so that $\frac{S_n^4}{n^4} \xrightarrow{a.s.} 0$ and $\frac{S_n}{n} \xrightarrow{a.s.} 0$. □

9.2 Convergence in Probability

9.2.1 Probabilistic Formulation

Although we saw in Remark 9.1.2(3) that a.s. convergence may be a weak concept in the sense that certain properties of the converging sequence are not shared by the limit, there are examples where this concept turns out to be still too strong.

Example 9.2.1 Let $\Omega := [0, 1]$, $\mathcal{F} := \mathcal{B}([0, 1])$, and let $\mathbb{P} = \mathcal{U}_{[0,1]} = \lambda$ be the uniform distribution on $\mathcal{B}([0, 1])$. Define (Fig. 9.2)

$$f_1(\omega) := \mathbb{1}_{\left[0, \frac{1}{2}\right)}(\omega), \quad f_2(\omega) := \mathbb{1}_{\left[\frac{1}{2}, 1\right)}(\omega),$$

$$f_3(\omega) := \mathbb{1}_{\left[0, \frac{1}{4}\right)}(\omega), \quad f_4(\omega) := \mathbb{1}_{\left[\frac{1}{4}, \frac{1}{2}\right)}(\omega), \ldots, f_6(\omega) := \mathbb{1}_{\left[\frac{3}{4}, 1\right)}(\omega),$$

$$f_7(\omega) := \mathbb{1}_{\left[0, \frac{1}{8}\right)}(\omega), \ldots$$

We do not have $f_n \xrightarrow{a.s.} 0$ as $\#\{n : f_n(\omega) = 1\} = \infty$ for all $\omega \in [0, 1)$.

Nevertheless, in Example 9.2.1 one has the feeling that $\lim_{n \to \infty} f_n = 0$, but in what sense? One solution consists in using *convergence in probability*:

Definition 9.2.2 Let $f, f_1, f_2, \ldots \in \mathcal{L}_0(\Omega, \mathcal{F}, \mathbb{P})$.

(1) The sequence $(f_n)_{n=1}^\infty$ **converges to f in probability** (we write $f_n \xrightarrow{\mathbb{P}} f$) if

$$\lim_{n \to \infty} \mathbb{P}(|f_n - f| > \varepsilon) = 0 \quad \text{for all} \quad \varepsilon > 0.$$

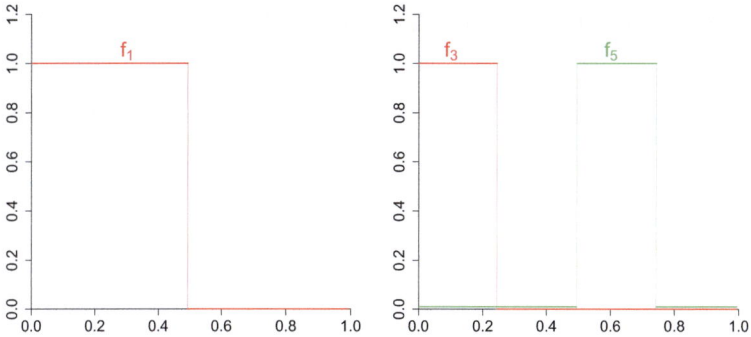

Fig. 9.2 The functions f_1, f_3 and f_5 considered in the example

(2) The sequence $(f_n)_{n=1}^\infty$ is a **Cauchy sequence in probability**[1] provided that for all $\varepsilon > 0$ there exists $n(\varepsilon) \geqslant 1$ such that for all $m, n \geqslant n(\varepsilon)$ one has that

$$\mathbb{P}(|f_m - f_n| > \varepsilon) \leqslant \varepsilon.$$

The following basic properties will be verified in Exercise 1 at the end of this chapter:

Proposition 9.2.3 *Let* $f, g, f_1, f_2, \ldots \in \mathcal{L}_0(\Omega, \mathcal{F}, \mathbb{P})$.

(1) *If* $f_n \xrightarrow{\mathbb{P}} f$, *then* $(f_n)_{n=1}^\infty$ *is a Cauchy sequence in probability.*
(2) **Uniqueness of the limit:** *If* $f_n \xrightarrow{\mathbb{P}} f$ *and* $f_n \xrightarrow{\mathbb{P}} g$, *then* $\mathbb{P}(f = g) = 1$.

Example (Example 9.2.1 Continued) We have that $f_n \xrightarrow{\lambda} 0$. In fact,

$$\lim_n \lambda(\{\omega \in [0,1] : |f_n(\omega)| > \varepsilon\}) \leq \lim_n \lambda(\{\omega \in [0,1] : f_n(\omega) \neq 0\}) = 0$$

since

$$\lambda(\{\omega \in [0,1] : f_n(\omega) \neq 0\}) = \begin{cases} \frac{1}{2} & \text{for } n = 1, 2 \\ \frac{1}{4} & \text{for } n = 3, 4, 5, 6 \\ \frac{1}{8} & \text{for } n = 7, \ldots \\ \vdots & \end{cases}.$$

Now we clarify the relation between the almost sure convergence and the convergence in probability, and verify the completeness and the algebraic properties of convergence in probability:

Theorem 9.2.4 *For* $f, g, f_1, g_1, f_2, g_2, \ldots \in \mathcal{L}_0(\Omega, \mathcal{F}, \mathbb{P})$ *one has:*

(1) *If* $f_n \xrightarrow{a.s.} f$, *then* $f_n \xrightarrow{\mathbb{P}} f$.
(2) **Completeness:** *If* $(f_n)_{n=1}^\infty$ *is a Cauchy sequence in probability, then there exists a subsequence* $1 \leqslant n_1 < n_2 < \cdots$ *and* $h \in \mathcal{L}_0(\Omega, \mathcal{F}, \mathbb{P})$ *with*

$$f_{n_k} \xrightarrow{a.s.} h.$$

Moreover, $f_n \xrightarrow{\mathbb{P}} h$ *as well.*
(3) *If* $f_n \xrightarrow{\mathbb{P}} f$, *then there exists a subsequence* $1 \leqslant n_1 < n_2 < n_3 < \cdots$ *such that* $f_{n_k} \xrightarrow{a.s.} f$ *as* $k \to \infty$.

---

[1] or **fundamental in probability**.

9.2 Convergence in Probability

(4) *One has* $f_n \xrightarrow{P} f$ *if and only if for all subsequences* $1 \leqslant n_1 < n_2 < n_3 < \ldots$ *there is a further subsequence* $1 \leqslant n_{k_1} < n_{k_2} < n_{k_3} < \cdots$ *such that*

$$f_{n_{k_l}} \xrightarrow{a.s.} f \quad \text{as} \quad l \to \infty.$$

(5) **Algebraic operations:** *If* $f_n \xrightarrow{P} f$ *and* $g_n \xrightarrow{P} g$ *and* $\alpha, \beta \in \mathbb{R}$, *then*

$$\alpha f_n + \beta g_n \xrightarrow{P} \alpha f + \beta g \quad \text{and} \quad f_n g_n \xrightarrow{P} fg.$$

Proof (1) The first assertion follows from Proposition 9.1.3 and

$$\mathbb{P}(|f_n - f| > \varepsilon) \leqslant \mathbb{P}\left(\sup_{k \geq n} |f_k - f| > \varepsilon\right).$$

(2) We let $\varepsilon_j := 2^{-j}$ for $j \geqslant 1$ and find $1 \leq n_1 < n_2 < \ldots$ such that

$$\mathbb{P}\left(\{\omega \in \Omega : |f_k(\omega) - f_l(\omega)| > \varepsilon_j\}\right) \leqslant \varepsilon_j$$

for $k, l \geq n_j$. For the sequence $(f_{n_j})_{j=1}^\infty$ we get that

$$\sum_{j=1}^\infty \mathbb{P}\left(\{\omega \in \Omega : |f_{n_{j+1}}(\omega) - f_{n_j}(\omega)| > \varepsilon_j\}\right) \leqslant \sum_{j=1}^\infty \varepsilon_j < \infty.$$

Applying the Lemma of Borel-Cantelli (Theorem 2.5.7) implies

$$\mathbb{P}\left(\{\omega \in \Omega : |f_{n_{j+1}}(\omega) - f_{n_j}(\omega)| > \varepsilon_j \text{ infinitely often}\}\right) = 0.$$

Hence

$$\mathbb{P}(\Omega_0) = 1 \quad \text{with} \quad \Omega_0 := \left\{\omega \in \Omega : \sum_{j=1}^\infty |f_{n_{j+1}}(\omega) - f_{n_j}(\omega)| < \infty\right\}.$$

We set

$$h(\omega) := \begin{cases} f_{n_1}(\omega) + \sum_{j=1}^\infty (f_{n_{j+1}}(\omega) - f_{n_j}(\omega)) & : \omega \in \Omega_0 \\ 0 & : \omega \notin \Omega_0 \end{cases}$$

and get that $f_{n_j} \xrightarrow{a.s.} h$. Finally, we check that $f_n \xrightarrow{\mathbb{P}} h$. For $\varepsilon > 0$ one gets

$$\mathbb{P}(|f_n - h| > \varepsilon) \leqslant \mathbb{P}(|f_n - f_{n_j}| + |f_{n_j} - h| > \varepsilon)$$
$$\leqslant \mathbb{P}\left(|f_n - f_{n_j}| > \frac{\varepsilon}{2}\right) + \mathbb{P}\left(|f_{n_j} - h| > \frac{\varepsilon}{2}\right),$$

where the last inequality follows from

$$\left\{|f_n - f_{n_j}| \leqslant \frac{\varepsilon}{2}\right\} \cap \left\{|f_{n_j} - h| \leqslant \frac{\varepsilon}{2}\right\} \subseteq \{|f_n - f_{n_j}| + |f_{n_j} - h| \leqslant \varepsilon\}.$$

Then $\lim_{j \to \infty} \mathbb{P}\left(|f_{n_j} - h| > \frac{\varepsilon}{2}\right) = 0$ follows from (1). We conclude by

$$\mathbb{P}\left(|f_n - f_{n_j}| > \frac{\varepsilon}{2}\right) \leqslant \eta$$

whenever $n, n_j \geqslant n(\eta, \varepsilon) \geqslant 1$ and $\eta > 0$ is arbitrary.

(3) As $(f_n)_{n=1}^{\infty}$ is a Cauchy sequence in probability by Proposition 9.2.3(1), by (2) there is a subsequence $(n_k)_{k=1}^{\infty}$ and a random variable h such that $f_{n_k} \xrightarrow{a.s.} h$ as $k \to \infty$. But this implies $f_{n_k} \xrightarrow{\mathbb{P}} h$ and $f_{n_k} \xrightarrow{\mathbb{P}} f$ as $k \to \infty$ so that $f = h$ a.s. by Proposition 9.2.3(2).

(4) Assume that $f_n \xrightarrow{\mathbb{P}} f$. Then $f_{n_k} \xrightarrow{\mathbb{P}} f$ as $k \to \infty$ and (3) implies the existence of one more subsequence $(n_{k_l})_{l=1}^{\infty}$ such that $f_{n_{k_l}} \xrightarrow{a.s.} f$ as $l \to \infty$.

To show the opposite implication we assume that f_n does not converge to f in probability. We find an $\varepsilon > 0$ and a subsequence $(n_k)_{k=1}^{\infty}$ such that $\mathbb{P}(|f_{n_k} - f| > \varepsilon) > \varepsilon$. However, by assumption there is a subsequence $(n_{k_l})_{l=1}^{\infty}$ such that $f_{n_{k_l}} \xrightarrow{a.s.} f$ as $l \to \infty$ and consequently also in probability. But this implies that $\mathbb{P}(|f_{n_{k_l}} - f| > \varepsilon) \to 0$ as $l \to \infty$ which is a contradiction.

(5) This item is the subject of Exercise 2. □

Example (Example 9.2.1 Continued) What is a possible subsequence for a.s. convergence? One can take

$$f_1 = \mathbb{1}_{[0,\frac{1}{2})}, \quad f_3 = \mathbb{1}_{[0,\frac{1}{4})}, \quad f_7 = \mathbb{1}_{[0,\frac{1}{8})}, \ldots.$$

Theorem 9.2.5 (Continuous Mapping Theorem) *Assume that for random variables* $f, f_1, f_2, \ldots \in \mathcal{L}_0(\Omega, \mathcal{F}, \mathbb{P})$ *we have* $f_n \xrightarrow{\mathbb{P}} f$. *If* $\varphi : \mathbb{R} \to \mathbb{R}$ *is continuous, then*

$$\varphi(f_n) \xrightarrow{\mathbb{P}} \varphi(f).$$

Proof The assertion follows directly from Theorem 9.2.4(4). □

9.2 Convergence in Probability

We conclude this subsection with a weak law of large numbers. Weak laws of large numbers are sometimes called Khinchin's laws of large numbers. Here the notion *weak* refers to *convergence in probability*. Note that in the *strong* law of large numbers Theorem 9.1.5 we used *almost sure convergence*. The remarkable fact is that the condition (2) below means that the random variables are only *uncorrelated* which is a considerably weaker assumption than *independence*:

Theorem 9.2.6 (Weak Law of Large Numbers) Let $f_1, f_2, \ldots \in \mathcal{L}_0(\Omega, \mathcal{F}, \mathbb{P})$ be such that

(1) $\mathbb{E} f_n^2 < \infty$ and $\mathbb{E} f_n = m \in \mathbb{R}$ for all $n \in \mathbb{N}$,
(2) $\text{cov}(f_k, f_l) := \mathbb{E}(f_k - \mathbb{E} f_k)(f_l - \mathbb{E} f_l) = 0$ for $k \neq l$,
(3) $\lim_n \frac{1}{n^2}[\text{var}(f_1) + \cdots + \text{var}(f_n)] = 0$.

Then $\frac{f_1 + \cdots + f_n}{n} \xrightarrow{\mathbb{P}} m$ as $n \to \infty$.

Proof By Chebyshev's inequality Corollary 8.12.7 we have that

$$\mathbb{P}\left(\left|\frac{f_1 + \cdots + f_n - nm}{n}\right| \geq \varepsilon\right) \leq \frac{1}{\varepsilon^2} \frac{\mathbb{E}|f_1 + \cdots + f_n - nm|^2}{n^2}$$

$$= \frac{1}{\varepsilon^2} \frac{\mathbb{E}\left(\sum_{k=1}^n (f_k - m)\right)^2}{n^2}$$

$$\leq \frac{1}{\varepsilon^2} \frac{\sum_{k=1}^n \text{var}(f_k)}{n^2} \xrightarrow{n \to \infty} 0.$$

□

9.2.2* Analytical Formulation

By the Ky-Fan distance we can express the convergence in probability:

Definition 9.2.7 (Ky Fan[2]-Metric) Assume that $(\Omega, \mathcal{F}, \mathbb{P})$ is a probability space. For $f, g \in \mathcal{L}_0(\Omega, \mathcal{F}, \mathbb{P})$ we define the distance

$$d(f, g) := \inf\{\varepsilon > 0 : \mathbb{P}(|f - g| > \varepsilon) \leq \varepsilon\}.$$

The distance $d(\cdot, \cdot)$ shares the following properties:

Lemma 9.2.8 For $f, g, h \in \mathcal{L}_0(\Omega, \mathcal{F}, \mathbb{P})$ one has:

(1) $d(f, g) = 0$ if and only $\mathbb{P}(f = g) = 1$.
(2) **Symmetry:** $d(f, g) = d(g, f)$.

---

[2] Ky Fan, 19/09/1914 (Hangzhou, China)–22/03/2010 (Santa Barbara, USA).

(3) **Triangle inequality:** $d(f, h) \leq d(f, g) + d(g, h)$.
(4) **Translation invariance:** $d(f, g) = d(f + h, g + h)$.

The proof is the subject of Exercise 4. The link to Sect. 9.2.1 is as follows:

Lemma 9.2.9 *For* $f, f_1, f_2, \ldots \in \mathcal{L}_0(\Omega, \mathcal{F}, \mathbb{P})$ *the following holds:*

(1) $\lim_n d(f_n, f) = 0$ *if and only if* $f_n \xrightarrow{\mathbb{P}} f$.
(2) *For all* $\varepsilon > 0$ *there is an* $n(\varepsilon) \geq 1$ *such that for all* $m, n \geq n(\varepsilon)$ *one has* $d(f_n, f_m) \leq \varepsilon$ *if and only if* $(f_n)_{n=1}^{\infty}$ *is a Cauchy sequence in probability.*

The proof is the subject of Exercise 5. We conclude by formulating the convergence in probability in terms of metric spaces. For this we need the notion of equivalence classes from Appendix A.7.

Definition 9.2.10

(1) For $f, g \in \mathcal{L}_0(\Omega, \mathcal{F}, \mathbb{P})$ we define the equivalence relation $f \sim g$ by $\mathbb{P}(f = g) = 1$. We denote by $[f]$ the class of all random variables equivalent to f.
(2) $L_0(\Omega, \mathcal{F}, \mathbb{P})$ is the space of all equivalence classes from $\mathcal{L}_0(\Omega, \mathcal{F}, \mathbb{P})$.
(3) For $f, g \in \mathcal{L}_0(\Omega, \mathcal{F}, \mathbb{P})$ and $\lambda \in \mathbb{R}$ we introduce the linear operations

$$\lambda[f] := [\lambda f] \quad \text{and} \quad [f] + [g] := [f + g].$$

(4) We define $d([f], [g]) := d(f, g)$.

The above lemmas immediately imply the following result.

Proposition 9.2.11 *The linear space* $[L_0(\Omega, \mathcal{F}, \mathbb{P}), d]$ *is a complete metric space, where the metric* $d(\cdot, \cdot)$ *is translation invariant.*

9.3 Convergence in L_p

9.3.1 Probabilistic Formulation

The following convergence in L_p is a quantitative notion of convergence, where one can naturally speak about a speed of convergence. This notion is (for example) used to describe the accuracy of approximation and simulation schemes, but also to build up function spaces in probability and analysis.

Definition 9.3.1 Let $(\Omega, \mathcal{F}, \mathbb{P})$ be a probability space and $f, f_1, f_2, \ldots \in \mathcal{L}_0(\Omega, \mathcal{F}, \mathbb{P})$.

(1) If $f \in \mathcal{L}_0(\Omega, \mathcal{F}, \mathbb{P})$, then we let

$$\operatorname{ess\,sup}_{\omega \in \Omega} |f(\omega)| := \inf \left\{ \sup_{\omega \in \Omega \setminus N} |f(\omega)| : N \in \mathcal{F}, \mathbb{P}(N) = 0 \right\}.$$

9.3 Convergence in L_p

(2) For $0 < p \leq \infty$ and $f \in \mathcal{L}_0(\Omega, \mathcal{F}, \mathbb{P})$ we let

$$\|f\|_{L_p} := \begin{cases} \left(\int_\Omega |f(\omega)|^p \, d\mathbb{P}(\omega)\right)^{\frac{1}{p}} & : p \in (0, \infty) \\ \operatorname{ess\,sup}_{\omega \in \Omega} |f(\omega)| & : p = \infty \end{cases} \in [0, \infty].$$

Moreover, we define the corresponding space by

$$\mathcal{L}_p(\Omega, \mathcal{F}, \mathbb{P}) := \{f \in \mathcal{L}_0(\Omega, \mathcal{F}, \mathbb{P}) : \|f\|_{L_p} < \infty\}.$$

(3) For $0 < p \leq \infty$ we say that f_n **converges to** f **in** L_p ($f_n \xrightarrow{L_p} f$) provided that

$$\lim_{n \to \infty} \|f_n - f\|_{L_p} = 0.$$

For the above definition we use in the case $0 < p < \infty$ that $x \mapsto |x|^p$ is a continuous function so that $|f|^p$ is a random variable whenever f is. In the following we will concentrate on the convergence in L_p if $0 < p < \infty$. But before doing so, we summarize some facts about $L_\infty(\Omega, \mathcal{F}, \mathbb{P})$:

Proposition 9.3.2 *Let $0 < p < q < \infty$ and $f \in \mathcal{L}_0(\Omega, \mathcal{F}, \mathbb{P})$. Then the following holds:*

(1) $\|f\|_{L_p} \leq \|f\|_{L_q} \leq \|f\|_{L_\infty}$.
(2) $\lim_{q \uparrow \infty} \|f\|_{L_q} = \|f\|_{L_\infty}$.

Proof (1) Corollary 8.12.6 gives $\|f\|_{L_p} \leq \|f\|_{L_q}$. If $\|f\|_{L_\infty} = \infty$, then $\|f\|_{L_q} \leq \|f\|_{L_\infty}$ is evident. If $\|f\|_{L_\infty} < \infty$, then for all $c > \|f\|_{L_\infty}$ there is a null set $N \in \mathcal{F}$ such that $|f(\omega)| \leq c$ for $\omega \in \Omega \setminus N$. This implies $|f(\omega)|^q \leq c^q$ for $\omega \in \Omega \setminus N$ and $\mathbb{E}|f|^q \leq c^q$. As this holds for all $c > \|f\|_{L_\infty(\mathbb{P})}$, we have $\mathbb{E}|f|^q \leq \|f\|_{L_\infty}^q$.

(2) If $\|f\|_{L_\infty} = 0$, then there is nothing to prove. So we assume $\|f\|_{L_\infty} \in (0, \infty]$. For any $0 \leq c < \|f\|_{L_\infty}$ there is a set $A_c \in \mathcal{F}$ of positive measure such that $|f(\omega)| \geq c$ for $\omega \in A_\varepsilon$. This implies $\|f\|_{L_q} \geq c \mathbb{P}(A_\varepsilon)^{\frac{1}{q}}$. By $q \uparrow \infty$ this gives $\lim_{q \uparrow \infty} \|f\|_{L_q} \geq c$. Finally, $c \uparrow \|f\|_{L_\infty}$ yields the statement. □

Now we continue with the case of the convergence in L_p if $0 < p < \infty$ and summarize some basic properties:

Proposition 9.3.3 *Let $0 < p < q < \infty$ and $f, g, f_1, g_1, f_2, g_2, \ldots \in \mathcal{L}_0(\Omega, \mathcal{F}, \mathbb{P})$.*

(1) *If $f_n \xrightarrow{L_p} f$, then $f_n \xrightarrow{\mathbb{P}} f$.*
(2) *If $f_n \xrightarrow{L_p} f$ and $g_n \xrightarrow{L_p} g$, then $f_n + g_n \xrightarrow{L_p} f + g$.*
(3) *If $f_n \xrightarrow{L_q} f$, then $f_n \xrightarrow{L_p} f$.*
(4) *If $f_n \xrightarrow{\mathbb{P}} f$ and $\mathbb{E} \sup_{n \in \mathbb{N}} |f_n|^p < \infty$, then $f \in \mathcal{L}_p(\Omega, \mathcal{F}, \mathbb{P})$ and*

$$\lim_{n \to \infty} \|f_n - f\|_{L_p} = 0.$$

Proof (1) Let $\varepsilon > 0$. Then by Markov's inequality Theorem 8.12.1,

$$\mathbb{P}(\{\omega \in \Omega : |f_n(\omega) - f(\omega)| > \varepsilon\}) = \mathbb{P}(\{\omega \in \Omega : |f_n(\omega) - f(\omega)|^p > \varepsilon^p\})$$
$$\leq \frac{1}{\varepsilon^p} \mathbb{E}|f_n - f|^p \xrightarrow[n]{} 0.$$

(2) The assertion follows from Minkowski's inequality (Theorem 8.12.9) as

$$\|(f_n + g_n) - (f + g)\|_{L_p} \leq c_p \left(\|f_n - f\|_{L_p} + \|g_n - g\|_{L_p}\right) \to 0 \text{ as } n \to \infty.$$

(3) This is an application of Corollary 8.12.6.
(4) We assume that the claim fails to be true and find a subsequence $n_1 < n_2 < \cdots$ and a $\delta > 0$ such that

$$\mathbb{E}|f_{n_k} - f|^p \geq \delta > 0. \tag{9.1}$$

By (1) and Theorem 9.2.4(3) there is one more subsequence $(n_{k_l})_{l=1}^{\infty}$ such that $f_{n_{k_l}} \xrightarrow{a.s.} f$ as $l \to \infty$. By Lebesgue's theorem on dominated convergence Theorem 8.5.2 we get that

$$\mathbb{E}|f|^p = \lim_{l \to \infty} \mathbb{E}|f_{n_{k_l}}|^p < \infty.$$

Hence

$$\mathbb{E} \sup_{n \in \mathbb{N}} |f_n - f|^p \leq c_p \mathbb{E} \left(\sup_{n \in \mathbb{N}} |f_n|^p + |f|^p\right) < \infty$$

and, again by dominated convergence,

$$0 = \mathbb{E} \lim_{l \to \infty} |f_{n_{k_l}} - f|^p = \lim_{l \to \infty} \mathbb{E}|f_{n_{k_l}} - f|^p.$$

This is a contradiction to (9.1). □

9.3.2* Analytical Formulation

Similarly to the case of convergence in probability in Sect. 9.2.2 we can translate the convergence in L_p into a concept described by analysis. For this reason we use the following spaces:

Definition 9.3.4 For $p \in (0, \infty]$ and a probability space $(\Omega, \mathcal{F}, \mathbb{P})$ we define

$$L_p(\Omega, \mathcal{F}, \mathbb{P}) := \{[f] : f \in \mathcal{L}_p(\Omega, \mathcal{F}, \mathbb{P})\} \text{ with } \|[f]\|_{L_p} := \|f\|_{L_p}.$$

9.4 Uniform Integrability

We recalled in Sect. 4.2 the notion of a Banach space, a p-normed Banach space, and a quasi-Banach space. The following Theorem 9.3.5 is a special case of Theorem 20.2.2 that we prove later:

Theorem 9.3.5 *The space $\left[L_p(\Omega, \mathcal{F}, \mathbb{P}), \|\cdot\|_{L_p}\right]$ is a Banach space if $p \in [1, \infty]$, a p-normed Banach space if $p \in (0, 1)$, and a quasi-Banach space if $p \in (0, \infty]$.*

9.4 Uniform Integrability

In this section we weaken the condition $\mathbb{E} \sup_{n \in \mathbb{N}} |f_n|^p < \infty$ for $p = 1$ from Proposition 9.3.3(4). But first we recall that $\int_\Omega |f| d\mathbb{P} < \infty$ implies by dominated convergence

$$\lim_{c \to \infty} \int_{\{|f| \geq c\}} |f| d\mathbb{P} = 0.$$

This leads to the following definition of uniform integrability:

Definition 9.4.1 Let $(f_i)_{i \in I} \subseteq \mathcal{L}_0(\Omega, \mathcal{F}, \mathbb{P})$, where $I \neq \emptyset$ is an arbitrary index set. The family $(f_i)_{i \in I}$ is called **uniformly integrable** (u.i.) provided that for all $\varepsilon > 0$ there is a constant $c > 0$ such that

$$\sup_{i \in I} \int_{\{|f_i| \geq c\}} |f_i| d\mathbb{P} \leq \varepsilon \quad \text{or equivalently} \quad \lim_{c \uparrow \infty} \sup_{i \in I} \int_{\{|f_i| \geq c\}} |f_i| d\mathbb{P} = 0.$$

Example 9.4.2 Let $(\Omega, \mathcal{F}, \mathbb{P}) = ([0, 1], \mathcal{B}([0, 1]), \lambda)$ and $f_n(t) := n \mathbb{1}_{\left[0, \frac{1}{n}\right]}(t)$, $n \in I = \{1, 2, 3, \ldots\}$. This family is not uniformly integrable because for any $c > 0$ we have that

$$\sup_{n \in \mathbb{N}} \int_{\{|f_n| \geq c\}} |f_n(t)| dt = 1.$$

The name *uniformly integrable* suggests that the expected values of a uniform integrable family of random variables is uniformly bounded. In fact, this is true:

Lemma 9.4.3 *Let $(f_i)_{i \in I} \subseteq \mathcal{L}_0(\Omega, \mathcal{F}, \mathbb{P})$ be uniformly integrable, then*

$$\sup_{i \in I} \mathbb{E}|f_i| < \infty.$$

Proof We choose an $\varepsilon > 0$ and find a $c > 0$ such that

$$\sup_{i \in I} \int_{\{|f_i| \geq c\}} |f_i| d\mathbb{P} \leq \varepsilon.$$

Then, for all $i \in I$,

$$\mathbb{E}|f_i| = \int_{\{|f_i|\geq c\}} |f_i| d\mathbb{P} + \int_{\{|f_i|<c\}} |f_i| d\mathbb{P} \leq \varepsilon + \int_{\{|f_i|\leq c\}} c\, d\mathbb{P} \leq \varepsilon + c.$$

□

An important sufficient criteria for uniform integrability is the following:

Lemma 9.4.4 *Let $G : [0, \infty) \to [0, \infty)$ be nondecreasing such that*

$$\lim_{y\to\infty, y>0} \frac{G(y)}{y} = \infty$$

and let $(f_i)_{i\in I} \subseteq \mathcal{L}_0(\Omega, \mathcal{F}, \mathbb{P})$ be such that

$$\sup_{i\in I} \mathbb{E}G(|f_i|) < \infty.$$

Then $(f_i)_{i\in I}$ is uniformly integrable.

Proof In Exercise 9 we show that G is $(\mathcal{B}([0, \infty)), \mathcal{B}(\mathbb{R}))$ measurable. Since $|f_i|$ is $(\mathcal{F}, \mathcal{B}([0, \infty)))$ measurable we conclude that $G(|f_i|)$ is $(\mathcal{F}, \mathcal{B}(\mathbb{R}))$ measurable, so it is a random variable. We let $\varepsilon > 0$ and $M := \sup_{i\in I} \mathbb{E}G(|f_i|)$, and find a $c > 0$ such that

$$\frac{M}{\varepsilon} \leq \frac{G(y)}{y} \quad \text{for} \quad y \geq c.$$

Then

$$\int_{\{|f_i|\geq c\}} |f_i| d\mathbb{P} \leq \frac{\varepsilon}{M} \int_{\{|f_i|\geq c\}} G(|f_i|) d\mathbb{P} \leq \varepsilon.$$

□

Example 9.4.5

(1) Examples for functions in Lemma 9.4.4 are $G(y) := y^p$ with $1 < p < \infty$ and $G(y) := y \log(1 + y)$. Hence $\sup_{i\in I} \mathbb{E}|f_i|^p < \infty$ or $\sup_{i\in I} \mathbb{E}[|f_i| \log(1 + |f_i|)] < \infty$ imply that $(f_i)_{i\in I}$ is uniformly integrable.
(2) In Example 9.4.2 we have seen that $\mathbb{E}|f_n| = 1$ does not guarantee that $(f_n)_{n=1}^\infty$ is uniformly integrable. This confirms that one can not take the function $G(y) = y$ in the above lemma.

9.4 Uniform Integrability

Our main result is a generalization of Lebesgue's dominated convergence:

Theorem 9.4.6 (Vitali's[3] Convergence Theorem) *For $0 < p < \infty$ and $f, f_1, f_2, \ldots \in \mathcal{L}_p(\Omega, \mathcal{F}, \mathbb{P})$ with $f_n \xrightarrow{\mathbb{P}} f$ the following assertions are equivalent:*

(1) $f_n \xrightarrow{L_p} f$.
(2) $(|f_n|^p)_{n=1}^\infty$ *is u.i.*
(3) $\lim_{n \to \infty} \|f_n\|_{L_p} = \|f\|_{L_p}$.

We prove Theorem 9.4.6 after Lemma 9.4.10. An application of Theorem 9.4.6 is the following:

Corollary 9.4.7 *Assume that $f, f_1, f_2, \ldots \in \mathcal{L}_0(\Omega, \mathcal{F}, \mathbb{P})$ are such that $(f_n)_{n=1}^\infty$ is u.i. and $f_n \xrightarrow{\mathbb{P}} f$. Then $\mathbb{E}|f| < \infty$ and*

$$\lim_{n \to \infty} |\mathbb{E} f_n - \mathbb{E} f| \leq \lim_{n \to \infty} \mathbb{E}|f_n - f| = 0.$$

Proof By Theorem 9.2.4 there is a subsequence $1 \leq n_1 < n_2 < \cdots$ with $f_{n_k} \xrightarrow{a.s.} f$ as $k \to \infty$. Applying Fatou's Lemma Theorem 8.5.1 and Lemma 9.4.3 we get that

$$\mathbb{E}|f| \leq \liminf_{k \to \infty} \mathbb{E}|f_{n_k}| \leq \sup_{n \in \mathbb{N}} \mathbb{E}|f_n| < \infty.$$

Hence we can apply for $f'_n := |f_n - f|$ and $f' := 0$ the implication (2) $\Rightarrow$ (1) of Theorem 9.4.6, where we use Exercise 11. □

The proof of Theorem 9.4.6 requires two statements. The first one shows that Lebesgue's dominated convergence in Theorem 8.5.2 holds also for convergence in probability instead of almost sure convergence:

Theorem 9.4.8 (Lebesgue's Theorem, Dominated Convergence II) *Assume that $f, f_1, f_2, \ldots \in \mathcal{L}_0(\Omega, \mathcal{F}, \mathbb{P})$ and $g \in \mathcal{L}_1(\Omega, \mathcal{F}, \mathbb{P})$ such that $f_n \xrightarrow{\mathbb{P}} f$, $|f_n| \leq g$ a.s., and $|f| \leq g$ a.s.. Then one has that*

$$\lim_{n \to \infty} \mathbb{E} f_n = \mathbb{E} f.$$

Proof Assume that the conclusion is not true. Then there is an $\varepsilon > 0$ and a subsequence $n_1 < n_2 < n_3 < \cdots$ such that

$$|\mathbb{E} f_{n_k} - \mathbb{E} f| \geq \varepsilon.$$

---

[3] Giuseppe Vitali, 26/08/1875 (Ravenna, Italy)–29/02/1932 (Bologna, Italy).

But we can find a subsequence n_{k_l} such that

$$f_{n_{k_l}} \xrightarrow{a.s.} f \quad \text{as} \quad l \to \infty$$

by Theorem 9.2.4. Applying dominated convergence yields to a contradiction because

$$\lim_{l \to \infty} \mathbb{E} f_{n_{k_l}} = \mathbb{E} f.$$

□

The next lemma shows that the absolute continuity of measures (which we introduce formally later) implies a continuity property:

Lemma 9.4.9 *Assume finite measures μ and ν on a measurable space $(\Omega, \mathcal{F})$ such that for all $N \in \mathcal{F}$ with $\nu(N) = 0$ one has $\mu(N) = 0$. Then, for all $\varepsilon > 0$ there is an $\eta > 0$ such that for all $A \in \mathcal{F}$ with $\nu(A) < \eta$ one has $\mu(A) < \varepsilon$.*

Proof We assume that the claim is not true. Then there is an $\varepsilon_0 > 0$ such that for all $n \in \mathbb{N}$ there is an $A_n \in \mathcal{F}$ with

$$\nu(A_n) < 2^{-n} \quad \text{and} \quad \mu(A_n) \geq \varepsilon_0.$$

For $A := \limsup_{n \to \infty} A_n$ we get

$$\nu(A) = \nu\left(\bigcap_{n=1}^{\infty} \bigcup_{i=n}^{\infty} A_i\right) = \lim_{n \to \infty} \nu\left(\bigcup_{i=n}^{\infty} A_i\right) \leq \lim_{n \to \infty} \sum_{i=n}^{\infty} \nu(A_i) = 0.$$

On the other side, by the Lemma of Fatou (Theorem 2.5.3), which holds by rescaling for finite measures exactly in the same form, we have that

$$\mu(A) \geq \varepsilon_0 > 0.$$

But this is a contradiction to our assumption. □

We will apply the above lemma in the following form:

Lemma 9.4.10 *Assume a probability space $(\Omega, \mathcal{F}, \mathbb{P})$ and a random variable $f : \Omega \to \mathbb{R}$ with $\mathbb{E}|f| < \infty$. Then for all $\varepsilon > 0$ there is an $\eta > 0$ such that $\mathbb{P}(A) < \eta$ implies that $\int_A |f| d\mathbb{P} < \varepsilon$.*

Proof We let $\nu := \mathbb{P}$ and $\mu(A) := \int_A |f| d\mathbb{P}$. Then, Proposition 10.1.3, which we will prove later, shows that the assumptions of Lemma 9.4.9 are satisfied and our claim follows. □

Proof of Theorem 9.4.6 (1) $\Rightarrow$ (2) We use the inequality $|x+y|^p \leq d_p[|x|^p + |y|^p]$ for $x, y \in \mathbb{R}$, where $d_p := 1$ for $0 < p \leq 1$ and $d_p := 2^{p-1}$ for $1 < p < \infty$. Then

9.4 Uniform Integrability

we get, for $c > 0$,

$$\int_{\{|f_n|^p \geq c\}} |f_n|^p \, d\mathbb{P} \leq d_p \left(\int_{\{|f_n|^p \geq c\}} |f_n - f|^p \, d\mathbb{P} + \int_{\{|f_n|^p \geq c\}} |f|^p \, d\mathbb{P} \right)$$

$$\leq d_p \left(\|f_n - f\|_{L_p}^p + \int_{\{|f_n|^p \geq c\}} |f|^p \, d\mathbb{P} \right).$$

We fix $\varepsilon, c > 0$ and find an $n(\varepsilon) \in \mathbb{N}$ such that

$$d_p \|f_n - f\|_{L_p}^p < \frac{\varepsilon}{2} \quad \text{for} \quad n \geq n(\varepsilon).$$

Moreover, by Minkowski's inequality (Theorem 8.12.9) we may estimate

$$\|f_n\|_{L_p} \leq c_p \left(\|f_n - f\|_{L_p} + \|f\|_{L_p} \right)$$

$$\leq M := c_p \left(\sup_{m \geq 1} \|f_m - f\|_{L_p} + \|f\|_{L_p} \right) < \infty,$$

so that by Markov's inequality,

$$\mathbb{P}(|f_n|^p \geq c) \leq \frac{M^p}{c}.$$

Now we use Lemma 9.4.10 and choose $c(\varepsilon) > 0$ such that

$$d_p \int_{\{|f_n|^p \geq c\}} |f|^p \, d\mathbb{P} < \frac{\varepsilon}{2} \quad \text{for all} \quad n \geq 1 \quad \text{and} \quad c \geq c(\varepsilon).$$

Summarizing, we get

$$\int_{\{|f_n|^p \geq c\}} |f_n|^p \, d\mathbb{P} < \varepsilon \quad \text{for all} \quad n \geq n(\varepsilon) \quad \text{and} \quad c \geq c(\varepsilon).$$

By Exercise 10 we find an $c'(\varepsilon) \geq c(\varepsilon)$ such that

$$\int_{\{|f_n|^p \geq c\}} |f_n|^p \, d\mathbb{P} < \varepsilon \quad \text{for all} \quad n = 1, \ldots, n(\varepsilon) - 1 \quad \text{and} \quad c \geq c'(\varepsilon).$$

(2) $\Rightarrow$ (3) We set $g_n := |f_n|^p$ and $g := |f|^p$. By Theorem 9.4.8 we get, using $a \wedge b := \min\{a, b\}$, that

$$\int_\Omega (g_n \wedge c) \, d\mathbb{P} \to \int_\Omega (g \wedge c) \, d\mathbb{P}$$

because $(g_n \wedge c) \xrightarrow{\mathbb{P}} (g \wedge c)$ as $x \to |x|^p \wedge c$ is a continuous function so that the continuous mapping Theorem 9.2.5 applies. But then

$$|\mathbb{E}g_n - \mathbb{E}g|$$
$$\leq |\mathbb{E}g_n - \mathbb{E}(g_n \wedge c)| + |\mathbb{E}(g_n \wedge c) - \mathbb{E}(g \wedge c)| + |\mathbb{E}(g \wedge c) - \mathbb{E}g|$$
$$\leq \int_{\{g_n \geq c\}} g_n d\mathbb{P} + |\mathbb{E}(g_n \wedge c) - \mathbb{E}(g \wedge c)| + \int_{\{g \geq c\}} g d\mathbb{P}.$$

Given $\varepsilon > 0$ we choose $c(\varepsilon) > 0$ such that for all $c \geq c(\varepsilon)$ one has

$$\int_{\{g \geq c\}} g d\mathbb{P} < \frac{\varepsilon}{3} \quad \text{and} \quad \int_{\{g_n \geq c\}} g_n d\mathbb{P} < \frac{\varepsilon}{3} \quad \text{for all } n \in \mathbb{N},$$

the latter is possible because of our assumption (2). In the next step we choose $n(\varepsilon) = n(\varepsilon, c(\varepsilon)) \in \mathbb{N}$ such that $|\mathbb{E}(g_n \wedge c(\varepsilon)) - \mathbb{E}(g \wedge c(\varepsilon))| < \varepsilon/3$ for $n \geq n(\varepsilon)$ which finally implies $|\mathbb{E}g_n - \mathbb{E}g| < \varepsilon$ for $n \geq n(\varepsilon)$.

(3) $\Rightarrow$ (1) Again we set $g_n := |f_n|^p$ and $g := |f|^p$. Then we have that

$$||g_n - g| - g_n + g| \leq 2g$$

and

$$|g_n - g| - g_n + g \xrightarrow{\mathbb{P}} 0$$

because $g_n \xrightarrow{\mathbb{P}} g$ implies $|g_n - g| \xrightarrow{\mathbb{P}} 0$ and $g - g_n \xrightarrow{\mathbb{P}} 0$. Hence, by Theorem 9.4.8,

$$\lim_{n \to \infty} \mathbb{E}[|g_n - g| - g_n + g] = 0.$$

By $\lim_{n \to \infty} \mathbb{E}g_n = \mathbb{E}g$, which is our assumption, this implies

$$\lim_{n \to \infty} \mathbb{E}\big||f_n|^p - |f|^p\big| = \lim_{n \to \infty} \mathbb{E}|g_n - g| = 0.$$

By (1) $\Rightarrow$ (2) for $p = 1$ and Exercise 11 we derive that $(|f_n|^p)_{n=1}^\infty$ is u.i. Finally we observe that

$$|f_n - f|^p \leq d_p[|f_n|^p + |f|^p]$$

so that $(|f_n - f|^p)_{n=1}^\infty$ is u.i. where we again use Exercise 11. Moreover, we have $|f_n - f|^p \xrightarrow{\mathbb{P}} 0$, so that (2) $\Rightarrow$ (3) for $p = 1$ implies $\lim_{n \to \infty} \mathbb{E}|f_n - f|^p = 0$. □

9.5 Summary

Lebesgue spaces

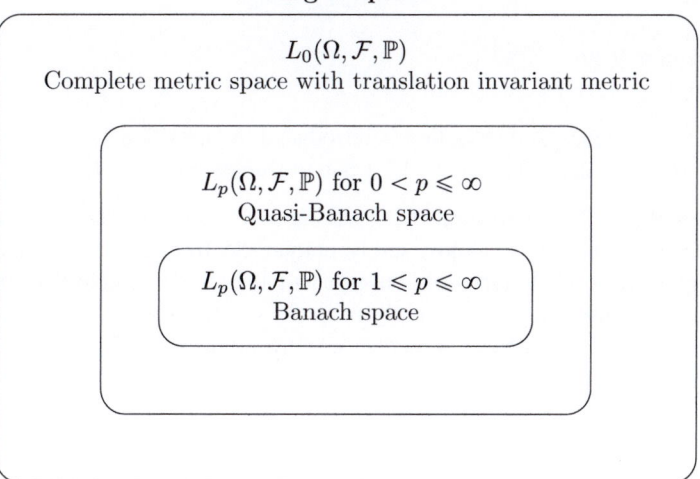

Types of convergence

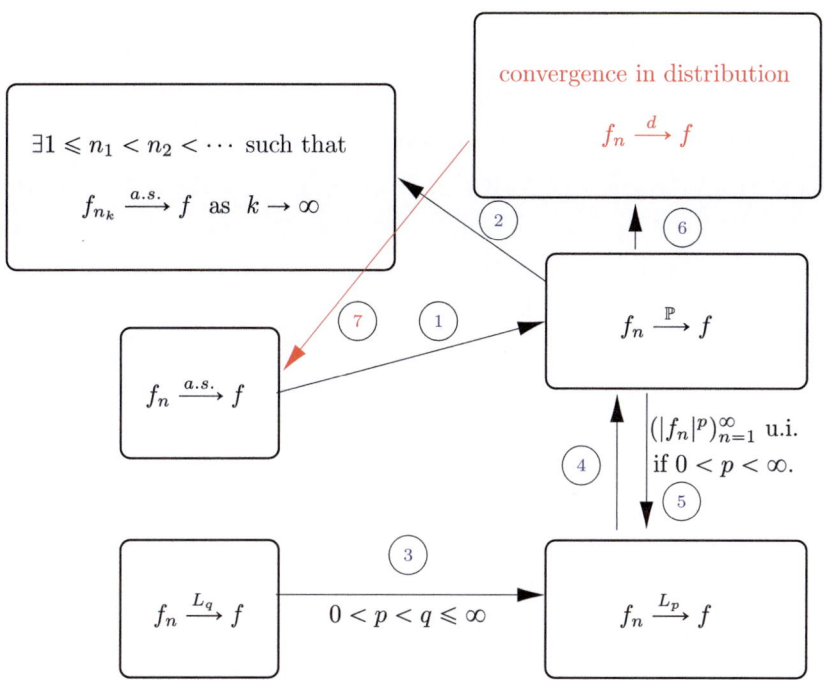

Types of convergence considered in Chap. 9:

① Theorem 9.2.4
② Theorem 9.2.4
③ Proposition 9.3.3
④ Proposition 9.3.3
⑤ Theorem 9.4.6
⑥ Corollary 12.1.9

⑦ Convergence in distribution is introduced in Definition 12.1.2. The implication ⑦ is in the sense Theorem 12.3.1 due to Dudley and Skorohod.

Remark 9.5.1 By Proposition 9.2.3 we know that the limit of a sequence converging in probability is almost surely unique. So from the above picture about the implications of the types of convergences it becomes visible that the limit of a sequence converging almost surely or in L_p is almost surely unique, regardless of the type of convergence. A sequence converging in distribution has a unique distribution.

9.6 Exercises

Ex 1: Prove Proposition 9.2.3: Assume $f, g, f_1, f_2, \ldots \in \mathcal{L}_0(\Omega, \mathcal{F}, \mathbb{P})$. Show the following:

(a) If $f_n \xrightarrow{\mathbb{P}} f$, then $(f_n)_{n=1}^\infty$ is a Cauchy sequence in probability,

(b) If $f_n \xrightarrow{\mathbb{P}} f$ and $f_n \xrightarrow{\mathbb{P}} g$, then $\mathbb{P}(f = g) = 1$.

Ex 2: Prove Theorem 9.2.4(5): If $f, g, f_1, g_1, f_2, g_2, \ldots \in \mathcal{L}_0(\Omega, \mathcal{F}, \mathbb{P})$ such that $f_n \xrightarrow{\mathbb{P}} f$, $g_n \xrightarrow{\mathbb{P}} g$ and $\mu, \lambda \in \mathbb{R}$, then $\lambda f_n + \mu g_n \xrightarrow{\mathbb{P}} \lambda f + \mu g$ and $f_n g_n \xrightarrow{\mathbb{P}} fg$.
Hint: You can prove it directly, but also easily with Theorem 9.2.4(4).

Ex 3: Let $f, f_1, f_2, \ldots \in \mathcal{L}_0(\Omega, \mathcal{F}, \mathbb{P})$ such that for all $\varepsilon > 0$ it holds

$$\sum_{n=1}^\infty \mathbb{P}(|f_n - f| \geq \varepsilon) < \infty.$$

Show that $f_n \xrightarrow{a.s.} f$.
Hint: You might use the lemma of Borel-Cantelli.

Ex 4: Recall for $f, g \in \mathcal{L}_0(\Omega, \mathcal{F}, \mathbb{P})$ the Ky-Fan metric $d(f, g) = \inf\{\varepsilon > 0 : \mathbb{P}(|f - g| > \varepsilon) \leq \varepsilon\}$. Verify Lemma 9.2.8, i.e. let $f, g, h \in \mathcal{L}_0(\Omega, \mathcal{F}, \mathbb{P})$ and prove the following:

(a) $d(f, g) = 0$ if and only if $\mathbb{P}(f = g) = 1$.
(b) $d(f, g) = d(g, f)$.

9.6 Exercises

(c) $d(f, h) \leq d(f, g) + d(g, h)$.
(d) $d(f, g) = d(f + h, g + h)$.

Ex 5: Verify Lemma 9.2.9, i.e. let $f, f_1, f_2, \ldots \in \mathcal{L}_0(\Omega, \mathcal{F}, \mathbb{P})$ and prove the following:

(a) $\lim_n d(f_n, f) = 0$ if and only if $f_n \xrightarrow{\mathbb{P}} f$.
(b) For all $\varepsilon > 0$ there is an $n(\varepsilon) \geq 1$ such that for all $m, n \geq n(\varepsilon)$ one has $d(f_n, f_m) \leq \varepsilon$ if and only if $(f_n)_{n=1}^\infty$ is a Cauchy sequence in probability.

Ex 6: For $f, g \in \mathcal{L}_0(\Omega, \mathcal{F}, \mathbb{P})$ define

$$D(f, g) := \int_\Omega \frac{|f - g|}{1 + |f - g|} d\mathbb{P}.$$

Prove that D is a translation invariant metric on $L_0(\Omega, \mathcal{F}, \mathbb{P})$, and that $f_n \xrightarrow{\mathbb{P}} f$ if and only if $\lim_{n \to \infty} D(f_n, f) = 0$.

Ex 7: Show that $f_n \xrightarrow{L_p} f$ does not imply $f_n \xrightarrow{a.s.} f$: Assume that $f_1, f_2, \ldots \in \mathcal{L}_0(\Omega, \mathcal{F}, \mathbb{P})$ are independent random variables such that $\mathbb{P}(f_n = 1) = \frac{1}{n} = 1 - \mathbb{P}(f_n = 0)$. Show that $f_n \xrightarrow{L_p} 0$ for any $0 < p < \infty$ and use the Lemma of Borel-Cantelli (Theorem 2.5.7) to derive that $\mathbb{P}(\lim_{n \to \infty} f_n \text{ does not exist}) = 1$.

Ex 8: Show that $f_n \xrightarrow{a.s.} f$ does not imply $f_n \xrightarrow{L_p} f$ for any $0 < p < \infty$: Use the probability space $([0, 1], \mathcal{B}([0, 1]), \lambda)$ and put $f_n := e^n \mathbb{1}_{[0, \frac{1}{n}]}$ for $n \in \mathbb{N}$.

Ex 9: Let $G : [0, \infty) \to [0, \infty)$ be a monotone function. Show that G is $(\mathcal{B}([0, \infty)), \mathcal{B}(\mathbb{R}))$ measurable.

Ex 10: Let $N \in \mathbb{N}$ and $f_1, \ldots, f_N \in \mathcal{L}_1(\Omega, \mathcal{F}, \mathbb{P})$. Show that the family $(f_n)_{n=1}^N$ is uniformly integrable.

Ex 11: Assume a probability space $(\Omega, \mathcal{F}, \mathbb{P})$ and two sequences of random variables $f_1, f_2, \ldots : \Omega \to \mathbb{R}$ and $g_1, g_2, \ldots : \Omega \to \mathbb{R}$. Assume that $(f_n)_{n=1}^\infty$ is u.i. and that $(g_n)_{n=1}^\infty$ is u.i. Given $\alpha, \beta \in \mathbb{R}$, prove that $(\alpha f_n + \beta g_n)_{n=1}^\infty$ is u.i.

Hint: For $c > 0$ and two random variables $f, g : \Omega \to \mathbb{R}$ verify and use the following inequality

$$\int_{\{|f+g| \geq c\}} |f + g| d\mathbb{P} \leq 2 \left[\int_{\{|f| \geq \frac{c}{2}\}} |f| d\mathbb{P} + \int_{\{|g| \geq \frac{c}{2}\}} |g| d\mathbb{P} \right],$$

i.e. you might want to check that

$$\mathbb{1}_{\{|f+g| \geq c\}} |f + g| \leq 2 \left[\mathbb{1}_{\{|f| \geq \frac{c}{2}\}} |f| + \mathbb{1}_{\{|g| \geq \frac{c}{2}\}} |g| \right].$$

Ex 12: Assume that $f_1, f_2, \cdots \in \mathcal{L}_1(\Omega, \mathcal{F}, \mathbb{P})$ and $g_1, g_2, \cdots \in \mathcal{L}_1(\Omega, \mathcal{F}, \mathbb{P})$ are uniformly integrable. Prove that

$$\lim_{n \to \infty} \mathbb{E}\left[f_n \mathbf{1}_{\{|g_n|>n\}}\right] = 0.$$

Ex 13: Let $(\Omega, \mathcal{F}, \mathbb{P})$ be a probability space and denote by $\mathcal{L}_0(\Omega, \mathcal{F}, \mathbb{P}; [-\infty, \infty])$ the collection of all extended measurable maps $f : \Omega \to [-\infty, \infty]$. Prove that for each nonempty set $H \subseteq \mathcal{L}_0(\Omega, \mathcal{F}, \mathbb{P}; [-\infty, \infty])$ there exists a countable $H_0 \subseteq H$ such that for

$$h(\omega) := \sup_{h_0 \in H_0} h_0(\omega)$$

the following holds:

(1) $h \in \mathcal{L}_0(\Omega, \mathcal{F}, \mathbb{P}; [-\infty, \infty])$.
(2) $\mathbb{P}(h \geq f) = 1$ for all $f \in H$.
(3) If g is another element of $\mathcal{L}_0(\Omega, \mathcal{F}, \mathbb{P}; [-\infty, \infty])$ such that $\mathbb{P}(g \geq f) = 1$ for all $f \in H$, then $\mathbb{P}(g \geq h) = 1$.

The map h is called essential supremum of H.
Hint: Check that w.l.o.g. we can assume that $f : \Omega \to [-1, 1]$ for all $f \in H$. Now let $\mathcal{C}$ be the collection of countable subsets $S \subseteq H$. Set

$$f_S(\omega) := \sup_{f \in S} f(\omega) \quad \text{for} \quad S \in \mathcal{C}.$$

Define $M := \sup\{\mathbb{E} f_S : S \in \mathcal{C}\}$ and find an $S_0 \in \mathcal{S}$ for which the supremum is attained.

Chapter 10
The Theorem of Radon-Nikodym and Conditional Expectation

So far, given a measurable space $(\Omega, \mathcal{F})$, we mainly did consider one measure on this space. However, in many situations the consideration of different measures and their relation to each other is of interest. In fact, we have been in this situation already as we defined the Gaussian distribution $\mathcal{N}_{m,\sigma^2}$ for $\sigma^2 > 0$ by its density p_{m,σ^2} with respect to the Lebesgue measure on $\mathbb{R}$. Could one introduce the Dirac measure by a density as well, i.e. does there exist a $(\mathcal{B}(\mathbb{R}), \mathcal{B}(\mathbb{R}))$-measurable function $f : \mathbb{R} \to \mathbb{R}$ satisfying $\int_\mathbb{R} |f(x)| d\lambda(x) < \infty$ such that

$$\delta_0((-\infty, b]) = \int_{(-\infty, b]} f(x) d\lambda(x) \quad \text{for all} \quad b \in \mathbb{R}?$$

We will see that the answer is *no*. What is the intrinsic difference between the Gaussian distribution with positive variance and the delta distribution? This section will provide us with the theory to answer this question.

The central concept will be *absolute continuity*: assuming two measures μ and ν on $(\Omega, \mathcal{F})$, we say that ν is absolutely continuous with respect to μ if and only if $\mu(A) = 0$ implies that $\nu(A) = 0$, and we write $\nu \ll \mu$. At first glance this concept seems to be rather weak as it only concerns null sets, however the Radon-Nikodym theorem (see Theorem 10.2.1 below) demonstrates that this notion is powerful and the right one. The Radon-Nikodym theorem makes it possible to relate different measures to each other by densities. An application is the construction of conditional expectations, one of the fundamental concepts in stochastic process theory and filtering. To prove the Radon-Nikodym theorem we use the Hahn-Jordan decomposition.

We continue with the existence and definition of the conditional expectation, its properties, and its connection to conditional probabilities. The chapter ends with a closed form formula for the conditional expectation in the case when the integrand consists of a measurable function of two random variables where one is measurable and the other one is independent from the σ-algebra used for conditioning.

10.1 Signed Measures

In this section we introduce and investigate signed measures, which generalize the notion of a finite measure. Signed measures are used to prove the existence of conditional expectations in this chapter and appear as dual of the space of continuous functions on a compactum in Sect. 18.4.

Definition 10.1.1 (Signed Measures and Absolute Continuity) Let $(\Omega, \mathcal{F})$ be a measurable space.

(1) A map $\nu : \mathcal{F} \to \mathbb{R}$ is called a (finite) **signed measure** if
   (a) $\nu(\emptyset) = 0$,
   (b) σ-**additivity** holds, that is for all $A_1, A_2, \ldots \in \mathcal{F}$ with $A_i \cap A_j = \emptyset$ for $i \neq j$ one has
   $$\nu\left(\bigcup_{i=1}^{\infty} A_i\right) = \sum_{i=1}^{\infty} \nu(A_i).$$

(2) Assume that μ is a measure on the measurable space $(\Omega, \mathcal{F})$ and ν is a signed measure on $(\Omega, \mathcal{F})$. Then ν is **absolutely continuous** with respect to μ if for all $A \in \mathcal{F}$ it holds that
   $$\mu(A) = 0 \quad \text{implies} \quad \nu(A) = 0.$$
   We shall write $\nu \ll \mu$.

Remark 10.1.2 Item (1b) implies that $\sum_{i=1}^{\infty} |\nu(A_i)| < \infty$. Indeed, any permutation of the sets $(A_i)_{i=1}^{\infty}$ yields to the same left-hand side $\nu\left(\bigcup_{i=1}^{\infty} A_i\right)$. So the right-hand side does not depend on the order of summation which implies $\sum_{i=1}^{\infty} |\nu(A_i)| < \infty$.

The basic example for a signed measure is
$$\nu := \nu^+ - \nu^-$$
for finite measures ν^+ and ν^-. The Hahn-Jordan decomposition, which we prove in this section, shows that this is actually already the general form.

Let us start with an example for a signed measure to which we return later in the Radon-Nikodym theorem:

Proposition 10.1.3 Let $(\Omega, \mathcal{F}, \mu)$ be a measure space, $L : \Omega \to \mathbb{R}$ be integrable, and
$$\nu(A) := \int_A L d\mu.$$

10.1 Signed Measures

Then ν is a signed measure and $\nu \ll \mu$. Moreover, if L is non-negative, then ν is a finite measure, and for any $(\mathcal{F}, \mathcal{B}(\mathbb{R}))$-measurable and nonnegative $f : \Omega \to \mathbb{R}$ it holds that

$$\int_\Omega f \, d\nu = \int_\Omega f L \, d\mu. \tag{10.1}$$

Proof We let $L^+ := \max\{L, 0\}$ and $L^- := \max\{-L, 0\}$ so that $L = L^+ - L^-$. Define

$$\nu^\pm(A) := \int_\Omega \mathbb{1}_A L^\pm \, d\mu.$$

Now we check that $\nu^\pm$ are finite measures. First we have that

$$\nu^\pm(\Omega) = \int_\Omega \mathbb{1}_\Omega L^\pm \, d\mu \leqslant \int_\Omega |L| \, d\mu < \infty.$$

Assume $(A_n)_{n=1}^\infty \subseteq \mathcal{F}$ to be pairwise disjoint sets. Then

$$\nu^+\left(\bigcup_{n=1}^\infty A_n\right) = \int_\Omega \mathbb{1}_{\bigcup_{n=1}^\infty A_n} L^+ \, d\mu = \int_\Omega \left(\sum_{n=1}^\infty \mathbb{1}_{A_n}(\omega)\right) L^+ \, d\mu$$

$$= \int_\Omega \lim_{N \to \infty} \left(\sum_{n=1}^N \mathbb{1}_{A_n}\right) L^+ \, d\mu = \lim_{N \to \infty} \int_\Omega \left(\sum_{n=1}^N \mathbb{1}_{A_n}\right) L^+ \, d\mu = \sum_{n=1}^\infty \nu^+(A_n),$$

where we have used the theorem about monotone convergence (Theorem 8.2.2). The same can be done for L^-.

To prove (10.1) for $L \geqslant 0$, by monotone convergence we only need to check the case $f = \mathbb{1}_A$ for $A \in \mathcal{F}$. But this is in fact the definition of ν. □

Next we prove the Hahn and the Hahn-Jordan decompositions of a signed measure which give an *optimal* decomposition $\nu = \nu^+ - \nu^-$ of a signed measure ν into finite measures ν^+, ν^-.

Theorem 10.1.4 (Hahn[1] Decomposition) *Let ν be a signed measure on $(\Omega, \mathcal{F})$. Then there exists a measurable partition $\Omega = \Omega^- \cup \Omega^+$ such that*

(1) $\nu(A) \geqslant 0$ *for all $A \in \mathcal{F}$ with $A \subseteq \Omega^+$,*
(2) $\nu(A) \leqslant 0$ *for all $A \in \mathcal{F}$ with $A \subseteq \Omega^-$.*

Given another partition $\Omega = \tilde{\Omega}^- \cup \tilde{\Omega}^+$ with these properties, one has that

$$\nu(A) = 0 \quad \text{for all} \quad A \in \mathcal{F} \quad \text{with} \quad A \subseteq \Omega^+ \cap \tilde{\Omega}^- \quad \text{or} \quad A \subseteq \Omega^- \cap \tilde{\Omega}^+.$$

---

[1] Hans Hahn, 27/09/1879 (Vienna, Austria)–24/07/1934 (Vienna, Austria).

Proof Let $\kappa := \sup_{B \in \mathcal{F}} \nu(B)$. Then there exists a sequence $(\Omega_n)_{n=1}^{\infty} \subseteq \mathcal{F}$ such that $\lim_{n \to \infty} \nu(\Omega_n) = \kappa$. We will construct by the help of this sequence a set $\Omega^+ \in \mathcal{F}$ for which ν attains the supremum κ. Let $\overline{\Omega} := \bigcup_{k=1}^{\infty} \Omega_k$ and consider for each $n \in \mathbb{N}$ the partition $\mathcal{C}_n$ of $\overline{\Omega}$ given by

$$\mathcal{C}_n = \{C_1 \cap \ldots \cap C_n : C_k \in \{\Omega_k, \overline{\Omega} \setminus \Omega_k\}\}.$$

Define

$$\mathcal{C}_n^+ := \{A \in \mathcal{C}_n : \nu(A) \geq 0\} \quad \text{and} \quad \Omega_n^+ := \bigcup_{A \in \mathcal{C}_n^+} A.$$

Since for all sets $A \in \mathcal{C}_n^+$ the measure ν is nonnegative, we have that

$$\nu(\Omega_n^+) = \sum_{A \in \mathcal{C}_n^+} \nu(A) \geq \sum_{A \in \mathcal{C}_n^+, A \subseteq \Omega_n} \nu(A) \geq \sum_{A \in \mathcal{C}_n, A \subseteq \Omega_n} \nu(A) = \nu(\Omega_n)$$

and

$$\nu\left(\bigcup_{k=n}^{\infty} \Omega_k^+\right) = \nu(\Omega_n^+) + \sum_{k=n+1}^{\infty} \nu\left(\Omega_k^+ \setminus \left(\bigcup_{l=n}^{k-1} \Omega_l^+\right)\right) \geq \nu(\Omega_n^+) \geq \nu(\Omega_n).$$

Setting

$$\Omega^+ := \limsup_{n \to \infty} \Omega_k^+ = \bigcap_{n=1}^{\infty} \bigcup_{k=n}^{\infty} \Omega_k^+$$

we conclude that

$$\nu(\Omega^+) = \nu\left(\bigcap_{n=1}^{\infty} \bigcup_{k=n}^{\infty} \Omega_k^+\right) = \lim_{n \to \infty} \nu\left(\bigcup_{k=n}^{\infty} \Omega_k^+\right) \geq \lim_{n \to \infty} \nu(\Omega_n) = \kappa.$$

Proof of (1) and (2): Assuming a measurable subset $A \subseteq \Omega^+$ with $\nu(A) < 0$ would yield to a contradiction because in this case,

$$\kappa = \nu(\Omega^+) < \nu(\Omega^+ \setminus A) \leq \kappa.$$

Set $\Omega^- := \Omega \setminus \Omega^+$ and assume a measurable $A \subseteq \Omega^-$ with $\nu(A) > 0$. This would again yield a contradiction because

$$\kappa = \nu(\Omega^+) < \nu(\Omega^+ \cup A) \leq \kappa.$$

The statement about the uniqueness is obvious. $\square$

10.1 Signed Measures

From the Hahn decomposition we derive the Hahn-Jordan decomposition:

Definition 10.1.5 (Hahn-Jordan[2] Decomposition) Given a signed measure ν on $(\Omega, \mathcal{F})$ and $\Omega^{\pm}$ from Theorem 10.1.4, we call $\nu := \nu_J^+ - \nu_J^-$ with

$$\nu_J^+(A) := \nu(A \cap \Omega^+) \quad \text{and} \quad \nu_J^-(A) := -\nu(A \cap \Omega^-)$$

the Hahn-Jordan decomposition of ν.

The Hahn-Jordan decomposition is unique in the following sense:

Theorem 10.1.6 *Assume a signed measure ν on $(\Omega, \mathcal{F})$ with decomposition $\nu = \nu^+ - \nu^-$. Then one has*

$$\nu_J^+(A) + \nu_J^-(A) \leq \nu^+(A) + \nu^-(A) \quad \text{for all} \quad A \in \mathcal{F},$$

where the equality holds if and only if $\nu_J^+ = \nu^+$ and $\nu_J^- = \nu^-$.

Proof We assume a Hahn decomposition $\Omega = \Omega^- \cup \Omega^+$ for ν. With this decomposition we derive from $\nu = \nu^+ - \nu^-$ that

$$\nu^+(A) + \nu^-(A) = \nu^+(A \cap \Omega^+) + \nu^+(A \cap \Omega^-) + \nu^-(A \cap \Omega^+) + \nu^-(A \cap \Omega^-)$$
$$\geq \nu^+(A \cap \Omega^+) - \nu^+(A \cap \Omega^-) - \nu^-(A \cap \Omega^+) + \nu^-(A \cap \Omega^-)$$
$$= \nu(A \cap \Omega^+) - \nu(A \cap \Omega^-)$$
$$= \nu_J^+(A \cap \Omega^+) + \nu_J^-(A \cap \Omega^-).$$

Therefore the inequality is verified and we have an equality if and only if $\nu^+(A \cap \Omega^-) = \nu^-(A \cap \Omega^+) = 0$. However this condition implies that

$$\nu^+(A) = \nu^+(A \cap \Omega^+) \geq \nu(A \cap \Omega^+) = \nu_J^+(A)$$

and

$$\nu^-(A) = \nu^-(A \cap \Omega^-) \geq -\nu(A \cap \Omega^-) = \nu_J^-(A).$$

Conversely, we have

$$\nu_J^+(A) = \nu(A \cap \Omega^+) \leq \nu^+(A \cap \Omega^+) \leq \nu^+(A)$$

---

[2] Camille Jordan, 05/01/1838 (Lyon, France)–22/01/1922 (Paris, France).

and
$$\nu_J^-(A) = -\nu(A \cap \Omega^-) \leqslant \nu^-(A \cap \Omega^-) \leqslant \nu^-(A).$$

This concludes the proof of $\nu_J^\pm = \nu^\pm$. □

Definition 10.1.7 Given a signed measure ν on $(\Omega, \mathcal{F})$ with Hahn-Jordan decomposition $\nu = \nu_J^+ - \nu_J^-$ we let
$$|\nu| := \nu_J^+ + \nu_J^- \quad \text{and} \quad |\nu|_{\text{TV}} := |\nu|(\Omega)$$
be the **total variation** of ν.

The name *total variation* is due to the fact that the Hahn-Jordan decomposition implies that
$$|\nu|(\Omega) = \sup \left\{ \sum_{i=1}^{\infty} |\nu(\Omega_i)| \right\},$$
where the supremum is taken over all partitions $\Omega = \bigcup_{i=1}^{\infty} \Omega_i$ with $\Omega_i \in \mathcal{F}$. By the next corollary we compute the total variation of a signed measure by integrating continuous functions. This will be useful for us in Sect. 18.4. We recall that for a metric space $[M, d]$ the space $C(M)$ consists of all continuous and bounded functions $f: M \to \mathbb{R}$ and is equipped with the norm $\|f\|_{C(M)} := \sup_{x \in M} |f(x)|$.

Corollary 10.1.8 *Let $[M, d]$ be a metric space and μ be a signed measure on $(M, \mathcal{B}(M))$. Then one has that*
$$|\mu|_{\text{TV}} = \sup \left\{ \left| \int_M \varphi(x) \mathrm{d}\mu(x) \right| : \|\varphi\|_{C(M)} \leqslant 1 \right\}.$$

Proof The inequality $\geqslant$ is obvious, we only need to verify $\leqslant$. We consider the Hahn-Jordan decomposition $\mu = \mu^+ - \mu^-$ with a partition $M = M^+ \cup M^-$. We choose closed sets $F^\pm \subseteq M^\pm$ that are necessarily disjoint. By Urysohn's Lemma 4.1.7 we find a continuous function $\varphi: M \to [-1, 1]$ such that $\varphi|_{M^\pm} = \pm 1$. Then
$$\int_M \varphi(x) \mathrm{d}\mu(x) \geqslant \mu^+(F^+) + \mu^-(F^-) - [\mu^+(M^+ \setminus F^+) + \mu^-(M^- \setminus F^-)]$$
$$= \mu^+(M^+) + \mu^-(M^-) - 2[\mu^+(M^+ \setminus F^+) + \mu^-(M^- \setminus F^-)].$$

Given $\varepsilon > 0$, by the inner regularity we can achieve that $2[\mu^+(M^+ \setminus F^+) + \mu^-(M^- \setminus F^-)] < \varepsilon$ so that the proof is complete. □

10.2 Theorem of Radon-Nikodym

Now we formulate and prove the first main result of this chapter, the Radon-Nikodym theorem, announced in the beginning of this chapter:

Theorem 10.2.1 (Radon[3]-Nikodym[4] Theorem) *Let $(\Omega, \mathcal{F}, \mu)$ be a σ-finite measure space and ν be a signed measure with $\nu \ll \mu$. Then there exists a measurable and integrable $L: \Omega \to \mathbb{R}$ such that*

$$\nu(A) = \int_A L \, \mathrm{d}\mu \quad \text{for all} \quad A \in \mathcal{F}. \tag{10.2}$$

The map L is unique in the following sense: If L and L' are maps as above satisfying (10.2), then $\mu(L \neq L') = 0$.

Definition 10.2.2 The measurable map L is called **Radon-Nikodym derivative** and we write

$$L = \frac{\mathrm{d}\nu}{\mathrm{d}\mu}.$$

Proof of Theorem 10.2.1

Reduction 1: We may assume that μ is a finite measure, and by scaling, a probability measure $\mu = \mathbb{P}$. Indeed, in the general case we can use a partition $\Omega = \bigcup_{n=1}^{\infty} \Omega_n$ with $\mu(\Omega_n) < \infty$, get $L_n : \Omega_n \to \mathbb{R}$, define $L = \sum_n \mathbb{1}_{\Omega_n} L_n$, and observe that

$$\int_\Omega |L| \mathrm{d}\mu = \sum_{n=1}^{\infty} \int_{\Omega_n} |L_n| \mathrm{d}\mu = \sum_{n=1}^{\infty} (\nu^+(\Omega_n) + \nu^-(\Omega_n)) = \nu^+(\Omega) + \nu^-(\Omega) < \infty,$$

where $\nu = \nu^+ - \nu^-$ is the Jordan decomposition.

Reduction 2: We may assume that ν is a nonnegative finite measure. In fact, by the Hahn-Jordan decomposition we have $\nu = \nu^+ - \nu^-$ with $\nu^+ \ll \mathbb{P}$ and $\nu^- \ll \mathbb{P}$ because, for example, $\mathbb{P}(A) = 0$ implies $\mathbb{P}(A \cap \Omega^+) = 0$ and therefore $0 = \nu(A \cap \Omega^+) = \nu^+(A)$. Hence we can consider ν^+ and ν^- separately, i.e. we may assume that ν is a nonnegative finite measure.

---

[3] Johann Radon, 16/12/1887 (Tetschen, Bohemia; now Decin, Czech Republic)–25/05/1956 (Vienna, Austria).

[4] Otton Marcin Nikodym, 13/08/1887 (Zablotow, Galicia, Austria-Hungary, now Ukraine)–04/05/1974 (Utica, USA).

The actual proof: We start our proof by defining

$$\mathcal{M} := \left\{ L \in \mathcal{L}_1(\Omega, \mathcal{F}, \mathbb{P}) : L \geq 0, \int_A L d\mathbb{P} \leq \nu(A) \text{ for all } A \in \mathcal{F} \right\}.$$

Because $0 \in \mathcal{M}$ the set $\mathcal{M}$ is nonempty. Moreover, $L_1, L_2 \in \mathcal{M}$ implies (recall that $a \vee b = \max\{a, b\}$)

$$L_1 \vee L_2 \leq L_1 + L_2 \in \mathcal{L}_1(\Omega, \mathcal{F}, \mathbb{P})$$

and, with $\Omega_1 := \{L_1 \geq L_2\}$ and $\Omega_2 := \{L_2 > L_1\}$, the estimate

$$\int_A L_1 \vee L_2 d\mathbb{P} = \int_{A \cap \Omega_1} L_1 d\mathbb{P} + \int_{A \cap \Omega_2} L_2 d\mathbb{P} \leq \nu(A \cap \Omega_1) + \nu(A \cap \Omega_2) = \nu(A).$$

Consequently, $L_1 \vee L_2 \in \mathcal{M}$. Observe that

$$\kappa := \sup_{L \in \mathcal{M}} \int_\Omega L d\mathbb{P} \leq \nu(\Omega) < \infty$$

and find $L_n \in \mathcal{M}$, $n \in \mathbb{N}$, such that

$$\sup_{n \in \mathbb{N}} \int_\Omega L_n d\mathbb{P} = \kappa.$$

Letting $K_n := \max\{L_1, \ldots, L_n\}$, we obtain $K_n \in \mathcal{M}$, $0 \leq K_1 \leq K_2 \leq \cdots$, and

$$\lim_{n \to \infty} \int_\Omega K_n d\mathbb{P} = \kappa.$$

Define the extended random variable

$$L(\omega) := \lim_{n \to \infty} K_n(\omega) \in [0, \infty].$$

From monotone convergence we get that

$$\int_\Omega L d\mathbb{P} = \kappa < \infty$$

and may redefine L on a null set such that $L : \Omega \to \mathbb{R}$, $L \in \mathcal{M}$, and

$$\int_A L d\mathbb{P} \leq \nu(A) \quad \text{for} \quad A \in \mathcal{F}.$$

10.2 Theorem of Radon-Nikodym

We show that $\int_A L d\mathbb{P} = \nu(A)$ for $A \in \mathcal{F}$: Assume that there is an $A_0 \in \mathcal{F}$ with

$$\int_{A_0} L d\mathbb{P} < \nu(A_0). \tag{10.3}$$

Since $\nu(A_0) > 0$ we conclude from $\nu \ll \mathbb{P}$ that $\mathbb{P}(A_0) > 0$. Choose an $\varepsilon > 0$ such that for

$$\nu'(A) := \nu(A) - \int_A (L + \varepsilon) d\mathbb{P} = \nu(A) - \int_A L d\mathbb{P} - \varepsilon \mathbb{P}(A), \quad A \in \mathcal{F},$$

one has $\nu'(A_0) > 0$. The Hahn decomposition with respect to ν' yields to a partition $\Omega = \Omega^+ \cup \Omega^-$ with $\nu'(\Omega^+) > 0$. Because $\nu' \ll \mathbb{P}$ as well, we obtain that $\mathbb{P}(\Omega^+) > 0$. Then for

$$L' := L + \varepsilon \mathbb{1}_{\Omega^+}$$

we have $L' \geq 0$ and $L' \in \mathcal{L}_1(\Omega, \mathcal{F}, \mathbb{P})$. To show that $L' \in \mathcal{M}$, we let $A \in \mathcal{F}$ and get that

$$\int_A L' d\mathbb{P} = \int_{A \cap \Omega^+} (L + \varepsilon) d\mathbb{P} + \int_{A \cap \Omega^-} L d\mathbb{P}$$

$$= \nu(A \cap \Omega^+) - \nu'(A \cap \Omega^+) + \int_{A \cap \Omega^-} L d\mathbb{P}$$

$$\leq \nu(A \cap \Omega^+) + \nu(A \cap \Omega^-)$$

$$= \nu(A).$$

But this leads to a contradiction since

$$\kappa = \int_\Omega L d\mathbb{P} < \int_\Omega L d\mathbb{P} + \varepsilon \mathbb{P}(\Omega^+) = \int_\Omega L' d\mathbb{P} \leq \sup_{L'' \in \mathcal{M}} \int_\Omega L'' d\mathbb{P} = \kappa$$

so that a set $A_0 \in \mathcal{F}$ satisfying (10.3) does not exist.

Now we show the uniqueness: Like above we may assume that $\mu = \mathbb{P}$ is a probability measure. Let $g' \in \mathcal{L}_1(\Omega, \mathcal{G}, \mathbb{P})$ with

$$\int_A g d\mathbb{P} = \int_A g' d\mathbb{P}$$

for all $A \in \mathcal{G}$ such that $\mathbb{P}(g \neq g') > 0$. Hence we find a set $A \in \mathcal{G}$ with $\mathbb{P}(A) > 0$ and real numbers $\alpha < \beta$ such that

$$g(\omega) \leq \alpha < \beta \leq g'(\omega) \quad \text{for} \quad \omega \in A$$

or
$$g'(\omega) \leq \alpha < \beta \leq g(\omega) \quad \text{for} \quad \omega \in A.$$

Consequently (for example in the first case)
$$\int_A g \, d\mathbb{P} \leqslant \alpha \mathbb{P}(A) < \beta \mathbb{P}(A) \leqslant \int_A g' \, d\mathbb{P}$$

which is a contradiction. □

10.3 Conditional Expectation

Given a probability space $(\Omega, \mathcal{F}, \mathbb{P})$ and a sub-σ-algebra $\mathcal{G} \subseteq \mathcal{F}$, a random variable $f : \Omega \to \mathbb{R}$ that is $\mathcal{G}$-measurable is automatically $\mathcal{F}$-measurable. The converse is not true in general. Roughly speaking, one can say that an $\mathcal{F}$-measurable random variable is more complex or does contain more information than a $\mathcal{G}$-measurable random variable does. There is a canonical way to project an $\mathcal{F}$-measurable random variable onto a subspace of $\mathcal{G}$-measurable random variables, that is called *conditional expectation*. The notion of conditional expectation plays a central role in the theory of stochastic processes and their applications. In particular, the notion of a martingale is derived from this. Martingales are used, for example, to describe fair games, for the definition of a stochastic integral, or to show why one can solve partial differential equations by Monte-Carlo methods.

Let us explain the conditional expectation in an easy, but typical, situation as an average procedure. Assume that the σ-algebra $\mathcal{G}$ is given by all possible unions of the elements of a partition

$$\Omega = \bigcup_{i=1}^{n} \Omega_i$$

where the Ω_i are pairwise disjoint and of positive measure. Given an $\mathcal{F}$-measurable map $f : \Omega \to \mathbb{R}$ such that $\int_\Omega |f| \, d\mathbb{P} < \infty$, we average f over the partition and obtain a new random variable

$$g(\omega) := \frac{1}{\mathbb{P}(\Omega_i)} \int_{\Omega_i} f \, d\mathbb{P} \quad \text{if} \quad \omega \in \Omega_i.$$

10.3 Conditional Expectation

Because g is constant on the Ω_i's it is measurable with respect to $\mathcal{G}$. Moreover, one can easily see that one cannot distinguish between f and g if one integrates over sets from $\mathcal{G}$, i.e.

$$\int_A f \, d\mathbb{P} = \int_A g \, d\mathbb{P} \quad \text{for all} \quad A \in \mathcal{G}.$$

The random variable g will be called *conditional expectation of f given $\mathcal{G}$*. The following proposition guarantees the existence of a conditional expectation in the general case, i.e. when $\mathcal{G}$ is not generated by a countable partition of Ω:

Theorem 10.3.1 (Conditional Expectation) *Let $(\Omega, \mathcal{F}, \mathbb{P})$ be a probability space, $\mathcal{G} \subseteq \mathcal{F}$ be a sub-σ-algebra, and $f \in \mathcal{L}_1(\Omega, \mathcal{F}, \mathbb{P})$. Then the following assertions hold:*

(1) There exists a $g \in \mathcal{L}_1(\Omega, \mathcal{G}, \mathbb{P})$ such that

$$\int_A f \, d\mathbb{P} = \int_A g \, d\mathbb{P} \text{ for all } A \in \mathcal{G}.$$

(2) If g and g' are as in (1), then $\mathbb{P}(g \neq g') = 0$.

Proof
(1) Define

$$\nu(A) := \int_A f \, d\mathbb{P} \quad \text{for} \quad A \in \mathcal{G},$$

so that ν is a signed measure on $\mathcal{G}$ by Proposition 10.1.3. Applying the Theorem of Radon-Nikodym gives a $g \in \mathcal{L}_1(\Omega, \mathcal{G}, \mathbb{P})$ such that, for $A \in \mathcal{G}$,

$$\nu(A) = \int_A g \, d\mathbb{P} \quad \text{so that} \quad \int_A g \, d\mathbb{P} = \nu(A) = \int_A f \, d\mathbb{P}.$$

(2) The uniqueness follows directly from the Theorem of Radon-Nikodym. □

This leads to the following definition:

Definition 10.3.2 (Conditional Expectation and Conditional Probability) The $\mathcal{G}$-measurable and integrable random variable g from Theorem 10.3.1 is called **conditional expectation of f given $\mathcal{G}$** and is denoted by

$$g = \mathbb{E}[f \mid \mathcal{G}].$$

Moreover, for $A \in \mathcal{F}$ we define the **conditional probability of A given $\mathcal{G}$** by

$$\mathbb{P}[A \mid \mathcal{G}] := \mathbb{E}[\mathbb{1}_A \mid \mathcal{G}].$$

Remark 10.3.3 One has to keep in mind that the conditional expectation is only unique up to null sets from $\mathcal{G}$ so that the notation $g = \mathbb{E}[f \mid \mathcal{G}]$ needs an explanation. Writing $g = \mathbb{E}[f \mid \mathcal{G}]$ we mean that $g \in \mathcal{L}_1(\Omega, \mathcal{G}, \mathbb{P})$ is a **version of the conditional expectation of f given $\mathcal{G}$**.

We continue with some basic properties of conditional expectations similar to Proposition 8.3.1:

Proposition 10.3.4 (Properties of the Conditional Expectation) *Let $f, f_1, f_2, \ldots \in \mathcal{L}_1(\Omega, \mathcal{F}, \mathbb{P})$ and let $\mathcal{H} \subseteq \mathcal{G} \subseteq \mathcal{F}$ be sub σ-algebras of $\mathcal{F}$. Then the following holds true:*

(1) *Linearity: For $\alpha_1, \alpha_2 \in \mathbb{R}$ one has*

$$\mathbb{E}[\alpha_1 f_1 + \alpha_2 f_2 \mid \mathcal{G}] = \alpha_1 \mathbb{E}[f_1 \mid \mathcal{G}] + \alpha_2 \mathbb{E}[f_2 \mid \mathcal{G}] \text{ a.s.}$$

(2) *Monotonicity: If $f_1 \leq f_2$ a.s., then $\mathbb{E}[f_1 \mid \mathcal{G}] \leq \mathbb{E}[f_2 \mid \mathcal{G}]$ a.s.*
(3) *Positivity: If $f \geq 0$ a.s., then $\mathbb{E}[f \mid \mathcal{G}] \geq 0$ a.s.*
(4) *Convexity: One has that $|\mathbb{E}[f \mid \mathcal{G}]| \leq \mathbb{E}[|f| \mid \mathcal{G}]$ a.s.*
(5) *Projection property: If f is $\mathcal{G}$-measurable, then $\mathbb{E}[f \mid \mathcal{G}] = f$ a.s.*
(6) *Tower property: $\mathbb{E}[\mathbb{E}[f \mid \mathcal{G}] \mid \mathcal{H}] = \mathbb{E}[\mathbb{E}[f \mid \mathcal{H}] \mid \mathcal{G}] = \mathbb{E}[f \mid \mathcal{H}]$ a.s.*
(7) *If $\mathcal{G} = \{\emptyset, \Omega\}$, then $\mathbb{E}[f \mid \mathcal{G}] = \mathbb{E}f$.*
(8) *If for all $B \in \mathcal{B}(\mathbb{R})$ and all $A \in \mathcal{G}$ one has that*

$$\mathbb{P}(\{f \in B\} \cap A) = \mathbb{P}(f \in B)\mathbb{P}(A),$$

i.e. if f is independent from $\mathcal{G}$, then $\mathbb{E}[f \mid \mathcal{G}] = \mathbb{E}f$ a.s.
(9) *Monotone convergence: If $0 \leq f_n \uparrow f$ a.s., then*

$$\lim_n \mathbb{E}[f_n \mid \mathcal{G}] = \mathbb{E}[f \mid \mathcal{G}] \text{ a.s..}$$

(10) *If $g: \Omega \to \mathbb{R}$ is $\mathcal{G}$-measurable and $gf \in \mathcal{L}_1(\Omega, \mathcal{F}, \mathbb{P})$, then*

$$\mathbb{E}[gf \mid \mathcal{G}] = g\mathbb{E}[f \mid \mathcal{G}] \text{ a.s.}$$

Proof
(1), (8) and (9) are exercises (see Exercise 1).
(2) Set $A := \{\omega \in \Omega : \mathbb{E}[f_1 \mid \mathcal{G}](\omega) > \mathbb{E}[f_2 \mid \mathcal{G}](\omega)\}$. The relation $f_1 \leq f_2$ a.s. implies $\mathbb{E}[\mathbb{1}_A(f_1 - f_2)] \leq 0$. Noticing that $A \in \mathcal{G}$, by the definition of conditional expectation we conclude

$$0 \geq \mathbb{E}\mathbb{1}_A f_1 - \mathbb{E}\mathbb{1}_A f_2 = \mathbb{E}\mathbb{1}_A \mathbb{E}[f_1 \mid \mathcal{G}] - \mathbb{E}\mathbb{1}_A \mathbb{E}[f_2 \mid \mathcal{G}]$$
$$= \mathbb{E}\mathbb{1}_A \big(\mathbb{E}[f_1 \mid \mathcal{G}] - \mathbb{E}[f_2 \mid \mathcal{G}]\big).$$

Therefore $\mathbb{P}(A) = 0$.

10.3 Conditional Expectation

(3) We apply (2) to $0 =: f_1 \leq f_2 := f$.

(4) The inequality $f \leq |f|$ gives $\mathbb{E}[f \mid \mathcal{G}] \leq \mathbb{E}[|f| \mid \mathcal{G}]$ a.s. and $-f \leq |f|$ gives $-\mathbb{E}[f \mid \mathcal{G}] \leq \mathbb{E}[|f| \mid \mathcal{G}]$ a.s., so that $-\mathbb{E}[|f| \mid \mathcal{G}] \leq \mathbb{E}[f \mid \mathcal{G}] \leq \mathbb{E}[|f| \mid \mathcal{G}]$ a.s..

(5) follows directly from the definition.

(6) Since $\mathbb{E}[f \mid \mathcal{H}]$ is $\mathcal{H}$-measurable and hence $\mathcal{G}$-measurable, item (5) implies that $\mathbb{E}[\mathbb{E}[f \mid \mathcal{H}] \mid \mathcal{G}] = \mathbb{E}[f \mid \mathcal{H}]$ a.s. so that one equality is shown. For the other equality we have to show that

$$\int_A \mathbb{E}[\mathbb{E}[f \mid \mathcal{G}] \mid \mathcal{H}] d\mathbb{P} = \int_A f \, d\mathbb{P}$$

for $A \in \mathcal{H}$. Letting $g := \mathbb{E}[f \mid \mathcal{G}]$ and $A \in \mathcal{H} \subseteq \mathcal{G}$, this follows from

$$\int_A \mathbb{E}[\mathbb{E}[f \mid \mathcal{G}] \mid \mathcal{H}] d\mathbb{P} = \int_A \mathbb{E}[g \mid \mathcal{H}] d\mathbb{P} = \int_A g \, d\mathbb{P} = \int_A \mathbb{E}[f \mid \mathcal{G}] d\mathbb{P} = \int_A f \, d\mathbb{P}.$$

(7) follows from

$$\int_\emptyset f \, d\mathbb{P} = 0 = \int_\emptyset (\mathbb{E}f) d\mathbb{P} \quad \text{and} \quad \int_\Omega f \, d\mathbb{P} = \mathbb{E}f = \int_\Omega (\mathbb{E}f) d\mathbb{P}.$$

(10) Assume first that $g = \sum_{n=1}^{N} \alpha_n \mathbb{1}_{A_n}$, where $\bigcup_{n=1}^{N} A_n = \Omega$ is a partition with $A_n \in \mathcal{G}$. For $A \in \mathcal{G}$ we get that

$$\int_A g f \, d\mathbb{P} = \sum_{n=1}^{N} \alpha_n \int_A \mathbb{1}_{A_n} f \, d\mathbb{P} = \sum_{n=1}^{N} \alpha_n \int_{A \cap A_n} f \, d\mathbb{P}$$

$$= \sum_{n=1}^{N} \alpha_n \int_{A \cap A_n} \mathbb{E}[f \mid \mathcal{G}] d\mathbb{P} = \int_A \left(\sum_{n=1}^{N} \alpha_n \mathbb{1}_{A_n} \right) \mathbb{E}[f \mid \mathcal{G}] d\mathbb{P}$$

$$= \int_A g \mathbb{E}(f \mid \mathcal{G}) d\mathbb{P}.$$

Hence $\mathbb{E}[gf \mid \mathcal{G}] = g\mathbb{E}[f \mid \mathcal{G}]$ a.s. For the general case we can assume that $f, g \geq 0$ since we can decompose $f = f^+ - f^-$ and $g = g^+ - g^-$ with $f^+ := \max\{f, 0\}$ and $f^- := \min\{-f, 0\}$ (and in the same way we proceed with g). We find simple functions $0 \leq g_n \leq g$ such that $g_n(\omega) \uparrow g(\omega)$. Then, by our first step, we get that

$$g_n \mathbb{E}[f \mid \mathcal{G}] = \mathbb{E}[g_n f \mid \mathcal{G}] \text{ a.s.}$$

By $n \to \infty$ the left-hand side follows. The right-hand side is a consequence of the monotone convergence treated in (9). $\square$

Theorem 10.3.5 (Conditional Inequalities) *Assume a probability space $(\Omega, \mathcal{F}, \mathbb{P})$ and a sub-σ-algebra $\mathcal{G} \subseteq \mathcal{F}$.*

(1) *Hölder's inequality:* If $1 < p, q < \infty$ with $\frac{1}{p} + \frac{1}{q} = 1$, $f \in \mathcal{L}_p(\Omega, \mathcal{F}, \mathbb{P})$, and $h \in \mathcal{L}_q(\Omega, \mathcal{F}, \mathbb{P})$. Then

$$\mathbb{E}[|fh| \mid \mathcal{G}] \leq (\mathbb{E}[|f|^p \mid \mathcal{G}])^{\frac{1}{p}} (\mathbb{E}[|h|^q \mid \mathcal{G}])^{\frac{1}{q}} \text{ a.s.}$$

(2) *Minkowski's inequality:* If $1 \leq p < \infty$ and $f, h \in \mathcal{L}_p(\Omega, \mathcal{F}, \mathbb{P})$, then

$$(\mathbb{E}[|f + h|^p \mid \mathcal{G}])^{\frac{1}{p}} \leq (\mathbb{E}[|f|^p \mid \mathcal{G}])^{\frac{1}{p}} + (\mathbb{E}[|h|^p \mid \mathcal{G}])^{\frac{1}{p}} \text{ a.s.}$$

(3) *Jensen's inequality:* If $f \in \mathcal{L}_1(\Omega, \mathcal{F}, \mathbb{P})$ and if $g : \mathbb{R} \to \mathbb{R}$ is convex such that $g(f) \in \mathcal{L}_1(\Omega, \mathcal{F}, \mathbb{P})$, then one has

$$g(\mathbb{E}[f \mid \mathcal{G}]) \leq (\mathbb{E}[g(f) \mid \mathcal{G}]) \text{ a.s.}$$

Proof
(1) For $N \in \mathbb{N}$ we set

$$\hat{f}(\omega) := \frac{|f(\omega)|}{(\mathbb{E}[|f|^p \mid \mathcal{G}](\omega))^{\frac{1}{p}} + \frac{1}{N}} \quad \text{and} \quad \hat{h}(\omega) := \frac{|h(\omega)|}{(\mathbb{E}[|h|^q \mid \mathcal{G}](\omega))^{\frac{1}{q}} + \frac{1}{N}},$$

where we agree about nonnegative versions of the conditional expectations. Using Proposition 10.3.4(10) this gives

$$\frac{\mathbb{E}[|fh| \mid \mathcal{G}]}{\left((\mathbb{E}[|f|^p \mid \mathcal{G}](\omega))^{\frac{1}{p}} + \frac{1}{N}\right)\left((\mathbb{E}[|h|^q \mid \mathcal{G}](\omega))^{\frac{1}{q}} + \frac{1}{N}\right)}$$

$$= \mathbb{E}[|\hat{f}\hat{h}| \mid \mathcal{G}] \leq \frac{\mathbb{E}[|\hat{f}|^p \mid \mathcal{G}]}{p} + \frac{\mathbb{E}[|\hat{h}|^q \mid \mathcal{G}]}{q} \leq \frac{1}{p} + \frac{1}{q} = 1 \text{ a.s.}$$

where we used $xy \leq \frac{x^p}{p} + \frac{y^q}{q}$ for $x, y \geq 0$. By rearranging the terms and $N \to \infty$ the assertion follows. (2) is deduced from (1) exactly as in Theorem 8.12.9.

(3) Let L_x be a support line of g in $x \in \mathbb{Q}$ and define $L(y) = \sup_{x \in \mathbb{Q}} L_x(y) \leq g(y)$. Then one has $L(x) = g(x)$ for $x \in \mathbb{Q}$, and since L and g are convex they are continuous. This implies that $L = g$ on $\mathbb{R}$ and a.s.,

$$g(\mathbb{E}[f \mid \mathcal{G}]) = \sup_{x \in \mathbb{Q}} L_x(\mathbb{E}[f \mid \mathcal{G}]) = \sup_{x \in \mathbb{Q}}(\mathbb{E}[L_x(f) \mid \mathcal{G}]) \leq (\mathbb{E}[g(f) \mid \mathcal{G}]).$$

□

10.4 A Representation of Conditional Expectations

We did start in Sect. 10.3 with the example that $\Omega = \bigcup_{i=1}^{n} \Omega_i$ is a measurable partition with $\mathbb{P}(\Omega_i) > 0$ and $\mathcal{G} := \sigma(\Omega_1, \ldots, \Omega_n)$. Then we can compute for $f \in \mathcal{L}_1(\Omega, \mathcal{F}, \mathbb{P})$ the conditional expectation $\mathbb{E}[f \mid \mathcal{G}]$ as

$$\mathbb{E}[f \mid \mathcal{G}] = \sum_{i=1}^{n} \mathbb{1}_{\Omega_i} \frac{1}{\mathbb{P}(\Omega_i)} \int_{\Omega_i} f \, d\mathbb{P}. \tag{10.4}$$

Let $\gamma : \Omega \to \{1, \ldots, n\}$ be given by $\gamma(\omega) := i$ if $\omega \in \Omega_i$. So, if we introduce the notation

$$\mathbb{E}[f \mid \gamma = i] := \frac{1}{\mathbb{P}(\Omega_i)} \int_{\Omega_i} f \, d\mathbb{P},$$

then we get that

$$\int_{\gamma^{-1}(B)} f \, d\mathbb{P} = \int_B \mathbb{E}[f \mid \gamma = i] d\mathbb{P}_\gamma(i) \tag{10.5}$$

where $B \subseteq E := \{1, \ldots, n\}$ and $\mathbb{P}_\gamma$ is the image measure of $\mathbb{P}$ on $\mathcal{E} := 2^E$. In this case we get $\mathbb{P}_\gamma(\{i\}) = \mathbb{P}(\Omega_i)$. If one would like to extend formula (10.5) to more general cases one problem consists in giving an expression $\mathbb{E}[f \mid \gamma = y]$ the correct meaning when $\mathbb{P}(\gamma = y) = 0$, in other words, to solve the question whether we are allowed to divide by zero in (10.4). The setting we consider is as follows:

We consider a probability space $(\Omega, \mathcal{F}, \mathbb{P})$, $f \in \mathcal{L}_1(\Omega, \mathcal{F}, \mathbb{P})$, and a sub-$\sigma$-algebra $\mathcal{G} \subseteq \mathcal{F}$ obtained by a measurable map $\gamma : \Omega \to E$, where $(E, \mathcal{E})$ is a measurable space, i.e.

$$\mathcal{G} := \sigma(\gamma) = \{\gamma^{-1}(B) : B \in \mathcal{E}\}.$$

We let $\mathbb{P}_\gamma$ be the image measure of $\mathbb{P}$ with respect to γ and will give a sense to

$$\int_{\gamma^{-1}(B)} f \, d\mathbb{P} = \int_B \mathbb{E}[f \mid \gamma = y] d\mathbb{P}_\gamma(y)$$

in Proposition 10.4.4 below. To do so, we need two statements before. We start with a factorisation lemma, which is of independent interest:

Lemma 10.4.1 (Factorization Lemma, Doob[5]-Dynkin[6] Lemma) *Assume $\Omega \neq \emptyset$, a measurable space $(E, \mathcal{E})$, $\gamma : \Omega \to E$, and let $\mathcal{G}$ be the σ-algebra on Ω such that*

$$\mathcal{G} := \sigma(\gamma) = \{\gamma^{-1}(B) : B \in \mathcal{E}\}.$$

Then for a map $g : \Omega \to \mathbb{R}$ the following assertions are equivalent:

(1) *The map g is $(\mathcal{G}, \mathcal{B}(\mathbb{R}))$ measurable.*
(2) *There exists an $(\mathcal{E}, \mathcal{B}(\mathbb{R}))$-measurable $h : E \to \mathbb{R}$ such that $g = h \circ \gamma$.*

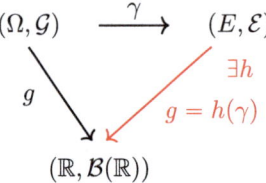

Proof
(2) $\Rightarrow$ (1) We have that γ is $(\sigma(\gamma), \mathcal{E})$ measurable and h is $(\mathcal{E}, \mathcal{B}(\mathbb{R}))$-measurable, so that the composition $g = h \circ \gamma$ is $(\sigma(\gamma), \mathcal{B}(\mathbb{R}))$-measurable.
(1) $\Rightarrow$ (2) We first assume that $\alpha_1, \ldots, \alpha_n \in \mathbb{R}$, $A_1, \ldots, A_n \in \sigma(\gamma)$, and

$$g = \sum_{i=1}^{n} \alpha_i \mathbb{1}_{A_i}.$$

We put $h := \sum_{i=1}^{n} \alpha_i \mathbb{1}_{B_i}$ where $B_i \in \mathcal{E}$ is such that $A_i = \gamma^{-1}(B_i)$. Then it holds that $h : E \to \mathbb{R}$ is $(\mathcal{E}, \mathcal{B}(\mathbb{R}))$-measurable and

$$g = \sum_{i=1}^{n} \alpha_i \mathbb{1}_{A_i} = \sum_{i=1}^{n} \alpha_i \mathbb{1}_{\gamma^{-1}(B_i)} = \sum_{i=1}^{n} \alpha_i \mathbb{1}_{B_i}(\gamma(\cdot)) = h \circ \gamma.$$

If $g \geq 0$, then we find simple functions g_n such that $0 \leq g_n(\omega) \uparrow g(\omega)$ for all $\omega \in \Omega$. For example, one can use $g_n := \sum_{k=0}^{4^n - 1} \frac{k}{2^n} \mathbb{1}_{g^{-1}([k/2^n, (k+1)/2^n))}$. For each n we construct an $(\mathcal{E}, \mathcal{B}(\mathbb{R}))$-measurable function $h_n : E \to [0, \infty)$ with $g_n = h_n \circ \gamma$ as above. Let $h : (E, \mathcal{E}) \to (\mathbb{R}, \mathcal{B}(\mathbb{R}))$ be defined by

$$h := \left(\sup_{n \in \mathbb{N}} h_n\right) \mathbb{1}_{\{\sup_{n \in \mathbb{N}} h_n < \infty\}}.$$

---

[5] Joseph Leo Doob, 27/02/1910 (Cincinnati, Ohio, USA)–07/06/2004 (Clark-Lindsey Village, Urbana, Illinois, USA).
[6] Eugene Borisovich Dynkin, 11/05/1924 (Leningrad, USSR, today St Petersburg, Russia)–14/11/2014 (Ithaca, New York, USA).

10.4 A Representation of Conditional Expectations

Then it holds that

$$h(\gamma(\omega)) = \left(\sup_{n\in\mathbb{N}} h_n(\gamma(\omega))\right) \mathbb{1}_{\{\sup_{n\in\mathbb{N}} h_n(\gamma(\omega))<\infty\}}$$

$$= \left(\sup_{n\in\mathbb{N}} g_n(\omega)\right) \mathbb{1}_{\{\sup_{n\in\mathbb{N}} g_n(\omega)<\infty\}} = g(\omega)\mathbb{1}_{\{g(\omega)<\infty\}} = g(\omega).$$

For a general g we write $g = g^+ - g^-$ with $g^+ := \max\{g, 0\}$ and $g^- := \max\{-g, 0\}$. We construct $(\mathcal{E}, \mathcal{B}(\mathbb{R}))$-measurable $h^+, h^- : E \to \mathbb{R}$ with $g^+ = h^+ \circ \gamma$ and $g^- = h^- \circ \gamma$ as before and obtain $g = (h^+ - h^-) \circ \gamma$. □

The next lemma applies the factorization lemma to conditional expectations:

Lemma 10.4.2 *Assume that $f \in \mathcal{L}_1(\Omega, \mathcal{F}, \mathbb{P})$ and let $(E, \mathcal{E})$ be a measurable space, and $\gamma : \Omega \to E$ be $(\mathcal{F}, \mathcal{E})$-measurable. Put*

$$\mathcal{G} := \sigma(\gamma) = \{\gamma^{-1}(B) : B \in \mathcal{E}\}.$$

Consider a version g of $\mathbb{E}[f \mid \mathcal{G}]$ and a factorization $g = h \circ \gamma$ according to Lemma 10.4.1. If g' is another version with a corresponding factorization $g' = h' \circ \gamma$, then

$$\mathbb{P}_\gamma(h = h') = 1,$$

where $\mathbb{P}_\gamma$ is the image measure of $\mathbb{P}$ with respect to γ.

Proof For $B \in \mathcal{E}$ we obtain by the change of variable formula Proposition 8.7.1 that

$$\mathbb{E}_{\mathbb{P}_\gamma}(h - h')\mathbb{1}_B = \int_\Omega (h - h')(\gamma(\omega))\mathbb{1}_B(\gamma(\omega))\mathrm{d}\mathbb{P}(\omega)$$

$$= \int_{\gamma^{-1}(B)} (\mathbb{E}[f \mid \mathcal{G}] - \mathbb{E}[f \mid \mathcal{G}])\mathrm{d}\mathbb{P} = 0$$

because $\gamma^{-1}(B) \in \mathcal{G}$. Testing with $B := \{y \in E : (h - h')(y) > 0\}$ and $B := \{y \in E : (h - h')(y) < 0\}$ leads to $h = h'$ $\mathbb{P}_\gamma$-a.s. □

Lemma 10.4.2 enables us to define the conditional probability we are interested in:

Definition 10.4.3 Assuming the setting as in Lemma 10.4.2 we let

$$\mathbb{E}[f \mid \gamma = y] := h(y) \quad \text{and} \quad \mathbb{P}[A \mid \gamma = y] := \mathbb{E}[\mathbb{1}_A \mid \gamma = y].$$

Again, it should be noted that $\mathbb{E}[f \mid \gamma = y]$ is only unique as equivalence class with respect to $\mathbb{P}_\gamma$. Now we obtain the general counterpart to relation (10.5) we were looking for:

Proposition 10.4.4 *For a set $B \in \mathcal{E}$ it holds that*

$$\int_{\gamma^{-1}(B)} f \, \mathrm{d}\mathbb{P} = \int_B \mathbb{E}[f \mid \gamma = y] \mathrm{d}\mathbb{P}_\gamma(y).$$

In particular, for $f = \mathbb{1}_A$ with $A \in \mathcal{F}$ we get that

$$\mathbb{P}(A \cap \gamma^{-1}(B)) = \int_B \mathbb{P}[A \mid \gamma = y] \mathrm{d}\mathbb{P}_\gamma(y).$$

Proof Because $\gamma^{-1}(B) \in \sigma(\gamma)$ and the definition of $\mathbb{E}[f \mid \gamma = y]$ we have that

$$\int_{\gamma^{-1}(B)} f \, \mathrm{d}\mathbb{P} = \int_{\gamma^{-1}(B)} \mathbb{E}[f \mid \sigma(\gamma)] \mathrm{d}\mathbb{P}$$

$$= \int_{\gamma^{-1}(B)} h \circ \gamma \, \mathrm{d}\mathbb{P}$$

$$= \int_B h(y) \mathrm{d}\mathbb{P}_\gamma(y)$$

$$= \int_B \mathbb{E}[f \mid \gamma = y] \mathrm{d}\mathbb{P}_\gamma(y).$$

□

With respect to densities we obtain the following representation:

Proposition 10.4.5 *Let $f : \Omega \to \mathbb{R}$ and $\gamma : \Omega \to \mathbb{R}^d$ be random variables whose joint distribution has a density $p_{(f,\gamma)}$, i.e. $p_{(f,\gamma)} : \mathbb{R}^{1+d} \to \mathbb{R}$ is $(\mathcal{B}(\mathbb{R}^{1+d}), \mathcal{B}(\mathbb{R}))$-measurable, nonnegative, and satisfies*

$$\mathbb{P}(\{\omega \in \Omega : (f(\omega), \gamma(\omega)) \in C\}) = \int_C p_{(f,\gamma)}(x, y) \mathrm{d}\lambda_{d+1}(x, y)$$

for $C \in \mathcal{B}(\mathbb{R}^{1+d})$. Then the following assertions hold:

(1) *The random variables f and γ have densities p_f and p_γ, respectively, given by*

$$p_f(x) := \int_{\mathbb{R}^d} p_{(f,\gamma)}(x, y) \mathrm{d}\lambda_d(y),$$

$$p_\gamma(y) := \int_{\mathbb{R}} p_{(f,\gamma)}(x, y) \mathrm{d}\lambda(x).$$

10.4 A Representation of Conditional Expectations

(2) If

$$p_{f|\gamma}(x|y) := \begin{cases} \frac{p_{(f,\gamma)}(x,y)}{p_\gamma(y)} & \text{if } p_\gamma(y) \neq 0 \\ 0 & \text{otherwise} \end{cases},$$

then for all $D \in \mathcal{B}(\mathbb{R})$ one has

$$\mathbb{P}[f \in D | \gamma = y] = \int_D p_{f|\gamma}(x|y) \, d\lambda(x) \quad \text{for} \quad \mathbb{P}_\gamma - \text{almost all } y \in \mathbb{R}^d.$$

Proof (1) is obvious. (2) First we observe that

$$\int_B \int_D p_{f|\gamma}(x|y) \, d\lambda(x) d\mathbb{P}_\gamma(y)$$

$$= \int_B \int_D p_{f|\gamma}(x|y) \, d\lambda(x) p_\gamma(y) d\lambda_d(y)$$

$$= \int_B \int_D p_{f|\gamma}(x|y) \, d\lambda(x) \mathbb{1}_{\{p_\gamma(y) > 0\}} p_\gamma(y) d\lambda_d(y)$$

$$= \int_B \int_D p_{(f,\gamma)}(x, y) \, d\lambda(x) \mathbb{1}_{\{p_\gamma(y) > 0\}} d\lambda_d(y)$$

$$= \int_B \int_D p_{(f,\gamma)}(x, y) \, d\lambda(x) d\lambda_d(y)$$

$$= \mathbb{P}(\{f \in D\} \cap \{\gamma \in B\}).$$

On the other hand, we have by Proposition 10.4.4 that

$$\mathbb{P}(f^{-1}(D) \cap \gamma^{-1}(B)) = \int_B \mathbb{P}[f \in D \mid \gamma = y] d\mathbb{P}_\gamma(y),$$

so that

$$\int_B \left[\int_D p_{f|\gamma}(x|y) \, d\lambda(x) \right] d\mathbb{P}_\gamma(y) = \int_B \mathbb{P}[f \in D \mid \gamma = y] d\mathbb{P}_\gamma(y)$$

for all $B \in \mathcal{B}(\mathbb{R}^d)$ which implies the assertion. □

We close this section with a generalization of Proposition 10.3.4(10), where in order to compute the conditional expectation of fg we can take out the $\mathcal{G}$-measurable factor g. If the connection between f and g is not given by a product but by a measurable function φ, under suitable integrability conditions, one can similarly compute the conditional expectation of $\varphi(f, g)$ provided that additionally to g being $\mathcal{G}$-measurable we have that f and $\mathcal{G}$ are independent.

Proposition 10.4.6 Assume a probability space $(\Omega, \mathcal{F}, \mathbb{P})$, a σ-algebra $\mathcal{G} \subseteq \mathcal{F}$, and random variables $f, g : \Omega \to \mathbb{R}$, where f is independent from $\mathcal{G}$ and g is $\mathcal{G}$-measurable. If $\varphi : \mathbb{R}^2 \to \mathbb{R}$ is $(\mathcal{B}(\mathbb{R}^2), \mathcal{B}(\mathbb{R}))$-measurable such that $\mathbb{E}|\varphi(f, g)| < \infty$ and $\mathbb{E}|\varphi(f, y)| < \infty$ for all $y \in \mathbb{R}$, then

$$\mathbb{E}[\varphi(f, g) \mid \mathcal{G}] = (\mathbb{E}\varphi(f, y))\big|_{y=g} \text{ a.s.}$$

Proof Applying the argument to prove Theorem 8.9.1(2) to $(\omega, y) \mapsto \varphi(f(\omega), y) \vee 0$ and $(\omega, y) \mapsto (-\varphi(f(\omega), y)) \vee 0$, and using that $\mathbb{E}|\varphi(f, y)| < \infty$ for all $y \in \mathbb{R}$, we get that the map $y \mapsto \mathbb{E}\varphi(f, y)$ is $(\mathcal{B}(\mathbb{R}), \mathcal{B}(\mathbb{R}))$-measurable. Now we have to prove that

$$\int_A \varphi(f, g) \mathrm{d}\mathbb{P} = \int_A (\mathbb{E}\varphi(f, y))\big|_{y=g} \mathrm{d}\mathbb{P}$$

for all $A \in \mathcal{G}$. To verify this equation, we start with a change of variables to get for $A \in \mathcal{G}$ that

$$\mathbb{E}(\varphi(f, g) \mathbb{1}_A) = \int_{\mathbb{R}^3} \varphi(x, y) z \mathrm{d}\mathbb{P}_{(f,g,\mathbb{1}_A)}(x, y, z).$$

Since g is $\mathcal{G}$-measurable and f is independent from $\mathcal{G}$ we get that f and $(g, \mathbb{1}_A)$ are independent, which means $\mathbb{P}_{(f,g,\mathbb{1}_A)} = \mathbb{P}_f \otimes \mathbb{P}_{(g,\mathbb{1}_A)}$. By Fubini's theorem and another change of variables,

$$\int_{\mathbb{R}^3} \varphi(x, y) z \mathrm{d}\mathbb{P}_{(f,g,\mathbb{1}_A)}(x, y, z) = \int_{\mathbb{R}^2} \left[\int_{\mathbb{R}} \varphi(x, y) \mathrm{d}\mathbb{P}_f(x)\right] z \mathrm{d}\mathbb{P}_{(g,\mathbb{1}_A)}(y, z)$$

$$= \int_{\mathbb{R}^2} [\mathbb{E}\varphi(f, y)] z \mathrm{d}\mathbb{P}_{(g,\mathbb{1}_A)}(y, z)$$

$$= \mathbb{E}\big[(\mathbb{E}\varphi(f, y))\big|_{y=g} \mathbb{1}_A\big]$$

which implies the assertion. □

10.5 Exercises

Ex 1: Assume $f, f_1, f_2, \ldots \in \mathcal{L}_1(\Omega, \mathcal{F}, \mathbb{P})$ and let $\mathcal{G} \subseteq \mathcal{F}$ be a sub-σ-algebra of $\mathcal{F}$.

(a) Prove Proposition 10.3.4(1): If $\alpha_1, \alpha_2 \in \mathbb{R}$, then

$$\mathbb{E}[\alpha_1 f_1 + \alpha_2 f_2 \mid \mathcal{G}] = \alpha_1 \mathbb{E}[f_1 \mid \mathcal{G}] + \alpha_2 \mathbb{E}[f_2 \mid \mathcal{G}] \text{ a.s.}$$

10.5 Exercises

(b) Prove Proposition 10.3.4(8): If for all $B \in \mathcal{B}(\mathbb{R})$ and all $A \in \mathcal{G}$ one has that

$$\mathbb{P}(\{f \in B\} \cap A) = \mathbb{P}(f \in B)\mathbb{P}(A),$$

then $\mathbb{E}[f \mid \mathcal{G}] = \mathbb{E}f$ a.s.

(c) Prove Proposition 10.3.4(9): If $0 \leqslant f_n \uparrow f$ a.s., then

$$\lim_n \mathbb{E}[f_n \mid \mathcal{G}] = \mathbb{E}[f \mid \mathcal{G}] \text{ a.s.}$$

(d) Derive from Proposition 10.3.4 that $\mathbb{E}f = \mathbb{E}\mathbb{E}[f|\mathcal{G}]$.

Ex 2: Assume that f and g are random variables on $(\Omega, \mathcal{F}, \mathbb{P})$ and $\mathcal{G} \subseteq \mathcal{F}$ is a sub σ-algebra such that f is independent from $\mathcal{G}$ and g is $\mathcal{G}$-measurable. Furthermore let f have uniform distribution on $[-1, 1]$. Prove that

$$\mathbb{E}[\sin(f+g)|\mathcal{G}] = \tfrac{1}{2}(\cos(g-1) - \cos(g+1)) \text{ a.s.}$$

Ex 3: For any $f, g \in L_2(\Omega, \mathcal{F}, \mathbb{P})$ define

$$\langle f, g \rangle := \mathbb{E}fg.$$

Then $\langle \, , \, \rangle : L_2(\Omega, \mathcal{F}, \mathbb{P}) \times L_2(\Omega, \mathcal{F}, \mathbb{P}) \to \mathbb{R}$ is the inner product of the Hilbert space $L_2(\Omega, \mathcal{F}, \mathbb{P})$. Let $\mathcal{G} \subseteq \mathcal{F}$ be a sub σ-algebra. Use Proposition 10.3.4 to show that the conditional expectation seen as map

$$f \mapsto \mathbb{E}[f \mid \mathcal{G}]$$

is the orthogonal projection from $L_2(\Omega, \mathcal{F}, \mathbb{P})$ onto $L_2(\Omega, \mathcal{G}, \mathbb{P})$. To be precise: We consider $L_2(\Omega, \mathcal{G}, \mathbb{P}) \subseteq L_2(\Omega, \mathcal{F}, \mathbb{P})$ in the sense that an equivalence class $[f] \in L_2(\Omega, \mathcal{F}, \mathbb{P})$ belongs to $L_2(\Omega, \mathcal{G}, \mathbb{P})$ if there is an $(\mathcal{G}, \mathcal{B}(\mathbb{R}))$-measurable representative $g \in [f]$.

Ex 4: Assume a probability space $(\Omega, \mathcal{F}, \mathbb{P})$ and independent $X, Y \in \mathcal{L}_0(\Omega, \mathcal{F}, \mathbb{P})$ such that the law of X and Y is the uniform distribution on $[0, 1]$. Let $f := X$, $g := X + Y$. In the notation of Proposition 10.4.5 compute $\mathbb{P}[f \in A | g = y]$ and $h_{f|g}(x|y)$.

Ex 5: Cantor set and devil's staircase: Let $C \subseteq [0, 1]$ denote the Cantor set, which can be generated as follows:

(1) $C_0 := [0, 1]$
(2) C_{n+1} is obtained from C_n by removing from each interval the middle third part (as an open interval) so that one gets

$$C_1 := [0, 1/3] \cup [2/3, 1],$$

$$C_2 := [0, 1/9] \cup [2/9, 3/9] \cup [6/9, 7/9] \cup [8/9, 1], \text{ and so on.}$$

(3) $C := \bigcap_{n \in \mathbb{N}} C_n$.

Problems

(a) Argue why we have $C \in \mathcal{B}([0, 1])$ and show that $\lambda(C) = 0$, where λ denotes the Lebesgue measure.

(b) Prove that $\#C = \#\mathbb{R}$.

   Hint: It is obvious that $\#C \leq \#\mathbb{R}$. For the other inequality describe the construction process by a sequence $(\varepsilon_1, \varepsilon_2, \ldots)$ with $\varepsilon_n \in \{-1, 1\}$ and use the cardinality of the set of all these sequences.

(c) Construct a continuous function $F : [0, 1] \to [0, 1]$ which is nondecreasing and which satisfies $F(0) = 0$ and $F(1) = 1$ as follows:

   - Step 1: On $[1/3, 2/3]$ take the value $1/2$.
   - Step 2: On $[1/9, 2/9]$ take the value $1/4$ and on $[7/9, 8/9]$ the value $3/4$.
   - In step n you get 2^{n-1} new intervals which get the values $k2^{-n}$ with $k = 1, 3, \ldots, 2^n - 1$.

   This function has a unique monotone extension to $[0, 1]$ and is called the Cantor-function or the **devil's staircase function.**

(d) Extend the constructed function F to a distribution function on $\mathbb{R}$ and consider the corresponding measure μ. Show that $\mu(C) = 1$, i.e. the Lebesgue-measure and μ are singular.

10.6 Comments

Section 10.1: The Hahn-Jordan decomposition Theorem 10.1.4 and Theorem 10.1.6 is due to Hahn [75, Chapters VI.1/2, Satz XIV, XV, and XVI]. Regarding the proof of Theorem 10.1.4 we followed the presentation in [16, Theorem 32.1].

Section 10.2: The general form of the Radon-Nikodym Theorem 10.2.1 was given by Nikodym [136]. As explained there, it is a generalization of Radon's result for the Euclidean space [151]. Daniell [43, page 445] proved a version for the Euclidean space as well.

Section 10.4: Lemma 10.4.1 can be found in, for example, in [4, Section II.11.7] or [97, p. 18]. Proposition 10.4.6 we learnt from [113, Proposition A.2.5].

Chapter 11
Fourier Transform and Gaussian Distributions

The concept of characteristic functions is a fundamental tool in probability theory. From the perspective of analysis it is the concept of Fourier transforms of measures. The idea is to describe properties of random variables $f : \Omega \to \mathbb{R}^d$ by the Fourier transform of their laws. This method is, at the same time, very elegant and extremely useful. Moreover, it will provide us with unexpected insights into probability theory.

After presenting basic properties of the Fourier transform, the Riemann-Lebesgue lemma and the uniqueness theorem are proven. As applications, the computation of moments of measures, the characterization of Gaussian random vectors, and the independence of random variables are discussed.

11.1 Preliminaries

For notation and basic facts about complex numbers the reader is referred to Appendix A.1. Let us begin by extending some concepts we did already work with:

(1) *Measurable maps:* For measurable spaces $(\Omega, \mathcal{F})$ and $(E, \mathcal{E})$ we denote by $\mathcal{L}_0(\Omega, \mathcal{F}; E)$ the space of all $(\mathcal{F}, \mathcal{E})$ measurable maps $f : \Omega \to E$, where we drop the σ-algebra $\mathcal{E}$ in the notation. In case the σ-algebra $\mathcal{F}$ is known from the context, we may drop it in the notation as well.

(2) *Borel σ-algebra:* If E is a real or complex normed space, then we use, if nothing to the contrary is stated, the following conventions: as σ-algebra we choose $\mathcal{E} := \mathcal{B}(E)$, the Borel σ-algebra generated by the norm-open sets from E, and for $E = \mathbb{R}^d$ and $E = \mathbb{C} \cong \mathbb{R}^2$ (with the identification from Appendix A.1) we use the Euclidean norm, i.e.

$$|x| := |(x_1, \ldots, x_d)| := (|x_1|^2 + \cdots + |x_d|^2)^{\frac{1}{2}} \quad \text{for} \quad x = (x_1, \ldots, x_d) \in \mathbb{R}^d.$$

A map $f = (f_1, \ldots, f_d) : \Omega \to \mathbb{R}^d$ is $(\mathcal{F}, \mathcal{B}(\mathbb{R}^d))$-measurable if and only if $f_1, \ldots, f_d : \Omega \to \mathbb{R}$ are $(\mathcal{F}, \mathcal{B}(\mathbb{R}))$-measurable. Indeed, if $f_1, \ldots, f_d$ are measurable, then the measurability of f can be verified by Lemma 6.2.3 by taking the generating system $\mathcal{G}$ of $\mathcal{B}(\mathbb{R}^d)$ consisting of all $B_1 \times \cdots \times B_d$ for $B_1, \ldots, B_d \in \mathcal{B}(\mathbb{R})$. Conversely, if f is measurable and $\pi_j : \mathbb{R}^d \to \mathbb{R}$ is the projection onto the j-th coordinate, then $f_j = \pi_j \circ f$ and as $\pi_j : \mathbb{R}^d \to \mathbb{R}$ is $(\mathcal{B}(\mathbb{R}^d), \mathcal{B}(\mathbb{R}))$-measurable, the composition is measurable as well.

(3) *Lebesgue spaces:* Given a measure space $(\Omega, \mathcal{F}, \mu)$, where either $E := \mathbb{C}$ or $E := \mathbb{R}^d$ with $d \in \mathbb{R}$ and the euclidean norm $|\cdot|$, and $p \in (0, \infty)$, we let $\mathcal{L}_p(\Omega, \mathcal{F}, \mu; E) = \mathcal{L}_p(\Omega, \mu; E)$ be the space of all $f \in \mathcal{L}_0(\Omega, \mathcal{F}; E)$ such that

$$\|f\|_{L_p(\Omega, \mu; E)} = \|f\|_{L_p} := \left(\int_\Omega |f|^p \mathrm{d}\mu \right)^{\frac{1}{p}} < \infty.$$

As in Definition 9.3.1 we may leave out E if $E = \mathbb{R}$, i.e. we use for example $\mathcal{L}_p(\Omega, \mathcal{F}, \mu) = \mathcal{L}_p(\Omega, \mathcal{F}, \mu; \mathbb{R})$. In case that $(\Omega, \mathcal{F}, \mu) = (\mathbb{R}^d, \mathcal{B}(\mathbb{R}^d), \lambda_d)$, we simply write $\mathcal{L}_p(\mathbb{R}^d; E)$ or even $\mathcal{L}_p(\mathbb{R}^d)$ if $E = \mathbb{R}$.

(4) *Independence:* For a probability space $(\Omega, \mathcal{F}, \mathbb{P})$ and a measurable space $(E, \mathcal{E})$ a family $(f_i)_{i \in I} \subseteq \mathcal{L}_0(\Omega, \mathcal{F}, \mathbb{P}; E)$ is called **independent** if the family of σ-algebras $(\sigma(f_i))_{i \in I}$ is independent, where $\sigma(f_i) := \{f_i^{-1}(B) : B \in \mathcal{E}\}$.

(5) *Integration:* Given a measure space $(\Omega, \mathcal{F}, \mu)$ and a measurable vector $f = (f_1, \ldots, f_d) : \Omega \to \mathbb{R}^d$, then $\int_\Omega f \mathrm{d}\mu := (\int_\Omega f_1 \mathrm{d}\mu, \ldots, \int_\Omega f_d \mathrm{d}\mu)$ where the component-wise integrals are supposed to exist. This applies to $\mathbb{C} \cong \mathbb{R}^2$ as well.

(6) *Integration and signed measures:* Assume a measurable space $(\Omega, \mathcal{F})$ and a (finite) signed measure μ. For a bounded and measurable $f : \Omega \to \mathbb{R}$ we let

$$\int_\Omega f \mathrm{d}\mu := \int_\Omega f \mathrm{d}\mu^+ - \int_\Omega f \mathrm{d}\mu^-$$

where $\mu^\pm$ are finite measures with $\mu = \mu^+ - \mu^-$. The integral $\int_\Omega f \mathrm{d}\mu$ does not depend on the particular choice of the decomposition $\mu = \mu^+ - \mu^-$. This extends to integrals over bounded and measurable vectors $f = (f_1, \ldots, f_d) : \Omega \to \mathbb{R}^d$.

(7) *Complex valued measurable maps:* Again using $\mathbb{C} \cong \mathbb{R}^2$ as in Appendix A.1, we apply the above concepts to complex valued maps to introduce for a measurable space $(\Omega, \mathcal{F})$ the following definitions:

   (a) A map $f : \Omega \to \mathbb{C}$ is measurable if $f : \Omega \to \mathbb{R}^2$ is $(\mathcal{F}, \mathcal{B}(\mathbb{R}^2))$-measurable.

   (b) Given a measure μ on $(\Omega, \mathcal{F})$ a measurable map $f : \Omega \to \mathbb{C}$ is integrable if

   $$\int_\Omega |f(\omega)| \mathrm{d}\mu(\omega) < \infty.$$

11.2 Definition and Basic Properties

In this case we let

$$\int_\Omega f(\omega)\mathrm{d}\mu(\omega) := \int_\Omega \mathrm{Re}(f(\omega))\mathrm{d}\mu(\omega) + i\int_\Omega \mathrm{Im}(f(\omega))\mathrm{d}\mathbb{P}(\omega).$$

We will make use of the following lemma, which seems to be obvious, but requires a proof:

Lemma 11.1.1 *Assume a measure space* (M, σ, μ). *If* $\psi \in \mathcal{L}_1(M, \mathcal{F}, \mu; \mathbb{C})$, *then one has*

$$\left|\int_M \psi \mathrm{d}\mu\right| \leqslant \int_M |\psi|\mathrm{d}\mu.$$

Proof For $\varepsilon > 0$ we obtain by Hölder's inequality Theorem 8.12.5 that

$$\left|\int_M \psi \mathrm{d}\mu\right|^2 = \left(\int_M \mathrm{Re}(\psi)\mathrm{d}\mu\right)^2 + \left(\int_M \mathrm{Im}(\psi)\mathrm{d}\mu\right)^2$$
$$\leqslant \left(\int_M \frac{|\mathrm{Re}(\psi)|^2}{|\psi|+\varepsilon}\mathrm{d}\mu + \int_M \frac{|\mathrm{Im}(\psi)|^2}{|\psi|+\varepsilon}\mathrm{d}\mu\right)\int_M [|\psi|+\varepsilon]\mathrm{d}\mu$$

and therefore

$$\left|\int_M \psi \mathrm{d}\mu\right|^2 \leqslant \int_M |\psi|\mathrm{d}\mu \int_M [|\psi|+\varepsilon]\mathrm{d}\mu.$$

By $\varepsilon \downarrow 0$ our assertion follows. $\square$

11.2 Definition and Basic Properties of Characteristic Functions

We start with introducing the main object of the chapter, the *Fourier[1]-transform of a measure* and the *characteristic function of a random variable*:

Definition 11.2.1 Given $d \in \mathbb{N}$, we denote the collection of *finite* signed measures on $(\mathbb{R}^d, \mathcal{B}(\mathbb{R}^d))$ (see Definition 10.1.1) by $\mathcal{M}(\mathbb{R}^d)$. The subset of all *finite* measures on $(\mathbb{R}^d, \mathcal{B}(\mathbb{R}^d))$ is denoted by $\mathcal{M}^+(\mathbb{R}^d)$, and the subset of probability measures by $\mathcal{M}_1^+(\mathbb{R}^d)$.

---

[1] Jean Baptiste Joseph Fourier, 21/03/1768 (Auxerre, Bourgogne, France)–16/05/1830 (Paris, France).

The notation '1' stands for $\mu(\mathbb{R}^d) = 1$ and '+' for $\mu(A) \geq 0$.

Definition 11.2.2

(1) For $\mu \in \mathcal{M}(\mathbb{R}^d)$ we let

$$\widehat{\mu}(x) := \int_{\mathbb{R}^d} e^{i\langle x, y \rangle} d\mu(y), \quad x \in \mathbb{R}^d,$$

where $\langle x, y \rangle = \sum_{k=1}^{d} x_k y_k$ and call $\widehat{\mu}$ **Fourier transform** of μ.

(2) Let $(\Omega, \mathcal{F}, \mathbb{P})$ be a probability space and $f \in \mathcal{L}_0(\Omega, \mathcal{F}, \mathbb{P}; \mathbb{R}^d)$. Then

$$\varphi_f(x) := \mathbb{E} e^{i\langle x, f \rangle} = \int_{\Omega} e^{i\langle x, f(\omega) \rangle} d\mathbb{P}(\omega), \quad x \in \mathbb{R}^d,$$

is called **characteristic function** of f.

Remark 11.2.3

(1) In harmonic analysis, typically the Fourier transform is defined with the exponent $-i\langle x, y \rangle$ instead of $i\langle x, y \rangle$, in particular the Fourier transform for functions like in Sect. 17.1 later. Obviously, the related mathematics does not change.

(2) The transforms $\widehat{\mu}$ and φ_f do exist, because $|e^{i\langle x,y \rangle}| = |e^{i\langle x,f \rangle}| = 1$ and $y \mapsto e^{i\langle x,y \rangle}$ is continuous, so that $y \mapsto e^{i\langle x,y \rangle}$ and $\omega \mapsto e^{i\langle x,f(\omega) \rangle}$ are measurable and bounded.

(3) If $\mathbb{P}_f$ is the law of $f : \Omega \to \mathbb{R}^d$, then $\varphi_f(x) = \widehat{\mathbb{P}_f}(x)$ for $x \in \mathbb{R}^d$. In fact, this follows from the change of variable formula Proposition 8.7.1 where we get, for $\psi(y) = e^{i\langle x,y \rangle}$,

$$\int_{\Omega} e^{i\langle x,f \rangle} d\mathbb{P} = \int_{\Omega} \psi(f) d\mathbb{P} = \int_{\mathbb{R}^d} \psi(y) d\mathbb{P}_f(y) = \int_{\mathbb{R}^d} e^{i\langle x,y \rangle} d\mathbb{P}_f(y).$$

(4) One clearly has $\widehat{\mu} + \widehat{\nu} = \widehat{\mu + \nu}$.

Let us consider some first examples:

Example 11.2.4

(a) *Dirac-measure:* Let $a \in \mathbb{R}^d$ and

$$\delta_a(B) := \begin{cases} 1 & : a \in B \\ 0 & : a \notin B \end{cases}.$$

Then

$$\widehat{\delta_a}(x) = \int_{\mathbb{R}^d} e^{i\langle x,y \rangle} d\delta_a(y) = e^{i\langle x,a \rangle} = \cos(\langle x, a \rangle) + i \sin(\langle x, a \rangle).$$

11.2 Definition and Basic Properties

(b) More generally, let $a_1, \ldots, a_n \in \mathbb{R}^d$, $\theta_1, \ldots, \theta_n \in \mathbb{R}$, and $\mu = \sum_{k=1}^n \theta_k \delta_{a_k}$. Then

$$\widehat{\mu}(x) = \sum_{k=1}^n \theta_k e^{i\langle x, a_k\rangle} = \sum_{k=1}^n \theta_k \big(\cos(\langle x, a_k\rangle) + i \sin(\langle x, a_k\rangle)\big),$$

which is a trigonometric polynomial.

(c) *Binomial distribution:* Let $0 < p < 1$, $d = 1$, $n \in \mathbb{N}$, and

$$\text{Bin}_{n,p}(\{k\}) := \binom{n}{k} p^k (1-p)^{n-k}, \quad \text{for } k = 0, \ldots, n.$$

Seeing $\text{Bin}_{n,p}$ as measure in $\mathcal{M}(\mathbb{R})$ by $\text{Bin}_{n,p} = \sum_{k=0}^n \binom{n}{k} p^k (1-p)^{n-k} \delta_k$, we get

$$\widehat{\text{Bin}_{n,p}}(x) = \int_{\mathbb{R}} e^{ixy} d\mu(y)$$

$$= \sum_{k=0}^n \binom{n}{k} p^k (1-p)^{n-k} e^{ixk}$$

$$= \sum_{k=0}^n \binom{n}{k} (pe^{ix})^k (1-p)^{n-k}$$

$$= \Big(pe^{ix} + (1-p)\Big)^n.$$

(d) *Poisson distribution:* Let $\lambda > 0$ and consider Pois_λ as measure in $\mathcal{M}(\mathbb{R})$ by $\text{Pois}_\lambda = \sum_{k=0}^\infty e^{-\lambda} \frac{\lambda^k}{k!} \delta_k$. Then we get

$$\widehat{\text{Pois}_\lambda}(x) = \sum_{k=0}^\infty e^{-\lambda} \frac{\lambda^k}{k!} e^{ixk} = e^{-\lambda} e^{\lambda e^{ix}} = e^{\lambda(e^{ix}-1)}.$$

(e) *Geometric distribution:* For $0 < p < 1$ we recall that $\text{Geom}_p = \sum_{k=0}^\infty (1-p)^k p \delta_k$ and consider this as a measure in $\mathcal{M}(\mathbb{R})$. Then

$$\widehat{\text{Geom}_p}(x) = \sum_{k=0}^\infty e^{ixk}(1-p)^k p = p \sum_{k=0}^\infty [e^{ix}(1-p)]^k = \frac{p}{1 - e^{ix}(1-p)}.$$

(f) *Exponential distribution:* For a parameter $\gamma > 0$ and $\text{Exp}_\gamma(A) = \int_A \mathbb{1}_{[0,\infty)}(x) e^{-\gamma x} d\lambda(x)$ we get that $\widehat{\text{Exp}_\gamma}(x) = \frac{\gamma}{\gamma - ix}$. This relation is the subject of Exercise 3.

(g) *Uniform distribution* on $[c,d]$ for $-\infty < c < d < \infty$, where the distribution is considered as measure in $\mathcal{M}(\mathbb{R})$ by $\mathcal{U}_{[c,d]}(B) = \frac{\lambda(B \cap [c,d])}{d-c}$: Here we get for $x \neq 0$ that

$$\widehat{\mathcal{U}_{[c,d]}}(x) = \frac{1}{d-c}\int_{[c,d]} e^{ixy}\,\mathrm{d}\lambda(y) = \frac{1}{d-c}\frac{e^{ixd} - e^{ixc}}{ix}.$$

We continue with some basic properties of the Fourier transform on $\mathcal{M}^+(\mathbb{R}^d)$ regarding affine transformations:

Proposition 11.2.5 *For $\mu, \nu \in \mathcal{M}^+(\mathbb{R}^d)$ one has the following:*
(1) *If $A = (a_{kl})_{k,l=1}^d : \mathbb{R}^d \to \mathbb{R}^d$ is a linear transformation, then*

$$\widehat{\mu_A}(x) = \widehat{\mu}(A^\mathrm{T} x),$$

where $\mu_A(B) := \mu(\{x \in \mathbb{R}^d : Ax \in B\})$ for $B \in \mathcal{B}(\mathbb{R}^d)$.
(2) *If $S_a : \mathbb{R}^d \to \mathbb{R}^d$ is the shift $S_a x := x + a$ by $a \in \mathbb{R}^d$, then*

$$\widehat{\mu_{S_a}} = \bar{\delta}_a \widehat{\mu}.$$

Proof (1) is Exercise 4 and (2) follows from

$$\widehat{\mu_{S_a}}(x) = \int_{\mathbb{R}^d} e^{i\langle x, y+a\rangle}\,\mathrm{d}\mu(y) = e^{i\langle x,a\rangle}\widehat{\mu}(x)$$

and Example 11.2.4. □

The Fourier transform maps a probability measure to a function from $\mathbb{R}^d$ into $\mathbb{C}$. If we are given a function from $\mathbb{R}^d$ into $\mathbb{C}$, how can we recognize whether this function is a Fourier transform of a probability measure? As we will see later, the next theorem provides the key to answer this question.

Theorem 11.2.6 *For $\mu, \mu_1, \mu_2 \in \mathcal{M}_1^+(\mathbb{R}^d)$ the following is true:*
(1) *The function $\widehat{\mu} : \mathbb{R}^d \to \mathbb{C}$ is uniformly continuous, i.e. for all $\varepsilon > 0$ there exists a $\delta > 0$ such that $|\widehat{\mu}(x) - \widehat{\mu}(y)| \leq \varepsilon$ whenever $|x - y| \leq \delta$.*
(2) *For all $x \in \mathbb{R}^d$ one has $|\widehat{\mu}(x)| \leq \widehat{\mu}(0) = 1$.*
(3) *The function $\widehat{\mu}$ is positive semidefinite, i.e. for all $x_1, \ldots, x_n \in \mathbb{R}^d$ and $\lambda_1, \ldots, \lambda_n \in \mathbb{C}$ it follows that*

$$\sum_{k,l=1}^n \lambda_k \bar{\lambda}_l \widehat{\mu}(x_k - x_l) \geq 0.$$

11.2 Definition and Basic Properties

Proof
(1) Let $\varepsilon > 0$. Choose a ball of radius $R > 0$ such that $\mu(\mathbb{R}^d \setminus B_R(0)) \leq \frac{\varepsilon}{3}$, where

$$B(0, R) := \left\{ x \in \mathbb{R}^d : |x| \leq R \right\}.$$

Take $\delta := \frac{\varepsilon}{3R}$. Then, since $|e^{i\alpha} - e^{i\beta}| \leq |\alpha - \beta|$, and if $|x_1 - x_2| \leq \delta$,

$$|\widehat{\mu}(x_1) - \widehat{\mu}(x_2)|$$

$$\leq \int_{B(0,R)} \left| e^{i\langle x_1, y\rangle} - e^{i\langle x_2, y\rangle} \right| d\mu(y) + \int_{\mathbb{R}^d \setminus B(0,R)} \left| e^{i\langle x_1, y\rangle} - e^{i\langle x_2, y\rangle} \right| d\mu(y)$$

$$\leq \int_{B(0,R)} |\langle x_1 - x_2, y\rangle| \, d\mu(y) + \int_{\mathbb{R}^d \setminus B(0,R)} 2 d\mu(y)$$

$$\leq |x_1 - x_2| \int_{B(0,R)} |y| d\mu(y) + 2\frac{\varepsilon}{3}$$

$$\leq \frac{\varepsilon}{3R} R + 2\frac{\varepsilon}{3} = \varepsilon.$$

(2) This part follows from Lemma 11.1.1 as

$$\left| \int_{\mathbb{R}^d} e^{i\langle x, y\rangle} d\mu(y) \right| \leq \int_{\mathbb{R}^d} |e^{i\langle x, y\rangle}| d\mu(y) = 1.$$

(3) Here we have that

$$\sum_{k,l=1}^n \lambda_k \bar{\lambda}_l \widehat{\mu}(x_k - x_l) = \sum_{k,l=1}^n \lambda_k \bar{\lambda}_l \int_{\mathbb{R}^d} e^{i\langle x_k - x_l, y\rangle} d\mu(y)$$

$$= \sum_{k,l=1}^n \lambda_k \bar{\lambda}_l \int_{\mathbb{R}^d} e^{i\langle x_k, y\rangle} e^{-i\langle x_l, y\rangle} d\mu(y).$$

Since $\overline{e^{i\alpha}} = \overline{\cos\alpha + i\sin\alpha} = \cos\alpha - i\sin\alpha = \cos(-\alpha) + i\sin(-\alpha) = e^{-i\alpha}$, we may write

$$\sum_{k,l=1}^n \lambda_k \bar{\lambda}_l \widehat{\mu}(x_k - x_l) = \int_{\mathbb{R}^d} \left[\sum_{k,l=1}^n \lambda_k \bar{\lambda}_l e^{i\langle x_k, y\rangle} \overline{e^{i\langle x_l, y\rangle}} \right] d\mu(y)$$

$$= \int_{\mathbb{R}^d} \left[\sum_{k=1}^n \lambda_k e^{i\langle x_k, y\rangle} \right] \left[\sum_{l=1}^n \bar{\lambda}_l \overline{e^{i\langle x_l, y\rangle}} \right] d\mu(y)$$

$$= \int_{\mathbb{R}^d} \left| \sum_{k=1}^n \lambda_k e^{i\langle x_k, y \rangle} \right|^2 d\mu(y) \geq 0.$$

□

Now we establish the connection between the Fourier transform for measures and the Fourier transform for functions $\psi : \mathbb{R}^d \to \mathbb{C}$. In the special case that $\psi \in \mathcal{L}_1(\mathbb{R}^d)$ takes only nonnegative values, the function ψ can be interpreted as the density of a measure.

Definition 11.2.7 For $\psi \in \mathcal{L}_1(\mathbb{R}^d; \mathbb{C})$ we define the **Fourier transform** $\widehat{\psi} : \mathbb{R}^d \to \mathbb{C}$ by

$$\widehat{\psi}(x) := \int_{\mathbb{R}^d} e^{i\langle x, y \rangle} \psi(y) d\lambda_d(y).$$

Proposition 11.2.8 *Let $\psi \in \mathcal{L}_1(\mathbb{R}^d; \mathbb{R})$ and define*

$$\mu(B) := \int_{\mathbb{R}^d} \mathbb{1}_B(x) \psi(x) d\lambda_d(x).$$

Then one has the following:

(1) $\mu \in M(\mathbb{R}^d)$ *with* $\mu = \mu^+ - \mu^-$, *where*

$$\mu^\pm(B) := \int_{\mathbb{R}^d} \mathbb{1}_B(x) \psi^\pm(x) d\lambda_d(x),$$

$\psi^+ := \max\{\psi, 0\}$, *and* $\psi^- := \max\{-\psi, 0\}$.
(2) $\widehat{\psi}(x) = \widehat{\mu}(x)$ *for all* $x \in \mathbb{R}^d$.

Proof (1) is obvious, cf. Proposition 10.1.3. (2) follows from

$$\widehat{\mu}(x) = \int_{\mathbb{R}^d} e^{i\langle x, y \rangle} d\mu^+(y) - \int_{\mathbb{R}^d} e^{i\langle x, y \rangle} d\mu^-(y)$$

$$= \int_{\mathbb{R}^d} e^{i\langle x, y \rangle} \psi^+(y) d\lambda(y) - \int_{\mathbb{R}^d} e^{i\langle x, y \rangle} \psi^-(y) d\lambda(y)$$

$$= \int_{\mathbb{R}^d} e^{i\langle x, y \rangle} \psi(y) d\lambda(y).$$

□

11.3 Convolutions of Measures

The problem to obtain information about the distribution of $f_1 + \cdots + f_n$ where $f_1, \ldots, f_n : \Omega \to \mathbb{R}^d$ are independent random variables occurs frequently, for example in central limit theorems and strong laws of large numbers. Now we introduce a concept especially designed for this purpose: the convolution.

Definition 11.3.1 (Convolution of Finite Measures) For finite measures $\mu_1, \ldots, \mu_n \in \mathcal{M}^+(\mathbb{R}^d)$ on $(\mathbb{R}^d, \mathcal{B}(\mathbb{R}^d))$ we define $\mu_1 * \cdots * \mu_n : \mathcal{B}(\mathbb{R}^d) \to [0, \infty)$ by

$$(\mu_1 * \cdots * \mu_n)(B) = (\mu_1 \otimes \cdots \otimes \mu_n)\left(\left\{(x_1, \ldots, x_n) \in (\mathbb{R}^d)^n : x_1 + \cdots + x_n \in B\right\}\right).$$

The map $\mu_1 * \cdots * \mu_n$ is called **convolution of** $\mu_1, \ldots, \mu_n$.

One reason for the above definition is the following relation:

Proposition 11.3.2 *Assume a probability space $(\Omega, \mathcal{F}, \mathbb{P})$ and independent random variables $f_1, \ldots, f_n \in \mathcal{L}_0(\Omega, \mathcal{F}, \mathbb{P}; \mathbb{R}^d)$. Then*

$$\mathbb{P}_{f_1} * \cdots * \mathbb{P}_{f_n} = \mathbb{P}_{f_1 + \cdots + f_n}.$$

Proof Analogously to Proposition 7.6.1, where real-valued random variables were considered, we have that by independence, the distribution of $(f_1, \ldots, f_n)$ is given by

$$\mathbb{P}_{(f_1, \ldots, f_n)} = \mathbb{P}_{f_1} \otimes \cdots \otimes \mathbb{P}_{f_n}.$$

Consequently, by a change of variables, for any $B \in \mathcal{B}(\mathbb{R}^d)$,

$$\mathbb{P}(f_1 + \cdots + f_n \in B)$$
$$= \mathbb{P}_{(f_1, \ldots, f_n)}\left(\left\{(x_1, \ldots, x_n) \in (\mathbb{R}^d)^n : x_1 + \cdots + x_n \in B\right\}\right)$$
$$= \mathbb{P}_{f_1} \otimes \cdots \otimes \mathbb{P}_{f_n}\left(\left\{(x_1, \ldots, x_n) \in (\mathbb{R}^d)^n : x_1 + \cdots + x_n \in B\right\}\right)$$
$$= \mathbb{P}_{f_1} * \cdots * \mathbb{P}_{f_n}(B).$$

□

We list some basic properties of the convolution.

Proposition 11.3.3 *For* $\mu, \mu_1, \mu_2, \mu_3 \in \mathcal{M}^+(\mathbb{R}^d)$ *the following holds:*

(1) $\mu_1 * \mu_2 \in \mathcal{M}(\mathbb{R}^d)$.
(2) $\mu_1 * \mu_2 = \mu_2 * \mu_1$.
(3) $\mu_1 * (\mu_2 * \mu_3) = (\mu_1 * \mu_2) * \mu_3 = \mu_1 * \mu_2 * \mu_3$.
(4) $(\delta_a * \mu)(B) = \mu(B - a)$.
(5) $\delta_0 * \mu = \mu$.
(6) $\delta_{a_1} * \cdots * \delta_{a_n} = \delta_{a_1 + \cdots + a_n}$ *for* $a_1, \ldots, a_n \in \mathbb{R}^d$.
(7) $\widehat{\mu_1 * \mu_2} = \widehat{\mu_1}\widehat{\mu_2}$.

Proof By a re-normalization we may assume for the proof that $\mu, \mu_1, \mu_2, \mu_3 \in \mathcal{M}_1^+(\mathbb{R}^d)$.

(1), (2), and (3) follow from Proposition 11.3.2 as we can find always independent random variables f_1, f_2, f_3 such that $\text{law}(f_j) = \mu_j$ for $j = 1, 2, 3$.
(4) By Fubini's theorem,

$$(\delta_a * \mu)(B) = (\delta_a \otimes \mu)(\{(x, y) : x + y \in B\})$$
$$= \int_{\mathbb{R}^d} \left[\int_{\mathbb{R}^d} \mathbb{1}_B(x + y) \mathrm{d}\delta_a(x) \right] \mathrm{d}\mu(y)$$
$$= \int_{\mathbb{R}^d} \mathbb{1}_B(a + y) \mathrm{d}\mu(y)$$
$$= \mu(B - a).$$

(5) and (6) follow directly from (4). (7) follows from

$$\widehat{\mu_1 * \mu_2}(x) = \int_{\mathbb{R}^d} \int_{\mathbb{R}^d} e^{i\langle x, y+z \rangle} \mathrm{d}\mu_1(y) \mathrm{d}\mu_2(z)$$
$$= \int_{\mathbb{R}^d} e^{i\langle x, y \rangle} \mathrm{d}\mu_1(y) \int_{\mathbb{R}^d} e^{i\langle x, z \rangle} \mathrm{d}\mu_2(z)$$
$$= \widehat{\mu_1}(x)\widehat{\mu_2}(x).$$

$\square$

11.4 Densities and Convolutions of Functions

Let us start with the notion of a density:

Definition 11.4.1 A measure $\mu \in \mathcal{M}^+(\mathbb{R}^d)$ has a **density** $\psi \in \mathcal{L}_1(\mathbb{R}^d)$ provided that $\psi(x) \geq 0$ for all $x \in \mathbb{R}^d$ and μ can be represented by

$$\mu(B) = \int_B \psi(x) \mathrm{d}\lambda_d(x) \quad \text{for all} \quad B \in \mathcal{B}(\mathbb{R}^d).$$

11.4 Densities and Convolutions of Functions

In other words, we have that $\mu \ll \lambda_d$, i.e. μ is absolutely continuous with respect to λ_d, with the Radon-Nikodym derivative

$$\psi = \frac{d\mu}{d\lambda_d}.$$

More generally, if we fix the 'reference measure' λ_d and allow signed measures, then we can interpret signed measures as 'generalized functions' by the formal embedding

$$J : \mathcal{L}_1(\mathbb{R}^d) \hookrightarrow \mathcal{M}(\mathbb{R}^d) \quad \text{given by} \quad \mu(B) := \int_B \psi(x) d\lambda_d(x)$$

for $B \in \mathcal{B}(\mathbb{R}^d)$.

In view of Proposition 11.3.2 the following question appears to be relevant: If we start with two probability measures μ_1 and μ_2 that have densities $p_1, p_2 \in \mathcal{L}_1(\mathbb{R}^d)$, does the convolution $\mu_1 * \mu_2$ have a density p and if *yes*, what is this density? In order to get an intuition for this density, we use Tonelli's theorem and compute, for $B \in \mathcal{B}(\mathbb{R}^d)$, that

$$(\mu_1 * \mu_2)(B) = (\mu_1 \otimes \mu_2)(\{(x_1, x_2) : x_1 + x_2 \in B\})$$

$$= \int_{\mathbb{R}^d} \left[\int_{\mathbb{R}^d} \mathbb{1}_B(x_1 + x_2) p_1(x_1) d\lambda_d(x_1) \right] p_2(x_2) d\lambda_d(x_2)$$

$$= \int_{\mathbb{R}^d} \left[\int_{\mathbb{R}^d} \mathbb{1}_B(x) p_1(x - x_2) d\lambda_d(x) \right] p_2(x_2) d\lambda_d(x_2)$$

$$= \int_{\mathbb{R}^d} \mathbb{1}_B(x) \left[\int_{\mathbb{R}^d} p_1(x - x_2) p_2(x_2) d\lambda_d(x_2) \right] d\lambda_d(x).$$

So our candidate is

$$p(x) := \int_{\mathbb{R}} p_1(x - y) p_2(y) d\lambda_d(y).$$

Let us make this precise and start with the definition of the convolution of two nonnegative functions:

Definition 11.4.2 (Convolution of Nonnegative Functions) Let $\psi_1, \psi_2 \in \mathcal{L}_0(\mathbb{R}^d, \mathcal{B}(\mathbb{R}^d))$ be nonnegative. Then $\psi_1 * \psi_2 : \mathbb{R}^d \to [0, \infty]$ with

$$(\psi_1 * \psi_2)(x) := \int_{\mathbb{R}^d} \psi_1(x - y) \psi_2(y) d\lambda_d(y)$$

is called **convolution** of ψ_1 and ψ_2.

This convolution has the following properties:

Proposition 11.4.3 *For nonnegative functions $\psi_1, \psi_2 \in \mathcal{L}_0(\mathbb{R}^d)$ one has:*

(1) $\psi_1 * \psi_2 : \mathbb{R}^d \to [0, \infty]$ *is an extended measurable map.*
(2) $\lambda_d(\{x \in \mathbb{R}^d : (\psi_1 * \psi_2)(x) = \infty\}) = 0$ *if* $\psi_1, \psi_2 \in \mathcal{L}_1(\mathbb{R}^d)$.
(3) $\int_{\mathbb{R}^d} (\psi_1 * \psi_2)(x) d\lambda_d(x) = \int_{\mathbb{R}^d} \psi_1(x) d\lambda_d(x) \int_{\mathbb{R}^d} \psi_2(x) d\lambda_d(x)$.

Proof Consider the map $g : \mathbb{R}^d \times \mathbb{R}^d \to \mathbb{R}$ defined by

$$g(x, y) := \psi_1(x - y)\psi_2(y).$$

We get a nonnegative measurable function $g : \mathbb{R}^d \times \mathbb{R}^d \to \mathbb{R}$ and can apply Tonelli's theorem. We observe that

$$\int_{\mathbb{R}^d} \int_{\mathbb{R}^d} g(x, y) d\lambda_d(x) d\lambda_d(y) = \int_{\mathbb{R}^d} \left[\int_{\mathbb{R}^d} \psi_1(x - y) \psi_2(y) d\lambda_d(x) \right] d\lambda_d(y)$$

$$= \int_{\mathbb{R}^d} \left[\int_{\mathbb{R}^d} \psi_1(x - y) d\lambda_d(x) \right] \psi_2(y) d\lambda^d(y)$$

$$= \int_{\mathbb{R}^d} \psi_2(y) d\lambda_d(y) \int_{\mathbb{R}^d} \psi_1(x) d\lambda_d(x)$$

implies (3). Moreover, (1) and (2) follow as a by-product from Tonelli's theorem, where (2) follows from $\int_{\mathbb{R}^d} (\psi_1 * \psi_2)(x) d\lambda_d(x) < \infty$ as a consequence of (3). □

We can extend this definition to integrable functions, where we have to be careful with null sets on which the convolution might not be defined:

Definition 11.4.4 (Convolution of $\mathcal{L}_1$-Functions) For $x \in \mathbb{R}^d$ and $\psi_1, \psi_2 \in \mathcal{L}_1(\mathbb{R}^d)$ we let $(\psi_1 * \psi_2)(x)$ be given by

$$\begin{cases} (\psi_1^+ * \psi_2^+)(x) - (\psi_1^+ * \psi_2^-)(x) - (\psi_1^- * \psi_2^+)(x) + (\psi_1^- * \psi_2^-)(x) & : x \in F, \\ 0 & : x \notin F \end{cases},$$

where F is the set of $x \in \mathbb{R}^d$ such that all terms

$$(\psi_1^+ * \psi_2^+)(x), \quad (\psi_1^+ * \psi_2^-)(x), \quad (\psi_1^- * \psi_2^+)(x), \quad (\psi_1^- * \psi_2^-)(x)$$

are finite.

From Proposition 11.4.3(2) one gets that $\lambda_d(\mathbb{R}^d \setminus F) = 0$ which makes Definition 11.4.4 meaningful in the sense that $\psi_1 * \psi_2$ is defined as a convolution as expected outside a set of measure zero.

Let us conclude by summarizing the relations between the convolutions of measures and functions, and the law of the sum of independent random variables:

Proposition 11.4.5

(1) If the probability measures $\mu_1, \mu_2 \in \mathcal{M}_1^+(\mathbb{R}^d)$ have densities $p_1, p_2 \in \mathcal{L}_1(\mathbb{R}^d)$, then $\mu_1 * \mu_2$ has the density $(p_1 * p_2)\mathbb{1}_{\{\psi_1 * \psi_2 < \infty\}}$.
(2) If $f_1, f_2 \in \mathcal{L}_0(\Omega, \mathcal{F}, \mathbb{P})$ are independent random variables such that the laws of f_1 and f_2 have the densities $p_1, p_2 \in \mathcal{L}_1(\mathbb{R}^d)$, then the law of $f_1 + f_2$ has the density $(p_1 * p_2)\mathbb{1}_{\{p_1 * p_2 < \infty\}}$.

11.5 The theorem of Riemann-Lebesgue and the Uniqueness Theorem

Let $(\Omega, \mathcal{F}, \mathbb{P})$ be a probability space and assume independent and identically distributed random variables $f_1, f_2, \ldots \in \mathcal{L}_2(\Omega, \mathcal{F}, \mathbb{P})$. In addition to the strong law of large numbers, we already discussed, the following different scaling of partial sums is of interest:

$$S_n := \frac{1}{\sqrt{n}}(f_1 + \cdots + f_n).$$

This scaling has the property that the variance is retained, i.e. $\mathrm{var}(S_n) = \mathrm{var}(f_1)$. Interpreting the variance as energy of a scaled random walk

$$B_t^n := \frac{1}{\sqrt{n}}(f_1 + \cdots + f_k) \quad \text{for} \quad t = \frac{k}{n} \quad \text{with} \quad k = 1, \ldots, n$$

after n steps, then we have some kind of a conservation property at time $t = 1$. We are interested in the convergence of S_n. The central idea is to use the characteristic function of S_n:

$$\varphi_{S_n}(x) = \mathbb{E}e^{ix\frac{1}{\sqrt{n}}(f_1 + \cdots + f_n)} = \mathbb{E}\prod_{j=1}^{n} e^{i\frac{x}{\sqrt{n}}f_j} = \left(\varphi_{f_1}\left(\frac{x}{\sqrt{n}}\right)\right)^n.$$

Here the $f_1, \ldots, f_n$ are independent and identically distributed. Now we may ask:

(Q1) Under which conditions does $(\varphi_{f_1}(\frac{x}{\sqrt{n}}))^n$ converge pointwise to a function $\varphi(x)$ as $n \to \infty$?
(Q2) If $(\varphi_{f_1}(\frac{x}{\sqrt{n}}))^n$ converges for any x, is the limit φ a characteristic function, i.e. does there exist a probability measure μ such that $\widehat{\mu} = \varphi$?
(Q3) If there is such a measure μ, what is its connection to the distribution of S_n?

Question (Q2) yields to Lévy's continuity theorem Theorem 12.1.3 and (Q3) to the notion of *weak convergence* introduced in Chap. 12. Prominent examples for (Q1) we will find in Sect. 15.1.

We begin with the first main result of this section, the lemma of Riemann and Lebesgue. For this we introduce the Banach space $[C_0(\mathbb{R}^d; \mathbb{C}), \|\cdot\|_{C_0}]$:

Definition 11.5.1 Let

$$C_0(\mathbb{R}^d; \mathbb{C}) := \left\{ \varphi : \mathbb{R}^d \to \mathbb{C} \text{ continuous and } \lim_{|x| \to \infty} |\varphi(x)| = 0 \right\}$$

with $\|\varphi\|_{C_0} := \sup_{x \in \mathbb{R}^d} |\varphi(x)|$.

Theorem 11.5.2 (Riemann-Lebesgue Lemma) *For all* $\psi \in \mathcal{L}_1(\mathbb{R}^d; \mathbb{C})$ *one has* $\widehat{\psi} \in C_0(\mathbb{R}^d; \mathbb{C})$.

Proof First of all we remark that by decomposing ψ into the positive and negative parts of the real and imaginary part (so that we have four parts) and applying Theorem 11.2.6 (which is obviously true for finite measures as well by renormalization) we get that $\widehat{\psi}$ is continuous. Next we recall that

$$\left| \widehat{\psi_1}(x) - \widehat{\psi_2}(x) \right| = \left| \int_{\mathbb{R}^d} [\psi_1(y) - \psi_2(y)] e^{i\langle x, y \rangle} \,\mathrm{d}\lambda_d(y) \right|$$
$$\leq \int_{\mathbb{R}^d} |\psi_1(y) - \psi_2(y)| \,\mathrm{d}\lambda_d(y) \qquad (11.1)$$

for $\psi_1, \psi_2 \in \mathcal{L}_1(\mathbb{R}^d; \mathbb{C})$. We define the set $E \subseteq \mathcal{L}_1(\mathbb{R}^d; \mathbb{C})$ as all finite linear combinations over $\mathbb{C}$ of indicator functions

$$\psi_0(x_1, \ldots, x_d) = \mathbb{1}_{(a_1, b_1)}(x_1) \cdots \mathbb{1}_{(a_d, b_d)}(x_d) \quad \text{for} \quad -\infty < a_k < b_k < \infty.$$

We obtain that $\widehat{\psi_0}(x_1, \ldots, x_d) = \widehat{\mathbb{1}}_{(a_1, b_1)}(x_1) \cdots \widehat{\mathbb{1}}_{(a_d, b_d)}(x_d)$ and

$$\widehat{\mathbb{1}}_{(a_k, b_k)}(x_k) = \int_{(a_k, b_k)} e^{i x_k y_k} \,\mathrm{d}\lambda(y_k) = \frac{e^{i x_k b_k} - e^{i x_k a_k}}{i x_k} \to 0 \quad \text{as} \quad |x_k| \to \infty.$$

Let us fix $\psi \in \mathcal{L}_1(\mathbb{R}^d; \mathbb{C})$. In Exercise 5 we show that there are $\psi_n \in E, n \in \mathbb{N}$, such that

$$\int_{\mathbb{R}^d} |\psi(y) - \psi_n(y)| \,\mathrm{d}\lambda_d(y) \leq \frac{1}{n}$$

11.5 Main theorems

so that $\sup_{x \in \mathbb{R}^d} |\widehat{\psi}(x) - \widehat{\psi_n}(x)| \leq \frac{1}{n}$ by (11.1). Then we get

$$\limsup_{|x| \to \infty} |\widehat{\psi}(x)| = \limsup_{|x| \to \infty} |\widehat{\psi}(x) - \widehat{\psi_n}(x)| \leq \frac{1}{n}.$$

As $n \in \mathbb{N}$ was arbitrary, $\lim_{|x| \to \infty} |\widehat{\psi}(x)| = 0$. □

Next we prepare with two lemmas the proof of the uniqueness theorem about the Fourier transform of probability measures:

Lemma 11.5.3 *The set* $\{\widehat{\psi} : \mathbb{R}^d \to \mathbb{C} : \psi \in \mathcal{L}_1(\mathbb{R}^d; \mathbb{C})\}$ *is dense in* $C_0(\mathbb{R}^d; \mathbb{C})$.

Proof Let $\mathcal{A} := \{\widehat{\psi} : \mathbb{R}^d \to \mathbb{C} : \psi \in \mathcal{L}_1(\mathbb{R}^d; \mathbb{C})\}$. Theorem 11.5.2 implies that $\mathcal{A} \subseteq C_0(\mathbb{R}^d; \mathbb{C})$. To verify the lemma we only need to check the conditions of the Stone-Weierstrass Theorem A.3.1. Given $\varepsilon > 0$ and $x \in \mathbb{R}^d$, we use

$$B(x, \varepsilon) := \{y \in \mathbb{R}^d : |x - y| \leq \varepsilon\}.$$

(1) is obvious as $\mathcal{L}_1(\mathbb{R}^d; \mathbb{C})$ is a linear space.

(2) follows from $\widehat{\psi_1 \psi_2} = \widehat{\psi_1} * \widehat{\psi_2}$ and

$$\|\psi_1 * \psi_2\|_{L_1(\mathbb{R}^d;\mathbb{C})} \leq \|\psi_1\|_{L_1(\mathbb{R}^d;\mathbb{C})} \|\psi_2\|_{L_1(\mathbb{R}^d;\mathbb{C})}.$$

(3) follows from

$$\overline{\widehat{\psi}(x)} = \overline{\int_{\mathbb{R}^d} e^{i\langle x, y \rangle} \psi(y) d\lambda_d(y)} = \int_{\mathbb{R}^d} e^{-i\langle x, y \rangle} \overline{\psi}(y) d\lambda_d(y) = \widehat{\overline{\psi}}(-x).$$

(4) For $x_0 \in \mathbb{R}^d$ and $\psi(x) := \mathbb{1}_{B(x_0, 1)}(x) e^{-i\langle x_0, x \rangle} \in \mathcal{L}_1(\mathbb{R}^d; \mathbb{C})$ we have

$$\widehat{\psi}(x_0) = \int_{\mathbb{R}^d} e^{i\langle x_0, x \rangle} \mathbb{1}_{B(x_0, 1)}(x) e^{-i\langle x_0, x \rangle} d\lambda_d(x) = \lambda_d(B(x_0, 1)) > 0.$$

(5) Let $x_0 \neq x_1$ and $z_0 := (x_0 - x_1)/|x_0 - x_1|^2$. Then $e^{i\langle x_0, z_0 \rangle} \neq e^{i\langle x_1, z_0 \rangle}$ because of $e^{i\langle x_0 - x_1, z_0 \rangle} = e^i \neq 1$ and $e^{i\langle x_0, z \rangle} \neq e^{i\langle x_1, z \rangle}$ for all $z \in B(z_0, \varepsilon)$ for some $\varepsilon > 0$. Hence we get for $\psi(x) := \mathbb{1}_{B(z_0, \varepsilon)}(x)(e^{-i\langle x_0, x \rangle} - e^{-i\langle x_1, x \rangle})$ that

$$\widehat{\psi}(x_0) - \widehat{\psi}(x_1) = \int_{B(z_0, \varepsilon)} \left[e^{-i\langle x_0, y \rangle} - e^{-i\langle x_1, y \rangle} \right] \left[e^{i\langle x_0, y \rangle} - e^{i\langle x_1, y \rangle} \right] d\lambda_d(y)$$

$$= \int_{B(z_0, \varepsilon)} \left| e^{i\langle x_0, y \rangle} - e^{i\langle x_1, y \rangle} \right|^2 d\lambda_d(y) > 0.$$

□

The second lemma is a simple approximation of open sets:

Lemma 11.5.4 *Let $B \subseteq \mathbb{R}^d$ be open. Then there exist continuous functions $0 \leq \varphi_1 \leq \varphi_2 \leq \cdots \leq \mathbb{1}_B$ with $\lim_{n \to \infty} \varphi_n(x) = \mathbb{1}_B(x)$ for $x \in \mathbb{R}^d$.*

Proof We can assume that $B \neq \mathbb{R}^d$ and $B \neq \emptyset$, and use $\varphi_n(x) := \min\{n\,d(x, B^c), 1\}$ with $d(x, B^c) := \inf\{|x - y| : y \in B^c\}$. □

Now we turn to the second main theorem in the section, the theorem about uniqueness which states that a probability measure is uniquely characterized by its Fourier transform:

Theorem 11.5.5 (Uniqueness) *Let $\mu, \nu \in \mathcal{M}_1^+(\mathbb{R}^d)$. Then $\mu = \nu$ if and only if $\widehat{\mu} = \widehat{\nu}$.*

Proof As it is clear that $\mu = \nu$ implies $\widehat{\mu} = \widehat{\nu}$ we only need to verify the other direction and assume that $\widehat{\mu} = \widehat{\nu}$. For any $\psi \in \mathcal{L}_1(\mathbb{R}^d; \mathbb{C})$ and $\mu \in \mathcal{M}_1^+(\mathbb{R}^d)$ we get by Fubini's theorem that

$$\int_{\mathbb{R}^d} \widehat{\mu}(x)\psi(x)\mathrm{d}\lambda_d(x) = \int_{\mathbb{R}^d} \left[\int_{\mathbb{R}^d} e^{i\langle x,y\rangle} \mathrm{d}\mu(y)\right] \psi(x)\mathrm{d}\lambda_d(x)$$

$$= \int_{\mathbb{R}^d} \left[\int_{\mathbb{R}^d} e^{i\langle y,x\rangle} \psi(x)\mathrm{d}\lambda_d(x)\right] \mathrm{d}\mu(y)$$

$$= \int_{\mathbb{R}^d} \widehat{\psi}(y)\mathrm{d}\mu(y).$$

Given $\varphi \in C_0(\mathbb{R}^d; \mathbb{C})$, we apply Lemma 11.5.3 and find $\psi_n \in \mathcal{L}_1(\mathbb{R}^d; \mathbb{C})$ such that $\|\varphi - \widehat{\psi}_n\|_{C_0} \xrightarrow[n]{} 0$. Then the previous computation implies that

$$\left| \int_{\mathbb{R}^d} \varphi(x)\mathrm{d}\mu(x) - \int_{\mathbb{R}^d} \varphi(x)\mathrm{d}\nu(x) \right|$$

$$= \left| \int_{\mathbb{R}^d} \left[\varphi(x) - \widehat{\psi}_n(x)\right] \mathrm{d}\mu(x) + \int_{\mathbb{R}^d} \widehat{\psi}_n(x)[\mathrm{d}\mu(x) - \mathrm{d}\nu(x)] \right.$$

$$\left. - \int_{\mathbb{R}^d} \left[\varphi(x) - \widehat{\psi}_n(x)\right] \mathrm{d}\nu(x) \right|$$

$$\leq 2\|\varphi - \widehat{\psi}_n\|_{C_0} + \left| \int_{\mathbb{R}^d} \widehat{\psi}_n(x)[\mathrm{d}\mu(x) - \mathrm{d}\nu(x)] \right|$$

$$= 2\|\varphi - \widehat{\psi}_n\|_{C_0} + \left| \int_{\mathbb{R}^d} [\widehat{\mu} - \widehat{\nu}](x)\psi_n(x)\mathrm{d}\lambda_d(x) \right|.$$

By $n \to \infty$ and $\widehat{\mu} = \widehat{\nu}$ we get

$$\int_{\mathbb{R}^d} \varphi(x) d\mu(x) = \int_{\mathbb{R}^d} \varphi(x) d\nu(x) \quad \text{for} \quad \varphi \in C_0(\mathbb{R}^d; \mathbb{C}).$$

By Lemma 11.5.4 and monotone convergence we deduce $\mu(B) = \nu(B)$ for all bounded open sets $B \subseteq \mathbb{R}^d$. With the monotonicity of the measures this extends to all open sets $B \subseteq \mathbb{R}^d$ which form a π-system containing $\mathbb{R}^d$. Now Theorem 3.2.4 implies $\mu = \nu$. □

The next corollary is a typical example how to code properties of a measure μ into its Fourier transform:

Corollary 11.5.6 *Let $\mu \in \mathcal{M}_1^+(\mathbb{R}^d)$. Then $\mu(B) = \mu(-B)$ for all $B \in \mathcal{B}(\mathbb{R}^d)$ if and only if $\widehat{\mu}(x) \in \mathbb{R}$ for all $x \in \mathbb{R}^d$.*

The proof is the subject of Exercise 6.

11.6 Moments of Measures

There are different types of moments of a measure $\mu \in \mathcal{M}^+(\mathbb{R}^d)$. For example, given integers $l_1, \ldots, l_d \geq 0$,

$$\int_{\mathbb{R}^d} \left| x_1^{l_1} \cdots x_d^{l_d} \right| d\mu(x_1, \ldots, x_d) \text{ is called } \textbf{absolute moment of order } (l_1, \ldots, l_d),$$

$$\int_{\mathbb{R}^d} x_1^{l_1} \cdots x_d^{l_d} d\mu(x_1, \ldots, x_d) \text{ is called } \textbf{moment of order } (l_1, \ldots, l_d).$$

We are interested in the second type and show that one can use the Fourier transform to compute these moments. The result, we are interested in, is the following:

Proposition 11.6.1 *Let $\mu \in \mathcal{M}^+(\mathbb{R}^d)$ and assume $k_1, \ldots, k_d \in \mathbb{N}_0$ such that for all integers $0 \leq l_j \leq k_j$ it holds that*

$$\int_{\mathbb{R}^d} \left| x_1^{l_1} \cdots x_d^{l_d} \right| d\mu(x_1, \ldots, x_d) < \infty.$$

Then, for all integers $l_j = 0, \ldots, k_j$, one has the following:

(1) $\frac{\partial^{l_1 + \cdots + l_d}}{\partial x_1^{l_1} \cdots \partial x_d^{l_d}} \widehat{\mu} : \mathbb{R}^d \to \mathbb{C}$ *is bounded and uniformly continuous.*

(2) $\frac{\partial^{l_1 + \cdots + l_d}}{\partial x_1^{l_1} \cdots \partial x_d^{l_d}} \widehat{\mu}(x) = i^{l_1 + \cdots + l_d} \int_{\mathbb{R}^d} e^{i\langle x, y \rangle} y_1^{l_1} \cdots y_d^{l_d} d\mu(y).$

(3) $\frac{\partial^{l_1 + \cdots + l_d}}{\partial x_1^{l_1} \cdots \partial x_d^{l_d}} \widehat{\mu}(0) = i^{l_1 + \cdots + l_d} \int_{\mathbb{R}^d} y_1^{l_1} \cdots y_d^{l_d} d\mu(y).$

For the proof of Proposition 11.6.1 we need the observation that under certain conditions a differential operator commutes with an integral:

Lemma 11.6.2 *Let* $(\Omega, \mathcal{F}, \mu)$ *be a measure space and let* $f = (\operatorname{Re}(f), \operatorname{Im}(f)) : (a, b) \times \Omega \to \mathbb{C}$ *with* $-\infty < a < b < \infty$ *be such that the following holds:*

(1) $\frac{\partial f}{\partial x}(\cdot, \omega) : (a, b) \to \mathbb{C}$ *is continuous for all* $\omega \in \Omega$ *with* $\frac{\partial f}{\partial x} := \left(\frac{\partial \operatorname{Re}(f)}{\partial x}, \frac{\partial \operatorname{Im}(f)}{\partial x} \right)$.
(2) $\frac{\partial f}{\partial x}(x, \cdot), f(x, \cdot) : \Omega \to \mathbb{C}$ *are* $(\mathcal{F}, \mathcal{B}(\mathbb{C}))$-*measurable for all* $x \in (a, b)$.
(3) *There is a* $g \in \mathcal{L}_1(\Omega, \mathcal{F}, \mu)$ *such that* $\left| \frac{\partial f}{\partial x}(x, \omega) \right| \leq g(\omega)$ *for* $(x, \omega) \in (a, b) \times \Omega$.
(4) $\int_\Omega |f(x, \omega)| d\mu(\omega) < \infty$ *for all* $x \in (a, b)$.

Then

$$\frac{\partial}{\partial x} \int_\Omega f(x, \omega) d\mu(\omega) = \int_\Omega \frac{\partial f}{\partial x}(x, \omega) d\mu(\omega).$$

Proof For $x, x + h \in (a, b)$ with $h \neq 0$ we have

$$\frac{\int_\Omega f(x+h, \cdot) d\mu - \int_\Omega f(x, \cdot) d\mu}{h} = \int_\Omega \frac{f(x+h, \cdot) - f(x, \cdot)}{h} d\mu. \quad (11.2)$$

By the mean-value theorem we have for $\tilde{f} = \operatorname{Re}(f)$ and $\tilde{f} = \operatorname{Im}(f)$ that

$$\left| \frac{\tilde{f}(x+h, \cdot) - \tilde{f}(x, \cdot)}{h} \right| \leq g \in \mathcal{L}_1(\Omega, \mathcal{F}, \mu).$$

Then the assertion follows by Lebesgue's theorem about dominated convergence Theorem 8.5.2 applied to (11.2) with f replaced by $\operatorname{Re}(f)$ and $\operatorname{Im}(f)$. □

Proof of Proposition 11.6.1

(a) It is obvious that (2) $\Rightarrow$ (3).
(b) For $l_1 = \cdots = l_d = 0$ items (2) and (1) follow from the definition of $\hat{\mu}$ and from Theorem 11.2.6.
(c) Now let us fix $j \in \{1, \ldots, d\}$ and $x_1, \ldots, x_{j-1}, x_{j+1}, \ldots, x_d \in \mathbb{R}$ and define $f : \mathbb{R} \times \mathbb{R}^d \to \mathbb{C}$ by

$$f(x_j, y) := e^{i \langle (x_1, \ldots, x_d), y \rangle}.$$

For $g(y) := |y_j|$ this implies

$$\left| \frac{\partial f}{\partial x_j}(x_j, y) \right| = |y_j| = g(y)$$

11.6 Moments of Measures

where by assumption $g \in \mathcal{L}_1(\mathbb{R}^d, \mathcal{B}(\mathbb{R}^d), \mu)$. Applying Lemma 11.6.2 with $(\Omega, \mathcal{F}, \mu) = (\mathbb{R}^d, \mathcal{B}(\mathbb{R}^d), \mu)$ and any $(a, b) \ni x_j$ we derive that

$$\frac{\partial}{\partial x_j} \int_{\mathbb{R}^d} e^{i\langle x, y \rangle} \, d\mu(y) = \int_{\mathbb{R}^d} \frac{\partial}{\partial x_j} e^{i\langle x, y \rangle} \, d\mu(y) = i \int_{\mathbb{R}^d} e^{i\langle x, y \rangle} y_j \, d\mu(y).$$

Now we define $d\nu_+(y) = \mathbb{1}_{\{y_j \geq 0\}} y_j d\mu(y)$ and $d\nu_-(y) = -\mathbb{1}_{\{y_j < 0\}} y_j d\mu(y)$ and obtain finite measures, so that

$$x \mapsto \int_{\mathbb{R}^d} e^{i\langle x, y \rangle} \, d\nu_\pm(y)$$

are uniformly continuous and bounded by Theorem 11.2.6 and (1) and (2) follow.

(d) Assume now that $l_j \in \{0, \ldots, k_j\}$ and that (1) and (2) have been proven for $(l_1, \ldots, l_d)$. Fix $j \in \{1, \ldots, d\}$ and assume that $l_j + 1 \leq k_j$. Define the finite measures

$$d\tilde{\mu}_+(y) := \mathbb{1}_{\{y_1^{l_1} \cdots y_d^{l_d} \geq 0\}} y_1^{l_1} \cdots y_d^{l_d} \, d\mu(y),$$

$$d\tilde{\mu}_-(y) := -\mathbb{1}_{\{y_1^{l_1} \cdots y_d^{l_d} < 0\}} y_1^{l_1} \cdots y_d^{l_d} \, d\mu(y),$$

where $0^0 := 1$. Using step (c) for $\tilde{\mu}_+$ and $\tilde{\mu}_-$ we deduce and (1) and (2) for $(l_1, \ldots, l_{j-1}, l_j + 1, l_{j+1}, \ldots, l_d)$. □

Example 11.6.3 Binomial distribution: For $0 < p < 1$, $d = 1$, and $n \in \{1, 2, \ldots\}$ we recall that

$$\operatorname{Bin}_{n,p}(\{k\}) = \binom{n}{k} p^k (1-p)^{n-k} \quad \text{for} \quad k = 1, \ldots, n.$$

By Example 11.2.4 this implies that $\widehat{\operatorname{Bin}_{n,p}}(x) = \left(pe^{ix} + (1-p)\right)^n$ so that

$$\widehat{\operatorname{Bin}_{n,p}}'(x) = n\left(pe^{ix} + (1-p)\right)^{n-1} pie^{ix} \quad \text{so that} \quad \frac{\widehat{\operatorname{Bin}_{n,p}}'(0)}{i} = np.$$

11.7 Gaussian Distribution on $\mathbb{R}^d$

The Gaussian distribution with mean $m \in \mathbb{R}$ and variance $\sigma^2 > 0$ was given by

$$\mathcal{N}_{m,\sigma^2}(B) = \int_B e^{-\frac{(x-m)^2}{2\sigma^2}} \frac{1}{\sqrt{2\pi\sigma^2}} d\lambda(x) \quad \text{for} \quad B \in \mathcal{B}(\mathbb{R}).$$

If we take the product measure

$$\gamma_d := \otimes_{i=1}^d \mathcal{N}_{0,1}$$

on $(\mathbb{R}^d, \mathcal{B}(\mathbb{R}^d))$, then we obtain a first natural extension to $\mathbb{R}^d$, the standard Gaussian distribution γ_d on $\mathbb{R}^d$. In this section we will define that a measure $\mu \in \mathcal{M}_1^+(\mathbb{R}^d)$ is Gaussian provided that it is an affine image of γ_d. This approach does not require the existence of a density - actually there are Gaussian measures on $\mathbb{R}^d$ without density. Before we start with this approach we recall the case $d = 1$ and compute the corresponding Fourier transform. Here we use the notation

$$\gamma := \gamma_1 = \mathcal{N}_{0,1}.$$

Theorem 11.7.1 *For $m \in \mathbb{R}$ and $\sigma > 0$ one has:*

(1) $\mathcal{N}_{m,\sigma^2} \in \mathcal{M}_1^+(\mathbb{R})$.
(2) $\int_\mathbb{R} x d\mathcal{N}_{m,\sigma^2}(x) = m$ is the mean.
(3) $\int_\mathbb{R} (x-m)^2 d\mathcal{N}_{m,\sigma^2}(x) = \sigma^2$ is the variance.
(4) $\widehat{\mathcal{N}}_{m,\sigma^2}(x) = e^{imx} e^{-\frac{1}{2}\sigma^2 x^2}$.

Proof (1), (2), and (3) were the subject of Corollary 8.10.2. (4) First we show that one has that

$$\widehat{\gamma}(x) = e^{-\frac{x^2}{2}}.$$

By definition

$$\widehat{\gamma}(x) = \int_\mathbb{R} e^{ixy} e^{-\frac{y^2}{2}} \frac{d\lambda(y)}{\sqrt{2\pi}} = \int_\mathbb{R} \cos(xy) e^{-\frac{y^2}{2}} \frac{d\lambda(y)}{\sqrt{2\pi}}.$$

Using Lemma 11.6.2 this gives that

$$\widehat{\gamma}'(x) = -\int_\mathbb{R} \sin(xy) y e^{-\frac{y^2}{2}} \frac{d\lambda(y)}{\sqrt{2\pi}} = -x \int_\mathbb{R} \cos(xy) e^{-\frac{y^2}{2}} \frac{d\lambda(y)}{\sqrt{2\pi}} = -x\widehat{\gamma}(x).$$

11.7 Gaussian Distribution on $\mathbb{R}^d$

by integration by parts since we can interpret the integrals as Riemann integrals. For $y \geqslant 0$ we let $h(y) := e^y \widehat{\gamma}(\sqrt{2y})$, so that, for $y > 0$,

$$h'(y) = e^y \sqrt{2} \frac{1}{2} \frac{1}{\sqrt{y}} \widehat{\gamma}'(\sqrt{2y}) + e^y \widehat{\gamma}(\sqrt{2y}) = 0.$$

Since $h(0) = 1$, we deduce $h(y) = 1$ for $y \geqslant 0$ and $\widehat{\gamma}(x) = e^{-\frac{x^2}{2}}$ for $x \geqslant 0$ and by symmetry of $\widehat{\gamma}(x)$ for all $x \in \mathbb{R}$. Now we can finish the proof as

$$\begin{aligned}
\widehat{\mathcal{N}}_{m,\sigma^2}(x) &= \int_{\mathbb{R}} e^{ixy} e^{-\frac{1}{2}\frac{(y-m)^2}{\sigma^2}} \frac{d\lambda(y)}{\sqrt{2\pi\sigma^2}} \\
&= \int_{\mathbb{R}} e^{ix(\sigma z + m)} e^{-\frac{z^2}{2}} \frac{d\lambda(z)}{\sqrt{2\pi}} \\
&= e^{ixm} \widehat{\gamma}(x\sigma) \\
&= e^{imx} e^{-\frac{1}{2}\sigma^2 x^2}.
\end{aligned}$$

$\square$

For the following it is convenient to agree about

$$\mathcal{N}_{m,0} = \delta_m \in \mathcal{M}_1^+(\mathbb{R})$$

is the Dirac measure in $m \in \mathbb{R}$, i.e. we have a Gaussian distribution with variance $\sigma^2 = 0$.

Definition 11.7.2

(1) A matrix $R = (r_{kl})_{k,l=1}^d \in \mathbb{R}^{d \times d}$ is called **symmetric** provided that $r_{kl} = r_{lk}$ for all $k, l = 1, \ldots, d$.
(2) A matrix $R = (r_{kl})_{k,l=1}^d \in \mathbb{R}^{d \times d}$ is called **positive semidefinite** provided that

$$\langle Rx, x \rangle = \sum_{k,l=1}^d r_{kl} x_k x_l \geqslant 0 \quad \text{for all} \quad x = (x_1, \ldots, x_d) \in \mathbb{R}^d.$$

Now we prove an equivalence that will lead us to the definition of Gaussian measures in Definition 11.7.4:

Theorem 11.7.3 *For $\mu \in \mathcal{M}_1^+(\mathbb{R}^d)$ the following assertions are equivalent:*

(1) *Characterization as affine image: There exist a vector $a \in \mathbb{R}^d$ and a matrix $A = (a_{kl})_{k,l=1}^d$ such that μ is the image measure of γ_d with respect to the affine map $G(x) := a + Ax$, i.e.*

$$\mu(B) = \gamma_d\left(\{x \in \mathbb{R}^d : G(x) \in B\}\right) \quad \text{for all} \quad B \in \mathcal{B}(\mathbb{R}^d).$$

(2) *Characterization by Fourier transform:* There exist a vector $m \in \mathbb{R}^d$ and a positive semidefinite and symmetric matrix $R = (r_{kl})_{k,l=1}^d$ such that

$$\widehat{\mu}(x) = e^{i\langle x,m\rangle - \frac{1}{2}\langle Rx,x\rangle}.$$

(3) *Characterization by functionals:* For all $b = (b_1, \ldots, b_d) \in \mathbb{R}^d$ the law of $L_b : \mathbb{R}^d \to \mathbb{R}$, $L_b(x) := \langle x, b\rangle$, with respect to $(\mathbb{R}^d, \mathcal{B}(\mathbb{R}^d), \mu)$ is a Gaussian measure on $(\mathbb{R}, \mathcal{B}(\mathbb{R}))$.

In particular, we have that a, m, and R are unique and that

(a) $m_k = \int_{\mathbb{R}^d} x_k \mathrm{d}\mu(x)$,
(b) $r_{kl} = \int_{\mathbb{R}^d} (x_k - m_k)(x_l - m_l)\mathrm{d}\mu(x)$,
(c) $m = a$ and $R = AA^\mathrm{T}$.

Before we prove Theorem 11.7.3 we turn its equivalent assertions into a definition:

Definition 11.7.4 (Gaussian Measures on $\mathbb{R}^d$)

(1) A measure $\mu \in \mathcal{M}_1^+(\mathbb{R}^d)$ is called **Gaussian measure** with mean $m \in \mathbb{R}^d$ and (a symmetric and positive semidefinite) covariance $R \in \mathbb{R}^{d \times d}$ if

$$\widehat{\mu}(x) = e^{i\langle x,m\rangle - \frac{1}{2}\langle Rx,x\rangle}.$$

The measure is denoted by $\mathcal{N}_{m,R}$.

(2) A Gaussian measure $\mu \in \mathcal{M}_1^+(\mathbb{R}^d)$ with covariance $R \in \mathbb{R}^{d \times d}$ is called **nondegenerated** if $\mathrm{rank}(R) = d$, and **degenerated** if $\mathrm{rank}(R) < d$.

Proof of Theorem 11.7.3
(1) $\Rightarrow$ (2) follows from

$$\begin{aligned}
\widehat{\mu}(x) &= \int_{\mathbb{R}^d} e^{i\langle x,y\rangle}\mathrm{d}\mu(y) = \int_{\mathbb{R}^d} e^{i\langle x,a+Ay\rangle}\mathrm{d}\gamma_d(y) \\
&= e^{i\langle x,a\rangle}\int_{\mathbb{R}^d} e^{i\langle A^\mathrm{T} x,y\rangle}\mathrm{d}\gamma_d(y) = e^{i\langle x,a\rangle}\widehat{\gamma_d}\left(A^\mathrm{T} x\right) \\
&= e^{i\langle x,a\rangle}e^{-\frac{1}{2}\langle A^\mathrm{T} x, A^\mathrm{T} x\rangle} = e^{i\langle x,a\rangle}e^{-\frac{1}{2}\langle AA^\mathrm{T} x,x\rangle}
\end{aligned}$$

so that $R = AA^\mathrm{T}$ and $m = a$, where we have used that

$$\begin{aligned}
\widehat{\gamma_d}(x) &= \int_{\mathbb{R}^d} e^{i\langle x,y\rangle}\mathrm{d}\gamma_d(y) \\
&= \int_\mathbb{R} e^{ix_1 y_1}\mathrm{d}\gamma(y_1)\cdots\int_\mathbb{R} e^{ix_d y_d}\mathrm{d}\gamma(y_d) \\
&= e^{-\frac{x_1^2}{2}}\cdots e^{-\frac{x_d^2}{2}} = e^{-\frac{1}{2}\langle x,x\rangle}.
\end{aligned}$$

11.7 Gaussian Distribution on $\mathbb{R}^d$

(2) $\Rightarrow$ (3) We compute the Fourier transform of law(L_b): For $x \in \mathbb{R}$ we obtain

$$\widehat{\text{law}(L_b)}(x) = \int_{\mathbb{R}} e^{ixz} \text{dlaw}(L_b)(z)$$

$$= \int_{\mathbb{R}^d} e^{ix\langle b, y\rangle} d\mu(y) = \widehat{\mu}(xb)$$

$$= e^{i\langle xb, m\rangle - \frac{1}{2}\langle Rxb, xb\rangle} = e^{ix\langle b, m\rangle - \frac{1}{2}x^2 \langle Rb, b\rangle}.$$

But this is the Fourier transform of a Gaussian measure on $\mathbb{R}$ with mean $\langle b, m\rangle$ and variance $\langle Rb, b\rangle$.

(3) $\Rightarrow$ (1) We define $a \in \mathbb{R}^d$ and the symmetric matrix $R' = (r'_{kl})_{k,l=1}^d \in \mathbb{R}^{d\times d}$ by

$$a = (a_k)_{k=1}^d := \left(\int_{\mathbb{R}^d} y_k d\mu(y)\right)_{k=1}^d,$$

$$r'_{kl} := \int_{\mathbb{R}^d} (y_k - a_k)(y_l - a_l) d\mu(y).$$

The integrals exist because for $\pi_k : \mathbb{R}^d \to \mathbb{R}$ with $\pi_k((y_1, \ldots, y_d)) := y_k$ the image measures of μ with respect to the π_k are supposed to be Gaussian. The matrix R' is positive semidefinite because for $x \in \mathbb{R}^d$ we have

$$\langle R'x, x\rangle = \sum_{k,l=1}^d r'_{kl} x_k x_l$$

$$= \sum_{k,l=1}^d \int_{\mathbb{R}^d} (y_k - a_k)(y_l - a_l) d\mu(y) x_k x_l$$

$$= \int_{\mathbb{R}^d} \left(\sum_{k=1}^d (y_k - a_k) x_k\right)^2 d\mu(y)$$

$$= \int_{\mathbb{R}^d} |\langle y - a, x\rangle|^2 d\mu(y) \geq 0.$$

It is known from linear algebra that there is a matrix $A \in \mathbb{R}^{d\times d}$ such that $R' = AA^T$. Now we consider the affine map $G : \mathbb{R}^d \to \mathbb{R}^d$ with

$$G(x) := a + Ax.$$

Let ν be the image measure of γ_d with respect to G. From (1) $\Rightarrow$ (2) we know that

$$\widehat{\nu}(x) = e^{i\langle x, a\rangle} e^{-\frac{1}{2}\langle R'x, x\rangle}.$$

In order to verify (3) ⇒ (1) we show $\mu = \nu$ and do this by the uniqueness Theorem 11.5.5 by proving

$$\widehat{\mu}(x) = e^{i\langle x,a\rangle} e^{-\frac{1}{2}\langle R'x,x\rangle}.$$

By assumption (3) and Theorem 11.7.1, for all $b \in \mathbb{R}^d$ there are $m_b \in \mathbb{R}$ and $\sigma_b \geq 0$ such that

$$\int_{\mathbb{R}^d} e^{ix\langle b,y\rangle} d\mu(y) = e^{im_b x - \frac{1}{2}\sigma_b^2 x^2},$$

$$m_b = \int_{\mathbb{R}^d} \langle b, y\rangle \, d\mu(y),$$

$$\sigma_b^2 = \int_{\mathbb{R}^d} |\langle b, y\rangle - m_b|^2 d\mu(y).$$

For these m_b and σ_b we also get, in terms of a and R',

$$m_b = \int_{\mathbb{R}^d} \langle b, y\rangle \, d\mu(y) = \langle b, a\rangle,$$

$$\sigma_b^2 = \int_{\mathbb{R}^d} |\langle b, y - a\rangle|^2 d\mu(y) = \langle R'b, b\rangle.$$

Now we conclude by

$$\widehat{\mu}(x) = \widehat{\text{law}(L_x)}(1) = e^{im_x} e^{-\frac{1}{2}\sigma_x^2} = e^{i\langle x,m\rangle} e^{-\frac{1}{2}\langle R'x,x\rangle}.$$

Finally, (a) and (b) follow from Proposition 11.6.1 (where for (b) one may use $\int_{\mathbb{R}^d}(x_k - m_k)(x_l - m_l)d\mu(x) = \int_{\mathbb{R}^d} x_k x_l d\mu(x) - m_k m_l$), which also yields that m and R are unique in item (2). Moreover, the proof of the implication (1) ⇒ (2) verifies $m = a$ and $R = AA^T$. □

Now we return to the case that the Gaussian measure is nondegenerated and observe that we have a density in this case:

Proposition 11.7.5 *Assume that $\mu \in \mathcal{M}_1^+(\mathbb{R}^d)$ is a nondegenerated Gaussian measure with mean $m \in \mathbb{R}^d$ and covariance matrix $R \in \mathbb{R}^{d \times d}$. Then one has that*

$$\mu(B) = \int_B e^{-\frac{1}{2}\langle R^{-1}(x-m),x-m\rangle} \frac{d\lambda_d(x)}{(2\pi)^{\frac{d}{2}} |\det R|^{\frac{1}{2}}}.$$

The proof is the subject of Exercise 11.

11.8 Gaussian $\mathbb{R}^d$-Valued Random Variables

Now we translate the concept from Sect. 11.7 into the language of random variables:

Definition 11.8.1 A random variable $f : \Omega \to \mathbb{R}^d$ on a probability space $(\Omega, \mathcal{F}, \mathbb{P})$ is called **Gaussian** provided that the law of f is a Gaussian measure on $\mathbb{R}^d$. A random variable $f : \Omega \to \mathbb{R}^d$ on a probability space $(\Omega, \mathcal{F}, \mathbb{P})$ is called **standard normal distributed (or standard Gaussian)** provided that the law of f is γ_d.

Reformulating Theorem 11.7.3 yields to the following twin:

Theorem 11.8.2 *For a probability space $(\Omega, \mathcal{F}, \mathbb{P})$ and a random variable $f = (f_1, \ldots, f_d) : \Omega \to \mathbb{R}^d$ the following assertions are equivalent:*

(1) *There exist an $a \in \mathbb{R}^d$, a matrix $A = (a_{kl})_{k,l=1}^d$ and a standard normally distributed random variable $g : \Omega \to \mathbb{R}^d$ such that f and $a + Ag$ have the same distribution:*

$$f \stackrel{d}{=} a + Ag.$$

(2) *Characterization by characteristic function: There exist $m \in \mathbb{R}^d$ and a positive semidefinite symmetric matrix $R = (r_{kl})_{k,l=1}^d$ such that*

$$\varphi_f(x) = e^{i\langle x,m\rangle - \frac{1}{2}\langle Rx,x\rangle} \quad \text{for all} \quad x \in \mathbb{R}^d.$$

(3) *Characterization by functionals: For all $b_1, \ldots, b_d \in \mathbb{R}$ the random variable*

$$\sum_{k=1}^d b_k f_k : \Omega \to \mathbb{R}$$

is Gaussian.

11.9 A Characterization of Independent Random Variables

In Corollary 7.5.1 we proved the existence of independent random variables, now we prove a characterization of independence.

Theorem 11.9.1 *Let $f_1, \ldots, f_d \in \mathcal{L}_0(\Omega, \mathcal{F}, \mathbb{P})$ be random variables. Then the following assertions are equivalent:*

(1) $f_1, \ldots, f_d$ *are independent.*
(2) $\varphi_{(f_1,\ldots,f_d)}(x) = \varphi_{f_1}(x_1) \ldots \varphi_{f_d}(x_d)$ *for all* $x = (x_1 \ldots, x_d) \in \mathbb{R}^d$.

Proof
(1) $\Rightarrow$ (2) For $h : \Omega \to \mathbb{R}^d$ with $h(\omega) := (f_1(\omega), \ldots, f_d(\omega))$ we get

$$\varphi_h(x) = \int_\Omega e^{i(x_1 f_1(\omega) + \cdots + x_d f_d(\omega))} d\mathbb{P}(\omega)$$
$$= \int_\Omega e^{i x_1 f_1(\omega)} d\mathbb{P}(\omega) \cdots \int_\Omega e^{i x_d f_d(\omega)} d\mathbb{P}(\omega) = \varphi_{f_1}(x_1) \cdots \varphi_{f_d}(x_d),$$

where we use Corollary 8.10.4.

(2) $\Rightarrow$ (1) On the product space $(\bigtimes_{i=1}^d \Omega, \bigotimes_{i=1}^d \mathcal{F}, \bigotimes_{i=1}^d \mathbb{P})$ we define $H : \bigtimes_{i=1}^d \Omega \to \mathbb{R}^d$ by

$$H(\omega_1, \ldots \omega_d) := (f_1(\omega_1), \ldots, f_d(\omega_d)).$$

Then the coordinates of H are independent by construction and

$$\varphi_H(x_1, \ldots, x_d) = \varphi_{f_1}(x_1) \cdots \varphi_{f_d}(x_d) = \varphi_h(x_1, \ldots, x_d),$$

where the last equality is our assumption that (2) holds. Therefore the laws of H and h coincide which implies that

$$\mathbb{P}(f_1 \in B_1, \ldots, f_d \in B_d) = \mathbb{P}(h \in B_1 \times \cdots \times B_d)$$
$$= (\mathbb{P} \otimes \cdots \otimes \mathbb{P})(H \in B_1 \times \cdots \times B_d)$$
$$= \mathbb{P}(f_1 \in B_1) \cdots \mathbb{P}(f_d \in B_d).$$

□

As an application of Theorem 11.9.1 we derive that uncorrelated components of a Gaussian random vector are independent.

Definition 11.9.2 The random variables $f_1, \ldots f_d \in \mathcal{L}_2(\Omega, \mathcal{F}, \mathbb{P})$ are called **uncorrelated**, provided that $\mathbb{E}(f_k - \mathbb{E}f_k)(f_l - \mathbb{E}f_l) = 0$ for all $k, l = 1, \ldots, d$ with $k \neq l$.

If $f_1, \ldots f_d \in \mathcal{L}_2(\Omega, \mathcal{F}, \mathbb{P})$ are independent, then they are uncorrelated as by Corollary 8.10.4 we have that

$$\mathbb{E}(f_k - \mathbb{E}f_k)(f_l - \mathbb{E}f_l) = [\mathbb{E}(f_k - \mathbb{E}f_k)][\mathbb{E}(f_l - \mathbb{E}f_l)] = 0$$

for all $k, l = 1, \ldots, d$ with $k \neq l$. The converse fails to be true in general. But there is a remarkable converse for random variables $f_1, \ldots, f_d$ such that $(f_1, \ldots, f_d)$ forms a Gaussian random vector. We emphasize that the law$(f_1, \ldots, f_d) \in \mathcal{M}_1^+(\mathbb{R}^d)$ needs to be Gaussian, it is not sufficient that only $f_1, \ldots, f_n : \Omega \to \mathbb{R}$ are Gaussian as maps $f_k : \Omega \to \mathbb{R}$ (see Exercise 10).

Theorem 11.9.3 Let $f_1, \ldots, f_d \in \mathcal{L}_0(\Omega, \mathcal{F}, \mathbb{P})$ be random variables such that $(f_1, \ldots, f_d) : \Omega \to \mathbb{R}^d$ is a Gaussian random variable. Then the following assertions are equivalent:

(1) $f_1, \ldots, f_d$ are uncorrelated.
(2) $f_1, \ldots, f_d$ are independent.

Proof We only have to check (1) $\Rightarrow$ (2). We know from Theorem 11.7.3 that for $x = (x_1, \ldots, x_d)$ one has

$$\varphi_{(f_1,\ldots,f_d)}(x_1, \ldots, x_d) = e^{i(x_1 \mathbb{E} f_1 + \ldots + x_d \mathbb{E} f_d) - \frac{1}{2} \langle Rx, x \rangle},$$

with $R = (r_{k,l})_{k,l=1}^d$ and

$$r_{k,l} = \mathbb{E}(f_k - \mathbb{E} f_k)(f_l - \mathbb{E} f_l) = 0 \quad \text{for} \quad k \neq l.$$

Consequently,

$$\varphi_{(f_1,\ldots,f_d)}(x_1, \ldots, x_d) = e^{ix_1 \mathbb{E} f_1 - \frac{1}{2} x_1^2 \text{var}(f_1)} \ldots e^{ix_d \mathbb{E} f_d - \frac{1}{2} x_d^2 \text{var}(f_d)}$$
$$= \varphi_{f_1}(x_1) \ldots \varphi_{f_d}(x_d).$$

Applying Theorem 11.9.1 we get the independence of $f_1, \ldots, f_d$. □

11.10 Exercises

Ex 1: Compute $\mu_1 * \mu_2$, where μ_i is the uniform distribution on $[a_i, a_i + 1] \subseteq \mathbb{R}$, $a_i \in \mathbb{R}$, for $i = 1, 2$.

Ex 2: From Exercise 1 we conclude that for any $a > 0$ the convolution $\mathcal{U}_{[-a/2, a/2]} * \mathcal{U}_{[-a/2, a/2]}$ has a 'tent-shaped' density

$$h(x) = \begin{cases} \frac{a - |x|}{a^2} & |x| \leq a \\ 0 & |x| > a \end{cases}.$$

Compute $\widehat{\mathcal{U}}_{[-a/2, a/2]}$ and show that $\mu = \mathcal{U}_{[-a/2, a/2]} * \mathcal{U}_{[-a/2, a/2]}$ has the Fourier transform

$$\widehat{\mu}(t) = 4 \frac{\sin(at/2)^2}{a^2 t^2} = 2 \frac{1 - \cos(at)}{a^2 t^2} \quad \text{for} \quad t \neq 0.$$

Ex 3: For a parameter $\gamma > 0$ and the exponential distribution $\text{Exp}_\gamma(A) = \int_A \mathbb{1}_{[0,\infty)}(x) e^{-\gamma x} d\lambda(x)$, $A \in \mathcal{B}(\mathbb{R})$, prove that $\widehat{\text{Exp}_\gamma}(x) = \frac{\gamma}{\gamma - ix}$.

Hint: Use the relation $e^{ixy} = \cos(xy) + i\sin(xy)$ and the oscillatory integrals from Lemma A.4.1.

Ex 4: Prove Proposition 11.2.5(1): Assume $\mu \in \mathcal{M}^+(\mathbb{R}^d)$. Let $A = (a_{kl})_{k,l=1}^d : \mathbb{R}^d \to \mathbb{R}^d$ be a linear transformation. Show that

$$\widehat{\mu_A}(x) = \widehat{\mu}(A^T x),$$

where $\mu_A(B) := \mu(\{x \in \mathbb{R}^d : Ax \in B\})$ for $B \in \mathcal{B}(\mathbb{R}^d)$.

Ex 5: Finish the proof of Theorem 11.5.2: Show that the linear combinations of indicator functions $g(x_1, \ldots, x_d) = \mathbb{1}_{(a_1,b_1)}(x_1) \cdots \mathbb{1}_{(a_d,b_d)}(x_d)$ for $-\infty < a_k < b_k < \infty$ are dense in $L_1(\mathbb{R}^d; \mathbb{C})$.

Ex 6: Prove Corollary 11.5.6: Assume $\mu \in \mathcal{M}_1^+(\mathbb{R}^d)$. Prove that $\mu(B) = \mu(-B)$ for all $B \in \mathcal{B}(\mathbb{R}^d)$ if and only if $\widehat{\mu}(x) \in \mathbb{R}$ for all $x \in \mathbb{R}^d$.

Ex 7: Find random variables $f_1, f_2 : \Omega \to \mathbb{R}$ on some probability space $(\Omega, \mathcal{F}, \mathbb{P})$ that are uncorrelated, but not independent.

Ex 8: Let $(\Omega, \mathcal{F}, \mathbb{P})$ be a probability space. Do there exist random variables $f_1, f_2 : \Omega \to \mathbb{R}$ such that $\mathbb{P}_{f_1}$ and $\mathbb{P}_{f_2}$ are Gaussian measures, but the distribution $\mathbb{P}_{(f_1, f_2)}$ of the vector (f_1, f_2) is not a Gaussian measure?
Hint: Show that the function

$$h(x, y) := \frac{1}{\pi} e^{-\frac{x^2+y^2}{2}} \mathbb{1}_{xy \geq 0}(x, y)$$

is a density. Could it be used as density of (f_1, f_2)?

Ex 9: Let $f : \Omega \to \mathbb{R}$ be standard normal distributed and define

$$g = f \mathbb{1}_{\{|f| \leq 1\}} - f \mathbb{1}_{\{|f| > 1\}}.$$

Show that g is Gaussian, but (f, g) is not Gaussian.
Hint: Note that $|f + g| \leq 1$.

Ex 10: Let $f, g, h_0 : \Omega \to \mathbb{R}$ be i.i.d. random variables with law $\mathcal{N}(0, 1)$. Define

$$h := |h_0| \mathrm{sgn}(fg).$$

Show that

(a) h is standard normal distributed,
(b) f, g, h are pairwise independent,
(c) f, g, h are not independent.

Ex 11: Prove Proposition 11.7.5: Assume that $\mu \in \mathcal{M}_1^+(\mathbb{R}^d)$ is a nondegenerated Gaussian measure with mean $m \in \mathbb{R}^d$ and covariance matrix $R \in \mathbb{R}^{d \times d}$. Then one has that

$$\mu(B) = \int_B e^{-\frac{1}{2}\langle R^{-1}(x-m), x-m \rangle} \frac{d\lambda_d(x)}{(2\pi)^{\frac{d}{2}} |\det R|^{\frac{1}{2}}}.$$

11.11 List of Characteristic Functions

Distribution	Fourier transform	
Dirac measure δ_a, $a \in \mathbb{R}^d$	$e^{i\langle x,a \rangle}$	$x \in \mathbb{R}^d$
Binomial distribution $\mathrm{Bin}_{n,p} = \sum_{k=0}^{n} \binom{n}{k} p^k (1-p)^{n-k} \delta_k$ $0 < p < 1, n \in \mathbb{N}$	$\left(p e^{ix} + (1-p) \right)^n$	$x \in \mathbb{R}$
Poisson distribution $\mathrm{Pois}_\lambda = \sum_{k=0}^{\infty} e^{-\lambda} \frac{\lambda^k}{k!} \delta_k$, $0 < p < 1$	$e^{\lambda(e^{ix}-1)}$	$x \in \mathbb{R}$
Geometric distribution $\mathrm{Geom}_p = \sum_{k=0}^{\infty} (1-p)^k p \delta_k$, $0 < p < 1$	$\frac{p}{1-e^{ix}(1-p)}$	$x \in \mathbb{R}$
Exponential distribution $\mathrm{dExp}_\gamma(x) = \mathbb{1}_{[0,\infty)}(x) \gamma e^{-\gamma x} \mathrm{d}\lambda(x)$, $\gamma > 0$	$\frac{\gamma}{\gamma - ix}$	$x \in \mathbb{R}$
Uniform distribution $\mathrm{d}\mathcal{U}_{[c,d]}(x) = \frac{\mathbb{1}_{[c,d]}(x)}{d-c} \mathrm{d}\lambda(x)$, $-\infty < c < d < \infty$	$\frac{1}{d-c} \frac{e^{ixd} - e^{ixc}}{ix}$	$x \in \mathbb{R}$
Gaussian distribution mean $m \in \mathbb{R}^d$, covariance $R \in \mathbb{R}^{d \times d}$	$e^{i\langle x,m \rangle - \frac{1}{2} \langle Rx, x \rangle}$	$x \in \mathbb{R}^d$
Cauchy distribution $\mathrm{d}\mu_c(x) = \frac{c}{\pi} \frac{1}{c^2 + x^2} \mathrm{d}\lambda(x)$, $c > 0$	$e^{-c|x|}$	$x \in \mathbb{R}$

Chapter 12
Weak Convergence

So far almost sure convergence, convergence in probability, and convergence in L_p of random variables were considered. All these types of convergence require that the convergent sequence of random variables is defined on the same probability space. This is relaxed by the weak convergence, where one only considers the convergence of the laws of the random variables. Main results about weak convergence, like the so-called Portmanteau theorem, Levy's continuity theorem, Prokhorov's theorem, and the continuous mapping theorem are proven. Finally it is shown that the theorem of Dudley and Skorohod provides a way back to the notion of almost sure convergence.

First we will formulate the main results of this section and after that we turn to their proofs.

12.1 The Main Theorems

The weak convergence is based on the so-called *Portmanteau theorem* which provides a concept of the convergence of measures. This concept is called *weak convergence* as we test the measures by integrating continuous and bounded functions. The *Portmanteau theorem* leads to Definition 12.1.2 below.

Theorem 12.1.1 (Portmanteau Theorem) *Let $\mu, \mu_1, \mu_2, \ldots \in \mathcal{M}_1^+(\mathbb{R}^d)$. Then the following assertions are equivalent.*

(1) *For all continuous and bounded functions $\varphi : \mathbb{R}^d \to \mathbb{R}$ one has that $\int_{\mathbb{R}^d} \varphi(x) d\mu_n(x) \xrightarrow[n]{} \int_{\mathbb{R}^d} \varphi(x) d\mu(x)$.*
(2) *For all closed sets $A \in \mathcal{B}(\mathbb{R}^d)$ one has $\limsup_{n \to \infty} \mu_n(A) \leq \mu(A)$.*
(3) *For all open sets $B \in \mathcal{B}(\mathbb{R}^d)$ one has $\liminf_{n \to \infty} \mu_n(B) \geq \mu(B)$.*
(4) $\widehat{\mu}_n(x) \xrightarrow[n]{} \widehat{\mu}(x)$ *for all $x \in \mathbb{R}^d$.*

(5) If $d = 1$ and if $F_n(x) := \mu_n((-\infty, x])$ and $F(x) := \mu((-\infty, x])$, then $F_n(x) \xrightarrow[n]{} F(x)$ for all points $x \in \mathbb{R}$ of continuity of F.

Based on the previous theorem we define the notion of *weak convergence* and of *convergence in distribution*:

Definition 12.1.2

(1) For $\mu, \mu_1, \mu_2, \ldots \in \mathcal{M}_1^+(\mathbb{R}^d)$, we say that μ_n **converges weakly to** μ ($\mu_n \Rightarrow \mu$ or $\mu_n \xrightarrow{w} \mu$) provided that the conditions of Theorem 12.1.1 are satisfied.

(2) Let $f_n : \Omega_n \to \mathbb{R}^d$, $n \in \mathbb{N}$, and $f : \Omega \to \mathbb{R}^d$ be random variables defined on the probability spaces $(\Omega_n, \mathcal{F}_n, \mathbb{P}_n)$ and $(\Omega, \mathcal{F}, \mathbb{P})$, respectively. Then f_n converges to f **weakly** or **in distribution** ($f_n \xrightarrow{d} f$) if the corresponding laws $\mu_n(B) = \mathbb{P}_n(f_n \in B)$ and $\mu(B) = \mathbb{P}(f \in B)$ converge weakly.

The next theorem, Lévy's *continuity theorem*, is a statement about the completeness of weak convergence, i.e. the existence of a limit measure μ of a sequence $(\mu_n)_{n=1}^\infty$:

Theorem 12.1.3 (Levy's Continuity Theorem) *Assume $(\mu_n)_{n=1}^\infty \subseteq \mathcal{M}_1^+(\mathbb{R}^d)$ such that $(\widehat{\mu}_n)_{n=1}^\infty$ converges pointwise to $\varphi : \mathbb{R}^d \to \mathbb{C}$ which is continuous at $x = 0$. Then there is a $\mu \in \mathcal{M}_1^+(\mathbb{R}^d)$ such that $\varphi = \widehat{\mu}$.*

Finally, Prokhorov's theorem provides the connection between *tightness* and *sequentially relative compactness* with respect to the weak convergence.

Definition 12.1.4 (Tightness) A nonempty set of measures $M \subseteq \mathcal{M}_1^+(\mathbb{R}^d)$ is called **tight** provided that for all $\varepsilon > 0$ there is a compact set $K \subseteq \mathbb{R}^d$ with

$$\mu(\mathbb{R}^d \setminus K) \leqslant \varepsilon \quad \text{for all} \quad \mu \in M.$$

A simple example of a set of measures that is not tight is $M := \{\delta_{x_n} : n \in \mathbb{N}\}$ for $x_n \in \mathbb{R}^d$ with $\lim_{n \to \infty} |x_n| = \infty$. For such a sequence one also observes that

$$\lim_{n \to \infty} \int_{\mathbb{R}^d} \varphi(x) d\delta_{x_n}(x) = \lim_{n \to \infty} \varphi(x_n) = 0$$

for all continuous $\varphi : \mathbb{R}^d \to \mathbb{R}$ with compact support. Therefore there cannot exist a weak limit in $\mathcal{M}_1^+(\mathbb{R}^d)$ of the sequence $(\delta_{x_n})_{n \in \mathbb{N}}$: In fact, assuming a weak limit $\mu \in \mathcal{M}_1^+(\mathbb{R}^d)$, we would get that $\int_{\mathbb{R}^d} \varphi(x) d\mu(x) = 0$ for all continuous $\varphi : \mathbb{R}^d \to \mathbb{R}$ with compact support, which would imply that $\mu = 0$. This observation yields to a characterization of tightness, known as the theorem of Prokhorov:

12.1 The Main Theorems

Theorem 12.1.5 (Prokhorov[1]'s Theorem) *For a nonempty subset $M \subseteq \mathcal{M}_1^+(\mathbb{R}^d)$ the following assertions are equivalent:*

(1) *M is tight.*
(2) *For all sequences $(\mu_n)_{n=1}^\infty$ there is a subsequence $(\mu_{n_k})_{k=1}^\infty$ and some $\mu \in \mathcal{M}_1^+(\mathbb{R}^d)$ such that $\mu_{n_k} \Rightarrow \mu$ as $k \to \infty$.*

Before we prove the main theorems in Sect. 12.2 we give some corollaries. We start with the uniqueness of a weak limit:

Corollary 12.1.6 (Uniqueness of the Weak Limit) *Let $\mu_1, \mu_2, \ldots \in \mathcal{M}_1^+(\mathbb{R}^d)$. Assume that there are $\mu, \nu \in \mathcal{M}_1^+(\mathbb{R}^d)$ such that $\mu_n \Rightarrow \mu$ and $\mu_n \Rightarrow \nu$. Then $\mu = \nu$.*

Proof If $\mu_n \Rightarrow \mu$ and $\mu_n \Rightarrow \nu$, then by Theorem 12.1.1 $\hat{\mu}_n(x) \to \hat{\mu}(x)$ and $\hat{\mu}_n(x) \to \hat{\nu}(x)$ for all $x \in \mathbb{R}^d$. This implies $\hat{\mu}(x) = \hat{\nu}(x)$ for all $x \in \mathbb{R}^d$, so that $\mu = \nu$ by Theorem 11.5.5. □

The next corollary is the counterpart to the *Continuous Mapping Theorem* for the convergence in probability (Theorem 9.2.5):

Corollary 12.1.7 (Continuous Mapping Theorem) *Assume random variables $f_n \in \mathcal{L}_0(\Omega_n, \mathcal{F}_n, \mathbb{P}_n; \mathbb{R}^d)$, $n \in \mathbb{N}$, and $f \in \mathcal{L}_0(\Omega, \mathcal{F}, \mathbb{P}; \mathbb{R}^d)$ with $f_n \xrightarrow{d} f$ and a continuous function $H : \mathbb{R}^d \to \mathbb{R}^N$ for some $d, N \in \mathbb{N}$. Then $H(f_n) \xrightarrow{d} H(f)$.*

Proof Assume a continuous and bounded function $\varphi : \mathbb{R}^N \to \mathbb{R}$. Then the composition $\varphi \circ H : \mathbb{R}^d \to \mathbb{R}$ is continuous and bounded. Therefore,

$$\mathbb{E}\varphi(H(f_n)) = \mathbb{E}(\varphi \circ H)(f_n) \xrightarrow[n]{} \mathbb{E}(\varphi \circ H)(f) = \mathbb{E}\varphi(H(f)).$$

□

We continue with simplifying (2) and (3) of the Portmanteau Theorem 12.1.1 for sets that have a boundary which is a null set with respect to the limit measure. We recall that for a set $A \subseteq \mathbb{R}^d$ the boundary of A is given by

$$\partial A := \overline{A} \setminus \mathring{A},$$

where $\overline{A}$ is the closure of A and $\mathring{A}$ the largest open set that is contained in A.

Corollary 12.1.8 *Let $\mu, \mu_1, \mu_2, \ldots \in \mathcal{M}_1^+(\mathbb{R}^d)$ with $\mu_n \Rightarrow \mu$ and let $A \in \mathcal{B}(\mathbb{R}^d)$ with $\mu(\partial A) = 0$. Then one has*

$$\lim_{n \to \infty} \mu_n(A) = \mu(A).$$

---

[1] Yurii Vasilevich Prokhorov, 15/12/1929 (Moscow, USSR, now Russia)–16/07/2013 (Moscow, Russia).

The proof is the subject of Exercise 1. We turn to connections to our previous types of convergence (see Sect. 9.5):

Corollary 12.1.9

(1) If $f, f_1, f_2, \ldots \in \mathcal{L}_0(\Omega, \mathcal{F}, \mathbb{P})$ and $f_n \xrightarrow{\mathbb{P}} f$, then one has

$$f_n \xrightarrow{d} f.$$

(2) If $f_n \in \mathcal{L}_0(\Omega_n, \mathcal{F}_n, \mathbb{P}_n)$ for $n \in \mathbb{N}$, $c \in \mathbb{R}$, and $f_n \xrightarrow{d} c$, then one has

$$\lim_{n \to \infty} \mathbb{P}_n(|f_n - c| \geqslant \varepsilon) = 0 \quad \text{for all} \quad \varepsilon > 0.$$

Proof
(1) By Theorem 12.1.1(1), for a continuous and bounded $\varphi : \mathbb{R} \to \mathbb{R}$ we need to show that $\mathbb{E}\varphi(f_n) \xrightarrow[n]{} \mathbb{E}\varphi(f)$. By the continuity Theorem 9.2.5 we have that $\varphi(f_n) \xrightarrow{\mathbb{P}} \varphi(f)$. Knowing this, the statement follows from the dominated convergence Theorem 9.4.8.
(2) The proof is the subject of Exercise 4. □

Corollary 12.1.10 (Slutsky's[2] Theorem) Let $f \in \mathcal{L}_0(\Omega, \mathcal{F}, \mathbb{P})$, $f_n, g_n \in \mathcal{L}_0(\Omega_n, \mathcal{F}_n, \mathbb{P}_n)$, $n \in \mathbb{N}$, be such that $f_n \xrightarrow{d} f$ and $g_n \xrightarrow{d} c$ for some $c \in \mathbb{R}$. Then one has

(1) $f_n + g_n \xrightarrow{d} f + c$,
(2) $f_n g_n \xrightarrow{d} cf$,
(3) $\frac{f_n}{g_n} \xrightarrow{d} \frac{f}{c}$ if $g_n(\omega_n) \neq 0$ for all $\omega_n \in \Omega_n$ and $c \neq 0$.

The proof is the subject of Exercise 7. We close with an example that will be useful for us later:

Example 12.1.11 Let $m \in \mathbb{R}^d$, $a_n \downarrow 0$, and let $D_{a_n^2} \in \mathbb{R}^{d \times d}$ be the diagonal matrix that has $a_n^2 > 0$ on its diagonal. Then

$$\mathcal{N}_{m, D_{a_n^2}} \Rightarrow \delta_m,$$

where δ_m is the Dirac measure in $m \in \mathbb{R}^d$.

---

[2] Evgenii Evgenievich Slutsky, 19/04/1880 (Yaroslavl Governorate, Russia)–10/03/1948 (Moscow, USSR).

Proof For $n \in \mathbb{N}$ we have

$$\widehat{\mathcal{N}_{m,D_{a_n^2}}}(x) = e^{i\langle x,m\rangle}e^{-a_n^2\frac{\langle x,x\rangle}{2}} \to e^{i\langle x,m\rangle} = \widehat{\delta_m}(x)$$

for $n \to \infty$ by Theorem 11.7.3. By Theorem 12.1.1(4) we have $\mathcal{N}_{m,D_{a_n^2}} \Rightarrow \delta_m$. □

12.2 Proofs of the Main Theorems

Our first lemma deduces from the pointwise convergence of the Fourier transforms of measures $\mu_1, \mu_2, \ldots \in \mathcal{M}_1^+(\mathbb{R}^d)$ and the continuity of the limit at zero their tightness:

Lemma 12.2.1 *Assume that* $(\mu_n)_{n=1}^\infty \subseteq \mathcal{M}_1^+(\mathbb{R}^d)$ *such that* $(\widehat{\mu}_n)_{n=1}^\infty$ *converges pointwise to a function* $\varphi : \mathbb{R}^d \to \mathbb{C}$ *that is continuous at* $x = 0$. *Then the family* $(\mu_n)_{n=1}^\infty$ *is tight.*

Proof For $x \in \mathbb{R}^d$ denote $|x|_\infty := \max\{|x_1|, \ldots, |x_d|\}$ and let $K > 0$. By Fubini's theorem,

$$\left(\frac{K}{2}\right)^d \int_{\{|x|_\infty \leqslant 1/K\}} (1 - \widehat{\mu}_n(x))\,d\lambda_d(x)$$

$$= \left(\frac{K}{2}\right)^d \int_{\{|x|_\infty \leqslant 1/K\}} \left(\int_{\mathbb{R}^d} \left(1 - e^{i\langle x,y\rangle}\right) d\mu_n(y)\right) d\lambda_d(x)$$

$$= \left(\frac{K}{2}\right)^d \int_{\mathbb{R}^d} \int_{\{|x|_\infty \leqslant 1/K\}} \left(1 - e^{i\langle x,y\rangle}\right) d\lambda_d(x)\,d\mu_n(y)$$

$$= \int_{\mathbb{R}^d} \left(1 - \prod_{k=1}^d \left(\frac{K}{2}\int_{[-1/K,1/K]} e^{ix_k y_k}\,d\lambda(x_k)\right)\right) d\mu_n(y).$$

Since for $y_k \neq 0$ one has $\frac{K}{2}\int_{[-1/K,1/K]} e^{ix_k y_k}\,d\lambda(x_k) = \frac{\sin(y_k/K)}{y_k/K} \in \mathbb{R}$ and $\left|\frac{\sin u}{u}\right| \leqslant 1$ for $u \neq 0$, the integrand in the last expression is nonnegative. This implies

$$\left(\frac{K}{2}\right)^d \int_{\{|x|_\infty \leqslant 1/K\}} (1 - \widehat{\mu}_n(x))\,d\lambda_d(x)$$

$$\geqslant \int_{\{|y|_\infty \geqslant 2K\}} \left(1 - \prod_{k=1}^d \left(\frac{K}{2}\int_{[-1/K,1/K]} e^{ix_k y_k}\,d\lambda(x_k)\right)\right) d\mu_n(y).$$

But for $|y|_\infty \geqslant 2K$ we have that

$$\left|\prod_{k=1}^{d}\left(\frac{K}{2}\int_{[-1/K,1/K]}e^{ix_ky_k}\,\mathrm{d}\lambda(x_k)\right)\right| = \prod_{k=1}^{d}\left|\frac{\sin(y_k/K)}{y_k/K}\right| \leqslant \frac{1}{2}.$$

Indeed, if $|y|_\infty \geqslant 2K$, then it holds $|y_k| \geqslant 2K$ for at least one $k \in \{1,\ldots,d\}$ and $\left|\frac{\sin(y_k/K)}{y_k/K}\right| \leqslant \frac{1}{2}$, and the remaining factors we can estimate by one. Therefore,

$$\mu_n\left(\{y \in \mathbb{R}^d : |y|_\infty \geqslant 2K\}\right) \leqslant 2\left(\frac{K}{2}\right)^d \int_{\{|x|_\infty \leqslant 1/K\}} (1 - \widehat{\mu}_n(x))\,\mathrm{d}\lambda_d(x).$$

Since necessarily $\varphi(0) = 1$ and φ is continuous at $x = 0$, we find for any $\varepsilon > 0$ a $K_\varepsilon > 0$ such that

$$2\left(\frac{K_\varepsilon}{2}\right)^d \int_{\{|x|_\infty \leqslant 1/K_\varepsilon\}} |1 - \varphi(x)|\,\mathrm{d}\lambda_d(x) < \varepsilon.$$

Because of $|\widehat{\mu}_n(x)| \leqslant 1$ we may conclude by dominated convergence (applied to the real and imaginary parts of $\widehat{\mu}_n$ and φ) that

$$\mu_n\left(\{y \in \mathbb{R}^d : |y|_\infty \geqslant 2K_\varepsilon\}\right) \leqslant 2\left(\frac{K_\varepsilon}{2}\right)^d \int_{\{|x|_\infty \leqslant 1/K_\varepsilon\}} (1 - \widehat{\mu}_n(x))\,\mathrm{d}\lambda_d(x) < 2\varepsilon$$

for $n \geqslant n(\varepsilon)$. By choosing $K'_\varepsilon \geqslant K_\varepsilon$ large enough we can also arrange that $\mu_n\left(\{y \in \mathbb{R}^d : |y|_\infty \geqslant 2K'_\varepsilon\}\right) < 2\varepsilon$ for $n = 1,\ldots,n(\varepsilon)-1$. □

The second lemma extracts from a tight sequence of probability measures a subsequence that converges weakly:

Lemma 12.2.2 *Assume a family $(\mu_n)_{n=1}^{\infty} \subseteq \mathcal{M}_1^+(\mathbb{R}^d)$ which is tight. Then there exists a subsequence $(\mu_{n_k})_{k=1}^{\infty}$ and a measure $\mu \in \mathcal{M}_1^+(\mathbb{R}^d)$ such that*

$$\lim_{k \to \infty} \int_{\mathbb{R}^d} \varphi(x)\,\mathrm{d}\mu_{n_k}(x) = \int_{\mathbb{R}^d} \varphi(x)\,\mathrm{d}\mu(x)$$

for all continuous and bounded functions $\varphi : \mathbb{R}^d \to \mathbb{R}$.

Proof
(a) First we show that there is a subsequence $1 \leqslant n_1 < n_2 < \cdots$ such that we have for the distribution functions

$$F_n((x_1,\ldots,x_d)) := \mu_n((-\infty,x_1] \times \cdots \times (-\infty,x_d])$$

12.2 Proofs of the Main Theorems

that the limit

$$G(r) := \lim_{k \to \infty} F_{n_k}(r)$$

exists for all $r \in \mathbb{Q}^d$. Such a subsequence can be constructed by the following diagonal argument: We enumerate the points in $\mathbb{Q}^d$ by $\{r_1, r_2, \ldots\}$. Since $(F_n(r_1))_{n=1}^\infty \subseteq [0, 1]$, there exists a convergent subsequence $F_{n_k(1)}(r_1) \to G(r_1)$ as $k \to \infty$. Then we find a sub-subsequence $(n_k(2))_{k=1}^\infty$ of $(n_k(1))_{k=1}^\infty$ such that $F_{n_k(2)}(r_2) \to G(r_2)$ and continue in this way. For the diagonal sequence $(n_k(k))_{k \in \mathbb{N}}$ it holds that $F_{n_k(k)}(r) \to G(r)$ for all $r \in \mathbb{Q}^d$.

(b) We define $F : \mathbb{R}^d \to [0, 1]$ by

$$F((x_1, \ldots, x_d)) := \inf\{G((r_1, \ldots, r_d)) : x_j < r_j \in \mathbb{Q}\}$$

and verify that F satisfies the assumptions of Theorem 6.7.1 with $m = 1$. Before we start, we remark that the tightness assumption implies for any $\varepsilon \in (0, 1)$ the existence of an $L_\varepsilon > 0$ such that

$$\mu_n\left([-L_\varepsilon, L_\varepsilon]^d\right) > 1 - \varepsilon \quad \text{for all} \quad n \in \mathbb{N}.$$

Assumption (DF1) follows from

$$0 \leq G((\underline{r}_1, \ldots, \underline{r}_d)) \leq F((x_1, \ldots, x_d)) \leq G((\overline{r}_1, \ldots, \overline{r}_d)) \leq 1$$

when $\underline{r}_j \leq x_j < \overline{r}_j$ and from

$$F_n((x_1, \ldots, x_d)) > 1 - \varepsilon \quad \text{whenever} \quad x_1, \ldots, x_d \geq L_\varepsilon.$$

Assumption (DF2) follows similarly from

$$\sup_{x_1, \ldots, x_{j-1}, x_{j+1}, \ldots, x_d} F_n(x_1, \ldots, x_d)$$

$$\leq \mu_n(\mathbb{R} \times \cdots \times \mathbb{R} \times (-\infty, x_j] \times \mathbb{R} \times \cdots \times \mathbb{R})$$

$$= 1 - \mu_n(\mathbb{R} \times \cdots \times \mathbb{R} \times (x_j, \infty) \times \mathbb{R} \times \cdots \times \mathbb{R})$$

$$< \varepsilon$$

for $x_j < -L_\varepsilon$.

Assumption (DF4) follows from the coordinate-wise monotonicity of G and from the construction of F.

Assumption (DF5): By construction of G and because the functions F_n are obtained from measures, we get that

$$\Delta_{a_1, b_1} \circ \cdots \circ \Delta_{a_d, b_d} G \geq 0 \tag{12.1}$$

for $a_j, b_j \in \mathbb{Q}$ with $a_j \leqslant b_j$ for $j = 1, \ldots, d$. For each $x = (x_1, \ldots, x_d) \in \mathbb{R}^d$ there is a sequence of $r^\ell = (r_1^\ell, \ldots, r_d^\ell) \in \mathbb{Q}^d$ with $r_j^\ell \downarrow x_j$ as $\ell \to \infty$, such that $F(x) = \lim_{\ell \to \infty} G(r^\ell)$. Therefore (12.1) transfers to F for $a_j, b_j \in \mathbb{R}$.

Summarizing, we obtain a $\mu \in \mathcal{M}_1^+(\mathbb{R}^d)$ such that

$$F((x_1, \ldots, x_d)) := \mu((-\infty, x_1] \times \cdots \times (-\infty, x_d]).$$

(c) For $x \in \mathbb{R}^d$ we have that

$$F((x_1, \ldots, x_d)) = \inf \left\{ \lim_{k \to \infty} F_{n_k}((r_1, \ldots, r_d)) : x_j < r_j \in \mathbb{Q} \right\}$$

$$\geqslant \limsup_{k \to \infty} F_{n_k}((x_1, \ldots, x_d)).$$

On the other hand, if $y \in \mathbb{R}^d$ with $y_j < x_j$ for $j = 1, \ldots, d$, then we get that

$$F((y_1, \ldots, y_d)) = \inf \left\{ \lim_{k \to \infty} F_{n_k}((r_1, \ldots, r_d)) : y_j < r_j \in \mathbb{Q} \right\}$$

$$= \inf \left\{ \lim_{k \to \infty} F_{n_k}((r_1, \ldots, r_d)) : y_j < r_j < x_j \in \mathbb{Q} \right\}$$

$$\leqslant \liminf_{k \to \infty} F_{n_k}((x_1, \ldots, x_d)).$$

If x is a continuity point of F, then for all $\varepsilon > 0$ there is an $\eta > 0$ such that

$$F((y_1, \ldots, y_d)) \leqslant F((x_1, \ldots, x_d)) \leqslant F((y_1, \ldots, y_d)) + \varepsilon$$

if $x_j - \eta < y_j < x_j$ for $j = 1, \ldots, d$. Hence

$$F((x_1, \ldots, x_d)) - \varepsilon \leqslant F((y_1, \ldots, y_d))$$

$$\leqslant \liminf_{k \to \infty} F_{n_k}((x_1, \ldots, x_d))$$

$$\leqslant \limsup_{k \to \infty} F_{n_k}((x_1, \ldots, x_d))$$

$$\leqslant F((x_1, \ldots, x_d))$$

and $\varepsilon \downarrow 0$ yields $F((x_1, \ldots, x_d)) = \lim_{k \to \infty} F_{n_k}((x_1, \ldots, x_d))$.

(d) By the monotonicity of F we can find countable sets $D_1, \ldots, D_d \subseteq \mathbb{R}$ such that F is continuous on $C := D_1^c \times \ldots \times D_d^c$ (see Exercise 3). Then for all cuboids

$$Q := \underset{j=1}{\overset{d}{\times}} (a_j, b_j] \text{ with } -\infty < a_j < b_j < \infty \text{ and } a_j, b_j \in D_j^c \tag{12.2}$$

12.2 Proofs of the Main Theorems

we have $\lim_{k \to \infty} \mu_{n_k}(Q) = \mu(Q)$ by step (c).

(e) Finally we show the convergence of the integrals. By rescaling and adding a constant (i.e. we replace φ by $A + B\varphi$) we may assume a continuous function $\varphi : \mathbb{R}^d \to (0, 1)$. Given $\varepsilon \in [0, 1]$, we find $L_\varepsilon > 0$ such that $\mu_{n_k}([-L_\varepsilon, L_\varepsilon]^d) \geqslant 1 - \varepsilon$ and $\mu([-L_\varepsilon, L_\varepsilon]^d) \geqslant 1 - \varepsilon$. And we find a simple function

$$\psi_\varepsilon = \sum_{r=1}^{R} \alpha_r \mathbb{1}_{Q_r} : \mathbb{R}^d \to [0, 1]$$

with sets Q_r of form (12.2), $\alpha_1, \ldots, \alpha_R \in \mathbb{R}$, and

$$\sup_{x \in [-L_\varepsilon, L_\varepsilon]^d} |\psi_\varepsilon(x) - \varphi(x)| < \varepsilon.$$

Hence

$$\lim_{k \to \infty} \int_{\mathbb{R}^d} \psi_\varepsilon(x) \mathrm{d}\mu_{n_k}(x) = \int_{\mathbb{R}^d} \psi_\varepsilon(x) \mathrm{d}\mu(x).$$

Moreover,

$$\left| \int_{\mathbb{R}^d} \varphi(x) \mathrm{d}\mu_{n_k}(x) - \int_{\mathbb{R}^d} \varphi(x) \mathrm{d}\mu(x) \right|$$

$$\leqslant \left| \int_{\mathbb{R}^d} \varphi(x) \mathrm{d}\mu_{n_k}(x) - \int_{\mathbb{R}^d} \psi_\varepsilon(x) \mathrm{d}\mu_{n_k}(x) \right|$$

$$+ \left| \int_{\mathbb{R}^d} \psi_\varepsilon(x) \mathrm{d}\mu_{n_k}(x) - \int_{\mathbb{R}^d} \psi_\varepsilon(x) \mathrm{d}\mu(x) \right|$$

$$+ \left| \int_{\mathbb{R}^d} \psi_\varepsilon(x) \mathrm{d}\mu(x) - \int_{\mathbb{R}^d} \varphi(x) \mathrm{d}\mu(x) \right|$$

$$\leqslant 6\varepsilon + \left| \int_{\mathbb{R}^d} \psi_\varepsilon(x) \mathrm{d}\mu_{n_k}(x) - \int_{\mathbb{R}^d} \psi_\varepsilon(x) \mathrm{d}\mu(x) \right|,$$

which proves our statement. $\square$

Proof of Theorem 12.1.1

(2) $\Leftrightarrow$ (3) follows by considering $A = \mathbb{R}^d \setminus B$ and by using that $\mu(A) + \mu(B) = \mu_n(A) + \mu_n(B) = 1$.

(1) $\Rightarrow$ (4) We choose $\varphi(x) := \cos(\langle y, x \rangle)$ and $\varphi(x) := \sin(\langle y, x \rangle)$ for $y \in \mathbb{R}^d$.

(1) $\Rightarrow$ (3) One knows that there is a sequence of continuous functions $\varphi_l : \mathbb{R}^d \to [0, 1]$ such that $\varphi_l \uparrow \mathbb{1}_B$ (see Lemma 11.5.4). So we have

$$\lim_{n \to \infty} \int_{\mathbb{R}^d} \varphi_l(x) \mathrm{d}\mu_n(x) = \int_{\mathbb{R}^d} \varphi_l(x) \mathrm{d}\mu(x)$$

for all $l \in \mathbb{N}$, so that

$$\liminf_{n\to\infty} \int_{\mathbb{R}^d} 1_B(x)\mathrm{d}\mu_n(x) \geq \int_{\mathbb{R}^d} \varphi_l(x)\mathrm{d}\mu(x).$$

By monotone convergence,

$$\liminf_{n\to\infty} \int_{\mathbb{R}^d} 1_B(x)\mathrm{d}\mu_n(x) \geq \int_{\mathbb{R}^d} 1_B(x)\mathrm{d}\mu(x).$$

(4) $\Rightarrow$ (1) By Theorem 11.2.6(1) and Lemma 12.2.1 the family $(\mu_n)_{n=1}^\infty$ is tight. Using Lemma 12.2.2 there is a subsequence $(n_k)_{k=1}^\infty$ and a measure $\nu \in \mathcal{M}_1^+(\mathbb{R}^d)$ such that

$$\lim_{k\to\infty} \int_{\mathbb{R}^d} \varphi(x)\mathrm{d}\mu_{n_k}(x) = \int_{\mathbb{R}^d} \varphi(x)\mathrm{d}\nu(x)$$

for all continuous and bounded $\varphi : \mathbb{R}^d \to \mathbb{R}$. This implies $\lim_{k\to\infty} \widehat{\mu_{n_k}}(x) = \widehat{\nu}(x)$ for all $x \in \mathbb{R}^d$ so that the uniqueness Theorem 11.5.5 gives $\mu = \nu$ and

$$\lim_{k\to\infty} \int_{\mathbb{R}^d} \varphi(x)\mathrm{d}\mu_{n_k}(x) = \int_{\mathbb{R}^d} \varphi(x)\mathrm{d}\mu(x)$$

for all continuous and bounded $\varphi : \mathbb{R}^d \to \mathbb{R}$. Assume a continuous and bounded $\varphi : \mathbb{R}^d \to \mathbb{R}$ such that

$$\lim_{n\to\infty} \int_{\mathbb{R}^d} \varphi(x)\mathrm{d}\mu_n(x) = \int_{\mathbb{R}^d} \varphi(x)\mathrm{d}\mu(x)$$

does not hold. We find an $\varepsilon > 0$ and a subsequence $(n_k)_{k=1}^\infty$ such that

$$\left| \int_{\mathbb{R}^d} \varphi(x)\mathrm{d}\mu_{n_k}(x) - \int_{\mathbb{R}^d} \varphi(x)\mathrm{d}\mu(x) \right| > \varepsilon$$

for all $k \in \mathbb{N}$. But using the previous arguments we find a further subsequence $(n_{k_l})_{l=1}^\infty$ such that

$$\lim_{l\to\infty} \int_{\mathbb{R}^d} \varphi(x)\mathrm{d}\mu_{n_{k_l}}(x) = \int_{\mathbb{R}^d} \varphi(x)\mathrm{d}\mu(x),$$

which is a contradiction.

(2) and (3) imply (1): By rescaling and adding a constant we can again assume that $\varphi : \mathbb{R}^d \to (0, 1)$. For $t \in [0, 1]$ we define the open sets $A_t := \varphi^{-1}((t, \infty)) \in \mathcal{B}(\mathbb{R}^d)$. Then $\partial A_t \subseteq \{\varphi = t\}$ so that

$$\mu(\partial A_t) = 0$$

except for t from a countable set $\mathcal{N} \subseteq [0,1]$, see Exercise 6. Hence $\lim_{n\to\infty} \mu_n(A_t) = \mu(A_t)$ for $t \in [0,1] \setminus \mathcal{N}$ which can be deduced from (2) and (3) (see Corollary 12.1.8). Therefore we use Corollary 8.10.3 and dominated convergence to conclude

$$\int_{\mathbb{R}^d} \varphi(x) d\mu_n(x) = \int_{[0,1]} \mu_n(A_t) d\lambda(t) \xrightarrow[n]{} \int_{[0,1]} \mu(A_t) d\lambda(t) = \int_{\mathbb{R}^d} \varphi(x) d\mu(x).$$

(2) and (3) imply (5): For $\mu(\{x\}) = 0$ we have

$$\limsup_{n\to\infty} \mu_n((-\infty, x]) \leq \mu((-\infty, x]) = \mu((-\infty, x)) \leq \liminf_{n\to\infty} \mu_n((-\infty, x))$$

so that $\lim_{n\to\infty} \mu_n((-\infty, x]) = \mu((-\infty, x])$.

(5) $\Rightarrow$ (1) follows by step (e) of the proof of Lemma 12.2.2. $\square$

Proof of Theorem 12.1.3 By Lemma 12.2.1 the set of measures $(\mu_n)_{n=1}^\infty$ is tight. By Lemma 12.2.2 we find a subsequence $(\mu_{n_k})_{k=1}^\infty$ and a measure $\mu \in \mathcal{M}_1^+(\mathbb{R}^d)$ such that $\widehat{\mu_{n_k}}(x) \to \widehat{\mu}(x)$ for all $x \in \mathbb{R}^d$. But then $\varphi = \widehat{\mu}$ which proves the statement. $\square$

Proof of Theorem 12.1.5

(1) $\Rightarrow$ (2) follows from Lemma 12.2.2.

(2) $\Rightarrow$ (1) Assume that $M \subseteq \mathcal{M}_1^+(\mathbb{R}^d)$ is not tight. Then we find a sequence of measures $(\mu_n)_{n=1}^\infty \subseteq M$, $0 < a_1 < a_2 < a_3 < \cdots \to \infty$, and $\varepsilon \in (0,1)$ such that

$$\mu_n(\{x \in \mathbb{R}^d : |x| \geq a_n\}) \geq \varepsilon.$$

Assume that there is a subsequence $1 \leq n_1 < n_2 < \cdots$ such that $\mu_{n_k} \Rightarrow \mu$ for some $\mu \in \mathcal{M}_1^+(\mathbb{R}^d)$ as $k \to \infty$. For all $K > 0$ this gives

$$\mu(\{x \in \mathbb{R}^d : |x| \geq K\}) \geq \limsup_{k\to\infty} \mu_{n_k}(\{x \in \mathbb{R}^d : |x| \geq K\}) \geq \varepsilon$$

by Theorem 12.1.1(2). But this implies $\mu(\{x \in \mathbb{R}^d : |x| \geq K\}) \geq \varepsilon$ for all $K > 0$, which is impossible. $\square$

12.3 From Weak Convergence Back to Almost Sure Convergence

The following theorem closes the missing link from weak convergence back to almost sure convergence. It can be seen as a coupling result as we want to construct random variables that are coupled as close as possible to each other. The result goes back to Dudley and Skorohod.

Theorem 12.3.1 (Dudley and Skorohod) *Let $\mu_n, \mu \in \mathcal{M}_1^+(\mathbb{R}^d)$ be such that $\mu_n \Rightarrow \mu$. Then there exists a probability space $(\Omega, \mathcal{F}, \mathbb{P})$ and random variables $f_n, f \in \mathcal{L}_0(\Omega, \mathcal{F}, \mathbb{P})$ such that $\operatorname{law}(f) = \mu$ and $\operatorname{law}(f_n) = \mu_n$ for $n \in \mathbb{N}$, and such that $f_n \xrightarrow{a.s.} f$.*

Proof We fix a sequence $1 > \varepsilon_1 \geqslant \varepsilon_2 \geqslant \cdots > 0$ such that

$$\sum_{m=1}^{\infty} \varepsilon_m < \infty. \tag{12.3}$$

(a) Building of partitions: We put $F((x_1, \ldots, x_d)) := \mu((-\infty, x_1] \times \cdots \times (-\infty, x_d])$. By the proof of Lemma 12.2.2 we know that there are countable sets $D_1, \ldots, D_d \subseteq \mathbb{R}$ such that F is continuous on $C := D_1^c \times \cdots \times D_d^c$. We assume a sequence of partitions $\mathcal{P}_m = \{Q_{ml} : l \in \mathbb{N}\}$ of $\mathbb{R}^d$, $m \in \mathbb{N}$, such that each Q_{ml} is a half-open cube of form

$$(a_1, b_1] \times \cdots \times (a_d, b_d] \quad \text{with} \quad -\infty < a_j < b_j < \infty, \ a_j, b_j \notin D_j,$$

and $\sup_{x, y \in Q_{m,l}} |x - y| < \varepsilon_m$. We can arrange the Q_{ml} in l for fixed m such that

$$\mu(Q_{m1}) \geqslant \mu(Q_{m2}) \geqslant \cdots \geqslant 0$$

and find an $L_m \in \mathbb{N}$ such that

$$\mu\left(\bigcup_{l=1}^{L_m} Q_{ml}\right) > 1 - \varepsilon_m \quad \text{and} \quad \mu(Q_{mL_m}) > 0.$$

As by construction $\mu(\partial Q_{ml}) = 0$ we get from Corollary 12.1.8 that

$$\lim_{n \to \infty} \mu_n(Q_{ml}) = \mu(Q_{ml}) \quad \text{for all} \quad m, l \in \mathbb{N}.$$

Therefore, for each $m \in \mathbb{N}$ there is an $N_m \in \mathbb{N}$ such that

$$\mu_n(Q_{ml}) > (1 - \varepsilon_m)\mu(Q_{ml}) > 0 \quad \text{for} \quad l = 1, \ldots, L_m \text{ and } n \geqslant N_m.$$

By a possible enlargement of the N_m we can assume that

$$1 =: N_0 < N_1 < N_2 < N_3 < \cdots.$$

(b) Constructing the random variables: We put $Q_{m0} := \mathbb{R}^d \setminus \bigcup_{l=1}^{L_m} Q_{ml}$ so that $\mu(Q_{m0}) < \varepsilon_m$. For $n \in \mathbb{N}$ there is a unique $m(n) \in \{0, 1, 2, \ldots\}$ such that $n \in [N_{m(n)}, N_{m(n)+1})$. Assume a probability space $(\Omega, \mathcal{F}, \mathbb{P})$ carrying independent

12.3 Back to Almost Sure Convergence

random variables $f, \xi, (Y_n)_{n \geq 1}, (Y_{nl})_{n \geq 1, l=1,\ldots,L_{m(n)}}$, and $(Z_n)_{n \geq 1}$ such that

$$\text{law}(f) = \mu, \qquad \text{law}(\xi) = U_{[0,1]},$$
$$\text{law}(Y_n) = \mu_n, \qquad \text{law}(Y_{nl}) = \mu_n(\cdot | Q_{m(n)l}),$$
$$\text{law}(Z_n) = \nu_n,$$

where for $A \in \mathcal{B}(\mathbb{R}^d)$ we define

$$\nu_n(A) := \frac{1}{\varepsilon_{m(n)}} \left(\mu_n(A) - (1 - \varepsilon_{m(n)}) \sum_{l=0}^{L_{m(n)}} \mu(Q_{m(n)l}) \mu_n(A | Q_{m(n)l}) \right).$$

From $\mu_n(Q_{m(n)l}) > (1 - \varepsilon_{m(n)}) \mu(Q_{m(n)l})$ for $l = 1, \ldots, L_{m(n)}$ it follows that ν_n is a probability measure. Now we choose

$$f_n := \begin{cases} Y_n & : n < N_1 \\ \sum_{l=0}^{L_{m(n)}} \mathbb{1}_{\{\xi \in [\varepsilon_{m(n)}, 1], f \in Q_{m(n)l}\}} Y_{nl} + \mathbb{1}_{\{\xi \in [0, \varepsilon_{m(n)})\}} Z_n & : n \in [N_{m(n)}, N_{m(n)+1}) \end{cases}.$$

By construction we get that $\text{law}(f_n) = \mu_n$. Indeed, this is clear for $n < N_1$, and for $N_{m(n)} \leq n < N_{m(n)+1}$ we have by independence of ξ, f and Y_{nl} that

$$\mathbb{P}(f_n \in A)$$
$$= \sum_{l=0}^{L_{m(n)}} \mathbb{P}(\xi \in [\varepsilon_{m(n)}, 1], f \in Q_{m(n)l}, Y_{nl} \in A) + \mathbb{P}(\xi \in [0, \varepsilon_{m(n)})) \mathbb{P}(Z_n \in A)$$
$$= (1 - \varepsilon_{m(n)}) \sum_{l=0}^{L_{m(n)}} \mu(Q_{m(n)l}) \mu_n(A | Q_{m(n)l}) + \varepsilon_{m(n)} \nu_n(A)$$
$$= \mu_n(A).$$

For $\Omega_m := \{f \notin Q_{m0}, \xi \notin [0, \varepsilon_m)\}$ it holds $\mathbb{P}(\Omega_m^c) \leq \mathbb{P}(f \in Q_{m0}) + \mathbb{P}(\xi \in [0, \varepsilon_m)) < 2\varepsilon_m$. By Theorem 2.5.7 (Borel-Cantelli) and (12.3) we conclude that $\mathbb{P}(\limsup_{m \to \infty} \Omega_m^c) = 0$. Then $\Omega_0 = \liminf_{m \to \infty} \Omega_m$ has probability one. We observe that on Ω_m, provided that $N_{m(n)} \leq n < N_{m(n)+1}$, the values of f and f_n are in the same Q_{ml}, $(l = 1, \ldots, L_m)$ which means that their distance is less than ε_m. Hence we have $f_n(\omega) \to f(\omega)$ as $n \to \infty$ for all $\omega \in \Omega_0$. Finally, we define $f_n(\omega) := f(\omega)$ for $\omega \in \Omega_0^c$. This implies $f_n(\omega) \to f(\omega)$ for all $\omega \in \Omega$. □

12.4 Exercises

Ex 1: Verify Corollary 12.1.8: Let $\mu_n, \mu \in \mathcal{M}_1^+(\mathbb{R}^d)$, $n \in \mathbb{N}$, with $\mu_n \Rightarrow \mu$ and $A \in \mathcal{B}(\mathbb{R}^d)$ with $\mu(\partial A) = 0$. Then one has $\lim_{n \to \infty} \mu_n(A) = \mu(A)$.

Ex 2: Assume that $f_n, g_n, f, g \in \mathcal{L}_0(\Omega, \mathcal{F}, \mathbb{P})$, $n \in \mathbb{N}$, such that f_n and g_n are independent for each $n \in \mathbb{N}$, $f_n \xrightarrow{d} f$, and $g_n \xrightarrow{d} g$. Does this imply that f and g are independent?

Ex 3: Let $\mu \in \mathcal{M}_1^+(\mathbb{R}^d)$ and $F(x) := \mu((-\infty, x_1] \times \cdots \times (-\infty, x_d])$ for $x = (x_1, \ldots, x_d) \in \mathbb{R}^d$. Show that there are countable sets $D_1, \ldots, D_d \subseteq \mathbb{R}$ such that F is continuous on $C = D_1^c \times \ldots \times D_d^c$.
Hint: Show that if $H : \mathbb{R} \to [0, 1]$ is nondecreasing, then it can only have jump discontinuities, and the set of points, where H jumps, is countable. Deduce from this fact the statement.

Ex 4: Verify Corollary 12.1.9(2): If $f_n \in \mathcal{L}_0(\Omega_n, \mathcal{F}_n, \mathbb{P}_n)$ for $n \in \mathbb{N}$, $c \in \mathbb{R}$, and $f_n \xrightarrow{d} c$, then $\lim_{n \to \infty} \mathbb{P}_n(|f_n - c| \geq \varepsilon) = 0$ for all $\varepsilon > 0$.
Hint: Use $\mathbb{P}_n(|f_n - c| \geq \varepsilon) = \mathbb{P}_{f_n}((c - \varepsilon, c + \varepsilon)^c)$ and Theorem 12.1.1.

Ex 5: Assume that $f, f_1, f_2, \ldots \in \mathcal{L}_0(\Omega, \mathcal{F}, \mathbb{P})$ such that $f_n \xrightarrow{d} f$. If each random variable f_n is Gaussian with mean $m_n \in \mathbb{R}$ and variance $\sigma_n^2 > 0$, does it follow that f is Gaussian?

Ex 6: Assume a probability space $(\Omega, \mathcal{F}, \mathbb{P})$ and $(A_i)_{i \in I}$ with I being uncountable, $A_i \in \mathcal{F}$, and $A_i \cap A_j = \emptyset$ for $i \neq j$. Prove that the set $\{i \in I : \mathbb{P}(A_i) > 0\}$ is countable.

Ex 7: To verify Corollary 12.1.10 we proceed as follows:

(a) For $x, y \in \mathbb{R}$ prove that the assumptions imply that $(f_n, g_n) \xrightarrow{d} (f, c)$.
Hint: Using Corollary 12.1.9(2) and Theorem 9.4.8 prove first that
$$\lim_{n \to \infty} \left| \mathbb{E} e^{i(xf_n + yg_n)} - \mathbb{E} e^{i(xf_n + yc)} \right| = 0.$$

(b) Using the continuous functions $H_1, H_2 : \mathbb{R}^2 \to \mathbb{R}$ with $H_1(x, y) = x + y$ and $H_2(x, y) := xy$ prove (1) and (2) of Corollary 12.1.10 using Corollary 12.1.7.

(c) For part (3) observe that $g_n \xrightarrow{\mathbb{P}} c$ implies $1/g_n \xrightarrow{\mathbb{P}} 1/c$ using, for instance, the characterization of the convergence in probability by the almost sure convergence

12.5 Comments

Section 12.1: Theorem 12.1.1 is due to [17, Theorem 2.1., pp. 16], where Billingsley named Theorem 12.1.1 after Portmanteau and provided in [17] the reference 'Jean-Pierre Portmanteau: Espoir pour l'ensemble vide? *Annales de*

12.5 Comments

l'Université de Felletin CXLI(1915)322–325'. The title translates to 'Hope for the empty set?'. Apparently, this is a joke since Felletin is just a small village without university and a portmanteau is large travelling bag. There was no mathematician named Portmanteau in this connection. Parts of the Portmanteau theorem can be found in [150, Theorem 1.2], which is a variant of results from Aleksandrov [1]. Theorem 12.1.5 goes back to Prokhorov [150] (see also [15]).

Section 12.2: The main inequality in the proof of Lemma 12.2.1 is a d-dimensional version of [16, (26.22)]. There is a d-dimensional maximal version of this inequality, called *Lévy's truncation inequality*, see [168, Satz 7.11]). The proof of Lemma 12.2.2 provides a more general statement about vague sequential relative compactness (see [97, Theorem 6.20]).

Section 12.3: Theorem 12.3.1 was proved by Skorohod [175] for complete separable metric spaces (Polish spaces) and extended by Dudley [48] to separable metric spaces. Further extensions were given by Wichura [200] and again by Dudley [49]. For our presentations we used [50, Theorem 11.7.2, pp. 415] and [17, Theorem 6.7, pp. 70].

Chapter 13
Strong Law of Large Numbers

The Strong Law of Large Numbers (SLLN) is one of the basic limit theorems in probability. The SLLN will be proven in the Etemadi form which only requires pairwise independence for a sequence of integrable, identically distributed random variables. A remarkable converse statement due to Kolmogorov is also included in this chapter. As applications, a problem about normal numbers and the speed of convergence of Monte Carlo methods are considered, and the solution to the classical needle problem of Buffon is given.

13.1 Formulation of the Strong Law of Large Numbers

Given a sequence of independent random variables $f_1, f_2, \ldots : \Omega \to \mathbb{R}$ such that

$$\frac{1}{n}(f_1 + \cdots + f_n) \to c \in \mathbb{R} \quad \text{as} \quad n \to \infty$$

we speak about a strong law of large numbers (SLLN) if the convergence takes place almost surely, and about a Weak Law of Large Numbers (WLLN) if the convergence takes place in probability. We did already meet a SLLN under a fourth moment condition in Theorem 9.1.5 and a WLLN in Theorem 9.2.6. In the following we prove in Theorem 13.1.1 a SLLN which only needs a first moment condition, but the condition that the random variables are identically distributed.

In the following we use the standard notation that a family of random variables $(f_i)_{i \in I}$, $f_i : \Omega \to \mathbb{R}$, is i.i.d. (independent, identically distributed) provided that the random variables are independent and share the same law.

Now we formulate the SLLN in the Etemadi form:

Theorem 13.1.1 *Assume a probability space $(\Omega, \mathcal{F}, \mathbb{P})$ and $f_1, f_2, \ldots \in \mathcal{L}_0(\Omega, \mathcal{F})$ such that*

(1) $f_1, f_2, \ldots$ have the same distribution,
(2) $f_1, f_2, \ldots$ are pairwise independent, which means that f_n and f_m are independent whenever $m \neq n$,
(3) $\mathbb{E}|f_1| < \infty$.

Then one has for $S_n := f_1 + \cdots + f_n$ that

$$\lim_{n \to \infty} \frac{S_n}{n} = \mathbb{E}f_1 \text{ a.s.}$$

We say that the SLLN is in the Kolmogorov form, if condition (2) is replaced by the stronger condition:

(2') $f_1, f_2, \ldots$ are independent.

We will prove the SLNN in the Etemadi form in Sect. 13.2, but give an alternative proof for the SLNN in Kolmogorov form in terms of ergodic theory in Sect. 14.2, which is of independent interest. Before, we state a remarkable converse of the SLNN in the Kolmogorov form:

Theorem 13.1.2 (Kolmogorov) *Assume that $(\Omega, \mathcal{F}, \mathbb{P})$ is a probability space and that $f_1, f_2, \ldots : \Omega \to \mathbb{R}$ are i.i.d. random variables. If there is a constant $c \in \mathbb{R}$ such that*

$$\lim_{n \to \infty} \frac{S_n}{n} = c \text{ a.s.,}$$

then $\mathbb{E}|f_1| < \infty$ and $\mathbb{E}f_1 = c$.

The proof is the subject of Exercise 2.

13.2 Proof of the SLLN in Etemadi Form

In this section we prove Theorem 13.1.1. By decomposing $f_i = f_i^+ - f_i^- := \max\{f_i, 0\} - \max\{-f_i, 0\}$ and considering $(f_i^+)_{i=1}^\infty$ and $(f_i^-)_{i=1}^\infty$ separately, we may and do assume that $f_i \geq 0$ with $\mathbb{E}f_i > 0$. By truncation we define the random variables

$$g_i := \mathbb{1}_{\{f_i \leq i\}} f_i \quad \text{and the partial sums} \quad T_n := g_1 + \cdots + g_n.$$

We fix $\alpha > 1$ and let $n_k := \lceil \alpha^k \rceil$ for $k \in \mathbb{N}$ ($\lceil x \rceil = \min\{n \in \mathbb{Z} : n \geq x\}$), so that there is a constant $\kappa_\alpha > 0$ with

$$\sum_{\{k \in \mathbb{N} : n_k \geq i\}} \frac{1}{n_k^2} \leq \frac{\kappa_\alpha}{i^2} \quad \text{for} \quad i \in \mathbb{N}.$$

13.2 Proof of the SLLN in Etemadi Form

For $\varepsilon > 0$ by Chebychev's inequality (Corollary 8.12.7) and the orthogonality of $g_i - \mathbb{E}g_i$ and $g_j - \mathbb{E}g_j$ for $i \neq j$, which follows by the pairwise independence of the sequence $(f_n)_{n=1}^\infty$, we get that

$$\sum_{k=1}^\infty \mathbb{P}\left(\left|\frac{T_{n_k}}{n_k} - \mathbb{E}\frac{T_{n_k}}{n_k}\right| > \varepsilon\right) \leq \frac{1}{\varepsilon^2}\sum_{k=1}^\infty \frac{1}{n_k^2}\mathbb{E}\left|T_{n_k} - \mathbb{E}T_{n_k}\right|^2$$

$$= \frac{1}{\varepsilon^2}\sum_{k=1}^\infty \frac{1}{n_k^2}\sum_{i=1}^{n_k}\mathbb{E}|g_i - \mathbb{E}g_i|^2$$

$$\leq \frac{1}{\varepsilon^2}\sum_{k=1}^\infty \frac{1}{n_k^2}\sum_{i=1}^{n_k}\mathbb{E}|g_i|^2$$

$$= \frac{1}{\varepsilon^2}\sum_{i=1}^\infty \mathbb{E}|g_i|^2 \left[\sum_{\{k\in\mathbb{N}: n_k \geq i\}} \frac{1}{n_k^2}\right]$$

$$\leq \frac{K_\alpha}{\varepsilon^2}\sum_{i=1}^\infty \frac{\mathbb{E}|g_i|^2}{i^2}$$

$$= \frac{K_\alpha}{\varepsilon^2}\sum_{i=1}^\infty \frac{\mathbb{E}|f_i|^2 \mathbb{1}_{\{f_i \leq i\}}}{i^2}$$

$$= \frac{K_\alpha}{\varepsilon^2}\sum_{i=1}^\infty \sum_{j=1}^i \frac{\mathbb{E}|f_i|^2 \mathbb{1}_{\{j-1 < f_i \leq j\}}}{i^2}$$

$$= \frac{K_\alpha}{\varepsilon^2}\sum_{j=1}^\infty \left[\mathbb{E}|f_1|^2 \mathbb{1}_{\{j-1 < f_i \leq j\}}\right]\sum_{i=j}^\infty \frac{1}{i^2}$$

$$\leq \frac{K_\alpha}{\varepsilon^2}\sum_{j=1}^\infty \left[\mathbb{E}f_1 \mathbb{1}_{\{j-1 < f_i \leq j\}}\right]\sum_{i=j}^\infty \frac{j}{i^2}$$

$$\leq \frac{K_\alpha}{\varepsilon^2}\sup_{j\in\mathbb{N}}\left[j\sum_{i=j}^\infty \frac{1}{i^2}\right]\left[\sum_{j=1}^\infty \left[\mathbb{E}f_1 \mathbb{1}_{\{j-1 < f_i \leq j\}}\right]\right]$$

$$= \frac{K_\alpha}{\varepsilon^2}\sup_{j\in\mathbb{N}}\left[j\sum_{i=j}^\infty \frac{1}{i^2}\right]\mathbb{E}f_1$$

$$< \infty.$$

We set $\varepsilon_N := 1/N$ for $N \in \mathbb{N}$. The lemma of Borel-Cantelli (Theorem 2.5.7) implies that there is a set $\Omega_0^N \in \mathcal{F}$ of probability one, such that $\left|\frac{T_{n_k}(\omega)}{n_k} - \mathbb{E}\frac{T_{n_k}}{n_k}\right| >$

$1/N$ happens only for finitely many k for each $\omega \in \Omega_0^N$. Since $\Omega_0 = \bigcap_{N=1}^\infty \Omega_0^N$ has probability one as well we deduce that

$$\lim_{k\to\infty} \left| \frac{T_{n_k}}{n_k} - \mathbb{E}\frac{T_{n_k}}{n_k} \right| = 0 \text{ a.s.}$$

As we also have that $n_k \uparrow \infty$ as $k \to \infty$ implies that

$$\mathbb{E}f_1 = \lim_{n\to\infty} \mathbb{E}g_n = \lim_{k\to\infty} \frac{\mathbb{E}g_1 + \cdots + \mathbb{E}g_{n_k}}{n_k} = \lim_{k\to\infty} \frac{\mathbb{E}T_{n_k}}{n_k},$$

we arrive at

$$\lim_{k\to\infty} \frac{T_{n_k}}{n_k} = \mathbb{E}f_1 \text{ a.s.}$$

Next we observe that

$$\sum_{n=1}^\infty \mathbb{P}(f_n \neq g_n) = \sum_{n=1}^\infty \mathbb{P}(f_n > n) = \sum_{n=1}^\infty \mathbb{P}(f_1 > n) \leq \mathbb{E}f_1 < \infty.$$

Again using the lemma of Borel-Cantelli implies that, a.s., $f_n \neq g_n$ can only happen finitely many times. Consequently,

$$\lim_{k\to\infty} \frac{S_{n_k}}{n_k} = \mathbb{E}f_1 \text{ a.s.}$$

Finally, by the monotonicity $S_1 \leq S_2 \leq \cdots$ and $\lim_{k\to\infty} n_{k+1}/n_k = \alpha$ we get

$$\frac{1}{\alpha} \lim_{k\to\infty} \frac{S_{n_k}(\omega)}{n_k} \leq \liminf_{n\to\infty} \frac{S_n(\omega)}{n} \leq \limsup_{n\to\infty} \frac{S_n(\omega)}{n} \leq \alpha \lim_{k\to\infty} \frac{S_{n_k}(\omega)}{n_k}$$

for all $\omega \in \Omega$ where $\lim_{k\to\infty} \frac{S_{n_k}(\omega)}{n_k}$ exists. As $\alpha > 1$ was arbitrary, the proof is complete. $\square$

13.3 Theorem About Normal Numbers

We apply Theorem 13.1.1 to a problem from number theory. We consider expansions of numbers $t \in [0, 1)$ in terms of a basis $b \geq 2$, $b \in \mathbb{N}$. This means, we write each $t \in [0, 1)$ as

$$t = \sum_{n=1}^\infty \frac{t_n}{b^n} \quad \text{with} \quad t_n \in \{0, \ldots, b-1\}$$

13.4 Speed of Convergence in Monte-Carlo Methods

such that

$$\{t \in [0, 1) : t_1 = x_1, \ldots, t_n = x_n\}$$
$$= \left\{ t \in [0, 1) : \frac{x_1}{b} + \cdots + \frac{x_n}{b^n} \leqslant t < \frac{x_1}{b} + \cdots + \frac{x_n}{b^n} + \frac{1}{b^n} \right\}.$$

We look for patterns in this representation and fix $r \in \mathbb{N}$ and $x_1^0, \ldots, x_r^0 \in \{0, \ldots, b-1\}$. The function

$$Z_n(t) := \# \left\{ k \in \{1, \ldots, n\} : (t_k, \ldots, t_{k+r-1}) = (x_1^0, \ldots, x_r^0) \right\}$$

counts how many times this pattern occurs in the first n sectors of length r.

Example 13.3.1 If $b = 2$, $r = 1$, $x_1^0 = 1$, and $t = t_1 t_2 \ldots$ is the binary representation of $t \in [0, 1)$, then $Z_n(t)$ counts the number of $i \in \{1, \ldots, n\}$ with $t_i = 1$.

Let $\Omega = [0, 1)$, $\mathcal{F} = \mathbb{B}([0, 1))$, and λ be the Lebesgue measure, so that we may consider $Z_n : \Omega \to \mathbb{R}$ as random variable. Then the following statement shows that, a.s., this pattern occurs in average with probability b^{-r}. This is É. Borel's *normal number theorem*:

Corollary 13.3.2 *One has* $\frac{Z_n}{n} \xrightarrow{a.s.} b^{-r}$.

Proof Letting $f_n(t) := t_n$ we have $Z_n = F_1 + \cdots + F_n$ with

$$F_k := \mathbb{1}_{\{x_1^0, \ldots, x_r^0\}}(f_k, \ldots, f_{k+r-1}).$$

The random variables $f_1, f_2, \ldots : \Omega \to \mathbb{R}$ satisfy

$$\lambda(f_1 = x_1, \ldots, f_n = x_n) = \frac{1}{b^n} = \lambda(f_1 = x_1) \cdots \lambda(f_n = x_n)$$

for all $n \in \mathbb{N}$ and $x_1, \ldots, x_n \in \{0, \ldots, b-1\}$, and are therefore independent as one can replace the condition $\{f_k = x_k\}$ by $\{f \in B_k\}$ for $B_k \in \mathcal{B}(\mathbb{R})$. Hence we can apply Theorem 13.1.1 to conclude the proof. □

In Example 13.3.1 we get $b^{-r} = \frac{1}{2}$. So, roughly speaking, there is no preference for the digit to be 0 or 1 in the binary representation of randomly chosen $t \in [0, 1)$.

13.4 Speed of Convergence in Monte-Carlo Methods

In this section we show how the Chebyshev and the Hoeffding inequality (Corollary 8.12.7 and Theorem 8.12.10) can be used to get an information on how fast the convergence in the SLLN Theorem 13.1.1 actually is. As the SLLN is the key to classical Monte-Carlo methods the knowledge of the speed of convergence allows to stop a computer simulation when accomplishing a pre-given error bound.

To explain the idea we start with $f, f_1, f_2, \ldots \in \mathcal{L}_1(\Omega, \mathcal{F}, \mathbb{P})$ such that $f_1, f_2, \ldots$ are independent and have the same law as f. We set $S_n := f_1 + \cdots + f_n$ and consider

$$\frac{S_n}{n}$$

which can be seen as an example of a Monte-Carlo method because Theorem 13.1.1 implies that

$$\frac{S_n}{n} \to \mathbb{E}f \quad \text{a.s.}$$

To estimate which n will provide a good approximation we use the concept of convergence in probability:

Definition 13.4.1 Let $\varepsilon \in (0, 1)$ and $t > 0$. We define $n(t, \varepsilon) \in \mathbb{N}$ to be the smallest $n_0 \in \mathbb{N}$ such that

$$\mathbb{P}\left(\left|\frac{S_n}{n} - \mathbb{E}f\right| \geq t\right) \leq \varepsilon \quad \text{for} \quad n \geq n_0.$$

We obtain the following table:

Condition on f	Upper bound for $n(t, \varepsilon)$	Asymptotic of $n(t, \varepsilon)$ for $\varepsilon \downarrow 0$
$f \in L_2$	$\lceil \mathrm{var}(f)/(t^2\varepsilon) \rceil$	$1/\varepsilon$
$-\infty < a \leq f \leq b < \infty$	$\lceil \frac{|b-a|^2}{2t^2} \log(2/\varepsilon) \rceil$	$\log(1/\varepsilon)$

Because $\mathbb{E}S_n = \mathbb{E}f$ the table can be verified as follows:
For $f \in L_2$ we use Chebyshev's inequality Corollary 8.12.7 and get, for $t > 0$,

$$\mathbb{P}\left(\left|\frac{S_n}{n} - \mathbb{E}f\right| \geq t\right) \leq \frac{1}{t^2} \mathbb{E}\left|\frac{S_n}{n} - \mathbb{E}f\right|^2 = \frac{1}{t^2} \frac{\mathrm{var}(f)}{n} =: \varepsilon. \quad (13.1)$$

Hence by rearranging we see that $n(t, \varepsilon) \leq \lceil \mathrm{var}(f)/(t^2\varepsilon) \rceil$, where this upper bound behaves asymptotically like $1/\varepsilon$ for $\varepsilon \to 0$.

For $-\infty < a \leq f \leq b < \infty$ we remark that $-\infty < a' \leq f - \mathbb{E}f \leq b' < \infty$ with $a' := a - \mathbb{E}f$ and $b' := b - \mathbb{E}f$, so that $b - a = b' - a'$. Now Hoeffding's inequality Theorem 8.12.10 implies for $\lambda \geqslant 0$ that

$$\mathbb{P}\left(\left|\frac{S_n}{n} - \mathbb{E}f\right| \geqslant t\right) \leqslant 2e^{-2n\frac{t^2}{(b-a)^2}} =: \varepsilon.$$

We note that Definition 13.4.1 relates rather to the WLLN than to the SLLN: Using the setting and proof of Theorem 9.2.6 with constant variance $\sigma^2 = \text{var}(f_n)$ gives $\mathbb{P}\left(\left|\frac{S_n}{n} - \mathbb{E}f\right| \geqslant t\right) \leqslant \frac{1}{t^2}\frac{\sigma^2}{n}$ as in (13.1).

13.5 Buffon's Needle Problem

We come back to Buffon's needle problem from Chap. 1. Let $\Omega := [0, L) \times [0, 2\pi)$ and equip Ω with the σ-algebra $\mathcal{F} := \mathcal{B}([0, L)) \otimes \mathcal{B}([0, 2\pi))$ and the product measure $\mathbb{P} := \mu \otimes \nu$, where $\mu := \frac{1}{L}\lambda$ and $\nu := \frac{1}{2\pi}\lambda$

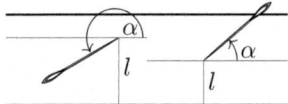

are the normalized restrictions of the Lebesgue measure to $[0, L)$ and $[0, 2\pi)$, respectively. In $(l, \alpha) \in \Omega = [0, L) \times [0, 2\pi)$ the parameter l denotes the distance of the needle's tip to the next line below and α denotes the angle between the parallel lines and the needle (measured counter clockwise at the needle's tip).

The event that the needle hits a line is

$$A := \left\{(l, \alpha) \in \Omega : l = 0 \text{ or for } l \in (0, L) \text{ one has} \right.$$
$$\left. \sin(\alpha) \geqslant 1 - \frac{l}{L} \text{ or } \sin(\alpha - \pi) \geqslant \frac{l}{L}\right\} \in \mathcal{F}.$$

Because $\lambda(\{0\}) = 0$ and because the distributions of $(\sin(\alpha), \frac{l}{L})$ on $(0, \pi) \times (0, L)$ and $(\sin(\alpha - \pi), 1 - \frac{l}{L})$ on $(\pi, 2\pi) \times (0, L)$ coincide we get by Tonelli's theorem that

$$\mathbb{P}(A) = 2\mathbb{P}\left(\left\{(\alpha, l) \in (0, \pi) \times (0, L) : \sin(\alpha) \geqslant \frac{l}{L}\right\}\right)$$
$$= 2\frac{1}{2\pi L}\int_{(0,\pi)\times(0,L)} \mathbb{1}_{\{\sin(\alpha) \geqslant \frac{l}{L}\}} d(\lambda \otimes \lambda)(\alpha, l)$$
$$= \frac{1}{\pi L}\int_{(0,\pi)} \left[\int_{(0,L)} \mathbb{1}_{\{\sin(\alpha) \geqslant \frac{l}{L}\}} d\lambda(l)\right] d\lambda(\alpha)$$

$$= \frac{1}{\pi L} \int_{(0,\pi)} L \sin(\alpha) d\lambda(\alpha)$$

$$= \frac{2}{\pi}.$$

Here we used that $\int_{(0,\pi)} \sin(\alpha) d\lambda_1(\alpha) = \int_0^\pi \sin(\alpha) d\alpha = 2$. Now we assume a probability space $(\Omega, \mathcal{F}, \mathbb{P})$ and independent random variables $f_1, f_2, \ldots : \Omega \to \mathbb{R}$ having the same distribution as $\mathbb{1}_A$ (on $(\Omega, \mathcal{F}, \mathbb{P})$) for $n = 1, 2, \ldots$. In this model $f_n(\omega)$ corresponds to the outcome of the n-th experiment. Applying Kolmogorov's SLLN (Theorem 13.1.1), we get that

$$\lim_{n\to\infty} \frac{S_n}{n} = \mathbb{E} f_1 = \mathbb{P}(A) = \frac{2}{\pi}.$$

13.6 Exercises

Ex 1: Monte Carlo method: Let $(\Omega, \mathcal{F}, \mathbb{P})$ be a probability space, and let $\xi_1, \eta_1, \xi_2, \eta_2, \cdots : \Omega \to \mathbb{R}$ be independent and uniformly on $[0, 1]$ distributed random variables. Let $g : [0, 1] \to [0, 1]$ be a continuous function. To use a Monte Carlo method for the computation of $\int_{[0,1]} g(x) d\lambda(x)$ define $f_n : \Omega \to \mathbb{R}$ by

$$f_n(\omega) := \mathbb{1}_{\{\eta_n(\omega) \leq g(\xi_n(\omega))\}}.$$

(a) Why are the $f_n : \Omega \to \mathbb{R}$ independent random variables?
(b) Prove by Tonelli's theorem that $\mathbb{E} f_n = \int_{[0,1]} g(x) d\lambda(x)$.
(c) Deduce that

$$\frac{1}{n}(f_1 + \cdots + f_n) \xrightarrow{a.s.} \int_{[0,1]} g(x) d\lambda(x).$$

(d) Run the R-programs from Appendices C.1 and C.2, where this Monte-Carlo method is applied to the cosine function and the computation of the number π, and experiment with the parameters.

Ex 2: Prove Theorem 13.1.2: Let $(\Omega, \mathcal{F}, \mathbb{P})$ be a probability space and $f_1, f_2, \ldots : \Omega \to \mathbb{R}$ be i.i.d. random variables. Assume that there is a constant $c \in \mathbb{R}$ such that

$$\frac{1}{n}(f_1 + \cdots + f_n) \xrightarrow{a.s.} c$$

as $n \to \infty$. Show that $\mathbb{E}|f_1| < \infty$ and $\mathbb{E} f_1 = c$.

Hint: Because of Theorem 13.1.1 we only need to show that $\mathbb{E}|f_1| < \infty$. According to Corollary 8.10.3 this is equivalent to $\sum_{n=1}^{\infty} \mathbb{P}(|f_n| \geq n) < \infty$. Using the Lemma of Borel-Cantelli, this is equivalent to

$$\mathbb{P}(\{\omega \in \Omega : \#\{n : |f_n(\omega)| \geq n\} = \infty\}) = 0. \tag{13.2}$$

Now use that

$$\frac{f_n}{n} = \frac{S_n}{n} - \frac{n-1}{n} \frac{S_{n-1}}{n-1} \xrightarrow{a.s.} 0$$

as $n \to \infty$ with $S_n := f_1 + \cdots + f_n$, which implies (13.2).

13.7 Comments

Section 13.1: The strong law of large numbers has a long history. The combination of Theorem 13.1.1 and Theorem 13.1.2 was stated by Kolomogorov in his monograph [105, page 59].

Section 13.2 is due to Etemadi [57]. The special case of the law of large numbers for nonnegative random variables is considered by Etemadi in [58].

Section 13.3: É. Borel's normal number theorem (Corollary 13.3.2) is due to Borel [24].

Section 13.4: For further reading on Monte-Carlo methods the reader is referred to [133].

Chapter 14
An Ergodic Theorem

In this chapter we consider measure preserving maps. These are measurable maps from the state space into the state space where the pre-image of any measurable set has the same probability as the set itself. If the map is such that any set coinciding with its pre-image has either probability one or zero we call the map ergodic. The Birkhoff-Khinchin theorem about ergodic maps is proven based on the maximal ergodic theorem, and a generalized form of the strong law of large numbers is deduced.

14.1 Birkhoff-Khinchin Ergodic Theorem

We start with some concepts from ergodic theory that we will need:

Definition 14.1.1 Let $(\Omega, \mathcal{F}, \mathbb{P})$ be a probability space.

(1) A measurable map $T : \Omega \to \Omega$ is called **measure preserving** provided that

$$\mathbb{P}(T^{-1}(A)) = \mathbb{P}(A) \quad \text{for all} \quad A \in \mathcal{F}.$$

(2) A measure preserving map $T : \Omega \to \Omega$ is called **ergodic** provided that, for $A \in \mathcal{F}$, the condition

$$T^{-1}(A) = A \quad \text{implies} \quad \mathbb{P}(A) \in \{0, 1\}.$$

(3) For a map $T : \Omega \to \Omega$ the σ-algebra $\mathcal{J}$, consisting of all sets $A \in \mathcal{F}$ such that $T^{-1}(A) = A$, is called σ-**algebra of T-invariant sets**.

While the notion *measure preserving* is self-explaining, the notion *ergodic* means that all sets that are invariant under T are trivial.

Example 14.1.2 On the probability space $(\{0, 1, \ldots, 7\}, 2^{\{0,1,\ldots,7\}}, \frac{1}{8}\sum_{k=0}^{7} \delta_k)$ we define for fixed $m \in \{1, \ldots, 7\}$ the map

$$T_m(k) := k + m \mod 8.$$

(1) Clearly, each T_m is measure preserving.
(2) T_m is ergodic if and only if m is an odd number: The nontrivial T_m-invariant sets for $m = 2, 6$ are $\{0, 2, 4, 6\}$ and $\{1, 3, 5, 7\}$, while T_4 has 4 nontrivial invariant sets. Consequently, these maps are not ergodic. If m is odd, one can see that $\{k, T_m(k), \ldots, T_m^7(k)\} = \{0, 1, \ldots, 7\}$ for any $k = 0, \ldots, 7$, so that $T_m^{-1}(A) = A$ implies $A = \{0, 1, \ldots, 7\}$ or $A = \emptyset$.

We continue with a *Maximal Ergodic Theorem*:

Theorem 14.1.3 *Let $(\Omega, \mathcal{F}, \mathbb{P})$ be a probability space, $T : \Omega \to \Omega$ be a measure preserving map, and $F \in \mathcal{L}_1(\Omega, \mathcal{F}, \mathbb{P})$. For $n \in \mathbb{N}$ we define*

$$S_n := F(T^0) + F(T) + \cdots + F(T^{n-1}) \quad \text{and} \quad M_n := \max\{0, S_1, \ldots, S_n\},$$

where $T^0 := I$ is the identical map, so that $F(T^0) = F$. Then $\mathbb{E}(F\mathbb{1}_{\{M_n > 0\}}) \geqslant 0$.

Proof First we observe that

$$F + M_n(T) \geqslant F + S_k(T) = S_{k+1} \quad \text{for} \quad 1 \leqslant k \leqslant n \in \mathbb{N}$$

and $F + M_n(T) \geqslant F = S_1$ so that

$$(F + M_n(T))^+ \geqslant M_{n+1} \geqslant M_n \quad \text{for} \quad n \in \mathbb{N}.$$

On $\{M_n > 0\}$ this implies $(F + M_n(T))^+ = F + M_n(T)$. Therefore,

$$\mathbb{E}F\mathbb{1}_{\{M_n > 0\}} \geqslant \mathbb{E}M_n\mathbb{1}_{\{M_n > 0\}} - \mathbb{E}M_n(T)\mathbb{1}_{\{M_n > 0\}}$$
$$\geqslant \mathbb{E}M_n\mathbb{1}_{\{M_n > 0\}} - \mathbb{E}M_n(T)\mathbb{1}_{\{M_n(T) > 0\}}$$
$$= 0,$$

where we used that

$$M_n(T)\mathbb{1}_{\{M_n > 0\}} \leqslant M_n(T) = M_n(T)\mathbb{1}_{\{M_n(T) > 0\}}$$

and that the latter expression has the same expectation as $M_n\mathbb{1}_{\{M_n > 0\}}$ since T is measure preserving. $\square$

Now we turn to the Birkhoff-Khinchin ergodic theorem:

14.1 Birkhoff-Khinchin Ergodic Theorem

Theorem 14.1.4 (Birkhoff[1] and Khinchin[2]) *Let $(\Omega, \mathcal{F}, \mathbb{P})$ be a probability space, $T : \Omega \to \Omega$ be ergodic, and $F : \Omega \to \mathbb{R}$ be a random variable such that $\mathbb{E}|F| < \infty$. Then one has that*

$$\lim_{n \to \infty} \frac{S_n}{n} = \mathbb{E}F \text{ a.s.} \quad \text{with} \quad S_n := \sum_{k=0}^{n-1} F(T^k).$$

Proof By normalization we can assume that $\mathbb{E}F = 0$. Let

$$\eta_- := \liminf_{n \to \infty} \frac{1}{n} \sum_{k=0}^{n-1} F(T^k) \leqslant \limsup_{n \to \infty} \frac{1}{n} \sum_{k=0}^{n-1} F(T^k) =: \eta_+.$$

We show that $0 \leqslant \eta_- \leqslant \eta_+ \leqslant 0$ a.s. By the symmetry of the problem (replace F by $-F$) it is sufficient to show that $\eta := \eta_+ \leqslant 0$ a.s. For $\varepsilon > 0$ and $n \in \mathbb{N}$ let

$$A_\varepsilon := \{\eta > \varepsilon\},$$
$$F_\varepsilon := (F - \varepsilon)\mathbb{1}_{A_\varepsilon},$$
$$S_{n,\varepsilon} := F_\varepsilon + F_\varepsilon(T) + \cdots + F_\varepsilon(T^{n-1}),$$
$$M_{n,\varepsilon} := \max\{0, S_{1,\varepsilon}, \ldots, S_{n,\varepsilon}\}.$$

If $\mathcal{J}$ is the σ-algebra of T-invariant sets, then we have $A_\varepsilon \in \mathcal{J}$ as $\limsup_{n \to \infty} \frac{1}{n} \sum_{k=0}^{n-1} F(T^k) > \varepsilon$ if and only if

$$\limsup_{n \to \infty} \frac{1}{n} \sum_{k=0}^{n-1} F(T^{k+1}) = \limsup_{n \to \infty} \frac{n+1}{n} \left[\frac{1}{n+1} \sum_{k=1}^{n} F(T^k) \right]$$
$$= \limsup_{n \to \infty} \left[\frac{1}{n+1} \sum_{k=0}^{n} F(T^k) \right] > \varepsilon.$$

---

[1] George David Birkhoff, 21/03/1884 (Overisel, Michigan, USA)–12/11/1944 (Cambridge, Massachusetts, USA).
[2] Aleksandr Yakovlevich Khinchin, 19/07/1894 (Kondrovo, Kaluzhskaya guberniya, Russia)–18/11/1959 (Moscow, USSR).

From Theorem 14.1.3 it follows that $\mathbb{E}(F_\varepsilon \mathbb{1}_{M_{n,\varepsilon}>0}) \geq 0$. Moreover,

$$\bigcup_{n=1}^{\infty}\{M_{n,\varepsilon} > 0\} = \left\{\sup_{n\in\mathbb{N}} S_{n,\varepsilon} > 0\right\} = \left\{\sup_{n\in\mathbb{N}} \frac{S_{n,\varepsilon}}{n} > 0\right\}$$

$$= \left\{\sup_{n\in\mathbb{N}} \frac{S_n}{n} > \varepsilon\right\} \cap A_\varepsilon = A_\varepsilon$$

because of $\sup_{n\in\mathbb{N}} \frac{S_n}{n} \geq \eta$ which follows from $\limsup_{n\to\infty} \frac{S_n}{n} = \eta$. By Lebesgue's dominated convergence,

$$0 \leq \mathbb{E}(F_\varepsilon \mathbb{1}_{\{M_{n,\varepsilon}>0\}}) \to \mathbb{E}(F_\varepsilon \mathbb{1}_{A_\varepsilon}) = \mathbb{E}(F \mathbb{1}_{A_\varepsilon}) - \varepsilon \mathbb{P}(A_\varepsilon)$$

as $n \to \infty$. Since we assumed that $\mathbb{E}F = 0$ and all sets from J have measure 0 or 1, it follows from Theorem 10.3.1 that $\mathbb{E}(F|J) = 0$ a.s. Because $A_\varepsilon \in J$ we get, using properties of the conditional expectation (Proposition 10.3.4), that

$$\mathbb{E}(F \mathbb{1}_{A_\varepsilon}) = \mathbb{E}(\mathbb{1}_{A_\varepsilon} \mathbb{E}(F|J)) = \mathbb{E}(\mathbb{1}_{A_\varepsilon} 0) = 0.$$

Therefore,

$$0 \leq \lim_{n\to\infty} \mathbb{E}(F_\varepsilon \mathbb{1}_{\{M_{n,\varepsilon}>0\}}) = -\varepsilon \mathbb{P}(A_\varepsilon)$$

so that $\mathbb{P}(A_\varepsilon) = 0$ for all $\varepsilon > 0$ which concludes the proof. □

14.2 A Generalized SLLN

The aim of this section is to apply the ergodic Theorem 14.1.4 in a setting with independent random variables.

Let μ be a probability measure on $(\mathbb{R}, \mathcal{B}(\mathbb{R}))$, $(\overline{\Omega}, \overline{\mathcal{F}}, \overline{\mathbb{P}}) := \otimes_{n=1}^{\infty}(\mathbb{R}, \mathcal{B}(\mathbb{R}), \mu)$, and define $T : \overline{\Omega} \to \overline{\Omega}$ by

$$T(x_1, x_2, x_3, \ldots) := (x_2, x_3, \ldots).$$

Then we have the following result:

Lemma 14.2.1 *The map T is ergodic.*

Proof
(a) T is measure-preserving which can be checked on the π-system of all $E_1 \times \cdots \times E_n \times \mathbb{R} \times \mathbb{R} \times \cdots$ with $n \in \mathbb{N}$ and $E_1, \ldots, E_n \in \mathcal{B}(\mathbb{R})$.

14.2 A Generalized SLLN

(b) Assume now that we have a T-invariant set $A \in \mathcal{F}$, that means it holds $T^{-1}(A) = A$. By iteration we get that $T^{-n}A = A$ for $n \in \mathbb{N}$. This implies that

$$A \in \bigcap_{n=2}^{\infty} \sigma(\pi_n, \pi_{n+1}, \ldots)$$

where the π_n are the coordinate functionals $\pi_n : \overline{\Omega} \to \mathbb{R}$ with $\pi_n(x_1, x_2, x_3, \ldots) := x_n$. By construction, the $(\pi_n)_{n \in \mathbb{N}}$ are i.i.d. where each random variable has the law μ. By the 0-1 law of Kolmogorov (Theorem 7.2.1) we get that $\overline{\mathbb{P}}(A) \in \{0, 1\}$. Hence T is ergodic. □

From the ergodic Theorem 14.1.4 we deduce the following version of a strong law of large numbers which cam be seen as a generalization of the SLLN in Kolmogorov form:

Theorem 14.2.2 *Let $(\Omega, \mathcal{F}, \mathbb{P})$ be a probability space, $f_1, f_2, \ldots : \Omega \to \mathbb{R}$ be i.i.d. random variables, and let $\Phi : \mathbb{R}^\mathbb{N} \to \mathbb{R}$ be $(\mathcal{B}(\mathbb{R}^\mathbb{N}), \mathcal{B}(\mathbb{R}))$-measurable such that $\mathbb{E}|\Phi(f_1, f_2, \ldots)| < \infty$. Then one has*

$$\lim_{n \to \infty} \frac{\Phi(f_1, f_2, \ldots) + \cdots + \Phi(f_n, f_{n+1}, \ldots)}{n} = \mathbb{E}\Phi(f_1, f_2, \ldots) \ a.s.$$

Proof We let μ be the law of f_1 and recall $(\overline{\Omega}, \overline{\mathcal{F}}, \overline{\mathbb{P}}) := \bigotimes_{n=1}^{\infty}(\mathbb{R}, \mathcal{B}(\mathbb{R}), \mu)$. We define $f : \overline{\Omega} \to \mathbb{R}$ by $f(x_1, x_2, \ldots) := x_1$. Then $f(T^n)(x_1, x_2, \ldots) = x_{n+1}$, so that the family $(f(T^n))_{n=0}^{\infty}$ forms an i.i.d. sequence of random variables, where the random variables have the same distribution as f_1. For $n \in \mathbb{N}$ we let

$$\overline{f}_n := f(T^{n-1}), \quad F_n := \Phi(f_n, f_{n+1}, \cdots), \quad \text{and} \quad \overline{F}_n := \Phi(\overline{f}_n, \overline{f}_{n+1}, \cdots).$$

Then for all $n \in \mathbb{N}$ and $B_1, \ldots, B_n \in \mathcal{B}(\mathbb{R})$ we have

$$\mathbb{P}(F_1 \in B_1, \ldots, F_n \in B_n) = \overline{\mathbb{P}}(\overline{F}_1 \in B_1, \ldots, \overline{F}_n \in B_n). \tag{14.1}$$

It follows by Theorem 14.1.4 that

$$\lim_{n \to \infty} \frac{1}{n} \left(\overline{F}_1(T^0) + \overline{F}_1(T) + \cdots + \overline{F}_1(T^{n-1}) \right) = \mathbb{E}\overline{F}_1 \ a.s.$$

which is by definition the same as

$$\lim_{n \to \infty} \frac{1}{n} \left(\overline{F}_1 + \cdots + \overline{F}_n \right) = \mathbb{E}\overline{F}_1 \ a.s.$$

The latter we can express as

$$1 = \overline{\mathbb{P}}\left(\bigcap_{N=1}^{\infty}\bigcup_{m=1}^{\infty}\bigcap_{n=m}^{\infty}\left\{\left|\frac{1}{n}\sum_{k=1}^{n}\overline{F}_k - \mathbb{E}\overline{F}_1\right| < \frac{1}{N}\right\}\right)$$

$$= \overline{\mathbb{P}}_{(\overline{F}_n)_{n=1}^{\infty}}\left(\bigcap_{N=1}^{\infty}\bigcup_{m=1}^{\infty}\bigcap_{n=m}^{\infty}\left\{(x_k)_{k=1}^{\infty} : \left|\frac{1}{n}\sum_{k=1}^{n}x_k - \mathbb{E}\overline{F}_1\right| < \frac{1}{N}\right\}\right)$$

where $\overline{\mathbb{P}}_{(\overline{F}_n)_{n=1}^{\infty}}$ is the law of $(\overline{F}_n)_{n=1}^{\infty}$ on $\otimes_{n=1}^{\infty}(\mathbb{R}, \mathcal{B}(\mathbb{R}))$. Because of (14.1) this implies

$$\lim_{n\to\infty}\frac{1}{n}(F_1 + \cdots + F_n) = \mathbb{E}F_1 \text{ a.s.}$$

which is the statement to prove. □

Finally we deduce the SLLN in Kolmogorov form:

Proof of Theorem 13.1.1 when $f_1, f_2, \ldots$ are independent. We apply Theorem 14.2.2 to the function $\Phi(x_1, x_2, \ldots) := x_1$ so that $\mathbb{E}\Phi(f_1, f_2, \ldots) = \mathbb{E}f_1$ and

$$\frac{\Phi(f_1, f_2, \ldots) + \cdots + \Phi(f_n, f_{n+1}, \ldots)}{n} = \frac{f_1 + \cdots + f_n}{n}.$$

□

14.3 Exercises

Ex 1: Given a measurable space $(\Omega, \mathcal{F})$ and a map $T : \Omega \to \Omega$, check that the collection of $A \in \mathcal{F}$ such that $T^{-1}(A) = A$ is a σ-algebra.

14.4 Comments

Section 14.1 follows the presentation in [173, Section V.3]. The maximal ergodic Theorem 14.1.3 goes back to Yosida and Kakutani [201] with a reference to Kolmogorov [106]. The proof we presented is due to A.M. Garsia [69]. The ergodic Theorem 14.1.4 originates from Birkhoff [19], the probabilistic formulation is due to Khinchin [102].

Chapter 15
Limit Theorems for Weak Convergence

In this chapter we present two basic limit theorems that use the weak convergence we studied in Chap. 12. The first one is the Central Limit Theorem (CLT) that explains why the Gaussian distribution plays such an exceptional role in probability and statistics. The second one is the limit theorem of Poisson, that shows accordingly the natural role of the Poisson distribution.

The plan of this chapter is as follows:

- In Theorem 15.1.1 we prove the classical central limit theorem.
- In Theorem 15.1.2 we prove the central limit theorem under the more general Lindeberg condition.
- In Theorem 15.2.1 we verify the Berry-Esseen theorem, which provides a speed of convergence for the classical central limit theorem (Theorem 15.1.1), we did start from.
- Finally in Theorem 15.4.1 we prove the Poisson limit theorem.

15.1 Central Limit Theorems

The classical central limit theorem states that, under certain prerequisites, the normalized sum of independent random variables converges weakly towards the standard normal distribution:

Theorem 15.1.1 (Classical CLT) *Let $(\Omega, \mathcal{F}, \mathbb{P})$ be a probability space and let $f_1, f_2, \ldots \Omega \to \mathbb{R}$ be a sequence of i.i.d. random variables such that $\mathbb{E} f_n^2 = 1$ and $\mathbb{E} f_n = 0$. If $S_n := f_1 + \cdots + f_n$, then one has*

$$\frac{S_n}{\sqrt{n}} \xrightarrow{d} g \quad \text{with} \quad \text{law}(g) = \mathcal{N}_{0,1}.$$

One interpretation and application is as follows: We wish to model the movement of a particle between time $t = 0$ and and time $t = 1$. For this we consider a time-discretization by n steps of length $1/n$. The change of the position of the particle from time $t = (k-1)/n$ to $t = k/n$ is modelled by $f_k/\sqrt{n}$ so that the particle at time $t = 1$ is at the position

$$\frac{1}{\sqrt{n}}(f_1 + \cdots + f_n).$$

What is the explanation of the factor $1/\sqrt{n}$? Naturally we want to have that the approximations preserve the energy, here expressed as the second moment. And, in fact, we have

$$\mathbb{E}\left(\frac{1}{\sqrt{n}}(f_1 + \cdots + f_n)\right)^2 = \frac{1}{n}\sum_{i=1}^{n}\mathbb{E}f_i^2 = 1.$$

Before we prove Theorem 15.1.1 we state a generalization of it, the CLT in Lindeberg[1] form:

Theorem 15.1.2 (CLT in Lindeberg Form) *Assume a probability space* $(\Omega, \mathcal{F}, \mathbb{P})$ *and independent random variables* $f_1, f_2, \ldots : \Omega \to \mathbb{R}$ *such that* $0 < \mathbb{E}f_n^2 < \infty$ *and* $\mathbb{E}f_n = 0$ *for* $n \in \mathbb{N}$.

(1) *Assume that*

$$\lim_{n\to\infty} \frac{1}{D_n^2} \sum_{k=1}^{n} \int_{\{|f_k|>\varepsilon D_n\}} f_k^2 \, d\mathbb{P} = 0 \quad \text{with} \quad D_n := \sqrt{\operatorname{var}(f_1 + \cdots + f_n)} \tag{15.1}$$

is satisfied for all $\varepsilon > 0$. *If* $S_n := f_1 + \cdots + f_n$, *then one has that*

$$\frac{S_n}{D_n} \xrightarrow{d} g \quad \text{with} \quad \operatorname{law}(g) = \mathcal{N}_{0,1}.$$

(2) *If* $\frac{S_n}{D_n} \xrightarrow{d} g$ *and* $\max_{1 \leqslant k \leqslant n} \frac{\operatorname{var}(f_k)}{D_n^2} \xrightarrow{n\to\infty} 0$, *then* (15.1) *holds.*

There are two ways to prove Theorem 15.1.1: directly, or by deducing Theorem 15.1.1 from Theorem 15.1.2 (see Exercise 1). The basic idea to prove the above theorems is as follows: Assume independent random variables $f_1, f_2, \ldots : \Omega \to \mathbb{R}$ such that $0 < \mathbb{E}f_n^2 < \infty$ and $\mathbb{E}f_n = 0$ for $n \in \mathbb{N}$ and define

$$D_n := \sqrt{\operatorname{var}(f_1 + \cdots + f_n)}, \tag{15.2}$$

---

[1] Jarl Waldemar Lindeberg, 04/08/1876 (Helsinki, Finland)–24/12/1932 (Helsinki, Finland).

15.1 Central Limit Theorems

$$S_n := f_1 + \cdots + f_n. \tag{15.3}$$

The random variable $\frac{S_n}{D_n}$ has variance one by definition. For its characteristic function we obtain

$$\varphi_{\frac{S_n}{D_n}}(x) = \prod_{k=1}^{n} \varphi_{\frac{f_k}{D_n}}(x) = \prod_{k=1}^{n} \varphi_{f_k}\left(\frac{x}{D_n}\right).$$

In order to prove the weak convergence $\frac{S_n}{D_n} \xrightarrow{d} g$ with $\text{law}(g) = \mathcal{N}_{0,1}$ we will check

$$\prod_{k=1}^{n} \varphi_{f_k}\left(\frac{x}{D_n}\right) \xrightarrow[n]{} e^{-\frac{x^2}{2}} \quad \text{for} \quad x \in \mathbb{R} \tag{15.4}$$

and then exploit Theorems 12.1.1 and 11.7.1. So, finally, we have to treat an analytical problem. This idea is made precise in Lemma 15.1.4. For $z \in \mathbb{C}$ with $|z| < 1$ we define a branch of the complex logarithm by the Newton[2]-Mercator[3] series as

$$\text{Log}(1+z) := \sum_{n=1}^{\infty} \frac{(-1)^{n+1}}{n} z^n.$$

Later we use that $e^{\text{Log}(1+z)} = 1 + z$ for $|z| < 1$ and the following estimate:

Lemma 15.1.3 *For $z \in \mathbb{C}$ with $|z| \leq \frac{1}{2}$ one has that*

$$|\text{Log}(1+z) - z| \leq |z|^2.$$

Proof Indeed, we get that

$$|\text{Log}(1+z) - z| = \left|\sum_{n=1}^{\infty} \frac{(-1)^{n+1}}{n} z^n - z\right| = |z|^2 \left|\sum_{n=2}^{\infty} \frac{(-1)^{n+1}}{n} z^{n-2}\right|$$

$$\leq \frac{|z|^2}{2} \left|\sum_{n=2}^{\infty} 2^{2-n}\right| = |z|^2.$$

□

---

[2] Isaac Newton, 04/01/1643 (Woolsthorpe, Lincolnshire, England)–31/03/1727 (London, England).

[3] Nicolaus Mercator, 1620 (Eutin, Holstein, Holy Roman Empire, now Germany)- –14/01/1687 (Paris, France).

Lemma 15.1.4 *Assume independent random variables $f_n : \Omega \to \mathbb{R}$ such that $0 < \mathbb{E} f_n^2 < \infty$ and $\mathbb{E} f_n = 0$ for $n \in \mathbb{N}$. Fix $x \in \mathbb{R}$, define D_n and S_n by (15.2) and (15.3), and $C_{kn} := \varphi_{f_k}\left(\frac{x}{D_n}\right) - 1$ for $k = 1, \ldots, n$. Assume that*

(1) $\lim_{n \to \infty} \sum_{k=1}^{n} |C_{kn}|^2 = 0$ *and*
(2) $\lim_{n \to \infty} \sum_{k=1}^{n} C_{kn} = -\frac{x^2}{2}$.

Then we get that $\lim_{n \to \infty} \varphi_{\frac{S_n}{D_n}}(x) = e^{-\frac{x^2}{2}}$.

Proof First we observe that

$$\varphi_{\frac{S_n}{D_n}}(x) = \prod_{k=1}^{n} \varphi_{f_k}\left(\frac{x}{D_n}\right) = \prod_{k=1}^{n}(1 + C_{kn}).$$

Using assumption (1) we find an $n_0 \in \mathbb{N}$ such that for $n \geq n_0$ we get

$$\max_{k=1,\ldots,n} |C_{kn}| \leq \frac{1}{2}.$$

For $n \geq n_0$ Lemma 15.1.3 implies

$$\left|\sum_{k=1}^{n} \operatorname{Log} \varphi_{f_k}\left(\frac{x}{D_n}\right) - \sum_{k=1}^{n} C_{kn}\right| \leq \sum_{k=1}^{n} \left|\operatorname{Log} \varphi_{f_k}\left(\frac{x}{D_n}\right) - C_{kn}\right| \leq \sum_{k=1}^{n} |C_{kn}|^2,$$

so that, by assumption (1) and (2),

$$\sum_{k=1}^{n} \operatorname{Log} \varphi_{f_k}\left(\frac{x}{D_n}\right) + \frac{x^2}{2} \xrightarrow[n]{} 0.$$

Hence

$$\varphi_{\frac{S_n}{D_n}}(x) = \exp\left(\sum_{k=1}^{n} \operatorname{Log} \varphi_{f_k}\left(\frac{x}{D_n}\right)\right) \xrightarrow[n]{} e^{-\frac{x^2}{2}}.$$

$\square$

Proof of Theorem 15.1.1 We have $D_n = \sqrt{n}$ and $\varphi := \varphi_{f_1} = \varphi_{f_2} = \cdots$. Since $\mathbb{E} f_1^2 < \infty$ Proposition 11.6.1 implies that φ, φ', and φ'' are continuous and bounded, so that the qualitative Taylor formula yields

$$\varphi(x) = \varphi(0) + x\varphi'(0) + \frac{x^2}{2}\varphi''(0) + o(x^2)$$

15.1 Central Limit Theorems

$$= \varphi(0) + xi\mathbb{E}f_1 + \frac{x^2}{2}i^2\mathbb{E}f_1^2 + o(x^2)$$

$$= 1 - \frac{x^2}{2} + o(x^2)$$

for $x \in \mathbb{R}$, where $o(x^2)$ stands for $\lim_{\xi \to 0, \xi \neq 0} \frac{o(\xi^2)}{\xi^2} = 0$ and $o(0) = 0$. So we get

$$\varphi\left(\frac{x}{\sqrt{n}}\right) = 1 - \frac{x^2}{2n} + o\left(\frac{x^2}{n}\right) = 1 + C_{kn} \quad \text{with} \quad C_{kn} := -\frac{x^2}{2n} + o\left(\frac{x^2}{n}\right),$$

and the assumptions of Lemma 15.1.4 are satisfied. We finish the proof with Theorem 12.1.1 which states that the pointwise convergence $\lim_{n \to \infty} \varphi_{\frac{S_n}{D_n}}(x) = e^{-\frac{x^2}{2}}$ is equivalent to $\frac{S_n}{D_n} \xrightarrow{d} g$, where g is a random variable with law$(g) = \mathcal{N}_{0,1}$. $\square$

Proof of Theorem 15.1.2 (1) Given $\varepsilon > 0$ we define

$$A_{kn}^\varepsilon := \frac{1}{D_n^2} \int_{\{|f_k| \leq \varepsilon D_n\}} f_k^2 d\mathbb{P} \quad \text{and} \quad B_{kn}^\varepsilon := \frac{1}{D_n^2} \int_{\{|f_k| > \varepsilon D_n\}} f_k^2 d\mathbb{P},$$

so that

$$A_{kn}^\varepsilon \leq \varepsilon^2, \quad \sum_{k=1}^n (A_{kn}^\varepsilon + B_{kn}^\varepsilon) = 1, \quad \text{and} \quad \sum_{k=1}^n B_{kn}^\varepsilon \xrightarrow[n]{} 0,$$

where the last convergence holds by assumption.
(a) First we show that

$$\left| C_{kn} + \frac{x^2}{2} A_{kn}^\varepsilon \right| \leq \frac{x^2}{2} B_{kn}^\varepsilon + \varepsilon \frac{|x|^3}{6} A_{kn}^\varepsilon$$

which follows with the notation $\varphi_{f_k}\left(\frac{x}{D_n}\right) = 1 + C_{kn}$ from

$$\left| C_{kn} + \frac{x^2}{2} A_{kn}^\varepsilon \right|$$

$$= \left| \varphi_{f_k}\left(\frac{x}{D_n}\right) - 1 + \frac{1}{2}\left(\frac{x}{D_n}\right)^2 \int_{\{|f_k| \leq \varepsilon D_n\}} f_k^2 d\mathbb{P} \right|$$

$$= \left| \int_\Omega e^{i\frac{x}{D_n} f_k} d\mathbb{P} - 1 + \frac{1}{2}\left(\frac{x}{D_n}\right)^2 \int_{\{|f_k| \leq \varepsilon D_n\}} f_k^2 d\mathbb{P} \right|$$

$$\leqslant \left| \int_{\{|f_k|\leqslant \varepsilon D_n\}} \left[1 + i\left(\frac{x}{D_n}f_k\right) - \frac{1}{2}\left(\frac{x}{D_n}f_k\right)^2 + R_3\left(\frac{1}{3!}\left(\frac{xf_k}{D_n}\right)^3\right)\right] d\mathbb{P}\right.$$

$$+ \int_{\{|f_k|>\varepsilon D_n\}} \left[1 + i\left(\frac{x}{D_n}f_k\right) + R_2\left(\frac{1}{2}\left(\frac{xf_k}{D_n}\right)^2\right)\right] d\mathbb{P}$$

$$\left. - 1 + \frac{1}{2}\left(\frac{x}{D_n}\right)^2 \int_{\{|f_k|\leqslant \varepsilon D_n\}} f_k^2 \, d\mathbb{P}\right|$$

$$= \left| \int_{\{|f_k|>\varepsilon D_n\}} R_2\left(\frac{1}{2}\left(\frac{xf_k}{D_n}\right)^2\right) d\mathbb{P} + \int_{\{|f_k|\leqslant \varepsilon D_n\}} R_3\left(\frac{1}{3!}\left(\frac{xf_k}{D_n}\right)^3\right) d\mathbb{P}\right|$$

$$\leqslant \frac{x^2}{2} B_{kn}^\varepsilon + \varepsilon \frac{|x|^3}{6} A_{kn}^\varepsilon,$$

where we used Proposition A.1.1(3) to obtain remainder terms $R_2, R_3 : \mathbb{R} \to \mathbb{C}$ satisfying $|R_i(x)| \leqslant |x|$.

(b) We prove that $\lim_{n\to\infty} \varphi_{\frac{S_n}{D_n}}(x) = e^{-\frac{x^2}{2}}$ by showing that the assumptions of Lemma 15.1.4 are satisfied: For Lemma 15.1.4(1) we observe

$$|C_{kn}| \leqslant \frac{x^2}{2} B_{kn}^\varepsilon + \left(\varepsilon \frac{|x|^3}{6} + \frac{x^2}{2}\right) A_{kn}^\varepsilon$$

so that

$$\sum_{k=1}^n |C_{kn}|^2 \leqslant \left(\sum_{k=1}^n |C_{kn}|\right)\left(\max_{k=1,\ldots,n} |C_{kn}|\right)$$

$$\leqslant \left(\frac{x^2}{2}\sum_{k=1}^n B_{kn}^\varepsilon + \left(\varepsilon\frac{|x|^3}{6} + \frac{x^2}{2}\right)\sum_{k=1}^n A_{kn}^\varepsilon\right)$$

$$\times \left(\frac{x^2}{2}\sum_{k=1}^n B_{kn}^\varepsilon + \left(\varepsilon\frac{|x|^3}{6} + \frac{x^2}{2}\right)\varepsilon^2\right).$$

This implies, thanks to $\sum_{k=1}^n A_{kn}^\varepsilon \leqslant 1$ and $\sum_{k=1}^n B_{kn}^\varepsilon \xrightarrow[n]{} 0$, that

$$\limsup_{n\to\infty} \sum_{k=1}^n |C_{kn}|^2 \leqslant \left(\varepsilon\frac{|x|^3}{6} + \frac{x^2}{2}\right)^2 \varepsilon^2.$$

15.1 Central Limit Theorems

As $\varepsilon > 0$ was arbitrary, we conclude that $\lim_{n\to\infty} \sum_{k=1}^{n} |C_{kn}|^2 = 0$. Regarding Lemma 15.1.4(2) we observe by step (a) that

$$\left| \frac{x^2}{2} + \sum_{k=1}^{n} C_{kn} \right| \leq \left| \frac{x^2}{2} - \frac{x^2}{2} \sum_{k=1}^{n} A_{kn}^{\varepsilon} \right| + \left| \frac{x^2}{2} \sum_{k=1}^{n} A_{kn}^{\varepsilon} + \sum_{k=1}^{n} C_{kn} \right|$$

$$\leq \frac{x^2}{2} \sum_{k=1}^{n} B_{kn}^{\varepsilon} + \sum_{k=1}^{n} \left(\frac{x^2}{2} B_{kn}^{\varepsilon} + \varepsilon \frac{|x|^3}{6} A_{kn}^{\varepsilon} \right)$$

$$\leq x^2 \sum_{k=1}^{n} B_{kn}^{\varepsilon} + \varepsilon \frac{|x|^3}{6}.$$

Therefore,

$$\limsup_{n\to\infty} \left| \frac{x^2}{2} + \sum_{k=1}^{n} C_{kn} \right| \leq \varepsilon \frac{|x|^3}{6}$$

for all $\varepsilon > 0$ and $\lim_{n\to\infty} \left| \frac{x^2}{2} + \sum_{k=1}^{n} C_{kn} \right| = 0$.

(2) We assume that

$$\lim_{n\to\infty} \varphi_{\frac{S_n}{D_n}}(x) = e^{-\frac{x^2}{2}}, \tag{15.5}$$

$$\lim_{n\to\infty} \left[\max_{1\leq k\leq n} \frac{\sigma_k^2}{D_n^2} \right] = 0. \tag{15.6}$$

Since Proposition A.1.1(3) applied to $e^{i\frac{x}{D_n} f_k}$ and $N = 1$ gives that $\left| \varphi_{f_k}\left(\frac{x}{D_n}\right) - 1 \right| \leq \frac{\sigma_k^2}{2D_n^2}$, (15.6) implies that

$$\max_{1\leq k\leq n} \left| \varphi_{f_k}\left(\frac{x}{D_n}\right) - 1 \right| \xrightarrow{n\to\infty} 0 \text{ and } \sum_{k=1}^{n} \left| \varphi_{f_k}\left(\frac{x}{D_n}\right) - 1 \right|^2 \xrightarrow{n\to\infty} 0. \tag{15.7}$$

We fix $x \in \mathbb{R}$ and choose $n_0 \in \mathbb{N}$ such that $\left| \varphi_{f_k}\left(\frac{x}{D_n}\right) - 1 \right| \leq 1/2$ for $n \geq n_0$. We get that $\text{Log}\,\varphi_{f_k}\left(\frac{x}{D_n}\right)$ exists for $n \geq n_0$ and (15.5) can be written as

$$\sum_{k=1}^{n} \text{Log}\,\varphi_{f_k}\left(\frac{x}{D_n}\right) \xrightarrow{n\to\infty} -\frac{x^2}{2}.$$

This implies

$$\left| \frac{x^2}{2} - \sum_{k=1}^{n}\left(1 - \varphi_{f_k}\left(\frac{x}{D_n}\right)\right) \right| \leq \left| \frac{x^2}{2} + \sum_{k=1}^{n} \operatorname{Log} \varphi_{f_k}\left(\frac{x}{D_n}\right) \right|$$
$$+ \left| -\sum_{k=1}^{n} \operatorname{Log} \varphi_{f_k}\left(\frac{x}{D_n}\right) - \sum_{k=1}^{n}\left(1 - \varphi_{f_k}\left(\frac{x}{D_n}\right)\right) \right| \stackrel{n\to\infty}{\to} 0, \quad (15.8)$$

where we used that the second summand can be estimated by Lemma 15.1.3 by

$$\left| \sum_{k=1}^{n}\left(\varphi_{f_k}\left(\frac{x}{D_n}\right) - 1\right) - \sum_{k=1}^{n} \operatorname{Log} \varphi_{f_k}\left(\frac{x}{D_n}\right) \right| \leq \sum_{k=1}^{n} \left|\varphi_{f_k}\left(\frac{x}{D_n}\right) - 1\right|^2$$

and therefore converges to zero by (15.7).

Considering the real part of the term on the left hand side of (15.8) and decomposing the integral, we have for $\varepsilon > 0$ that

$$\frac{x^2}{2} - \sum_{k=1}^{n} \int_{\{|z|\leq \varepsilon D_n\}} \left(1 - \cos\left(\frac{xz}{D_n}\right)\right) d\mathbb{P}_{f_k}(z)$$
$$= \sum_{k=1}^{n} \int_{\{|z|>\varepsilon D_n\}} \left(1 - \cos\left(\frac{xz}{D_n}\right)\right) d\mathbb{P}_{f_k}(z) + R_n(x)$$
$$\leq 2 \sum_{k=1}^{n} \int_{\{|z|>\varepsilon D_n\}} d\mathbb{P}_{f_k}(z) + R_n(x) \leq \frac{2}{\varepsilon^2} + R_n(x)$$

with $\lim_{n\to\infty} R_n(x) = 0$. We use $1 - \cos x \leq \frac{x^2}{2}$ to deduce

$$\sum_{k=1}^{n} \int_{\{|z|\leq \varepsilon D_n\}} \left(1 - \cos\left(\frac{xz}{D_n}\right)\right) d\mathbb{P}_{f_k}(z) \leq \frac{x^2}{2D_n^2} \sum_{k=1}^{n} \int_{\{|z|\leq \varepsilon D_n\}} z^2 d\mathbb{P}_{f_k}(z)$$
$$= \frac{x^2}{2}(1 - a_n(\varepsilon))$$

for $a_n(\varepsilon) := \frac{1}{D_n^2} \sum_{k=1}^{n} \int_{\{|f_k|>\varepsilon D_n\}} f_k^2 d\mathbb{P}$. This implies $\frac{x^2}{2} a_n(\varepsilon) \leq \frac{2}{\varepsilon^2} + R_n(x)$ and hence $\lim_{n\to\infty} a_n(\varepsilon) \leq \frac{4}{\varepsilon^2 x^2}$ for all $x \in \mathbb{R} \setminus \{0\}$. But this means that $\lim_{n\to\infty} a_n(\varepsilon) = 0$. □

15.2 The Berry-Esseen Theorem

In Theorem 15.1.1 we proved the classical central limit theorem, but without saying how fast the convergence is. The following theorem gives us an information about this:

Theorem 15.2.1 (Berry[4]-Esseen[5]) *Let $(\Omega, \mathcal{F}, \mathbb{P})$ be a probability space, and let $f_1, f_2, \ldots$ be i.i.d. random variables such that $\mathbb{E} f_1 = 0$ and $\mathbb{E} f_1^2 = 1$ and $\mathbb{E}|f_1|^{2+\alpha} < \infty$ for some $\alpha \in (0, 1]$. If $S_n := f_1 + \cdots + f_n$, then there is a $c > 0$ such that for all $n \in \mathbb{N}$ one has*

$$\sup_{-\infty < z < \infty} \left| \mathbb{P}\left(\frac{S_n}{\sqrt{n}} \leq z\right) - \mathcal{N}_{0,1}((-\infty, z]) \right| \leq c n^{-\frac{\alpha}{2}}.$$

For the proof of the Berry-Esseen theorem we use the following three lemmas:

Lemma 15.2.2 *Assume $f, g \in \mathcal{L}_0(\Omega, \mathcal{F}, \mathbb{P})$ such that*

$$\int_{\mathbb{R}} |F_f(y) - F_g(y)| \mathrm{d}\lambda(y) < \infty$$

and $x \neq 0$. Then one has

$$\int_{\mathbb{R}} e^{ixy} [F_f(y) - F_g(y)] \mathrm{d}\lambda(y) = \frac{\varphi_f(x) - \varphi_g(x)}{-ix}.$$

Proof Using Fubini's theorem we get that

$$\int_{\mathbb{R}} e^{ixy} [F_f(y) - F_g(y)] \mathrm{d}\lambda(y)$$

$$= \int_{\mathbb{R}} e^{ixy} [\mathbb{P}(f \leq y) - \mathbb{P}(g \leq y)] \mathrm{d}\lambda(y)$$

$$= \int_{\mathbb{R}} \int_{\Omega} e^{ixy} [\mathbb{1}_{\{f \leq y\}} - \mathbb{1}_{\{g \leq y\}}] \mathrm{d}\mathbb{P} \mathrm{d}\lambda(y)$$

$$= \int_{\mathbb{R}} \int_{\Omega} e^{ixy} \left[\mathbb{1}_{\{f < g\}} \mathbb{1}_{\{f \leq y < g\}} - \mathbb{1}_{\{g < f\}} \mathbb{1}_{\{g \leq y < f\}} \right] \mathrm{d}\mathbb{P} \mathrm{d}\lambda(y)$$

$$= \int_{\Omega} \left[\mathbb{1}_{\{f < g\}} \frac{e^{ixg} - e^{ixf}}{ix} - \mathbb{1}_{\{g < f\}} \frac{e^{ixf} - e^{ixg}}{ix} \right] \mathrm{d}\mathbb{P}$$

---

[4] Andrew Campbell Berry, 23/11/1906 (Somerville, Massachusetts, USA)–13/01/1998 (Appleton, Wisconsin, USA).
[5] Carl-Gustav Esseen, 13/09/1918 (Linköping, Sweden)–10/11/2001 (Linköping, Sweden).

$$= \frac{\varphi_f(x) - \varphi_g(x)}{-ix}.$$

□

Lemma 15.2.3 *For $f, g \in \mathcal{L}_2(\Omega, \mathcal{F}, \mathbb{P})$ one has*

$$\int_{\mathbb{R}} |F_f(y) - F_g(y)| \mathrm{d}\lambda(y) < \infty.$$

Proof For $y > 0$ we have

$$\begin{aligned}
|F_f(y) - F_g(y)| &= |\mathbb{P}(f \leqslant y) - \mathbb{P}(g \leqslant y)| \\
&= |\mathbb{P}(f > y) - \mathbb{P}(g > y)| \\
&\leqslant \mathbb{P}(|f| > y) + \mathbb{P}(|g| > y)| \\
&\leqslant \min\left\{y^{-2}\mathbb{E}f^2, 1\right\} + \min\left\{y^{-2}\mathbb{E}g^2, 1\right\},
\end{aligned}$$

whereas for $y < 0$ we have

$$\begin{aligned}
|F_f(y) - F_g(y)| &= |\mathbb{P}(f \leqslant y) - \mathbb{P}(g \leqslant y)| \\
&= |\mathbb{P}(-f \geqslant -y) - \mathbb{P}(-g \geqslant -y)| \\
&\leqslant \mathbb{P}(|f| \geqslant -y) + \mathbb{P}(|g| \geqslant -y)| \\
&\leqslant \min\left\{y^{-2}\mathbb{E}f^2, 1\right\} + \min\left\{y^{-2}\mathbb{E}g^2, 1\right\}.
\end{aligned}$$

Because $\int_{\mathbb{R}} \min\{y^{-2}, 1\} \mathrm{d}\lambda(y) < \infty$ the assertion follows. □

Lemma 15.2.4 *Let $f, g \in \mathcal{L}_2(\Omega, \mathcal{F}, \mathbb{P})$, where g has a standard normal law. Let F and G be the distribution functions of f and g, respectively. Then*

$$\sup_{x \in \mathbb{R}} |F(x) - G(x)| \leqslant c_{(15.9)} \left[\int_{(0,T]} \left| \frac{\varphi_f(x) - \varphi_g(x)}{x} \right| \mathrm{d}\lambda(x) + \frac{1}{T} \right] \quad (15.9)$$

for all $T > 0$, where $c_{(15.9)} > 0$ is an absolute constant not depending on (f, g).

Proof (a) We let $D(x) := F(x) - G(x)$, so that

$$M := \sup_{x \in \mathbb{R}} |D(x)| < \infty \quad \text{and} \quad \lim_{|x| \to \infty} |D(x)| = 0.$$

15.2 The Berry-Esseen Theorem

We assume $M > 0$, otherwise there is nothing to prove. From Lemmas 15.2.2 and 15.2.3 we know

$$\frac{\varphi_f(x) - \varphi_g(x)}{-ix} e^{-ixc} = \int_{\mathbb{R}} e^{ixy} D(y+c) d\lambda(y) \quad \text{for} \quad x, c \in \mathbb{R}. \tag{15.10}$$

Using (15.10), Fubini's theorem, Exercise 2 of Sect. 11.10 and the notation

$$[-T, T]_0 := [-T, T] \setminus \{0\},$$

we get

$$\int_{[-T,T]_0} \frac{\varphi_f(x) - \varphi_g(x)}{-ix} e^{-ixc} \frac{T - |x|}{T} d\lambda(x)$$

$$= \int_{[-T,T]_0} \int_{\mathbb{R}} e^{ixy} D(y+c) dy \frac{T - |x|}{T} d\lambda(x)$$

$$= \int_{\mathbb{R}} \int_{[-T,T]_0} e^{-ixy} \frac{T - |x|}{T} d\lambda(x) D(y+c) d\lambda(y)$$

$$= \frac{4}{T} \int_{\mathbb{R}} \frac{\sin^2(Tx/2)}{x^2} D(x+c) d\lambda(x).$$

(b) For $T > 0$ we define the probability measure (see Exercise (4))

$$d\mu_T(y) := \frac{2}{\pi T} \frac{|\sin(\frac{T}{2} y)|^2}{y^2} d\lambda(y).$$

For $c \in \mathbb{R}$ we get by (a) that

$$\left| \int_{\mathbb{R}} D(y+c) d\mu_T(y) \right| = \frac{1}{2\pi} \left| \int_{[-T,T]_0} \frac{\varphi_f(x) - \varphi_g(x)}{-ix} e^{-ixc} \frac{T - |x|}{T} d\lambda(x) \right|$$

$$\leq \frac{1}{2\pi} \int_{[-T,T]_0} \left| \frac{\varphi_f(x) - \varphi_g(x)}{x} \right| d\lambda(x)$$

$$= \frac{1}{\pi} \int_{(0,T]} \left| \frac{\varphi_f(x) - \varphi_g(x)}{x} \right| d\lambda(x).$$

For $\delta > 0$ we deduce that

$$\left| \int_{|y| < \delta} D(y+c) d\mu_T(y) \right|$$

$$\leq \frac{1}{\pi} \int_{(0,T]} \left| \frac{\varphi_f(x) - \varphi_g(x)}{x} \right| d\lambda(x) + M \mu_T((-\delta, \delta)^c)$$

$$\leq \frac{1}{\pi} \int_{(0,T]} \left| \frac{\varphi_f(x) - \varphi_g(x)}{x} \right| d\lambda(x) + 2M \int_{[\delta,\infty)} \frac{2}{\pi T} \frac{1}{y^2} d\lambda(y)$$

$$= \frac{1}{\pi} \int_{(0,T]} \left| \frac{\varphi_f(x) - \varphi_g(x)}{x} \right| d\lambda(x) + \frac{4M}{\pi T \delta}.$$

If $\kappa > 0$ and $\delta = \kappa M$ this implies that

$$\left| \int_{|y| < \kappa M} D(y+c) d\mu_T(y) \right| \leq \frac{1}{\pi} \int_{(0,T]} \left| \frac{\varphi_f(x) - \varphi_g(x)}{x} \right| d\lambda(x) + \frac{4}{\pi \kappa} \frac{1}{T}.$$

(c) There exists a sequence with $\lim y_n = y_0$ such that

$$M = \lim_{n \to \infty} |D(y_n)| = \left| \lim_{n \to \infty} F(y_n) - G(y_0) \right|.$$

Now there are two different cases:
(c1) We have $M = G(y_0) - F(y_0-)$. Let $|y| < \delta$. By the mean value theorem we have

$$G(y + y_0 - \delta) = G(y_0) + G'(\xi)(y - \delta) \quad \text{for some} \quad \xi \in [-2\delta + y_0, y_0],$$

so that for $\delta := \kappa M$ we get

$$D(y + y_0 - \delta) = F(y + y_0 - \delta) - G(y + y_0 - \delta)$$
$$\leq F(y_0-) - G(y_0) - G'(\xi)(y - \delta)$$
$$\leq -M + \frac{1}{\sqrt{2\pi}}(\delta - y) = -\left(1 - \frac{\kappa}{\sqrt{2\pi}}\right) M - \frac{y}{\sqrt{2\pi}}.$$

Choosing $\kappa := \sqrt{\pi/2}$ gives that

$$-D(y + y_0 - \delta) \geq \frac{M}{2} + \frac{y}{\sqrt{2\pi}} \quad \text{for} \quad |y| < \delta.$$

Setting $c := y_0 - \delta$ we derive

$$\left| \int_{|y| < \kappa M} D(y+c) d\mu_T(y) \right| \geq \int_{|y| < \kappa M} -D(y+c) d\mu_T(y) \geq \frac{M}{2} \mu_T((-\kappa M, \kappa M)),$$

since the integral of $y/\sqrt{2\pi}$ is zero. Hence

$$\frac{M}{2} \mu_T((-\kappa M, \kappa M)) \leq \frac{2}{\pi} \int_{(0,T]} \left| \frac{\varphi_f(x) - \varphi_g(x)}{x} \right| d\lambda(x) + \frac{4}{\pi \sqrt{\pi/2}} \frac{1}{T}.$$
(15.11)

15.2 The Berry-Esseen Theorem

For $M \leqslant \frac{8}{\pi\sqrt{\pi/2}}\frac{1}{T}$ we trivially have that

$$M \leqslant \frac{2}{\pi}\int_{(0,T]}\left|\frac{\varphi_f(x)-\varphi_g(x)}{x}\right|d\lambda(x) + \frac{8}{\pi\sqrt{\pi/2}}\frac{1}{T}.$$

Assume now that $M > \frac{\sqrt{28}}{\pi^{3/2}}\frac{1}{T} =: \frac{\alpha}{T}$. Then

$$\mu_T((-\kappa M, \kappa M)) \geqslant \mu_T\left(\kappa\alpha\left(-\frac{1}{T},\frac{1}{T}\right)\right)$$

$$= \frac{2}{\pi}\int_{[0,\kappa\alpha/2)}\frac{|\sin(z)|^2}{z^2}d\lambda(z)$$

$$=: c(\kappa\alpha) \in (0, 1)$$

since μ_T is a probability measure. Therefore, (15.11) implies

$$M \leqslant \frac{1}{c(\kappa\alpha)}\left[\frac{4}{\pi}\int_{(0,T]}\left|\frac{\varphi_f(x)-\varphi_g(x)}{x}\right|d\lambda(x) + \frac{8}{\pi\sqrt{\pi/2}}\frac{1}{T}\right].$$

(c2) We have $M = F(y_0) - G(y_0)$. Let $|y| < \delta$. Similar to the case (c1) we get, choosing again $\delta := \sqrt{\pi/2}M$, that

$$D(y + y_0 + \delta) = F(y + y_0 + \delta) - G(y + y_0 + \delta)$$
$$\geqslant F(y_0) - G(y_0) - G'(\xi)(y + \delta)$$
$$\geqslant M - \frac{1}{\sqrt{2\pi}}(y + \delta) = \frac{M}{2} - \frac{y}{\sqrt{2\pi}}.$$

We get

$$D(y + y_0 - \delta) \geqslant \frac{M}{2} - \frac{y}{\sqrt{2\pi}} \quad \text{for} \quad |y| < \delta$$

and for $c := y_0 - \delta$, exploiting that the integral of $y/\sqrt{2\pi}$ is zero,

$$\left|\int_{|y|<\kappa M}D(y+c)d\mu_T(y)\right| \geqslant \int_{|y|<\kappa M}D(y+c)d\mu_T(y) \geqslant \frac{M}{2}\mu_T((-\kappa M, \kappa M)).$$

Now we can conclude as in (c1). □

Proof of Theorem 15.2.1 We let $f := f_1 \in \mathcal{L}_{2+\alpha}(\Omega, \mathcal{F}, \mathbb{P})$ with $\mathbb{E}f = 0$ and $\mathbb{E}f^2 = 1$, where $\alpha \in (0, 1]$.

(a) For $x \in \mathbb{R}$ one can verify that

$$\left| e^{ix} - 1 - ix + \frac{x^2}{2} \right| \leq \frac{5}{2} \begin{cases} |x|^2 & : |x| \geq 1 \\ |x|^3 & : |x| < 1 \end{cases},$$

so that

$$\left| e^{ix} - 1 - ix + \frac{x^2}{2} \right| \leq \frac{5}{2} |x|^{2+\alpha} \quad \text{for all} \quad x \in \mathbb{R}.$$

Indeed, since $|e^{ix} - 1| \leq |x|$ it holds for $|x| \geq 1$ that

$$\left| e^{ix} - 1 - ix + \frac{x^2}{2} \right| \leq |x| + |x| + \frac{|x|^2}{2} \leq \frac{5}{2} |x|^2.$$

For $|x| < 1$ we use Proposition A.1.1(3) which implies

$$\left| e^{ix} - 1 - ix + \frac{x^2}{2} \right| \leq \frac{1}{3!} |x|^3.$$

Then, for $c_\alpha := \frac{5}{2} \mathbb{E}|f|^{2+\alpha}$ we get

$$\left| \varphi_f(x) - 1 + \frac{x^2}{2} \right| = \left| \mathbb{E}\left(e^{ixf} - 1 - i(xf) + \frac{(xf)^2}{2} \right) \right|$$

$$\leq \mathbb{E} \left| e^{ixf} - 1 - i(xf) + \frac{(xf)^2}{2} \right|$$

$$\leq \frac{5}{2} |x|^{2+\alpha} \mathbb{E}|f|^{2+\alpha}$$

$$= c_\alpha |x|^{2+\alpha}.$$

(b) For the following we need that

$$\frac{x^2}{n} + c_\alpha \left| \frac{x}{\sqrt{n}} \right|^{2+\alpha} \leq \frac{1}{2} \quad \text{so that} \quad \left| \varphi_f\left(\frac{x}{\sqrt{n}} \right) - 1 \right| \leq \frac{1}{2}. \tag{15.12}$$

To satisfy this assumption we find an $\varepsilon_\alpha \in (0, 1)$ such that

$$\frac{|x|}{\sqrt{n}} \leq \varepsilon_\alpha$$

15.2 The Berry-Esseen Theorem

guarantees (15.12) and assume this for the following. Now Lemma 15.1.3 implies that

$$\left|\text{Log}\,\varphi_f\left(\frac{x}{\sqrt{n}}\right) + \frac{x^2}{2n}\right|$$

$$\leqslant \left|\text{Log}\,\varphi_f\left(\frac{x}{\sqrt{n}}\right) - \varphi_f\left(\frac{x}{\sqrt{n}}\right) + 1\right| + \left|\varphi_f\left(\frac{x}{\sqrt{n}}\right) - 1 + \frac{x^2}{2n}\right|$$

$$\leqslant \left|\varphi_f\left(\frac{x}{\sqrt{n}}\right) - 1\right|^2 + c_\alpha \left|\frac{x}{\sqrt{n}}\right|^{2+\alpha}$$

$$\leqslant \left[\frac{1}{2}\left(\frac{x}{\sqrt{n}}\right)^2 + c_\alpha \left|\frac{x}{\sqrt{n}}\right|^{2+\alpha}\right]^2 + c_\alpha \left|\frac{x}{\sqrt{n}}\right|^{2+\alpha}$$

$$\leqslant \frac{1}{2}\left(\frac{x}{\sqrt{n}}\right)^4 + 2c_\alpha^2 \left|\frac{x}{\sqrt{n}}\right|^{4+2\alpha} + c_\alpha \left|\frac{x}{\sqrt{n}}\right|^{2+\alpha}$$

$$\leqslant \left[\frac{1}{2} + 2c_\alpha^2 + c_\alpha\right]\left|\frac{x}{\sqrt{n}}\right|^{2+\alpha}$$

$$=: c_\alpha' \left|\frac{x}{\sqrt{n}}\right|^{2+\alpha},$$

where we exploited that $\frac{x^2}{n} \leqslant 1$. Multiplying the above inequality by n gives

$$\left|n\text{Log}\,\varphi_f\left(\frac{x}{\sqrt{n}}\right) + \frac{x^2}{2}\right| \leqslant c_\alpha' \frac{|x|^{2+\alpha}}{n^{\frac{\alpha}{2}}}.$$

As for $z \in \mathbb{C}$ we have

$$|e^z - 1| \leqslant |z| \sum_{k=1}^{\infty} \frac{|z|^{k-1}}{k!} \leqslant |z|e^{|z|}$$

we conclude

$$\left|\varphi_{S_n}\left(\frac{x}{\sqrt{n}}\right) - e^{-\frac{x^2}{2}}\right| = e^{-\frac{x^2}{2}}\left|e^{n\text{Log}\,\varphi_f\left(\frac{x}{\sqrt{n}}\right)+\frac{x^2}{2}} - 1\right|$$

$$\leqslant e^{-\frac{x^2}{2}} c_\alpha' \frac{|x|^{2+\alpha}}{n^{\frac{\alpha}{2}}} \exp\left(c_\alpha' \frac{|x|^{2+\alpha}}{n^{\frac{\alpha}{2}}}\right)$$

$$= c_\alpha' \frac{|x|^{2+\alpha}}{n^{\frac{\alpha}{2}}} \exp\left(c_\alpha' \frac{|x|^{2+\alpha}}{n^{\frac{\alpha}{2}}} - \frac{x^2}{2}\right).$$

We find another $\varepsilon'_\alpha \in (0, \varepsilon_\alpha]$ such that for

$$\frac{|x|}{\sqrt{n}} \leqslant \varepsilon'_\alpha$$

we have $c'_\alpha \frac{|x|^{2+\alpha}}{n^{\frac{\alpha}{2}}} - \frac{x^2}{2} \leqslant -\frac{x^2}{3}$ so that

$$\left| \varphi_{S_n}\left(\frac{x}{\sqrt{n}}\right) - e^{-\frac{x^2}{2}} \right| \leqslant c'_\alpha \frac{|x|^{2+\alpha}}{n^{\frac{\alpha}{2}}} e^{-\frac{x^2}{3}} \quad \text{for} \quad |x| \leqslant \varepsilon'_\alpha \sqrt{n}.$$

(c) For $T := \varepsilon'_\alpha n^{\frac{\alpha}{2}} \leqslant \varepsilon'_\alpha \sqrt{n}$ Lemma 15.2.4 implies

$$\sup_{x \in \mathbb{R}} \left| \mathbb{P}\left(\frac{S_n}{\sqrt{n}} \leqslant x\right) - \int_{-\infty}^{x} \frac{1}{\sqrt{2\pi}} e^{-\frac{z^2}{2}} \, d\lambda(z) \right|$$

$$\leqslant c_{(15.9)} \left[\int_0^T \left| \frac{\varphi_{S_n}\left(\frac{x}{\sqrt{n}}\right) - e^{-\frac{x^2}{2}}}{x} \right| d\lambda(x) + \frac{1}{T} \right]$$

$$\leqslant c_{(15.9)} \left[\frac{c'_\alpha}{n^{\frac{\alpha}{2}}} \int_0^T |x|^{1+\alpha} e^{-\frac{x^2}{3}} \, d\lambda(x) + \frac{1}{T} \right]$$

$$\leqslant \frac{c_{(15.9)}}{n^{\frac{\alpha}{2}}} \left[c'_\alpha \int_0^\infty |x|^{1+\alpha} e^{-\frac{x^2}{3}} \, d\lambda(x) + \frac{1}{\varepsilon'_\alpha} \right].$$

□

15.3 Stein's Method

In this section we provide with Theorem 15.3.4 Stein's[6] method, an alternative for proving a bound for the normal approximation like in the Berry-Esseen Theorem 15.2.1.

There are two observations that lead to Stein's method. Firstly, the distance of two measures $\mu_1, \mu_2 \in \mathcal{M}_1^+(\mathbb{R})$ can be measured in various ways. Here the general approach is to consider

$$d_\mathcal{H}(\mu_1, \mu_2) := \sup_{h \in \mathcal{H}} \left| \int_\mathbb{R} h(x) d\mu_1(x) - \int_\mathbb{R} h(x) d\mu_2(x) \right|$$

---

[6] Charles Max Stein, 22/03/1920 (Brooklyn, New York, USA)–24/11/2016 (Fremont, California, USA).

15.3 Stein's Method

where $\mathcal{H}$ is an appropriate class of functions $h : \mathbb{R} \to \mathbb{R}$. We discuss below two important cases. The Kolmogorov distance measures the maximal distance of the distribution functions:

Definition 15.3.1 (Kolmogorov Distance) For $\mu_1, \mu_2 \in \mathcal{M}_1^+(\mathbb{R})$ their **Kolmogorov distance** is given by

$$d_K(\mu_1, \mu_2) := \sup_{x \in \mathbb{R}} |\mu_1((-\infty, x]) - \mu_2((-\infty, x])|.$$

Here the function class $\mathcal{H} := \{h = \mathbb{1}_{(-\infty, x]} : x \in \mathbb{R}\}$ is used. Looking from the perspective of optimal transport, the Wasserstein[7] or Kantorovich[8]-Rubinstein[9] distance is the natural object to take, that builds on the class $\mathcal{H}$ of all 1- Lipschitz[10] functions:

Definition 15.3.2 (Wasserstein or Kantorovich-Rubinstein Distance) For $\mu_1, \mu_2 \in \mathcal{M}_1^+(\mathbb{R})$ with $\int_\mathbb{R} |x| d\mu_j(x) < \infty$ for $j = 1, 2$ the **Wasserstein** or **Kantorovich-Rubinstein distance** is given by

$$d_W(\mu_1, \mu_2) := \sup_{h \in \text{Lip}(1)} \left| \int_\mathbb{R} h(x) d\mu_1(x) - \int_\mathbb{R} h(x) d\mu_2(x) \right|,$$

where $\text{Lip}(1) := \{h : \mathbb{R} \to \mathbb{R} : |h(x) - h(y)| \leq |x - y| \text{ for } x, y \in \mathbb{R}\}$.

For random variables $f_i \in \mathcal{L}_0(\Omega_i, \mathcal{F}_i, \mathbb{P}_i)$ the above distances are translated to

$$d_K(f_1, f_2) := d_K(\text{law}(f_1), \text{law}(f_2)) \quad \text{and} \quad d_W(f_1, f_2) := d_W(\text{law}(f_1), \text{law}(f_2)).$$

The relation between the distances is obtained by an optimization argument:

Proposition 15.3.3 *For $\mu_1, \mu_2 \in \mathcal{M}_1^+(\mathbb{R})$ with $\int_\mathbb{R} |x| d\mu_j(x) < \infty$ for $j = 1, 2$, and where μ_2 has a density bounded by $c > 0$, one has*

$$d_K(\mu_1, \mu_2) \leq 2\sqrt{c\, d_W(\mu_1, \mu_2)}.$$

---

[7] Leonid Vaserstein, born 15/09/1944.

[8] Leonid Vitalyevich Kantorovich, 19/01/1912 (St Petersburg, Russia)–07/04/1986 (Moscow, USSR).

[9] Gennadii Shlemovich Rubinstein, 26/04/1923 (Odessa, Ukraine)–02/05/2004.

[10] Rudolf Lipschitz, 14/05/1832 (Königsberg, East Prussia, now Kaliningrad, Russia)–07/10/1903 (Bonn, Germany).

Proof We fix $x \in \mathbb{R}$ and let $\varepsilon > 0$. We define the Lipschitz function $h_\varepsilon(y) := 1$ for $y \leqslant x$, $h_\varepsilon(y) := 0$ for $y \geqslant x + \varepsilon$, and otherwise linear. Then we get

$$\mu_1((-\infty, x]) - \mu_2((-\infty, x])$$

$$\leqslant \left(\int_\mathbb{R} h_\varepsilon(y) d\mu_1(y) - \int_\mathbb{R} h_\varepsilon(y) d\mu_2(y) \right)$$

$$+ \left(\int_\mathbb{R} h_\varepsilon(y) d\mu_2(y) - \int_\mathbb{R} \mathbb{1}_{\{y \leqslant x\}} d\mu_2(y) \right)$$

$$\leqslant \left| \int_\mathbb{R} h_\varepsilon(y) d\mu_1(y) - \int_\mathbb{R} h_\varepsilon(y) d\mu_2(y) \right| + c\varepsilon$$

$$\leqslant \frac{1}{\varepsilon} d_W(\mu_1, \mu_2) + c\varepsilon.$$

Minimizing over $\varepsilon > 0$ we arrive at

$$\mu_1((-\infty, x]) - \mu_2((-\infty, x]) \leqslant 2\sqrt{c}\, d_W(\mu_1, \mu_2).$$

For $\mu_2((-\infty, x]) - \mu_1((-\infty, x])$ we proceed similarly by using the function $h_\varepsilon(y) := 1$ for $y \leqslant x - \varepsilon$, $h_\varepsilon(y) := 0$ for $y \geqslant x$, and otherwise linear. $\square$

The main result of this section is:

Theorem 15.3.4 *Let $f_1, \ldots, f_n : \Omega \to \mathbb{R}$ be independent random variables such that $\mathbb{E}|f_i|^4 < \infty$, $\mathbb{E}|f_i|^2 = 1$, and $\mathbb{E}f_i = 0$ for $i = 1, \ldots, n$. Let*

$$S_n := f_1 + \cdots + f_n$$

and let g be a random variable such that $\text{law}(g) = \mathcal{N}_{0,1}$. *Then*

$$d_W\left(g, \frac{S_n}{\sqrt{n}}\right) \leqslant n^{-\frac{3}{2}} \sum_{j=1}^n \mathbb{E}|f_j|^3 + \sqrt{\frac{2}{\pi}} \frac{1}{n} \mathbb{E}\left| \sum_{j=1}^n (1 - f_j^2) \right|.$$

Note that for $\max_{j=1,\ldots,n} \|f_j\|_{L_4} \leqslant M < \infty$ we can use

$$\mathbb{E}\left| \sum_{j=1}^n (1 - f_j^2) \right| \leqslant \left(\mathbb{E}\left| \sum_{j=1}^n (1 - f_j^2) \right|^2 \right)^{\frac{1}{2}} = \left(\sum_{j=1}^n \|1 - f_j^2\|_{L_2}^2 \right)^{\frac{1}{2}}$$

$$\leqslant \sqrt{n} \max_{j=1,\ldots,n} \|1 - f_j^2\|_{L_2} \leqslant \sqrt{n} \max_{j=1,\ldots,n} (1 + \|f_j\|_{L_4}^2)$$

$$\leqslant \sqrt{n}(1 + M^2)$$

15.3 Stein's Method

and get

$$d_W\left(g, \frac{S_n}{\sqrt{n}}\right) \leq n^{-\frac{3}{2}} \sum_{i=1}^{n} \mathbb{E}|f_i|^3 + \sqrt{\frac{2}{\pi}} \frac{1}{n} \mathbb{E}\left|\sum_{j=1}^{n}(1-f_j^2)\right|$$

$$\leq n^{-\frac{1}{2}} \left[M^3 + \sqrt{\frac{2}{\pi}}(1+M^2)\right],$$

where we used $\mathbb{E}|f_i|^3 \leq (\mathbb{E}|f_i|^4)^{\frac{3}{4}} \leq M^3$.

For the proof of Theorem 15.3.4 two lemmas are required. The first one is about the Mills' ratio for a Gaussian random variable:

Lemma 15.3.5 *If $g : \Omega \to \mathbb{R}$ is a random variable with $\mathrm{law}(g) = \mathcal{N}_{0,1}$, then*

$$\frac{1}{\sqrt{2\pi}} \frac{x}{1+x^2} \leq \frac{\mathbb{P}(g > x)}{e^{-\frac{x^2}{2}}} \leq \min\left\{\frac{1}{2}, \frac{1}{\sqrt{2\pi}x}\right\} \quad \text{for} \quad x > 0.$$

Proof
(a) Right-hand side, second bound: We use

$$\mathbb{P}(g > x) = \int_{(x,\infty)} e^{-\frac{y^2}{2}} \frac{d\lambda(y)}{\sqrt{2\pi}} \leq \frac{1}{x} \int_{(x,\infty)} y e^{-\frac{y^2}{2}} \frac{d\lambda(y)}{\sqrt{2\pi}} = \frac{1}{x} \frac{1}{\sqrt{2\pi}} e^{-\frac{x^2}{2}}.$$

(b) Right-hand side, first bound: Because of (a) we only need to prove

$$\int_{(x,\infty)} e^{-\frac{y^2}{2}} \frac{d\lambda(y)}{\sqrt{2\pi}} \leq \frac{1}{2} e^{-\frac{x^2}{2}} \quad \text{for} \quad x \in \left[0, \sqrt{\frac{2}{\pi}}\right).$$

For $x = 0$ we have equality. Differentiating both sides it would be sufficient to have that

$$-e^{-\frac{x^2}{2}} \frac{1}{\sqrt{2\pi}} \leq -\frac{x}{2} e^{-\frac{x^2}{2}} \quad \text{for} \quad x \in \left[0, \sqrt{\frac{2}{\pi}}\right).$$

But this inequality is obviously satisfied.
(c) Left-hand side: For $x > 0$ integration by parts implies that

$$\int_{(x,\infty)} e^{-\frac{y^2}{2}} y^2 d\lambda(y) = \int_{(x,\infty)} \left(e^{-\frac{y^2}{2}} y\right) y d\lambda(y) = x e^{-\frac{x^2}{2}} + \int_{(x,\infty)} e^{-\frac{y^2}{2}} d\lambda(y),$$

so that
$$\int_{(x,\infty)} e^{-\frac{y^2}{2}}(y^2-1)d\lambda(y) = xe^{-\frac{x^2}{2}}.$$

Therefore the left-hand side is equivalent to
$$\int_{(x,\infty)} e^{-\frac{y^2}{2}}(y^2-1)d\lambda(y) \leqslant (1+x^2)\int_{(x,\infty)} e^{-\frac{y^2}{2}}d\lambda(y)$$

or
$$\int_{(x,\infty)} e^{-\frac{y^2}{2}}(y^2-x^2-2)d\lambda(y) \leqslant 0. \tag{15.13}$$

We also have that
$$\frac{d}{dx}\int_{(x,\infty)} e^{-\frac{y^2}{2}}(y^2-x^2-2)d\lambda(y) = 2e^{-\frac{x^2}{2}} - 2x\int_{(x,\infty)} e^{-\frac{y^2}{2}}d\lambda(y) \geqslant 0$$

where for the inequality we use (a). As
$$\lim_{x\to\infty}\int_{(x,\infty)} e^{-\frac{y^2}{2}}(y^2-x^2-2)d\lambda(y) = 0$$

the inequality (15.13), and therefore the left-hand side follows. □

Our next ingredient is based on Stein's lemma, which we will not discuss here in full detail, but only consider the case that by integration by parts applied to the Gaussian density one obtains the following: If $g : \Omega \to \mathbb{R}$ is a random variable with law$(g) = \mathcal{N}_{0,1}$ and $\varphi : \mathbb{R} \to \mathbb{R}$ is continuously differentiable with $\sup_{x\in\mathbb{R}} |\varphi'(x)| < \infty$, then

$$\mathbb{E}\varphi'(g) = \mathbb{E}g\varphi(g) \quad \text{or} \quad \mathbb{E}\varphi'(g) - \mathbb{E}g\varphi(g) = 0.$$

The next lemma is a converse to this observation:

Lemma 15.3.6 *If $g : \Omega \to \mathbb{R}$ is a random variable with law$(g) = \mathcal{N}_{0,1}$ and $f \in \mathcal{L}_1(\Omega, \mathcal{F}, \mathbb{P})$, then*

$$d_W(g, f) \leqslant \sup_{\varphi \in \mathcal{K}} |\mathbb{E}\varphi'(f) - \mathbb{E}f\varphi(f)|,$$

where $\mathcal{K}$ is the set of all $\varphi : \mathbb{R} \to \mathbb{R}$ such that φ, φ', φ'', and φ''' are continuous and such that $|\varphi(x)| \leqslant 1$, $|\varphi'(x)| \leqslant \sqrt{\frac{2}{\pi}}$, and $|\varphi''(x)| \leqslant 2$ for all $x \in \mathbb{R}$.

15.3 Stein's Method

Proof For $h \in \mathrm{Lip}(1)$, which is infinitely often differentiable, we define the test function

$$\varphi_h(x) := -\int_{(0,1)} \frac{1}{2\sqrt{t(1-t)}} \mathbb{E}[g\, h(\sqrt{t}x + \sqrt{1-t}g)]\mathrm{d}\lambda(t).$$

(a) First we convince ourselves that φ_h solves

$$\varphi_h'(x) - x\varphi_h(x) = h(x) - \mathbb{E}h(g). \tag{15.14}$$

(a1) Using Lemma 11.6.2 we can compute the derivative

$$\varphi_h'(x) = -\int_{(0,1)} \frac{1}{2\sqrt{1-t}} \mathbb{E}[g\, h'(\sqrt{t}x + \sqrt{1-t}g)]\mathrm{d}\lambda(t).$$

(a2) We re-write the right-hand side of (15.14) by Fubini's theorem as

$$h(x) - \mathbb{E}h(g) = \int_{(0,1)} \mathbb{E}\frac{\mathrm{d}}{\mathrm{d}t} h(\sqrt{t}x + \sqrt{1-t}g)\mathrm{d}\lambda(t)$$

$$= \int_{(0,1)} \mathbb{E}h'(\sqrt{t}x + \sqrt{1-t}g)\left(\frac{x}{2\sqrt{t}} - \frac{g}{2\sqrt{1-t}}\right)\mathrm{d}\lambda(t).$$

(a3) We transform the expectation in the definition of φ_h by integration by parts, where we exploit that $h(\sqrt{t}x + \sqrt{1-t}y)\frac{1}{\sqrt{1-t}}e^{-\frac{y^2}{2}} \to 0$ if $|y| \to \infty$, to get that

$$\mathbb{E}h'(\sqrt{t}x + \sqrt{1-t}g) = \frac{1}{\sqrt{2\pi}}\int_\mathbb{R} h'(\sqrt{t}x + \sqrt{1-t}y)e^{-\frac{y^2}{2}}\mathrm{d}\lambda(y)$$

$$= \frac{1}{\sqrt{2\pi}}\int_\mathbb{R} \left(\frac{\mathrm{d}}{\mathrm{d}y}h(\sqrt{t}x + \sqrt{1-t}y)\right)\frac{1}{\sqrt{1-t}}e^{-\frac{y^2}{2}}\mathrm{d}\lambda(y)$$

$$= \frac{1}{\sqrt{2\pi}}\int_\mathbb{R} h(\sqrt{t}x + \sqrt{1-t}y)\frac{y}{\sqrt{1-t}}e^{-\frac{y^2}{2}}\mathrm{d}\lambda(y)$$

$$= \frac{1}{\sqrt{1-t}}\mathbb{E}[g\, h(\sqrt{t}x + \sqrt{1-t}g)].$$

Combining (a1), (a2), and (a3) we obtain (15.14). Moreover, by (a3) we get that

$$\varphi_h(x) = -\int_{(0,1)} \mathbb{E}h'(\sqrt{t}x + \sqrt{1-t}g)\frac{1}{2\sqrt{t}}\mathrm{d}\lambda(t). \tag{15.15}$$

(b) We still need another representation of φ_h. We first define

$$\widetilde{\varphi}_h(x) := e^{\frac{x^2}{2}} \int_{(-\infty,x]} e^{-\frac{y^2}{2}} (h(y) - \mathbb{E}h(g)) d\lambda(y).$$

By differentiating we see that $\widetilde{\varphi}_h$ also solves the Eq. (15.14). We conclude that

$$\frac{d}{dx}\left(e^{-\frac{x^2}{2}}(\varphi_h(x) - \widetilde{\varphi}_h(x))\right) = 0$$

so that, for some $c \in \mathbb{R}$,

$$\varphi_h(x) - \widetilde{\varphi}_h(x) = c e^{\frac{x^2}{2}}.$$

Since for $x \to \infty$ it holds that

$$e^{-\frac{x^2}{2}} \widetilde{\varphi}_h(x) \to \int_{\mathbb{R}} e^{-\frac{y^2}{2}} (h(y) - \mathbb{E}h(g)) d\lambda(y) = 0$$

and that $e^{-\frac{x^2}{2}} \varphi_h(x) \to 0$, we have that $c = 0$ and

$$\varphi_h(x) = e^{\frac{x^2}{2}} \int_{(-\infty,x]} e^{-\frac{y^2}{2}} (h(y) - \mathbb{E}h(g)) d\lambda(y). \tag{15.16}$$

(c) $|\varphi_h(x)| \leq 1$ follows from (15.15) and from $|h'(y)| \leq 1$ for all $y \in \mathbb{R}$.
(d) $|\varphi_h'(x)| \leq \sqrt{\frac{2}{\pi}}$ follows from $|h'(y)| \leq 1$ for all $y \in \mathbb{R}$, from the fact that $\mathbb{E}|g| = \sqrt{\frac{2}{\pi}}$ (see Exercise (5)), and from

$$|\varphi_h'(x)| = \left|\int_{(0,1)} \frac{1}{2\sqrt{1-t}} \mathbb{E}[g\, h'(\sqrt{t}x + \sqrt{1-t}g)] d\lambda(t)\right|$$

$$\leq \int_{(0,1)} \frac{1}{2\sqrt{1-t}} d\lambda(t)\, \mathbb{E}|g| = \sqrt{\frac{2}{\pi}}.$$

(e) Proof of $|\varphi_h''(x)| \leq 2$: From (15.16) and (15.14) we get that

$$\varphi_h''(x) = x\varphi_h'(x) + \varphi_h(x) + h'(x)$$
$$= (1 + x^2)\varphi_h(x) + x(h(x) - \mathbb{E}h(g)) + h'(x)$$

so that

$$|\varphi_h''(x)| \leq 1 + \left|(1 + x^2)\varphi_h(x) + x(h(x) - \mathbb{E}h(g))\right|$$

15.3 Stein's Method

For
$$\gamma(x) := \frac{1}{\sqrt{2\pi}} \int_{(-\infty,x]} e^{-\frac{y^2}{2}} \, d\lambda(y)$$

we observe by Fubini's theorem that

$$h(x) - \mathbb{E}h(g)$$
$$= \frac{1}{\sqrt{2\pi}} \int_{\mathbb{R}} e^{-\frac{u^2}{2}} (h(x) - h(u)) \, d\lambda(u)$$
$$= \frac{1}{\sqrt{2\pi}} \int_{(-\infty,x]} \int_{[u,x]} h'(z) e^{-\frac{u^2}{2}} \, d\lambda(z) d\lambda(u)$$
$$\quad - \frac{1}{\sqrt{2\pi}} \int_{[x,\infty)} \int_{[x,u]} h'(z) e^{-\frac{u^2}{2}} \, d\lambda(z) d\lambda(u)$$
$$= \int_{(-\infty,x]} h'(z) \gamma(z) \, d\lambda(z) - \int_{[x,\infty)} h'(z)(1 - \gamma(z)) \, d\lambda(z),$$

which we insert into (15.16) to deduce

$$\varphi_h(x)$$
$$= e^{\frac{x^2}{2}} \int_{(-\infty,x]} e^{-\frac{y^2}{2}}$$
$$\quad \left(\int_{(-\infty,y]} h'(z) \gamma(z) \, d\lambda(z) - \int_{[y,\infty)} h'(z)(1 - \gamma(z)) \, d\lambda(z) \right) d\lambda(y)$$
$$= -\sqrt{2\pi} e^{\frac{x^2}{2}} (1 - \gamma(x)) \int_{(-\infty,x]} h'(z) \gamma(z) \, d\lambda(z)$$
$$\quad - \sqrt{2\pi} e^{\frac{x^2}{2}} \gamma(x) \int_{[x,\infty)} h'(z)(1 - \gamma(z)) \, d\lambda(z),$$

where the last equality follows by integration by parts (we use $e^{-\frac{y^2}{2}} = -\sqrt{2\pi}(1 - \gamma(y))' = \sqrt{2\pi}\gamma'(y)$). Now we come back to our estimate for $|\varphi_h''(x)|$. For the following computation we use the identities for $\varphi_h(x)$ and $h(x) - \mathbb{E}h(g)$, that $x - \sqrt{2\pi}(1+x^2)e^{\frac{x^2}{2}}(1-\gamma(x)) \leq 0$ (as consequence of Mills' ratio (Lemma 15.3.5)) to justify the second inequality, and $|h'(z)| \leq 1$:

$$|\varphi_h''(x)| \leq 1 + \left| (x - \sqrt{2\pi}(1+x^2)e^{\frac{x^2}{2}}(1-\gamma(x))) \int_{(-\infty,x]} h'(z)\gamma(z) \, d\lambda(z) \right|$$
$$\quad + \left| (x + \sqrt{2\pi}(1+x^2)e^{\frac{x^2}{2}}\gamma(x)) \int_{[x,\infty)} h'(z)(1-\gamma(z)) \, d\lambda(z) \right|$$

$$\leqslant 1 + (\sqrt{2\pi}(1+x^2)e^{\frac{x^2}{2}}(1-\gamma(x)) - x)\int_{(-\infty,x]} \gamma(z)\mathrm{d}\lambda(z)$$

$$+ (x + \sqrt{2\pi}(1+x^2)e^{\frac{x^2}{2}}\gamma(x))\int_{[x,\infty)} (1-\gamma(z))\mathrm{d}\lambda(z)$$

$$= 1 + (\sqrt{2\pi}(1+x^2)e^{\frac{x^2}{2}}(1-\gamma(x)) - x)\left(x\gamma(x) + \frac{e^{\frac{-x^2}{2}}}{\sqrt{2\pi}}\right)$$

$$+ (x + \sqrt{2\pi}(1+x^2)e^{\frac{x^2}{2}}\gamma(x))\left(-x(1-\gamma(x)) + \frac{e^{\frac{-x^2}{2}}}{\sqrt{2\pi}}\right) = 2.$$

The second last equality can bee seen by integration by parts.

(f) To conclude, let us assume $h \in \mathrm{Lip}(1)$ as in Definition 15.3.2. Let $S : \mathbb{R} \to [0,\infty)$ be infinitely differentiable and such that $S(x) = 0$ if $|x| \geqslant 1$ and $\int_{\mathbb{R}} S(x)\mathrm{d}\lambda(x) = 1$. Then

$$h_n(x) := \int_{\mathbb{R}} h\left(x - \frac{y}{n}\right) S(y)\mathrm{d}\lambda(y)$$

is infinitely differentiable, belongs to $\mathrm{Lip}(1)$, satisfies

$$|h_n(x)| \leqslant \sup_{\{|y-x|\leqslant 1\}} |h(y)| =: \overline{h}(x) \leqslant A + B|x|$$

for some $A, B \geqslant 0$, and satisfies $\lim_{n\to\infty} h_n(x) = h(x)$. Now, by dominated convergence it follows that

$$\lim_{n\to\infty} |\mathbb{E}h_n(f) - \mathbb{E}h_n(g)| = |\mathbb{E}h(f) - \mathbb{E}h(g)|.$$

In other words we can restrict us in Definition 15.3.2 to h that are infinitely differentiable, and the consideration of (15.14) under our assumptions is sufficient. $\square$

Proof of Theorem 15.3.4 We assume $\varphi \in \mathcal{K}$ as in Stein's Lemma 15.3.6. Let $S_n^{(j)} := \sum_{i\neq j} f_i$ for $j = 1, \ldots, n$. By independence of f_j and $\varphi\left(\frac{S_n^{(j)}}{\sqrt{n}}\right)$ we get $\mathbb{E}\left(f_j \varphi\left(\frac{S_n^{(j)}}{\sqrt{n}}\right)\right) = (\mathbb{E}f_j)\left(\mathbb{E}\varphi\left(\frac{S_n^{(j)}}{\sqrt{n}}\right)\right) = 0$ and deduce that

$$\left|\mathbb{E}\varphi'\left(\frac{S_n}{\sqrt{n}}\right) - \mathbb{E}\frac{S_n}{\sqrt{n}}\varphi\left(\frac{S_n}{\sqrt{n}}\right)\right|$$

$$= \left|\mathbb{E}\varphi'\left(\frac{S_n}{\sqrt{n}}\right) - \mathbb{E}\left[\frac{1}{\sqrt{n}}\sum_{j=1}^n f_j\left[\varphi\left(\frac{S_n}{\sqrt{n}}\right) - \varphi\left(\frac{S_n^{(j)}}{\sqrt{n}}\right) - \left(\frac{S_n}{\sqrt{n}} - \frac{S_n^{(j)}}{\sqrt{n}}\right)\varphi'\left(\frac{S_n}{\sqrt{n}}\right)\right]\right]\right|$$

$$-\mathbb{E}\left[\frac{1}{\sqrt{n}}\sum_{j=1}^n f_j\left[\left(\frac{S_n}{\sqrt{n}}-\frac{S_n^{(j)}}{\sqrt{n}}\right)\varphi'\left(\frac{S_n}{\sqrt{n}}\right)\right]\right]\Bigg|$$

$$\leqslant \left|\mathbb{E}\left[\frac{1}{\sqrt{n}}\sum_{j=1}^n f_j\left[\varphi\left(\frac{S_n}{\sqrt{n}}\right)-\varphi\left(\frac{S_n^{(j)}}{\sqrt{n}}\right)-\left(\frac{S_n}{\sqrt{n}}-\frac{S_n^{(j)}}{\sqrt{n}}\right)\varphi'\left(\frac{S_n}{\sqrt{n}}\right)\right]\right]\right|$$

$$+\left|\mathbb{E}\varphi'\left(\frac{S_n}{\sqrt{n}}\right)\left[1-\frac{1}{\sqrt{n}}\sum_{j=1}^n f_j\left(\frac{S_n}{\sqrt{n}}-\frac{S_n^{(j)}}{\sqrt{n}}\right)\right]\right|$$

$$\leqslant \left|\mathbb{E}\left[\frac{1}{\sqrt{n}}\sum_{j=1}^n |f_j|\left(\frac{S_n}{\sqrt{n}}-\frac{S_n^{(j)}}{\sqrt{n}}\right)^2\right]\right|+\sqrt{\frac{2}{\pi}}\mathbb{E}\left|1-\frac{1}{\sqrt{n}}\sum_{j=1}^n f_j\left(\frac{S_n}{\sqrt{n}}-\frac{S_n^{(j)}}{\sqrt{n}}\right)\right|$$

$$=n^{-\frac{3}{2}}\sum_{j=1}^n \mathbb{E}|f_j|^3+\sqrt{\frac{2}{\pi}}\frac{1}{n}\mathbb{E}\left|\sum_{j=1}^n (1-f_j^2)\right|$$

where we use Taylor's formula for the second inequality. Finally, the statement follows by an application of Stein's Lemma 15.3.6. □

Remark 15.3.7 From Proposition 15.3.3 and under the assumptions made in Theorem 15.3.4 one can recover a Berry-Esseen (Theorem 15.2.1) type result:

$$\sup_{-\infty<z<\infty}\left|\mathbb{P}\left(\frac{S_n}{\sqrt{n}}\leqslant z\right)-\mathcal{N}_{0,1}((-\infty,z])\right|=d_K\left(\frac{S_n}{\sqrt{n}},g\right)$$

$$\leqslant 2\sqrt{\frac{1}{\sqrt{2\pi}}\,d_W\left(\frac{S_n}{\sqrt{n}},g\right)}\leqslant Cn^{-\frac{1}{4}}.$$

15.4 The Poisson Limit Theorem

To motivate Poisson's limit theorem we consider the following model: Over the time interval $[0,1]$ we consider a counting process $(N_t)_{t\in[0,1]}$, $N_t:\Omega\to\{0,1,2,\ldots\}$, such that for each $\omega\in\Omega$ the map $t\mapsto N(t,\omega)$ starts at zero and has only jumps of size one with a certain intensity constant in time. The process has independent increments, i.e. for all $0=t_0<t_1<\cdots<t_n=1$ the random variables $N_{t_1}-N_{t_0},\ldots,N_{t_n}-N_{t_{n-1}}$ are independent. The expected frequency of jumps is expressed by $\mathbb{E}(N_t-N_s)=\lambda(t-s)$ for some intensity $\lambda>0$ and all $0\leqslant s<t\leqslant 1$. Such a process exists and is called Poisson process with intensity $\lambda>0$.

To approach the distribution of N_1 one can go different ways. We will do it by the following *approximation*: We divide the interval $(0,1]$ into the n intervals $\left(\frac{k-1}{n},\frac{k}{n}\right]$,

$k = 1, \ldots, n$, and assume exactly one jump in each interval, independently from the other intervals, with probability λ/n if $n \geq \lambda$. The expected value of jumps over the interval $(0, 1]$ would be λ. Is there a limit as $n \to \infty$? The answer gives the Poisson limit theorem:

Theorem 15.4.1 (Poisson Limit Theorem) *Let $(\Omega, \mathcal{F}, \mathbb{P})$ be a probability space and assume for each $n \in \mathbb{N}$ independent random variables $f_{n1}, \ldots, f_{nn} : \Omega \to \mathbb{R}$ such that $\mathbb{P}(f_{nk} = 1) = p_{nk}$ and $\mathbb{P}(f_{nk} = 0) = q_{nk}$ with $p_{nk} + q_{nk} = 1$, and such that*

$$\max_{1 \leq k \leq n} p_{nk} \xrightarrow[n]{} 0 \quad \text{and} \quad \sum_{k=1}^{n} p_{nk} \xrightarrow[n]{} \lambda > 0.$$

If $S_n := f_{n1} + \cdots + f_{nn}$ and if g is a random variable with $\operatorname{law}(g) = \operatorname{Pois}_\lambda$, then

$$S_n \xrightarrow{d} g.$$

Example 15.4.2 Our introducing motivation to Theorem 15.4.1 corresponds to

$$p_{n1} = \cdots = p_{nn} := \frac{\lambda}{n} \quad \text{for} \quad n > \lambda.$$

Proof of Theorem 15.4.1
(a) For $\theta := e^{ix} - 1$, $x \in \mathbb{R}$, and

$$b_{nl} := \frac{l!}{\lambda^l} \sum_{1 \leq k_1 < \cdots < k_l \leq n} p_{nk_1} \cdots p_{nk_l} \quad \text{for} \quad n \geq l \geq 1$$

we get that

$$\varphi_{S_n}(x) = \prod_{k=1}^{n} \varphi_{f_{nk}}(x) = \prod_{k=1}^{n} \left(p_{nk} e^{ix} + q_{nk} \right)$$

$$= \prod_{k=1}^{n} \left(1 + p_{nk} \left(e^{ix} - 1 \right) \right)$$

$$= \prod_{k=1}^{n} (1 + p_{nk}\theta)$$

$$= 1 + \theta \left(\sum_{k=1}^{n} p_{nk} \right) + \theta^2 \left(\sum_{1 \leq k_1 < k_2 \leq n} p_{nk_1} p_{nk_2} \right)$$

15.4 The Poisson Limit Theorem

$$+ \cdots + \theta^l \left(\sum_{1 \leq k_1 < \cdots < k_l \leq n} p_{nk_1} \cdots p_{nk_l} \right) + \cdots$$

$$+ \theta^n p_{n1} \cdots p_{nn}$$

$$= 1 + \sum_{l=1}^{n} \frac{(\theta \lambda)^l}{l!} b_{nl}.$$

(b) We prove $\lim_{n \to \infty, n \geq l} b_{nl} = 1$ for $l \in \mathbb{N}$: In fact, we have

$$b_{nl} = \frac{l!}{\lambda^l} \sum_{1 \leq k_1 < \cdots < k_l \leq n} p_{nk_1} \cdots p_{nk_l}$$

$$= \left(\frac{p_{n1} + \cdots + p_{nn}}{\lambda} \right)^l \frac{\sum_{\substack{1 \leq k_1, \ldots, k_l \leq n \\ \text{all indices are distinct}}} p_{nk_1} \cdots p_{nk_l}}{\sum_{1 \leq k_1, \ldots, k_l \leq n} p_{nk_1} \cdots p_{nk_l}}.$$

The first factor converges to 1 as $n \to \infty$ by assumption. For $l \geq 2$ we write the second factor as

$$\frac{\sum_{\substack{1 \leq k_1, \ldots, k_l \leq n \\ \text{all indices are distinct}}} p_{nk_1} \cdots p_{nk_l}}{\sum_{1 \leq k_1, \ldots, k_l \leq n} p_{nk_1} \cdots p_{nk_l}} = 1 - \frac{\sum_{\substack{1 \leq k_1, \ldots, k_l \leq n \\ \text{not all indices are distinct}}} p_{nk_1} \cdots p_{nk_l}}{\sum_{1 \leq k_1, \ldots, k_l \leq n} p_{nk_1} \cdots p_{nk_l}}$$

and observe that

$$\sum_{\substack{1 \leq k_1, \ldots, k_l \leq n \\ \text{not all indices are distinct}}} p_{nk_1} \cdots p_{nk_l} \leq \binom{l}{2} \left(\sum_{k=1}^{n} p_{nk}^2 \right) \left(\sum_{k=1}^{n} p_{nk} \right)^{l-2}$$

$$\leq \binom{l}{2} \left(\sum_{k=1}^{n} p_{nk} \right)^{l-1} \max_{1 \leq k \leq n} p_{nk}$$

which converges to zero as $n \to \infty$.

(c) From step (b) we also know that

$$b_{nl} \leq \left(\frac{p_{n1} + \cdots + p_{nn}}{\lambda} \right)^l.$$

Because our assumption implies that

$$M := \sup_{n \in \mathbb{N}} \frac{p_{n1} + \cdots + p_{nn}}{\lambda} < \infty,$$

we have the upper bound $b_{nl} \leq M^l$ for $n \geq l \geq 1$. With this in our hands we can finish our proof. Because

$$\sum_{l=1}^{\infty} \frac{|\theta\lambda|^l}{l!} M^l \leq e^{|\theta|\lambda M} < \infty$$

and $b_{nl} \leq M^l$ for $n \geq l \geq 1$ we may apply dominated convergence (regardless that the series is complex valued) to deduce

$$\lim_{n \to \infty} \varphi_{S_n}(x) = \lim_{n \to \infty} \left(1 + \sum_{l=1}^{n} \frac{(\theta\lambda)^l}{l!} b_{nl} \right)$$

$$= \lim_{n \to \infty} \left(1 + \sum_{l=1}^{\infty} \mathbb{1}_{\{l \leq n\}} \frac{(\theta\lambda)^l}{l!} b_{nl} \right)$$

$$= \left(1 + \sum_{l=1}^{\infty} \frac{(\theta\lambda)^l}{l!} \lim_{n \to \infty, n \geq l} b_{nl} \right)$$

$$= e^{\lambda\theta} = e^{\lambda(e^{ix}-1)} = \widehat{\text{Pois}}_\lambda(x),$$

where we use Example 11.2.4. Now Theorem 12.1.1 implies our assertion. □

15.5 Exercises

Ex 1: The classical CLT condition implies the Lindeberg condition: Verify that the Lindeberg condition is satisfied in the case

$$\frac{S_n}{\sqrt{n}} = \frac{f_1 + \cdots + f_n}{\sqrt{n}}$$

with

(a) $\mathbb{E} f_n^2 = 1$,
(b) $\mathbb{E} f_n = 0$,
(c) $f_1^2, f_2^2, f_3^2, \ldots$ is a uniformly integrable family of random variables.

Hint: For $f_{nk} := \frac{f_k}{\sqrt{n}}$ verify that

$$\sum_{k=1}^{n} \int_{\{|f_{nk}| > \varepsilon\}} |f_{nk}|^2 d\mathbb{P} \leq \sup_{k \geq 1} \int_{\{|f_k| > \sqrt{n}\varepsilon\}} f_k^2 d\mathbb{P}.$$

15.5 Exercises

Ex 2: The Lyapunov[11] condition implies the Lindeberg condition: The Lyapunov condition for the CLT is satisfied if $f_1, f_2, \ldots : \Omega \to \mathbb{R}$ are independent, such that $\mathbb{E}|f_n|^{2+\delta} < \infty$ for $n = 1, 2, \ldots$ for some $\delta > 0$, $\mathbb{E}f_n = 0$, and if

$$\lim_{n \to \infty} \sum_{k=1}^{n} \frac{\mathbb{E}|f_k|^{2+\delta}}{\operatorname{var}(S_n)^{\frac{2+\delta}{2}}} = 0$$

for $S_n := f_1 + \cdots + f_n$. Under these conditions show that the family

$$f_{nk} := \frac{f_k}{\sqrt{\operatorname{var}(S_n)}}$$

satisfies the Lindeberg condition.

Ex 3: The CLT can not be strengthened to convergence in probability:
Let $(\Omega, \mathcal{F}, \mathbb{P})$ be a probability space and let $f_1, f_2, \ldots \Omega \to \mathbb{R}$ be a sequence of i.i.d. random variables such that $\mathbb{E}f_n^2 = 1$ and $\mathbb{E}f_n = 0$. Explain why $\frac{S_n}{\sqrt{n}} \xrightarrow{d} g$ with $g \sim \mathcal{N}_{0,1}$. Assume that $\frac{S_n}{\sqrt{n}}$ converges in probability to a random variable f. Conclude that there exists a subsequence $(n_k)_{k=1}^{\infty}$ such that $\frac{S_{n_k}}{\sqrt{n_k}} \xrightarrow{a.s.} f$. Can you derive by the 0-1 law of Kolmogorov that f is almost surely constant? How are f and g related and what consequences does that have concerning the convergence in probability of the sequence?

Ex 4: A Dirichlet-like integral: In Lemma A.4.2 we did show that $\lim_{\substack{c \to \infty \\ c > 0}} \int_{(0,c]} \frac{\sin(y)}{y} d\lambda(y) = \frac{\pi}{2}$. Use this result and integration by parts to verify

$$\int_{(0,\infty)} \frac{\sin^2(y)}{y^2} d\lambda(y) = \frac{\pi}{2}.$$

Ex 5: The first absolute Gaussian moment: Show that for $g \sim \mathcal{N}_{0,1}$ it holds

$$\mathbb{E}|g| = \frac{2}{\sqrt{2\pi}} \int_{(0,\infty)} x e^{-\frac{x^2}{2}} d\lambda(x) = \sqrt{\frac{2}{\pi}}.$$

Ex 6: A sequence for which the CLT is not satisfied: Let $f_1, f_2, \ldots : \Omega \to \mathbb{R}$ be independent random variables such that

$$\mathbb{P}(f_1 = 1) = \mathbb{P}(f_1 = -1) = \frac{1}{2},$$

$$\mathbb{P}(f_n = \pm 1) = \frac{1}{4}, \quad \mathbb{P}(f_n = \pm n) = \frac{1}{4n^2}, \quad \mathbb{P}(f_n = 0) = \frac{1}{2}\left(1 - \frac{1}{n^2}\right)$$

---

[11] Aleksandr Mikhailovich Lyapunov, 06/06/1857 (Yaroslavl, Russia)–03/11/1918 (Odessa, Ukraine).

for $n \geq 2$. Check that $\mathbb{E} f_n = 0$ and $\text{var}(f_n) = 1$ for $n \in \mathbb{N}$. Show that the Lindeberg condition is not satisfied. Use Theorem 15.1.2(2) to conclude that $\frac{f_1 + \cdots + f_n}{\sqrt{n}}$ does not converge weakly to g with $g \sim \mathcal{N}_{0,1}$.

Ex 7: Poisson's theorem with two jumps: Assume parameters $\lambda, \mu > 0$ and that

$$\mathbb{P}(f_{nk} = 0) = (1 - \lambda/n)(1 - \mu/n),$$
$$\mathbb{P}(f_{nk} = 1) = (\lambda/n)(1 - \mu/n) + (\mu/n)(1 - \lambda/n),$$
$$\mathbb{P}(f_{nk} = 2) = \lambda\mu/n^2.$$

Show that the limit distribution is $\text{Pois}_{\lambda+\mu}$.

Ex 8: Experimenting with R: Run the R-program from Appendix C.3 and change the parameter for the SLLN and the CLT in the program to read from the histograms the different empirical distributions that appear for the SLLN and the CLT.

15.6 Comments

Section 15.1: For an account on the history of the central limit theorem we refer to [63]. The proof of Theorem 15.1.2(2) we learnt from [121, page 293].

Section 15.2: The Berry-Esseen theorem goes back to Berry [13] and Esseen [55, 56]. For the proof of Theorem 15.2.1 we used [36, Theorem 9.1.3 and Corollary 9.1.4]. The multivariate Berry-Esseen theorem is investigated also, where in particular the dependence of the error estimate on the dimension is of interest, see [10, 135, 152], where [152] gives an overview as well.

Section 15.3: Stein introduced his method in [177, 178]. A basic reference for this method is [35]. In Sect. 15.3 we also use the presentations in [32, 164]. The proof of Lemma 15.3.6 we learnt from [32]. Stein's method is developed further to infinite dimensional Gaussian measures and Malliavin calculus by D. Nualart, Pecatti, Nourdin, Chen, and others, see [34, 137–139]. Stein's method is also used to prove Berry-Esseen type theorems, see the overview in [152].

Section 15.4: One of the sources of Poisson's limit theorem (Theorem 15.4.1) is [149]. The speed of convergence of the Poisson approximation was first investigated by Le Cam [114, 115] and Hodges and LeCam [87], see also [176].

Section 15.5: Exercise 3 we learnt from an anonymous referee, Exercise 6 is taken from [182, page 184].

Chapter 16
Fourier Inversion Formulas

In this chapter we prove inversion formulas which provide the inverse of the Fourier transform of measures so that one can reconstruct from the Fourier transform the measure and possibly a density. After this investigation the Bochner-Khintchin theorem is verified by which one can deduce the existence of certain fundamental probability measures.

16.1 Inversion Formulas

Theorem 11.5.5 states that there is a one to one relation between a probability measure on $(\mathbb{R}^d, \mathcal{B}(\mathbb{R}^d))$ and its Fourier transform. This can be shown also with the help of explicit inversion formulas which are subject of this chapter. In the following we use for $c > 0$ the notation

$$[-c, c]_0 := [-c, c] \setminus \{0\}.$$

Theorem 16.1.1 *For $\mu \in \mathcal{M}_1^+(\mathbb{R}^d)$ and $-\infty < a_k < b_k < \infty$, $k = 1, \ldots, d$, it holds*

$$\int_{\mathbb{R}^d} \prod_{k=1}^d \left[\frac{1}{2} \mathbb{1}_{\{b_k\}}(x_k) + \mathbb{1}_{(a_k, b_k)}(x_k) + \frac{1}{2} \mathbb{1}_{\{a_k\}}(x_k) \right] d\mu(x)$$

$$= \lim_{c \to \infty} \frac{1}{(2\pi)^d} \int_{([-c,c]_0)^d} \prod_{k=1}^d \frac{e^{-iy_k a_k} - e^{-iy_k b_k}}{iy_k} \widehat{\mu}(y) d\lambda_d(y).$$

If we do not have any mass at the boundary of the cuboid $\bigtimes_{k=1}^{d}[a_k, b_k]$, then the above inversion formula simplifies:

Corollary 16.1.2 *For $\mu \in \mathcal{M}_1^+(\mathbb{R}^d)$ and $-\infty < a_k < b_k < \infty$ for $k = 1, \ldots, d$ such that $\mu\left(\bigtimes_{k=1}^{d}[a_k, b_k] \setminus \bigtimes_{k=1}^{d}(a_k, b_k)\right) = 0$ it holds*

$$\mu\left(\bigtimes_{k=1}^{d}(a_k, b_k)\right) = \lim_{c \to \infty} \frac{1}{(2\pi)^d} \int_{([-c,c]_0)^d} \prod_{k=1}^{d} \frac{e^{-iy_k a_k} - e^{-iy_k b_k}}{iy_k} \widehat{\mu}(y) d\lambda_d(y).$$

In particular, if $\mu \in \mathcal{M}_1^+(\mathbb{R})$, then for the distribution function $F(x) = \mu((-\infty, x])$, $x \in \mathbb{R}$, it holds

$$F(b) - F(a) = \lim_{c \to \infty} \frac{1}{2\pi} \int_{[-c,c]_0} \frac{e^{-iya} - e^{-iyb}}{iy} \widehat{\mu}(y) d\lambda(y)$$

if $-\infty < a < b < \infty$ are points of continuity of F.

One key to the proof of Theorem 16.1.1 is the value of the Dirichlet integral proven in Lemma A.4.2. The Dirichlet integral has two consequences that are important for us. Namely, for any $d \neq 0$ we have that

$$\sup_{c>0}\left|\int_{(0,c]} \frac{\sin(yd)}{y} d\lambda(y)\right| = \sup_{c>0}\left|\int_{(0,c|d|]} \frac{\sin(y)}{y} d\lambda(y)\right|$$

$$= \sup_{c>0}\left|\int_{(0,c]} \frac{\sin(y)}{y} d\lambda(y)\right| =: M < \infty \quad (16.1)$$

and

$$\lim_{c \to \infty, c>0} \int_{(0,c]} \frac{\sin(yd)}{y} d\lambda(y) = \operatorname{sgn}(d)\frac{\pi}{2}. \quad (16.2)$$

Proof of Theorem 16.1.1 For $c > 0$ the definition of $\widehat{\mu}$ gives

$$\frac{1}{(2\pi)^d} \int_{([-c,c]_0)^d} \prod_{k=1}^{d} \frac{e^{-iy_k a_k} - e^{-iy_k b_k}}{iy_k} \widehat{\mu}(y) d\lambda_d(y)$$

$$= \frac{1}{\pi^d} \int_{\mathbb{R}^d} \int_{([-c,c]_0)^d} \prod_{k=1}^{d} \left[\frac{e^{-iy_k a_k} - e^{-iy_k b_k}}{2iy_k} e^{ix_k y_k}\right] d\lambda_d(y) d\mu(x)$$

$$= \frac{1}{\pi^d} \int_{\mathbb{R}^d} \prod_{k=1}^{d} \left[\int_{[-c,c]_0} \left[\frac{e^{-iy_k a_k} - e^{-iy_k b_k}}{2iy_k} e^{ix_k y_k}\right] d\lambda(y_k)\right] d\mu(x)$$

16.1 Inversion Formulas

where we used Fubini's theorem. Indeed, the requirement of Fubini's theorem is satisfied because we have the integrable bound

$$\left| \frac{e^{-iy_k a_k} - e^{-iy_k b_k}}{iy_k} e^{ix_k y_k} \right| \leq |b_k - a_k| \quad \text{for} \quad y_k \neq 0.$$

By symmetry we have that

$$\int_{[-c,c]_0} \frac{e^{-iy_k a_k} - e^{-iy_k b_k}}{2iy_k} e^{ix_k y_k} d\lambda(y_k)$$
$$= \int_{(0,c]} \frac{\sin(y_k(x_k - a_k)) - \sin(y_k(x_k - b_k))}{y_k} d\lambda(y_k).$$

Exploiting (16.1), dominated convergence, and (16.2) we deduce that

$$\lim_{c \to \infty} \frac{1}{(2\pi)^d} \int_{([-c,c]_0)^d} \prod_{k=1}^d \frac{e^{-iy_k a_k} - e^{-iy_k b_k}}{iy_k} \widehat{\mu}(y) d\lambda_d(y)$$
$$= \lim_{c \to \infty} \frac{1}{\pi^d} \int_{\mathbb{R}^d} \prod_{k=1}^d \left[\int_{(0,c]} \frac{\sin(y_k(x_k - a_k)) - \sin(y_k(x_k - b_k))}{y_k} d\lambda(y_k) \right] d\mu(x)$$
$$= \frac{1}{\pi^d} \int_{\mathbb{R}^d} \prod_{k=1}^d \lim_{c \to \infty} \left[\int_{(0,c]} \frac{\sin(y_k(x_k - a_k)) - \sin(y_k(x_k - b_k))}{y_k} d\lambda(y_k) \right] d\mu(x)$$
$$= \frac{1}{\pi^d} \int_{\mathbb{R}^d} \prod_{k=1}^d \left[\frac{\pi}{2} \mathbb{1}_{\{b_k\}}(x_k) + \pi \mathbb{1}_{(a_k, b_k)}(x_k) + \frac{\pi}{2} \mathbb{1}_{\{a_k\}}(x_k) \right] d\mu(x)$$
$$= \int_{\mathbb{R}^d} \prod_{k=1}^d \left[\frac{1}{2} \mathbb{1}_{\{b_k\}}(x_k) + \mathbb{1}_{(a_k, b_k)}(x_k) + \frac{1}{2} \mathbb{1}_{\{a_k\}}(x_k) \right] d\mu(x).$$

$\square$

In Corollary 16.1.6 below we obtain a sufficient condition that a measure $\mu \in \mathcal{M}_1^+(\mathbb{R}^d)$ has a density, and we are able to compute this density by the inverse Fourier transform. For this we have some closer look at Fourier transforms of functions. Here one of the essential property of the Fourier transform is the fact that the inverse transform has exactly the same structure:

Definition 16.1.3 Let $\varphi \in \mathcal{L}_1(\mathbb{R}^d; \mathbb{C})$ be bounded and continuous. Then the **inverse Fourier transform** $\check{\varphi} : \mathbb{R}^d \to \mathbb{C}$ is defined by

$$\check{\varphi}(y) := \frac{1}{(2\pi)^d} \int_{\mathbb{R}^d} e^{-i\langle y, x \rangle} \varphi(x) d\lambda_d(x).$$

First we remark that the Riemann-Lebesgue Lemma (Theorem 11.5.2) implies $\check{\varphi} \in C_0(\mathbb{R}^d; \mathbb{C})$. The name *inverse Fourier transform* is justified by the following relation:

Theorem 16.1.4 (Inverse Fourier Transform I) *If $\varphi \in \mathcal{L}_1(\mathbb{R}^d; \mathbb{C})$ is bounded and continuous and $\check{\varphi} \in \mathcal{L}_1(\mathbb{R}^d; \mathbb{C})$, then*

$$\varphi(x) = \int_{\mathbb{R}^d} e^{i\langle x, y\rangle} \check{\varphi}(y) d\lambda_d(y) \quad \text{for all} \quad x \in \mathbb{R}^d.$$

Proof For a sequence $a_n \downarrow 0$ and $D_{a_n^2} \in \mathbb{R}^{d \times d}$ being the diagonal matrix that has $a_n^2 > 0$ on its diagonal we obtain from Example 12.1.11 and Theorem 11.7.3 that

$$\begin{aligned}
\varphi(x) &= \lim_{n \to \infty} \int_{\mathbb{R}^d} \varphi(y) d\mathcal{N}_{x, D_{a_n^2}}(y) \\
&= \lim_{n \to \infty} \int_{\mathbb{R}^d} \varphi(y) e^{-\frac{1}{2} \frac{\langle y-x, y-x\rangle}{a_n^2}} \frac{d\lambda_d(y)}{(2\pi a_n^2)^{\frac{d}{2}}} \\
&= \lim_{n \to \infty} \int_{\mathbb{R}^d} \varphi(y) \widehat{\mathcal{N}_{0, D_{a_n^{-2}}}}(x-y) \frac{d\lambda_d(y)}{(2\pi a_n^2)^{\frac{d}{2}}} \\
&= \lim_{n \to \infty} \int_{\mathbb{R}^d} \varphi(y) \left[\int_{\mathbb{R}^d} e^{i\langle x-y, z\rangle} e^{-\frac{1}{2} a_n^2 \langle z, z\rangle} \left(\frac{a_n^2}{2\pi}\right)^{\frac{d}{2}} d\lambda_d(z) \right] \frac{d\lambda_d(y)}{(2\pi a_n^2)^{\frac{d}{2}}} \\
&= \lim_{n \to \infty} \int_{\mathbb{R}^d} e^{i\langle x, z\rangle} \left[\frac{1}{(2\pi)^d} \int_{\mathbb{R}^d} \varphi(y) e^{-i\langle y, z\rangle} d\lambda_d(y) \right] e^{-\frac{1}{2} a_n^2 \langle z, z\rangle} d\lambda_d(z) \\
&= \lim_{n \to \infty} \int_{\mathbb{R}^d} e^{i\langle x, z\rangle} \check{\varphi}(z) e^{-\frac{1}{2} a_n^2 \langle z, z\rangle} d\lambda_d(z) \\
&= \int_{\mathbb{R}^d} e^{i\langle x, z\rangle} \check{\varphi}(z) d\lambda_d(z),
\end{aligned}$$

where we used that

$$\int_{\mathbb{R}^d \times \mathbb{R}^d} |\varphi(y)| e^{-\frac{1}{2} a_n^2 \langle z, z\rangle} d(\lambda_d \otimes \lambda_d)(y, z) < \infty,$$

$\check{\varphi} \in \mathcal{L}_1(\mathbb{R}^d; \mathbb{C})$, and dominated convergence. $\square$

In Theorem 16.1.4 the conclusion $\varphi(x) = \int_{\mathbb{R}^d} e^{i\langle x, y\rangle} \check{\varphi}(y) d\lambda_d(y)$ can be seen as a spectral decomposition of φ where the spectrum $\check{\varphi}$ is complex valued and Hermitian in the sense that $\overline{\check{\varphi}(y)} = \check{\varphi}(-y)$. In particular, when φ is real-valued, we get $\overline{\check{\varphi}(y)} = \check{\varphi}(-y)$.

In the case $\check{\varphi} \geq 0$ we can omit the condition $\check{\varphi} \in \mathcal{L}_1(\mathbb{R}^d; \mathbb{C})$:

Theorem 16.1.5 (Inverse Fourier Transform II) *If $\varphi \in \mathcal{L}_1(\mathbb{R}^d; \mathbb{C})$ is bounded and continuous and $\check{\varphi}(y) \geq 0$ for all $y \in \mathbb{R}^d$, then $\check{\varphi} \in \mathcal{L}_1(\mathbb{R}^d; \mathbb{C})$ with*

$$\int_{\mathbb{R}^d} \check{\varphi}(y) d\lambda_d(y) = \varphi(0).$$

Proof We proceed as in the proof of Theorem 16.1.4. We take a sequence $a_n \downarrow 0$ and $D_{a_n^2} \in \mathbb{R}^{d \times d}$ being the diagonal matrix that has $a_n^2 > 0$ on its diagonal and obtain

$$\begin{aligned}
\varphi(0) &= \lim_{n \to \infty} \int_{\mathbb{R}^d} \varphi(y) d\mathcal{N}_{0, D_{a_n^2}}(y) \\
&= \lim_{n \to \infty} \int_{\mathbb{R}^d} \varphi(y) e^{-\frac{1}{2} \frac{\langle y, y \rangle}{a_n^2}} \frac{d\lambda_d(y)}{(2\pi a_n^2)^{\frac{d}{2}}} \\
&= \lim_{n \to \infty} \int_{\mathbb{R}^d} \varphi(y) \widehat{\mathcal{N}_{0, D_{a_n^{-2}}}}(-y) \frac{d\lambda_d(y)}{(2\pi a_n^2)^{\frac{d}{2}}} \\
&= \lim_{n \to \infty} \int_{\mathbb{R}^d} \varphi(y) \left[\int_{\mathbb{R}^d} e^{i \langle -y, z \rangle} e^{-\frac{1}{2} a_n^2 \langle z, z \rangle} \left(\frac{a_n^2}{2\pi} \right)^{\frac{d}{2}} d\lambda_d(z) \right] \frac{d\lambda_d(y)}{(2\pi a_n^2)^{\frac{d}{2}}} \\
&= \lim_{n \to \infty} \int_{\mathbb{R}^d} \left[\int_{\mathbb{R}^d} e^{-i \langle z, y \rangle} \varphi(y) \frac{d\lambda_d(y)}{(2\pi)^d} \right] e^{-\frac{1}{2} a_n^2 \langle z, z \rangle} d\lambda_d(z) \\
&= \lim_{n \to \infty} \int_{\mathbb{R}^d} \check{\varphi}(z) e^{-\frac{1}{2} a_n^2 \langle z, z \rangle} d\lambda_d(z) \\
&= \int_{\mathbb{R}^d} \check{\varphi}(z) d\lambda_d(z),
\end{aligned}$$

where we used monotone convergence as $\check{\varphi} \geq 0$. This implies especially that $\check{\varphi} \in \mathcal{L}_1(\mathbb{R}^d; \mathbb{C})$. □

The following corollary is a direct consequence of the previous two statements:

Corollary 16.1.6 *Let $\varphi \in \mathcal{L}_1(\mathbb{R}^d; \mathbb{C})$ be bounded and continuous. Then $\varphi = \widehat{\mu}$ for some $\mu \in \mathcal{M}_1^+(\mathbb{R}^d)$ if and only if*

(1) $\varphi(0) = 1$,
(2) $\check{\varphi}(y) \geq 0$ for all $y \in \mathbb{R}^d$.

In this case, $\check{\varphi}$ is the density of μ.

16.2 The Theorem of Bochner and Khinchin

Since there is a unique relation between a measure $\mu \in \mathcal{M}_1^+(\mathbb{R}^d)$ and its Fourier transform $\widehat{\mu}$, properties of μ can be derived from properties of $\widehat{\mu}$. This suggests that one might go the other way round: We start with a candidate φ of a Fourier transform which would imply certain properties of the corresponding measure we wish to have—*given we know that this measure exists*. The celebrated Bochner-Khinchin theorem provides a criteria for a complex function being a Fourier transform of some $\mu \in \mathcal{M}_1^+(\mathbb{R}^d)$:

Theorem 16.2.1 (Bochner and Khinchin) *Assume that $\varphi : \mathbb{R}^d \to \mathbb{C}$ is continuous with $\varphi(0) = 1$. Then the following assertions are equivalent:*

(1) *The function φ is the Fourier transform of some $\mu \in \mathcal{M}_1^+(\mathbb{R}^d)$.*
(2) *The function φ is positive semidefinite, i.e. for all $n \in \mathbb{N}$, all $x_1, \ldots, x_n \in \mathbb{R}^d$, and all $\lambda_1, \ldots, \lambda_n \in \mathbb{C}$ it holds*

$$\sum_{k,l=1}^n \lambda_k \overline{\lambda_l} \, \varphi(x_k - x_l) \geq 0.$$

Proof (1) $\Rightarrow$ (2) follows from Theorem 11.2.6, so we only need to verify (2) $\Rightarrow$ (1).
(a) First we suppose additionally that $\varphi \in \mathcal{L}_1(\mathbb{R}^d; \mathbb{C})$. For $x \in \mathbb{R}^d$ we again use $|x|_\infty := \max\{|x_1|, \ldots, |x_d|\}$ and let $K > 0$. For $z \in \mathbb{R}^d$ consider

$$p_K(z) := \frac{1}{(4\pi K)^d} \int_{\{|x|_\infty \leq K\}} \int_{\{|y|_\infty \leq K\}} \varphi(x-y) e^{-i\langle x-y, z\rangle} \, d\lambda_d(x) d\lambda_d(y).$$

Since φ is continuous, by an approximation and the positive semidefiniteness of φ we deduce $p_K \geq 0$ (observe $e^{-i\langle x-y, z\rangle} = e^{-i\langle x, z\rangle} \overline{e^{-i\langle y, z\rangle}}$ and approximate $\int_{\{|x|_\infty \leq K\}} \int_{\{|y|_\infty \leq K\}} \varphi(x-y) e^{-i\langle x-y, z\rangle} d\lambda_d(x) d\lambda_d(y)$ by $\sum_{k,l=1}^n \varphi(x_k - x_l) \lambda_k \overline{\lambda_l}$). We substitute $v := x - y$ and $u := y$, where for the set we integrate over we have

$$\{(x, y) \in \mathbb{R}^d \times \mathbb{R}^d : |x|_\infty \leq K, |y|_\infty \leq K\}$$
$$= \{(v, u) \in \mathbb{R}^d \times \mathbb{R}^d : |v|_\infty \leq 2K,$$
$$\max\{-K - v_k, -K\} \leq u_k \leq \min\{K - v_k, K\}, \; k = 1, \ldots, d\},$$

so that

$$p_K(z)$$
$$= \frac{1}{(4\pi K)^d} \int_{\{|v|_\infty \leq 2K\}} \left(\prod_{k=1}^d \int_{\max\{-K-v_k, -K\}}^{\min\{K-v_k, K\}} d\lambda_1(u_k) \right) \varphi(v) e^{-i\langle v, z\rangle} d\lambda_d(v)$$
$$= \frac{1}{(2\pi)^d} \int_{\{|v|_\infty \leq 2K\}} \prod_{k=1}^d \left(1 - \frac{|v_k|}{2K}\right) \varphi(v) e^{-i\langle v, z\rangle} d\lambda_d(v).$$

16.2 The Theorem of Bochner and Khinchin

We apply Corollary 16.1.6 to

$$\varphi_K(v) := \mathbb{1}_{\{|v|_\infty \leqslant 2K\}} \prod_{k=1}^{d}\left(1 - \frac{|v_k|}{2K}\right)\varphi(v)$$

and conclude that $p_K = \check{\varphi}_K$ in the notation of Corollary 16.1.6, and that φ_K is the Fourier transform of some $\mu_K \in \mathcal{M}_1^+(\mathbb{R}^d)$. The sequence $(\varphi_K)_{K=1}^{\infty}$ converges uniformly on compacts to φ and hence φ is the Fourier transform of some $\mu \in \mathcal{M}_1^+(\mathbb{R}^d)$ by Lévy's continuity theorem Theorem 12.1.3.

(b) For the general case we first notice that the assumption that φ is positive semidefinite with $\varphi(0) = 1$ implies that

(i) $\varphi(-x) = \overline{\varphi(x)}$,
(ii) $|\varphi(x)| \leqslant 1$.

Indeed, using (2) for $n = 2$, $x_2 = 0$, $\lambda_2 = 1$, gives

$$1 + \lambda_1 \varphi(x_1) + \overline{\lambda}_1 \varphi(-x_1) + |\lambda_1|^2 \geqslant 0. \tag{16.3}$$

For $\lambda_1 = 1$ we conclude that $\varphi(x_1) + \varphi(-x_1)$ is real-valued. Assuming $\varphi(x_1) = a + ib$ and $\varphi(-x_1) = c + id$ implies that $b = -d$. Since for $\lambda_1 = i$ we have that $i\varphi(x_1) - i\varphi(-x_1)$ is real-valued we conclude that $\varphi(x_1) = a + ib$ implies that $\varphi(-x_1) = a - ib$ which means we have shown $\varphi(-x) = \overline{\varphi(x)}$.
Putting $\lambda_1 = -\overline{\varphi(x_1)}$ in (16.3) leads to

$$1 - \overline{\varphi(x_1)}\varphi(x_1) - \varphi(x_1)\varphi(-x_1) + |\varphi(x_1)|^2 \geqslant 0.$$

Hence by $\varphi(-x_1) = \overline{\varphi(x_1)}$ we conclude that $|\varphi(x_1)|^2 \leqslant 1$. Now we introduce for $k \in \mathbb{N}$ the functions

$$\varphi_k(x) := \varphi(x) e^{-\frac{|x|^2}{2k}}.$$

Obviously, each φ_k is continuous and satisfies $\varphi_k(0) = 1$, and is integrable because $|\varphi(x)| \leqslant 1$. Since $e^{-\frac{|x|^2}{2k}}$ is the Fourier transform of a Gaussian measure it is a positive semidefinite function and so is φ_k by the Schur product Theorem A.5.1. Hence by (a) there exists for each k a $\mu_k \in \mathcal{M}_1^+(\mathbb{R}^d)$ with $\varphi_k = \widehat{\mu}_k$. Since φ is continuous and $\lim_{k \to \infty} \varphi_k(x) = \varphi(x)$ for all $x \in \mathbb{R}^d$, Lévy's continuity theorem implies that φ is the Fourier transform of some $\mu \in \mathcal{M}_1^+(\mathbb{R}^d)$. □

The assumption *positive semidefinite* in Theorem 16.2.1 of Bochner and Khinchin is sometimes difficult to check. On the real line there is an easier sufficient condition:

Corollary 16.2.2 (Pólya) *Let* $\varphi : \mathbb{R} \to [0, \infty)$ *be*

(1) *continuous,*
(2) *even, i.e.* $\varphi(x) = \varphi(-x)$ *for* $x \in \mathbb{R}$,

(3) *convex on* $[0, \infty)$,
(4) *and assume that* $\varphi(0) = 1$ *and* $\lim_{x \to \infty} \varphi(x) = 0$.

Then there exists some $\mu \in \mathcal{M}_1^+(\mathbb{R})$ *such that* $\widehat{\mu} = \varphi$.

Proof We find a sequence of φ_n that are integrable, satisfy the above assumptions, and such that $\lim_n \varphi_n(x) = \varphi(x)$ for all $x \in \mathbb{R}$. By the convexity of φ_n we verify below that

$$\int_{\mathbb{R}} e^{-ixy} \varphi_n(x) d\lambda(x) = \int_{\mathbb{R}} \cos(yx) \varphi_n(x) d\lambda(x) \geq 0 \tag{16.4}$$

for all $y \in \mathbb{R}$ so that φ_n is the Fourier-transform of some $\mu_n \in \mathcal{M}_+^1(\mathbb{R})$ by Corollary 16.1.6. Since the φ_n converge pointwise, Theorem 12.1.3 implies that there exists some $\mu \in \mathcal{M}_1^+(\mathbb{R})$ such that $\widehat{\mu} = \varphi$.

Now let us prove (16.4): the inequality (16.4) is obvious for $y = 0$, moreover the left-hand side does not change when passing from y to $-y$. Therefore we may assume $y > 0$. By a change of variables we can restrict ourselves to $y = 1$ when φ_n is replaced by $x \mapsto \varphi_n(x/y)$. And for $y = 1$ we have

$$\int_{[0, 2\pi)} \cos(x) \varphi_n(x) d\lambda(x)$$
$$= \int_{[0, \frac{\pi}{2})} \cos(x) \left[\varphi_n(x) + \varphi_n(2\pi - x) - \varphi_n(\pi - x) - \varphi(\pi + x) \right] d\lambda(x)$$

and observe (Exercise 1) that $\varphi_n(x) + \varphi_n(2\pi - x) \geq \varphi_n(\pi - x) + \varphi(\pi + x)$ by the convexity of φ_n. The same argument is applied to all intervals $[2\pi k, 2\pi(k+1))$, $k \in \mathbb{Z}$, where we also use the symmetry of the function φ_n. □

16.3 Cauchy Distribution

We start with an experiment (Fig. 16.1). Take a point source which sends small particles to a wall. The distance between the point source and the wall is $c > 0$. The angle $\psi \in \left(-\frac{\pi}{2}, \frac{\pi}{2}\right)$ is uniformly distributed on $\left(-\frac{\pi}{2}, \frac{\pi}{2}\right)$. What is the probability μ_c that a particle hits a set $B \in \mathcal{B}(\mathbb{R})$ from this wall? Here we get for $x \in \mathbb{R}$ that

$$\mu_c((-\infty, x]) = \frac{1}{2} + \frac{\arctan \frac{x}{c}}{\pi} = \frac{c}{\pi} \int_{-\infty}^{x} \frac{d\lambda(y)}{c^2 + y^2}.$$

16.3 Cauchy Distribution

Fig. 16.1 Example: Cauchy distribution

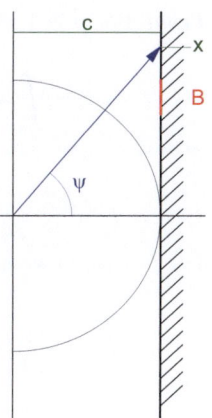

Definition 16.3.1 For $c > 0$ the distribution

$$\mathrm{d}\mu_c(x) := \frac{c}{\pi} \frac{1}{c^2 + x^2} \mathrm{d}\lambda(x) \in \mathcal{M}_1^+(\mathbb{R})$$

is called **Cauchy distribution with parameter** $c > 0$.

Proposition 16.3.2 *For $c > 0$ one has that $\widehat{\mu}_c(x) = e^{-c|x|}$.*

Proof We prove the statement for $c = 1$, the general case is done by a change of variables. By Lemma A.4.1 we get

$$\frac{1}{2\pi} \int_{\mathbb{R}} e^{-iyx} e^{-|x|} \mathrm{d}\lambda(x) = \frac{1}{\pi} \int_0^\infty \cos(yx) e^{-x} \mathrm{d}\lambda(x) = \frac{1}{\pi(1 + y^2)}.$$

Letting $\varphi(x) := e^{-|x|}$ and $\check{\varphi}(y) := \frac{1}{\pi(1+y^2)}$ we get $\varphi(0) = 1$ and $\check{\varphi}(y) \geq 0$ for all $y \in \mathbb{R}^d$. Therefore we can apply Corollary 16.1.6 to conclude the proof. □

The Cauchy distribution does not have any moments of order greater or equal to one:

Proposition 16.3.3 *For all $c > 0$ one has*

$$\int_{\mathbb{R}} |x|^p \mathrm{d}\mu_c(x) = \infty \quad \text{for} \quad p \geq 1,$$

$$\int_{\mathbb{R}} |x|^p \mathrm{d}\mu_c(x) < \infty \quad \text{for} \quad p \in (0, 1).$$

Proof For $p \geqslant 1$ we use

$$\int_\mathbb{R} \frac{|x|^p}{c^2+x^2} d\lambda(x) \geqslant \frac{1}{c^2+1} \int_{|x| \geqslant 1} \frac{|x|^p}{x^2} d\lambda(x)$$
$$= \frac{2}{c^2+1} \int_{[1,\infty)} x^{p-2} d\lambda(x) = \infty.$$

For $p \in (0,1)$ we have

$$\int_\mathbb{R} \frac{|x|^p}{c^2+x^2} d\lambda(x) = \int_{[-1,1]} \frac{|x|^p}{c^2+x^2} d\lambda(x) + 2 \int_{(1,\infty)} \frac{|x|^p}{c^2+x^2} d\lambda(x)$$
$$\leqslant \frac{2}{c^2} + 2 \int_{(1,\infty)} x^{p-2} d\lambda(x) < \infty.$$

□

Now let us come back to our introducing example. We think about two walls at the distances $0 < c_1 < c_1 + c_2 < \infty$ (Fig. 16.2). The particle will be emitted with a direction given by the angle ψ_1 at the origin and will be re-emitted at the wall at distance c_1 with the angle ψ_2. The angles ψ_1 and ψ_2 are independent and uniformly distributed on $\left(-\frac{\pi}{2}, \frac{\pi}{2}\right)$. If f_1 is the position at the first wall and $f_1 + f_2$ its position on the second wall, then one would expect the following:

(a) f_1 has the Cauchy distribution with parameter c_1.
(b) f_2 has the Cauchy distribution with parameter c_2.
(c) $f_1 + f_2$ has the Cauchy distribution with parameter $c_1 + c_2$.

As f_1 and f_2 are assumed to be independent, this is the so-called 1-*stability* and is confirmed by the following proposition:

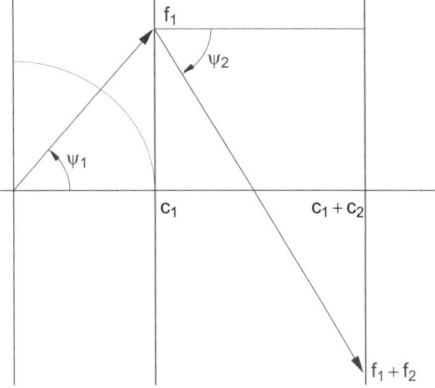

Fig. 16.2 Example: sum of Cauchy distributed random variables

Proposition 16.3.4 *Let $(\Omega, \mathcal{F}, \mathbb{P})$ be a probability space and let $f_1, f_2 : \Omega \to \mathbb{R}$ be independent, Cauchy distributed random variables with parameters $c_1 > 0$ and $c_2 > 0$, respectively. Then for all $\alpha_1, \alpha_2 \in \mathbb{R}$ the random variable $\alpha_1 f_1 + \alpha_2 f_2$ has a Cauchy distribution with parameter $|\alpha_1|c_1 + |\alpha_2|c_2$.*

Proof For the characteristic function $\varphi_{\alpha_1 f_1 + \alpha_2 f_2}$ we compute by independence of f_1 and f_2 that

$$\varphi_{\alpha_1 f_1 + \alpha_2 f_2}(x) = \mathbb{E}e^{x(\alpha_1 f_1 + \alpha_2 f_2)} = \mathbb{E}e^{x\alpha_1 f_1}\mathbb{E}e^{\alpha_2 f_2} = e^{-c_1|x\alpha_1|}e^{-c_2|x\alpha_2|}$$
$$= e^{-|x|(|\alpha_1|c_1 + |\alpha_2|c_2)}$$

which implies our assertion. □

16.4 Exercises

Ex 1: Show that for a convex function $\varphi : \mathbb{R} \to [0, \infty)$ as in Corollary 16.2.2 it holds that

$$\varphi(\pi - x) + \varphi(\pi + x) \leqslant \varphi(x) + \varphi(2\pi - x) \quad \text{for} \quad x \in \left[0, \frac{\pi}{2}\right).$$

Ex 2: Let $a > 0$ and $d = 1$. Using statements of this chapter prove that

$$\varphi(x) := \begin{cases} \frac{a - |x|}{a} & |x| \leqslant a \\ 0 & |x| > a \end{cases}$$

is the Fourier transform of the measure with density $\psi(x) := \frac{1 - \cos(ax)}{\pi a x^2}$.

Ex 3: The theorem of Bochner and Khinchin (Theorem 16.2.1) implies that the function $\varphi(x) = e^{-c|x|}$ is positive semidefinite for $c > 0$ since it is the Fourier transform of the Cauchy distribution. On the other hand, φ is integrable, bounded and continuous, so that by Corollary 16.1.6 the function

$$\check{\varphi}(y) := \frac{1}{2\pi}\int_{\mathbb{R}} e^{-iyx}\varphi(x)\mathrm{d}\lambda(x)$$

is the density of the Cauchy distribution, i.e. $\check{\varphi}(y) = \frac{c}{\pi}\frac{1}{c^2 + y^2}$. Since φ is symmetric, we have $\check{\varphi}(y) = \frac{1}{2\pi}\int_{\mathbb{R}} e^{iyx}\varphi(x)\mathrm{d}\lambda(x) = \frac{1}{2\pi}\hat{\varphi}(y)$. Consequently, $\check{\varphi}$ is again a positive semidefinite, integrable, bounded and continuous function.

For the examples below find $\hat{\varphi}$ and show that the relation $\check{\varphi}(y) = \frac{1}{2\pi}\hat{\varphi}(y)$ holds:

(a) $\varphi(x) := e^{-\frac{1}{2}\sigma^2 x^2}$ for $x \in \mathbb{R}$, where $\sigma > 0$ is fixed,
(b) $\varphi(x) := \max\left\{0, \frac{2-|x|}{2}\right\}$ for $x \in \mathbb{R}$.

Explain why $\varphi, \check{\varphi}$ are both positive semidefinite, integrable, bounded and continuous functions.

16.5 Comments

Section 16.1: For the proof of Corollary 16.1.2 we also used the presentations in [173, pp. 283] ($d = 1$) and [16, p. 382].

Section 16.2: A general form of the Bochner-Khinchin Theorem 16.2.1 uses locally compact abelian groups G. For G being the torus the result was proved by Herglotz [83] and for $G = \mathbb{R}$ by Bochner [21, Satz 19], see also [196, Chapitre VI]. A general account on the analysis on locally compact abelian groups can be found in Folland [64, Chapter 4]. For the proof of Theorem 16.2.1 we used the sources [72, 243–247] (here $d = 1$ is considered), [22, p. 121, Volume II], and [62, Lemma 3, page 622] to verify $\varphi(-x) = \overline{\varphi(x)}$ and $|\varphi(x)| \leq 1$. Corollary 16.2.2 is due to Pólya [148].

Section 16.4: Exercise 3 is taken from Engelbert and Schladitz [54].

Chapter 17
Norm Estimates for the Fourier Transform

In this section we look at the Fourier transform from a more functional analytic point of view and prove fundamental norm-estimates in terms of Plancherel's identity and the Hausdorff-Young inequality. As an application of a Fourier transform we briefly introduce the Hilbert transform.

17.1 Plancherel's Equality and Hausdorff-Young Inequality

We start with Plancherel's equality and generalize the equality to the Hausdorff-Young inequality. For the latter inequality the Riesz-Thorin Theorem A.8.1 is essential which is based on interpolation techniques from function space theory. Let us start with the Fourier transform of functions where we use a different normalization:

Definition 17.1.1 For $\psi \in \mathcal{L}_1(\mathbb{R}^d; \mathbb{C})$ we define

$$\mathcal{F}(\psi)(x) := (2\pi)^{-\frac{d}{2}} \int_{\mathbb{R}^d} e^{i\langle x, y\rangle} \psi(y) \mathrm{d}\lambda_d(y).$$

By Definition 10.2.7 we have

$$\mathcal{F}(\psi)(x) = (2\pi)^{-\frac{d}{2}} \widehat{\psi}(x).$$

We start with the theorem due to Plancherel, that says that the Fourier transform gives rise to an isometry on $L_2(\mathbb{R}^d; \mathbb{C})$:

Theorem 17.1.2 (Plancherel[1]) *One has*

$$\|\mathcal{F}(\psi)\|_{L_2(\mathbb{R}^d;\mathbb{C})} = \|\psi\|_{L_2(\mathbb{R}^d;\mathbb{C})} \quad \text{for} \quad \psi \in \mathcal{L}_1(\mathbb{R}^d;\mathbb{C}) \cap \mathcal{L}_2(\mathbb{R}^d;\mathbb{C}).$$

Moreover, the Fourier transform can be uniquely extended to a surjective isometric map $\widetilde{\mathcal{F}} : L_2(\mathbb{R}^d;\mathbb{C}) \to L_2(\mathbb{R}^d;\mathbb{C})$.

In the following we use for the extended Fourier transform $\widetilde{\mathcal{F}} : L_2(\mathbb{R}^d;\mathbb{C}) \to L_2(\mathbb{R}^d;\mathbb{C})$ also the notation $\mathcal{F}(\psi)$ regardless whether $\psi \in \mathcal{L}_1(\mathbb{R}^d;\mathbb{C})$ or $\psi \in \mathcal{L}_2(\mathbb{R}^d;\mathbb{C})$.

There is a technical aspect one has to keep in mind: For $\psi \in \mathcal{L}_1(\mathbb{R}^d;\mathbb{C})$ the original definition is pointwise, i.e for each $x \in \mathbb{R}^d$ we can compute $\mathcal{F}(\psi)(x) = (2\pi)^{-\frac{d}{2}} \int_{\mathbb{R}^d} e^{i\langle x,y\rangle} \psi(y) d\lambda_d(y)$. For $\psi \in \mathcal{L}_2(\mathbb{R}^d;\mathbb{C})$ this is not possible when ψ is not integrable, i.e. $\psi \notin \mathcal{L}_1(\mathbb{R}^d;\mathbb{C})$. Instead of a function $x \mapsto \mathcal{F}(\psi)(x)$ we only obtain an equivalence class $\mathcal{F}(\psi) \in L_2(\mathbb{R}^d;\mathbb{C})$, so that $\mathcal{F}(\psi)$ is only unique up to the behaviour on null sets. Finally, if $\psi \in \mathcal{L}_1(\mathbb{R}^d;\mathbb{C}) \cap \mathcal{L}_2(\mathbb{R}^d;\mathbb{C})$, then we can compute $\mathcal{F}(\psi)$ pointwise as $\psi \in \mathcal{L}_1(\mathbb{R}^d;\mathbb{C})$. Looking at the same function ψ as an element of $L_2(\mathbb{R}^d;\mathbb{C})$, then we obtain an equivalence class in $L_2(\mathbb{R}^d;\mathbb{C})$. It turns out that the pointwise computation is a representative of this equivalence class.

For the proof of Plancherel's theorem we use the following space of functions:

Definition 17.1.3 We let $C_c(\mathbb{R}^d;\mathbb{C})$ be the space of continuous functions with compact support, i.e. continuous $\varphi : \mathbb{R}^d \to \mathbb{C}$ such that there is an $N > 0$ with $f(x) = 0$ if $|x| > N$. Moreover, $C_c(\mathbb{R}^d) := C_c(\mathbb{R}^d;\mathbb{R})$.

The following lemma provides necessary approximation properties, exploiting the Gaussian density from Proposition 11.7.5, in order to prove Plancherel's theorem:

Lemma 17.1.4 *For $\varepsilon > 0$ let $v_\varepsilon : \mathbb{R}^d \to \mathbb{R}$ be the Gaussian density*

$$v_\varepsilon(x) := \frac{1}{(2\pi\varepsilon^2)^{\frac{d}{2}}} e^{-\frac{1}{2}\frac{|x|^2}{\varepsilon^2}}. \tag{17.1}$$

(1) *For $\psi \in \mathcal{L}_1(\mathbb{R}^d;\mathbb{C}) \cap \mathcal{L}_2(\mathbb{R}^d;\mathbb{C})$ and $\eta > 0$ there are $\psi_0 \in C_c(\mathbb{R}^d;\mathbb{C})$ and $\varepsilon > 0$ such that*

$$\|\psi - \psi_0 * v_\varepsilon\|_{L_1(\mathbb{R}^d;\mathbb{C})} < \eta \quad \text{and} \quad \|\psi - \psi_0 * v_\varepsilon\|_{L_2(\mathbb{R}^d;\mathbb{C})} < \eta.$$

(2) *For $\psi \in \mathcal{L}_2(\mathbb{R}^d;\mathbb{C})$ and $\eta > 0$ there are $\psi_0 \in C_c(\mathbb{R}^d;\mathbb{C})$ and $\varepsilon > 0$ such that*

$$\|\psi - \psi_0 * v_\varepsilon\|_{L_2(\mathbb{R}^d;\mathbb{C})} < \eta.$$

---

[1] Michel Plancherel, 16/01/1885 (Bussy, Kanton Freiburg, Switzerland)–04/03/1967 (Zürich, Switzerland).

17.1 Plancherel and Hausdorff-Young

Proof
(a) Reduction to nonnegative ψ: In both cases, item (1) and (2), we decompose ψ as

$$\psi = [\max\{\operatorname{Re}(\psi), 0\} - \max\{-\operatorname{Re}(\psi), 0\}] + i[\max\{\operatorname{Im}(\psi), 0\} - \max\{-\operatorname{Im}(\psi), 0\}].$$

As we can approximate each of the four functions on the right-hand side separately by functions of the form $\psi_0 * v_\varepsilon$ with $\psi_0 \in C_c(\mathbb{R}^d)$ and $\varepsilon > 0$, we may and do assume that $\psi : \mathbb{R}^d \to [0, \infty)$.

(b) Reduction to simple ψ: Similarly to the proof of Proposition 6.1.4 we find a sequence of simple functions $\psi_n : \mathbb{R}^n \to [0, \infty)$ with $0 \leqslant \psi_n(x) \uparrow \psi(x)$ for all $x \in \mathbb{R}^d$. By monotone convergence we have for (1), where by assumption $\psi \in \mathcal{L}_1(\mathbb{R}^d; \mathbb{C}) \cap \mathcal{L}_2(\mathbb{R}^d; \mathbb{C})$, that

$$\lim_{n \to \infty} \|\psi - \psi_n\|_{L_1(\mathbb{R}^d)} = 0, \quad \text{and} \quad \lim_{n \to \infty} \|\psi - \psi_n\|_{L_2(\mathbb{R}^d)} = 0.$$

For (2) we derive from $\psi \in \mathcal{L}_2(\mathbb{R}^d; \mathbb{C})$ that $\lim_{n \to \infty} \|\psi - \psi_n\|_{L_2(\mathbb{R}^d)} = 0$. So we may assume that ψ is simple and takes only nonnegative values.

Now we will approximate the simple function $\psi : \mathbb{R}^d \to [0, \infty)$ simultaneously in $L_1(\mathbb{R}^d)$ and $L_2(\mathbb{R}^d)$. For the remaining part of the proof we do not need to distinguish between (1) and (2) since any simple function in $L_2(\mathbb{R}^d)$ is also in $L_1(\mathbb{R}^d)$.

(c) Reduction to $\psi = \mathbb{1}_B$ with B of finite Lebesgue measure: In both cases we necessarily have $\|\psi_n\|_{L_2(\mathbb{R}^d)} \leqslant \|\psi\|_{L_2(\mathbb{R}^d)} < \infty$. If we have the representation $\psi_n(x) = \sum_{j=1}^{N_n} \beta_j^n \mathbb{1}_{B_j^n}(x)$ with $\beta_j^n > 0$ and $B_j^n \in \mathcal{B}(\mathbb{R}^d)$, then we automatically get that $\lambda_d(B_j^n) < \infty$ as $\|\psi_n\|_{L_2(\mathbb{R}^d)} < \infty$. For this reason it is sufficient to assume $\psi = \mathbb{1}_B$ where $B \in \mathcal{B}(\mathbb{R}^d)$ is of finite Lebesgue measure.

(d) Reduction to $\psi = \mathbb{1}_G$ where $G \subseteq \mathbb{R}^d$ is open and bounded: Exploiting the outer regularity from Theorem 5.2.4, given $\varepsilon > 0$ there is an open set $G \supseteq B$ such that $\lambda_d(G \setminus B) < \varepsilon$, which implies $\|\mathbb{1}_B - \mathbb{1}_G\|_{L_1(\mathbb{R}^d)} < \varepsilon$ and $\|\mathbb{1}_B - \mathbb{1}_G\|_{L_2(\mathbb{R}^d)} < \sqrt{\varepsilon}$. Moreover, by replacing G by $G_N := G \cap \{x \in \mathbb{R}^d : |x| < N\}$ for $N \in \mathbb{N}$ we can choose N large enough such that $\|\mathbb{1}_B - \mathbb{1}_{G_N}\|_{L_1(\mathbb{R}^d)} < \varepsilon$ and $\|\mathbb{1}_B - \mathbb{1}_{G_N}\|_{L_2(\mathbb{R}^d)} < \sqrt{\varepsilon}$ as well. Therefore, we can assume that the open set is bounded.

(e) Reduction to $\psi \in C_c(\mathbb{R}^d)$: Finally, let $\emptyset \neq G \subseteq \mathbb{R}^d$ be a bounded open set. Then the set $F := G^c$ is closed. Given $n \in \mathbb{N}$ we define the compact sets $K_n := \{x \in G : d(x, \partial G) \geqslant 1/n\}$. By Urysohn's Lemma 4.1.7 there is a continuous function $\psi_n : \mathbb{R}^d \to [0, 1]$ such that $\psi_n = 1$ on K_n and $\psi_n = 0$ on F. Moreover, we have $\lim_{n \to \infty} \psi_n(x) = \mathbb{1}_G(x)$. By dominated convergence, $\lim_{n \to \infty} \|\mathbb{1}_G - \psi_n\|_{L_2(\mathbb{R}^d)} = 0$ and $\lim_{n \to \infty} \|\mathbb{1}_G - \psi_n\|_{L_1(\mathbb{R}^d)} = 0$.

(f) Final approximation: Let $\psi_\varepsilon := \psi * v_\varepsilon$ with $\psi \in C_c(\mathbb{R}^d)$ and $\varepsilon \in (0, 1]$. Then $\lim_{\varepsilon \to 0} \|\psi - \psi_\varepsilon\|_{L_1(\mathbb{R}^d)} = \lim_{\varepsilon \to 0} \|\psi - \psi_\varepsilon\|_{L_2(\mathbb{R}^d)} = 0$ by dominated convergence. □

Proof of Theorem 17.1.2
(a) We define

$$A := \left\{ \psi = \psi_0 * v_\epsilon, \psi_0 \in C_c(\mathbb{R}^d; \mathbb{C}), \varepsilon > 0 \right\} \subseteq L_2(\mathbb{R}^d; \mathbb{C})$$

and the space of all finite linear combinations of elements of A,

$$\mathcal{H} := \mathrm{Lin}(A) \subseteq L_2(\mathbb{R}^d; \mathbb{C}).$$

If $\psi \in \mathcal{H}$, then $\psi \in L_1(\mathbb{R}^d; \mathbb{C})$, and it is bounded and continuous. It holds $\varphi := \mathcal{F}(\psi) \in L_1(\mathbb{R}^d; \mathbb{C})$. The latter follows from

$$\mathcal{F}(\psi_0 * v_\varepsilon)(x) = \mathcal{F}(\psi_0)(x)\, e^{-\varepsilon^2 \frac{|x|^2}{2}}$$

since $\mathcal{F}(\psi_0)$ is bounded as $\psi_0 \in C_c(\mathbb{R}^d; \mathbb{C}) \subseteq L_1(\mathbb{R}^d; \mathbb{C})$. By Lemma 17.1.4 the linear space $\mathcal{H}$ is dense in $L_2(\mathbb{R}^d; \mathbb{C})$.

(b) If $\psi \in L_1(\mathbb{R}^d; \mathbb{C})$ is bounded and continuous and if $\varphi := \mathcal{F}(\psi) \in L_1(\mathbb{R}^d; \mathbb{C})$, then Theorem 16.1.4 yields to

$$\mathcal{F}(\mathcal{F}(\psi))(x) = \psi(-x).$$

Since $\mathcal{F}(\psi)$ is bounded this implies that

$$\int_{\mathbb{R}^d} \mathcal{F}(\psi)(x) \overline{\mathcal{F}(\psi)(x)} \mathrm{d}\lambda_d(x) = \int_{\mathbb{R}^d} \varphi(x) \overline{\mathcal{F}(\psi)(x)} \mathrm{d}\lambda_d(x)$$

$$= \int_{\mathbb{R}^d} \varphi(x) \mathcal{F}(\overline{\psi})(-x) \mathrm{d}\lambda_d(x)$$

$$= \int_{\mathbb{R}^d} \psi(x) \mathcal{F}(\mathcal{F}(\overline{\psi}))(-x) \mathrm{d}\lambda_d(x)$$

$$= \int_{\mathbb{R}^d} \psi(x) \overline{\psi}(x) \mathrm{d}\lambda_d(x)$$

and

$$\int_{\mathbb{R}^d} |\mathcal{F}(\psi)(x)|^2 \mathrm{d}\lambda_d(x) = \int_{\mathbb{R}^d} |\psi(x)|^2 \mathrm{d}\lambda_d(x).$$

In particular, we have that

$$\|\mathcal{F}(\psi)\|_{L_2(\mathbb{R}^d; \mathbb{C})} = \|\psi\|_{L_2(\mathbb{R}^d; \mathbb{C})} \quad \text{for} \quad \psi \in \mathcal{H}$$

and get a unique linear extension $\widetilde{\mathcal{F}} : L_2(\mathbb{R}^d; \mathbb{C}) \to L_2(\mathbb{R}^d; \mathbb{C})$ from $\mathcal{F} : \mathcal{H} \to L_2(\mathbb{R}^d; \mathbb{C})$.

17.1 Plancherel and Hausdorff-Young

(c) For $\psi \in \mathcal{L}_1(\mathbb{R}^d; \mathbb{C}) \cap \mathcal{L}_2(\mathbb{R}^d; \mathbb{C})$ we find, by Lemma 17.1.4(1), a sequence $(\psi_n)_{n \in \mathbb{N}} \subseteq \mathcal{H}$ such that

$$\lim_{n \to \infty} \|\psi - \psi_n\|_{L_1(\mathbb{R}^d; \mathbb{C})} = 0 \quad \text{and} \quad \lim_{n \to \infty} \|\psi - \psi_n\|_{L_2(\mathbb{R}^d; \mathbb{C})} = 0.$$

This implies that

$$\lim_{n \to \infty} \mathcal{F}(\psi_n)(x) = \mathcal{F}(\psi)(x) \quad \text{for all} \quad x \in \mathbb{R}^d,$$

$$\lim_{n \to \infty} \mathcal{F}(\psi_n) = \widetilde{\mathcal{F}}(\psi) \quad \text{in} \quad L_2(\mathbb{R}^d; \mathbb{C}).$$

Therefore, $\mathcal{F}(\psi) = \widetilde{\mathcal{F}}(\psi)$ a.e.

(d) The map $\widetilde{\mathcal{F}}$ is surjective: We assume $\psi_0 \in C_c(\mathbb{R}^d; \mathbb{C})$ and $\varepsilon > 0$, and define $\psi := \psi_0 * v_\varepsilon$. We get

$$\check{\psi}(y) = e^{-\varepsilon^2 \frac{|y|^2}{2}} \check{\psi}_0(y) \quad \text{and} \quad \check{\psi} \in \mathcal{L}_1(\mathbb{R}^d; \mathbb{C}).$$

Therefore

$$\mathcal{F}((2\pi)^{\frac{d}{2}} \check{\psi}) = \mathcal{F}(\mathcal{F}(\psi(-\cdot))) = \psi.$$

Because the set $\{\psi_0 * v_\varepsilon : \psi_0 \in C_c(\mathbb{R}^d; \mathbb{C}), \varepsilon > 0\}$ is dense in $L_2(\mathbb{R}^d; \mathbb{C})$ by Lemma 17.1.4(2) the map $\widetilde{\mathcal{F}}$ is surjective. □

Now we go one step further and prove the Hausdorff-Young inequality that extents the Plancherel's equality to an inequality in the case when $\psi \in \mathcal{L}_p(\mathbb{R}^d; \mathbb{C})$ for $p \in (1, 2)$. We will prove this inequality by the Riesz-Thorin interpolation theorem which 'interpolates' between Plancherel's equality and the case $\psi \in \mathcal{L}_1(\mathbb{R}^d; \mathbb{C})$. Before we state and prove the Hausdorff-Young inequality we need some preparations:

Definition 17.1.5 For $p, q \in [1, \infty]$ we define the linear space

$$L_p(\mathbb{R}^d; \mathbb{C}) + L_q(\mathbb{R}^d; \mathbb{C})$$
$$:= \left\{ [\psi] : \psi = \psi_p + \psi_q : \psi_p \in \mathcal{L}_p(\mathbb{R}^d; \mathbb{C}), \psi_q \in \mathcal{L}_q(\mathbb{R}^d; \mathbb{C}) \right\}$$

of equivalence classes $[\cdot]$ from $L_0(\mathbb{R}^d, \mathcal{B}(\mathbb{R}^d), \lambda_d; \mathbb{C})$ equipped with the norm

$$\|[\psi]\|_{L_p + L_q} = \|\psi\|_{L_p + L_q}$$
$$:= \inf \left\{ \|\psi_p\|_{L_p(\mathbb{R}^d; \mathbb{C})} + \|\psi_q\|_{L_q(\mathbb{R}^d; \mathbb{C})} : \psi = \psi_p + \psi_q \right\}.$$

That $\|\cdot\|_{L_p+L_q}$ is actually a norm we leave as an exercise. Now the Fourier transform is a well-defined linear map

$$\mathcal{F}: L_1(\mathbb{R}^d; \mathbb{C}) + L_2(\mathbb{R}^d; \mathbb{C}) \to L_\infty(\mathbb{R}^d; \mathbb{C}) + L_2(\mathbb{R}^d; \mathbb{C}). \tag{17.2}$$

In fact, assume that $[\psi] = [\psi_1] + [\psi_2] = [\psi_1'] + [\psi_2']$ for $\psi_1, \psi_1' \in \mathcal{L}_1(\mathbb{R}^d; \mathbb{C})$ and $\psi_2, \psi_2' \in \mathcal{L}_2(\mathbb{R}^d; \mathbb{C})$. Then $\psi_1 - \psi_1' \stackrel{a.e.}{=} \psi_2' - \psi_2 \in \mathcal{L}_1(\mathbb{R}^d; \mathbb{C}) \cap \mathcal{L}_2(\mathbb{R}^d; \mathbb{C})$. This implies that $\mathcal{F}(\psi_1 - \psi_1')(x)$ and $\mathcal{F}(\psi_2' - \psi_2)(x)$ can be computed pointwise by Definition 17.1.1 and for these representatives one has $\mathcal{F}(\psi_1 - \psi_1')(x) = \mathcal{F}(\psi_2' - \psi_2)(x)$ for all $x \in \mathbb{R}^d$. Therefore, in terms of equivalence classes, $\mathcal{F}(\psi_1) - \mathcal{F}(\psi_1') = \mathcal{F}(\psi_2') - \mathcal{F}(\psi_2)$ and $\mathcal{F}(\psi_1) + \mathcal{F}(\psi_2) = \mathcal{F}(\psi_1') + \mathcal{F}(\psi_2')$. Moreover, we have that $\mathcal{F}(\psi_1), \mathcal{F}(\psi_1') \in L_\infty(\mathbb{R}^d; \mathbb{C})$ and $\mathcal{F}(\psi_2), \mathcal{F}(\psi_2') \in L_2(\mathbb{R}^d; \mathbb{C})$. So

$$\mathcal{F}(\psi) := \mathcal{F}(\psi_1) + \mathcal{F}(\psi_2) \quad \text{for} \quad [\psi] = [\psi_1] + [\psi_2]$$

yields to a well-defined operator in (17.2). We collect some properties of this approach:

Lemma 17.1.6

(1) *The map* $\mathcal{F}: L_1(\mathbb{R}^d; \mathbb{C}) + L_2(\mathbb{R}^d; \mathbb{C}) \to L_\infty(\mathbb{R}^d; \mathbb{C}) + L_2(\mathbb{R}^d; \mathbb{C})$ *is linear and continuous with*

$$\|\mathcal{F}(\psi)\|_{L_\infty+L_2} \leqslant \|\psi\|_{L_1+L_2}.$$

(2) *If* $p \in (1, 2)$, *then we have* $L_p(\mathbb{R}^d; \mathbb{C}) \subseteq L_1(\mathbb{R}^d; \mathbb{C}) + L_2(\mathbb{R}^d; \mathbb{C})$ *with*

$$\|\psi\|_{L_1+L_2} \leqslant 2\|\psi\|_{L_p(\mathbb{R}^d;\mathbb{C})}.$$

(3) *If* $q \in (2, \infty)$, *then we have* $L_q(\mathbb{R}^d; \mathbb{C}) \subseteq L_\infty(\mathbb{R}^d; \mathbb{C}) + L_2(\mathbb{R}^d; \mathbb{C})$ *with*

$$\|\psi\|_{L_\infty+L_2} \leqslant 2\|\psi\|_{L_q(\mathbb{R}^d;\mathbb{C})}.$$

Proof
(1) For $\psi = \psi_1 + \psi_2$ with $\psi_1 \in \mathcal{L}_1(\mathbb{R}^d; \mathbb{C})$ and $\psi_2 \in \mathcal{L}_2(\mathbb{R}^d; \mathbb{C})$ we get that

$$\begin{aligned}
\|\mathcal{F}(\psi)\|_{L_\infty+L_2} &= \|\mathcal{F}(\psi_1 + \psi_2)\|_{L_\infty+L_2} \\
&\leqslant \|\mathcal{F}(\psi_1)\|_{L_\infty(\mathbb{R}^d;\mathbb{C})} + \|\mathcal{F}(\psi_2)\|_{L_2(\mathbb{R}^d;\mathbb{C})} \\
&\leqslant \frac{1}{(2\pi)^{\frac{d}{2}}} \|\psi_1\|_{L_1(\mathbb{R}^d;\mathbb{C})} + \|\psi_2\|_{L_2(\mathbb{R}^d;\mathbb{C})} \\
&\leqslant \|\psi_1\|_{L_1(\mathbb{R}^d;\mathbb{C})} + \|\psi_2\|_{L_2(\mathbb{R}^d;\mathbb{C})}.
\end{aligned}$$

Taking the infimum over all decompositions $\psi = \psi_1 + \psi_2$ we arrive at (1).

17.1 Plancherel and Hausdorff-Young

(2) If $\psi \in \mathcal{L}_p(\mathbb{R}^d; \mathbb{C})$ with $\|\psi\|_{L_p(\mathbb{R}^d;\mathbb{C})} = 1$, then we define

$$\psi_1(x) := \psi(x)\mathbb{1}_{\{|\psi(x)|>1\}} \quad \text{and} \quad \psi_2(x) := \psi(x)\mathbb{1}_{\{|\psi(x)|\leqslant 1\}}$$

and get

$$\int_{\mathbb{R}^d} |\psi_1(x)| \mathrm{d}\lambda_d(x) = \int_{\mathbb{R}^d} |\psi(x)|\mathbb{1}_{\{|\psi(x)|>1\}} \mathrm{d}\lambda_d(x)$$
$$\leqslant \int_{\mathbb{R}^d} |\psi(x)|^p \mathbb{1}_{\{|\psi(x)|>1\}} \mathrm{d}\lambda_d(x) \leqslant 1$$

and

$$\int_{\mathbb{R}^d} |\psi_2(x)|^2 \mathrm{d}\lambda_d(x) = \int_{\mathbb{R}^d} |\psi(x)|^2 \mathbb{1}_{\{|\psi(x)|\leqslant 1\}} \mathrm{d}\lambda_d(x)$$
$$\leqslant \int_{\mathbb{R}^d} |\psi(x)|^p \mathbb{1}_{\{|\psi(x)|\leqslant 1\}} \mathrm{d}\lambda_d(x) \leqslant 1.$$

(3) We use the same argument as in (2) for $\psi \in \mathcal{L}_q(\mathbb{R}^d; \mathbb{C})$ with $\|\psi\|_{L_q(\mathbb{R}^d;\mathbb{C})} = 1$ and

$$\psi_\infty(x) := \psi(x)\mathbb{1}_{\{|\psi(x)|\leqslant 1\}} \quad \text{and} \quad \psi_2(x) := \psi(x)\mathbb{1}_{\{|\psi(x)|>1\}}.$$

□

Now we turn to the Hausdorff-Young inequality:

Theorem 17.1.7 (Hausdorff[2]-Young) *Let $p \in [1, 2]$ and $q \in [2, \infty]$ with $1 = \frac{1}{p} + \frac{1}{q}$. Then there is a constant $c_p > 0$ such that one has that*

$$\|\mathcal{F}(\psi)\|_{L_q(\mathbb{R}^d;\mathbb{C})} \leqslant c_p \|\psi\|_{L_p(\mathbb{R}^d;\mathbb{C})} \quad \text{for} \quad \psi \in \mathcal{L}_p(\mathbb{R}^d; \mathbb{C}).$$

Proof For $p = 1$ we know that

$$\|\mathcal{F}(\psi)\|_{L_\infty(\mathbb{R}^d;\mathbb{C})} \leqslant \frac{1}{(2\pi)^{\frac{d}{2}}} \|\psi\|_{L_1(\mathbb{R}^d;\mathbb{C})}.$$

For $p = 2$ Plancherel's Theorem 17.1.2 gives

$$\|\mathcal{F}(\psi)\|_{L_2(\mathbb{R}^d;\mathbb{C})} = \|\psi\|_{L_2(\mathbb{R}^d;\mathbb{C})}.$$

---

[2] Felix Hausdorff, 08/11/1868 (Breslau, now Wroclaw, Poland)–26/01/1942 (Bonn, Germany).

Now let $p \in (1, 2)$ and $\psi \in \mathcal{L}_p(\mathbb{R}^d; \mathbb{C})$. For $S(\mathbb{R}^d, \mathcal{B}(\mathbb{R}^d), \lambda_d; \mathbb{C})$ as defined in the Riesz-Thorin Theorem A.8.1 we find $\psi_n \in S(\mathbb{R}^d, \mathcal{B}(\mathbb{R}^d), \lambda_d; \mathbb{C})$ such that $\lim_{n \to \infty} \|\psi - \psi_n\|_{L_p(\mathbb{R}^d; \mathbb{C})} = 0$. Applying the Riesz-Thorin theorem gives

$$\|\mathcal{F}(\psi_n) - \mathcal{F}(\psi_m)\|_{L_q(\mathbb{R}^d; \mathbb{C})} \leqslant \left(\frac{1}{(2\pi)^{\frac{d}{2}}} \right)^{\frac{2-p}{p}} \|\psi_n - \psi_m\|_{L_p(\mathbb{R}^d; \mathbb{C})}.$$

Therefore $(\mathcal{F}(\psi_n))_{n=1}^{\infty}$ is a Cauchy sequence in $L_q(\mathbb{R}^d; \mathbb{C})$ that converges to some $\varphi \in L_q(\mathbb{R}^d; \mathbb{C})$. Using the Riesz-Thorin theorem in the form

$$\|\mathcal{F}(\psi_n)\|_{L_q(\mathbb{R}^d; \mathbb{C})} \leqslant \left(\frac{1}{(2\pi)^{\frac{d}{2}}} \right)^{\frac{2-p}{p}} \|\psi_n\|_{L_p(\mathbb{R}^d; \mathbb{C})}$$

yields to

$$\|\varphi\|_{L_q(\mathbb{R}^d; \mathbb{C})} \leqslant \left(\frac{1}{(2\pi)^{\frac{d}{2}}} \right)^{\frac{2-p}{p}} \|\psi\|_{L_p(\mathbb{R}^d; \mathbb{C})}.$$

It remains to show that $\varphi = \mathcal{F}(\psi)$. For this we observe that $(\psi_n)_{n=1}^{\infty}$ converges in $L_1(\mathbb{R}^d; \mathbb{C}) + L_2(\mathbb{R}^d; \mathbb{C})$ to ψ as well because of Lemma 17.1.6(2), hence $\mathcal{F}(\psi_n)$ converges to $\mathcal{F}(\psi)$ in $L_\infty(\mathbb{R}^d; \mathbb{C}) + L_2(\mathbb{R}^d; \mathbb{C})$ by Lemma 17.1.6(1). But at the same time $\mathcal{F}(\psi_n)$ converges to φ in $L_\infty(\mathbb{R}^d; \mathbb{C}) + L_2(\mathbb{R}^d; \mathbb{C})$ according to Lemma 17.1.6(3), so that the proof is complete. □

17.2 Hilbert Transform

The Hilbert transform has a central place in real and complex analysis. It can be seen as a Fourier multiplier in complex analysis or as a singular integral operator in real analysis. The transform has its applications, among others, in the theory of holomorphic functions, in the theory of function spaces, in probability theory, and in signal processing.

To start with, we remark that Plancherel's Theorem 17.1.2 allows us to introduce a class of operators, called Fourier multipliers:

Definition 17.2.1 For $m \in \mathcal{L}_\infty(\mathbb{R}^d; \mathbb{C})$ we define the **Fourier multiplier** $M_m : L_2(\mathbb{R}^d; \mathbb{C}) \to L_2(\mathbb{R}^d; \mathbb{C})$ by

$$M_m \psi := \mathcal{F}^{-1}(m\mathcal{F}(\psi)).$$

17.2 Hilbert Transform

By Plancherel's Theorem 17.1.2 we get the L_2-continuity of the Fourier multipliers:

$$\|M_m \psi\|_{L_2} = \|m \mathcal{F}(\psi)\|_{L_2} \leqslant \|m\|_{L_\infty} \|\mathcal{F}(\psi)\|_{L_2} = \|m\|_{L_\infty} \|\psi\|_{L_2}.$$

In the case $d = 1$ the Hilbert transform is the special multiplier with $m(y) := -i\,\text{sgn}(y)$:

Definition 17.2.2 For $d = 1$ and $\psi \in L_2(\mathbb{R}, \mathbb{C})$ we let

$$\mathcal{H}\psi := \mathcal{F}^{-1}(-i\,\text{sgn}(\cdot)\,\mathcal{F}(\psi))$$

be the **Hilbert[3]-transform**

Theorem 17.2.3 *If $\mathcal{H}$ is the Hilbert transform, then*
(1) $\|\mathcal{H}\psi\|_{L_2(\mathbb{R};\mathbb{C})} = \|\psi\|_{L_2(\mathbb{R};\mathbb{C})}$ for $\psi \in L_2(\mathbb{R}; \mathbb{C})$,
(2) $\mathcal{H}\psi \in L_2(\mathbb{R}; \mathbb{R})$ for $\psi \in L_2(\mathbb{R}; \mathbb{R})$,
(3) $\psi + i\mathcal{H}\psi = \mathcal{F}^{-1}(2\mathbb{1}_{[0,\infty)} \mathcal{F}(\psi))$ for $\psi \in L_2(\mathbb{R}; \mathbb{R})$.

Proof
(1) By Plancherel's Theorem 17.1.2 we have

$$\|\mathcal{H}\psi\|_{L_2(\mathbb{R};\mathbb{C})} = \|-i\,\text{sgn}(\cdot)\mathcal{F}(\psi)\|_{L_2(\mathbb{R};\mathbb{C})} = \|\mathcal{F}(\psi)\|_{L_2(\mathbb{R};\mathbb{C})} = \|\psi\|_{L_2(\mathbb{R};\mathbb{C})}.$$

(2) We consider $\psi = \psi_0 * v_\varepsilon$ for $\psi_0 \in C_c(\mathbb{R})$ (we suppose that ψ_0 is real-valued) and $\varepsilon > 0$, and get that

$$\overline{\mathcal{H}\psi} = \overline{\mathcal{F}^{-1}(-i\,\text{sgn}(\cdot)\,\mathcal{F}(\psi))}$$
$$= \mathcal{F}^{-1}(\overline{-i\,\text{sgn}(-\cdot)\,\mathcal{F}(\psi)(-\cdot)})$$
$$= \mathcal{F}^{-1}(i\,\text{sgn}(-\cdot)\,\overline{\mathcal{F}(\psi)(-\cdot)})$$
$$= \mathcal{F}^{-1}(-i\,\text{sgn}(\cdot)\,\mathcal{F}(\psi))$$
$$= \mathcal{H}\psi.$$

This means that $\mathcal{H}\psi$ is real-valued. Given a general real-valued $\psi \in \mathcal{L}_2(\mathbb{R}^d)$, we find by Lemma 17.1.4 (and its proof) a sequence $\psi_n * v_{\varepsilon_n}$ with $\psi_n \in C_c(\mathbb{R})$ and $\varepsilon_n > 0$, such that $\lim_{n \to \infty} \|\psi_n * v_{\varepsilon_n} - \psi\|_{L_2(\mathbb{R}^d)} = 0$. Therefore, $\lim_{n \to \infty} \|\mathcal{H}(\psi_n * v_{\varepsilon_n}) - \mathcal{H}\psi\|_{L_2(\mathbb{R}^d;\mathbb{C})} = 0$ as well. As all equivalence classes $\mathcal{H}(\psi_n * v_{\varepsilon_n})$ have a real-valued representative, $\mathcal{H}\psi$ has a real-valued representative as well.

---

[3] David Hilbert, 23/01/1862 (Wehlau near Königsberg, now Kaliningrad, Russia)–14/02/1943 (Göttingen, Germany).

(3) follows from

$$\begin{aligned}
\psi + i\mathcal{H}\psi &= \mathcal{F}^{-1}(\mathcal{F}(\psi) + i\mathcal{F}(\mathcal{H}\psi)) \\
&= \mathcal{F}^{-1}(\mathcal{F}(\psi) + i(-i)\operatorname{sgn}(\cdot)\mathcal{F}(\psi)) \\
&= \mathcal{F}^{-1}((\mathbb{1}_{\{0\}} + 2\mathbb{1}_{(0,\infty)})\mathcal{F}(\psi)) \\
&= \mathcal{F}^{-1}(2\mathbb{1}_{[0,\infty)}\mathcal{F}(\psi)).
\end{aligned}$$

$\square$

17.3 Comments

Section 17.1: The theory of Fourier series goes back to Fourier in 1907, see [74]. Theorem 17.1.2 is due to Plancherel [147] and Theorem 17.1.7 due to Young [202] and Hausdorff [81]. The best constant in Theorem 17.1.7 is $c_p = \sqrt{p^{\frac{1}{p}}/q^{\frac{1}{q}}}$ for $1 < p \leqslant 2 \leqslant q < \infty$ with $1 = (1/p) + (1/q)$. For $q = 2, 4, 6, \ldots$ this was proved by Babenko [5] and was extended to the full range $p \in (1, 2]$ by Beckner [6].

Section 17.2: The Hilbert transform can be dated back to Hilbert [86] to his work on the Riemann problem. This problem is an extension problem of a given function on the boundary of a bounded domain to holomorphic functions inside and outside the boundary. To solve this problem the Cauchy integral is used. The interested reader is referred to [107, Section 7]. The operator one obtains (that maps a function to its 'conjugate' function) in this context refers to the Hilbert transform on the torus. The Hilbert transform in $L_p(\mathbb{R})$ with $p \in (1, \infty)$, which corresponds to our setting, was investigated by M. Riesz [158]. The work of M. Riesz can be seen as a starting point to develop the theory of singular integrals, see the work of Calderon and Zygmund [28] on the occasion of the 65th birthday of M. Riesz.

Chapter 18
Riesz Representation Theorems

Riesz representation theorems belong to the family of results about duality in Banach space theory. Given a Banach space E, the identification of its dual space is needed and useful in various situations. For example, there are various problems that can be transformed into a 'dual problem' which is easier to understand and to handle. We will see in Theorem 18.2.1 that Hölder's inequality and the Radon-Nikodym theorem can be viewed as results about duality for L_p. Moreover, Theorem 18.4.1 shows that the concept of outer measures and the related construction of measures can be understood as the identification of the dual of the space of continuous functions on a compactum.

18.1 L_p Spaces Over σ-Finite Measures

In this section we show that—in certain respects—L_p spaces over σ-finite measures can be identified as L_p spaces over probability spaces we already considered. We start with the definition:

Definition 18.1.1 Assume a σ-finite measure space $(\Omega, \mathcal{F}, \mu)$ with $\mu(\Omega) > 0$.

(1) For $f \in \mathcal{L}_0(\Omega, \mathcal{F})$ we let

$$\mu - \operatorname{ess\,sup}_\Omega |f| := \inf \left\{ \sup_{\omega \in \Omega \setminus N} |f(\omega)| : N \in \mathcal{F}, \mu(N) = 0 \right\}.$$

(2) For $p \in [1, \infty]$ and $f \in \mathcal{L}_0(\Omega, \mathcal{F})$ we let

$$\|f\|_{L_p} = \|f\|_{L_p(\mu)} := \begin{cases} \left(\int_\Omega |f(\omega)|^p \mathrm{d}\mu(\omega) \right)^{\frac{1}{p}} & : p \in (0, \infty) \\ \mu - \operatorname{ess\,sup}_\Omega |f| & : p = \infty \end{cases}$$

and $\mathcal{L}_p(\Omega, \mathcal{F}, \mu) := \{f \in \mathcal{L}_0(\Omega, \mathcal{F}) : \|f\|_{L_p} < \infty\}$.
(3) For $p \in [1, \infty]$ and a σ-finite measure space $(\Omega, \mathcal{F}, \mu)$ we define

$$L_p(\Omega, \mathcal{F}, \mu) := \{[f] : f \in \mathcal{L}_p(\Omega, \mathcal{F}, \mu)\} \text{ with } \|[f]\|_{L_p} := \|f\|_{L_p}.$$

In case $\mu = \mathbb{P}$ is a probability measure we prove for $p \in [1, \infty]$ that $[L_p(\Omega, \mathcal{F}, \mathbb{P}), \|\cdot\|_{L_p(\mathbb{P})}]$ is a Banach space in Theorem 20.2.2. To transfer this statement to the case of σ-finite measure spaces $(\Omega, \mathcal{F}, \mu)$, where we assume that $\mu(\Omega) > 0$, we choose any measurable partition of $\Omega = \bigcup_{i \in I} \Omega_i$ such that $\mu(\Omega_i) \in (0, \infty)$ and where $I = \{1\}$ or $I = \mathbb{N}$. Then we find a random variable $D : \Omega \to (0, \infty)$ which is constant on all Ω_i and such that $\int_\Omega D d\mu = 1$, and let $d\mathbb{P} := D d\mu$. For $p \in [1, \infty]$ we define the map

$$I_p : \mathcal{L}_0(\Omega, \mathcal{F}) \to \mathcal{L}_0(\Omega, \mathcal{F}) \text{ by } I_p f := f D^{\frac{1}{p}},$$

with the convention $D^{\frac{1}{\infty}} = D^{-\frac{1}{\infty}} \equiv 1$. The map I_p satisfies the following properties:

(a) $(I_p)^{-1} : \mathcal{L}_0(\Omega, \mathcal{F}) \to \mathcal{L}_0(\Omega, \mathcal{F})$ exists and satisfies $(I_p)^{-1} f = f D^{-\frac{1}{p}}$.
(b) $\|I_p g\|_{L_p(\mu)} = \|g\|_{L_p(\mathbb{P})}$ for all $g \in \mathcal{L}_0(\Omega, \mathcal{F})$.
(c) $\|(I_p)^{-1} f\|_{L_p(\mathbb{P})} = \|f\|_{L_p(\mu)}$ for all $f \in \mathcal{L}_0(\Omega)$.
(d) The equivalence classes in $\mathcal{L}_0(\Omega, \mathcal{F})$ with respect to μ and P coincide.

In other words, the linear map $I_p : L_p(\Omega, \mathcal{F}, \mathbb{P}) \to L_p(\Omega, \mathcal{F}, \mu)$ is isometric and onto. As we stated in Theorem 9.3.5 that $L_p(\Omega, \mathcal{F}, \mathbb{P})$ is a Banach space, we also have:

Theorem 18.1.2 *For a σ-finite measure space $(\Omega, \mathcal{F}, \mu)$ and $p \in [1, \infty]$ the space $[L_p(\Omega, \mathcal{F}, \mu), \|\cdot\|_{L_p(\mu)}]$ is a Banach space.*

18.2 The Dual Space of $L_p(\Omega, \mathcal{F}, \mu)$

In this section we identify the dual space of $L_p(\Omega, \mathcal{F}, \mu)$ for σ-finite measures μ. To lighten the notation we shall write $L_p^*(\Omega, \mathcal{F}, \mu)$ for $(L_p(\Omega, \mathcal{F}, \mu))^*$. To avoid degenerated situations we recall that we exclude the case $\mu(\Omega) = 0$. Furthermore, for $p, q \in (1, \infty)$ with $1 = \frac{1}{p} + \frac{1}{q}$ we use the relations

$$(p-1)q = p \text{ and accordingly } (q-1)p = q.$$

The result is:

18.2 The Dual Space of $L_p(\Omega, \mathcal{F}, \mu)$

Theorem 18.2.1 *Let $(\Omega, \mathcal{F}, \mu)$ be a σ-finite measure space, $p \in [1, \infty)$, and $1 = \frac{1}{p} + \frac{1}{q}$.*

(1) *One has isometrically that*

$$L_p^*(\Omega, \mathcal{F}, \mu) = L_q(\Omega, \mathcal{F}, \mu)$$

in the following sense: Each $g \in L_q(\Omega, \mathcal{F}, \mu)$ defines an $a_g \in L_p^*(\Omega, \mathcal{F}, \mu)$ by

$$\langle f, a_g \rangle := \int_\Omega fg \, d\mu \quad \text{with} \quad \|a_g\|_{L_p^*(\mu)} = \|g\|_{L_q(\mu)},$$

and each $a \in L_p^*(\Omega, \mathcal{F}, \mu)$ can be represented by a $g \in L_q(\Omega, \mathcal{F}, \mu)$ in this form.

(2) *If $f \in L_p(\Omega, \mathcal{F}, \mu)$, then there is a $g \in L_q(\Omega, \mathcal{F}, \mu)$ such that*

$$\|g\|_{L_q(\Omega, \mathcal{F}, \mu)} = 1 \quad \text{and} \quad \|f\|_{L_p(\Omega, \mathcal{F}, \mu)} = \int_\Omega fg \, d\mu.$$

Proof
(1) **(a)** By Hölder's inequality we know that

$$\left| \int_\Omega fg \, d\mu \right| \leq \|f\|_{L_p(\mu)} \|g\|_{L_q(\mu)},$$

so that

$$\|g\|_{L_p^*(\mu)} \leq \|g\|_{L_q(\mu)}.$$

The equality that can be seen as follows: First assume that $p \in (1, \infty)$. We choose $f := \operatorname{sgn}(g)|g|^{q-1}$ so that $fg = |g|^q$ and

$$\|f\|_{L_p(\mu)} = \left(\int_\Omega |g|^{(q-1)p} d\mu \right)^{\frac{1}{p}} = \left(\int_\Omega |g|^q d\mu \right)^{\frac{1}{p}}.$$

Hence

$$\left| \int_\Omega fg \, d\mu \right| = \int_\Omega |g|^q d\mu = \|f\|_{L_p(\mu)} \|g\|_{L_q(\mu)}$$

and $\|g\|_{L_p^*(\mu)} \geq \|g\|_{L_q(\mu)}$. If $p = 1$, then we choose $f = \mathbb{1}_A$ and observe that

$$\sup_{\mu(A) > 0} \frac{1}{\mu(A)} \left| \int_A g \, d\mu \right| = \|g\|_{L_\infty(\mu)},$$

see Exercise 2. Summarizing (a) in other words, we proved that $L_q(\Omega, \mathcal{F}, \mu)$ is an isometric sub-space of $L_p^*(\Omega, \mathcal{F}, \mu)$.

(b) To verify $L_q(\Omega, \mathcal{F}, \mu) = L_p^*(\Omega, \mathcal{F}, \mu)$ we have to find for each $a \in L_p^*(\Omega, \mathcal{F}, \mu)$ an element $g \in L_q(\Omega, \mathcal{F}, \mu)$ that represents the functional a like in (a).

(b1) We first assume that $\mu(\Omega) = 1$ and a functional $a \in L_p^*(\Omega, \mathcal{F}, \mu)$ and define the map $\nu : \mathcal{F} \to \mathbb{R}$ by

$$\nu(A) := a(\mathbb{1}_A).$$

This defines a signed measure which is absolutely continuous with respect to μ so that by the theorem of Radon-Nikodym there is a $g \in \mathcal{L}_1(\Omega, \mathcal{F}, \mu)$ such that

$$\langle \mathbb{1}_A, a \rangle = \int_A g \, d\mu.$$

For any simple function f_0 it follows that

$$\langle f_0, a \rangle = \int_\Omega f_0 g \, d\mu$$

and

$$\left| \int_\Omega f_0 g \, d\mu \right| = |\langle f_0, a \rangle| \leq \|a\|_{L_p^*} \|f_0\|_{L_p}.$$

Replacing f_0 by $f_0 \operatorname{sgn}(f_0 g)$ we deduce

$$\int_\Omega |f_0 g| \, d\mu \leq \|a\|_{L_p^*} \|f_0\|_{L_p}$$

for any simple f_0. For a general $f \in L_p(\Omega, \mathcal{F}, \mu)$ we find simple f_n such that $0 \leq f_n \uparrow |f|$ and get

$$\int_\Omega |fg| \, d\mu \leq \|a\|_{L_p^*} \|f\|_{L_p} \tag{18.1}$$

by monotone convergence. If we set $f := \operatorname{sgn}(g)|g|^{q-1}\mathbb{1}_{\{|g| \leq N\}}$ for $N \in \mathbb{N}$, then

$$\int_\Omega |g|^q \mathbb{1}_{\{|g| \leq N\}} \, d\mu \leq \|a\|_{L_p^*(\mu)} \left(\int_\Omega |g|^q \mathbb{1}_{\{|g| \leq N\}} \, d\mu \right)^{\frac{1}{p}}$$

and $\|g \mathbb{1}_{\{|g| \leq N\}}\|_{L_q(\mu)} \leq \|a\|_{L_p^*(\mu)}$ for all $N \in \mathbb{N}$. By $N \to \infty$ we conclude $\|g\|_{L_q(\mu)} \leq \|a\|_{L_p^*(\mu)} < \infty$.

(b2) We remove the assumption $\mu(\Omega) = 1$ and assume that μ is σ-finite. Again we assume a functional $a \in L_p^*(\Omega, \mathcal{F}, \mu)$. By the map $I_p : L_p(\Omega, \mathcal{F}, \mathbb{P}) \to$

18.2 The Dual Space of $L_p(\Omega, \mathcal{F}, \mu)$

$L_p(\Omega, \mathcal{F}, \mu)$ we get a functional $b := a \circ I_p \in L_p^*(\Omega, \mathcal{F}, \mathbb{P})$. We apply part (b1) and get an element $h \in L_q(\Omega, \mathcal{F}, \mathbb{P})$ such that

$$\langle \psi, b \rangle = \int_\Omega \psi h \, d\mathbb{P} \quad \text{for all} \quad \psi \in L_p(\Omega, \mathcal{F}, \mathbb{P}).$$

By definition this means that

$$\langle I_p \psi, a \rangle = \int_\Omega \psi h D \, d\mu$$

or, for $\varphi \in L_p(\Omega, \mathcal{F}, \mu)$,

$$\langle \varphi, a \rangle = \int_\Omega (I_p^{-1}\varphi) h D \, d\mu = \int_\Omega \varphi D^{-\frac{1}{p}} h D \, d\mu = \int_\Omega D^{\frac{1}{q}} \varphi h \, d\mu = \int_\Omega (I_q h) \varphi \, d\mu.$$

Finally, for the representing function $I_q h$ we get $\|I_q h\|_{L_q(\mu)} = \|h\|_{L_q(\mathbb{P})} < \infty$.
(2) If $\|f\|_{L_p(\Omega, \mathcal{F}, \mu)} = 0$, then any $g \in L_q(\Omega, \mathcal{F}, \mu)$ with $\|g\|_{L_q(\Omega, \mathcal{F}, \mu)} = 1$ can be taken. So we assume $\|f\|_{L_p(\Omega, \mathcal{F}, \mu)} > 0$. If $q = \infty$, then we choose $g := \text{sgn}(f)$ so that $\|g\|_{L_\infty} = 1$ and $\int_\Omega f g \, d\mu = \int_\Omega |f| \, d\mu$. For the case $p, q \in (1, \infty)$ we put

$$g := \text{sgn}(f) |f|^{p-1} \|f\|_{L_p}^{-\frac{p}{q}},$$

so that

$$\int_\Omega f g \, d\mu = \int_\Omega f \, \text{sgn}(f) |f|^{p-1} \, d\mu \, \|f\|_{L_p}^{-\frac{p}{q}} = \int_\Omega |f|^p \, d\mu \, \|f\|_{L_p}^{-\frac{p}{q}} = \|f\|_{L_p}.$$

Furthermore, we compute

$$\|g\|_{L_q} = \left(\int_\Omega |\text{sgn}(f)|f|^{p-1}|^q \, d\mu \right)^{\frac{1}{q}} \|f\|_{L_p}^{-\frac{p}{q}}$$

$$= \left(\int_\Omega |f|^p \, d\mu \right)^{\frac{1}{q}} \left(\int_\Omega |f|^p \, d\mu \right)^{-\frac{1}{q}} = 1.$$

$\square$

In computations often iterated L_p-L_q-means occur, where one would like to change their order. To do so, the following corollary is useful:

Corollary 18.2.2 *Assume σ-finite measure spaces $(\Omega_1, \mathcal{F}_1, \mu_1)$ and $(\Omega_2, \mathcal{F}_2, \mu_2)$, an $\mathcal{F}_1 \otimes \mathcal{F}_2$-measurable function $f : \Omega_1 \times \Omega_2 \to \mathbb{R}$, and $0 < p \leq q \leq \infty$. Then one has*

$$\left\| \|f\|_{L_p(\Omega_1, \mu_1)} \right\|_{L_q(\Omega_2, \mu_2)} \leq \left\| \|f\|_{L_q(\Omega_2, \mu_2)} \right\|_{L_p(\Omega_1, \mu_1)}. \tag{18.2}$$

Proof We assume that $\left\|\|f\|_{L_q(\Omega_2,\mu_2)}\right\|_{L_p(\Omega_1,\mu_1)} < \infty$, otherwise there is nothing to prove.

(a) First we assume $q < \infty$. By Fubini-Tonelli's theorem the maps

$$\Omega_2 \ni \omega_2 \mapsto \|f(\cdot,\omega_2)\|_{L_p(\Omega_1,\mu_1)} \in [0,\infty],$$

$$\Omega_1 \ni \omega_1 \mapsto \|f(\omega_1,\cdot)\|_{L_q(\Omega_2,\mu_2)} \in [0,\infty]$$

are measurable. We also assume w.l.o.g. that $p < q$, otherwise there is nothing to prove by the Fubini-Tonelli theorem. Letting $r := q/p \in (1,\infty)$ and $g := |f|^p$, the inequality (18.2) is equivalent to

$$\left\|\|g\|_{L_1(\Omega_1,\mu_1)}\right\|_{L_r(\Omega_2,\mu_2)} \leqslant \left\|\|g\|_{L_r(\Omega_2,\mu_2)}\right\|_{L_1(\Omega_1,\mu_1)}.$$

According to Theorem 18.2.1(2) we choose a measurable function $h : \Omega_2 \to [0,\infty)$ with $\|h\|_{L_{r'}} = 1$, where $1 = (1/r) + (1/r')$, and

$$\left\|\|g\|_{L_1(\Omega_1,\mu_1)}\right\|_{L_r(\Omega_2,\mu_2)} = \int_{\Omega_2} h(\omega_2) \|g(\cdot,\omega_2)\|_{L_1(\Omega_1,\mu_1)} \, d\mu_2(\omega_2).$$

By this choice we obtain the desired statement because

$$\left\|\|g\|_{L_1(\Omega_1,\mu_1)}\right\|_{L_r(\Omega_2,\mu_2)} = \int_{\Omega_2} h(\omega_2) \|g(\cdot,\omega_2)\|_{L_1(\Omega_1,\mu_1)} \, d\mu_2(\omega_2)$$

$$= \int_{\Omega_2} \int_{\Omega_1} h(\omega_2) g(\omega_1,\omega_2) d\mu_1(\omega_1) d\mu_2(\omega_2)$$

$$= \int_{\Omega_1} \left[\int_{\Omega_2} h(\omega_2) g(\omega_1,\omega_2) d\mu_2(\omega_2)\right] d\mu_1(\omega_1)$$

$$\leqslant \int_{\Omega_1} \|g(\omega_1,\cdot)\|_{L_r(\Omega_2,\mu_2)} \, d\mu_1(\omega_1).$$

(b) We assume $q = \infty$. As μ_2 is σ-finite, we find a measurable map $D_2 : \Omega_2 \to (0,\infty)$ such that $\int_{\Omega_2} D_2 d\mu_2 = 1$ and define the probability measure $d\mathbb{P}_2 := D_2 d\mu_2$. We have that $\|h\|_{L_\infty(\Omega_2,\mathbb{P}_2)} = \|h\|_{L_\infty(\Omega_2,\mu_2)}$ for all $h \in \mathcal{L}_0(\Omega_2,\mathcal{F}_2)$. By the previous step we know for $0 < p < \overline{q} < \infty$ that

$$\left\|\|f\|_{L_p(\Omega_1,\mu_1)}\right\|_{L_{\overline{q}}(\Omega_2,\mathbb{P}_2)} \leqslant \left\|\|f\|_{L_{\overline{q}}(\Omega_2,\mathbb{P}_2)}\right\|_{L_p(\Omega_1,\mu_1)}.$$

By $\bar{q} \uparrow \infty$ and Proposition 9.3.2 we get that

$$\Omega_1 \ni \omega_1 \mapsto \|f(\omega_1, \cdot)\|_{L_\infty(\Omega_2, \mu_2)} = \|f(\omega_1, \cdot)\|_{L_\infty(\Omega_2, \mathbb{P}_2)}$$
$$= \lim_{\bar{q} \uparrow \infty} \|f(\omega_1, \cdot)\|_{L_{\bar{q}}(\Omega_2, \mathbb{P}_2)} \in [0, \infty].$$

So the final statement follows by $\bar{q} \uparrow \infty$.

□

18.3 The Spaces $C(K)$

In the following $[K, d]$ is a compact metric space and $C(K)$ is the space of all continuous functions $\varphi : K \to \mathbb{R}$. As K is compact, every continuous function $\varphi : K \to \mathbb{R}$ is automatically bounded. We equip $C(K)$ with the norm

$$\|\varphi\|_{C(K)} := \sup_{x \in K} |\varphi(x)|$$

and obtain a Banach space $[C(K), \|\cdot\|_{C(K)}]$. For its dual $(C(K))^*$ we use the shorter notation $C^*(K)$. Before we determine this dual space in Sect. 18.4 we prove a decomposition of an element $a \in C^*(K)$ fully in the spirit of the Hahn-Jordan decomposition of signed measures in Theorem 10.1.6. First we explain what does it mean that an $a \in C^*(K)$ is nonnegative:

Definition 18.3.1 A functional $a \in C^*(K)$ is called **nonnegative** (we write $a \geq 0$) provided that

$$\langle \varphi, a \rangle \geq 0 \quad \text{for} \quad \varphi \in C(K) \text{ with } \varphi \geq 0.$$

The following decomposition is a twin of the Hahn-Jordan decomposition of signed measures:

Proposition 18.3.2 *For each $a \in C^*(K)$ there is a decomposition $a = a^+ - a^-$ with $a^+, a^- \in C^*(K)$ and $a^+, a^- \geq 0$. The decomposition $a = a^+ - a^-$ can be chosen to be minimal in the following sense: Given another decomposition $a = b^+ - b^-$, then one has for all $\varphi \geq 0$ that*

$$\langle \varphi, a^\pm \rangle \leq \langle \varphi, b^\pm \rangle.$$

Proof Let $a \in C^*(K)$. First we define for $\varphi \geq 0$ the functionals

$$\langle \varphi, a^+ \rangle := \sup\{\langle \psi, a \rangle : 0 \leq \psi \leq \varphi\},$$
$$\langle \varphi, a^- \rangle := \sup\{\langle \psi - \varphi, a \rangle : 0 \leq \psi \leq \varphi\},$$

so that

$$|\langle\varphi, a^+\rangle| \leq \sup\{|\langle\psi, a\rangle| : 0 \leq \psi \leq \varphi\} \leq \|a\|_{C^*(K)} \sup\{\|\psi\|_{C(K)} : 0 \leq \psi \leq \varphi\}$$
$$= \|a\|_{C^*(K)}\|\varphi\|_{C(K)},$$

and similarly, $|\langle\varphi, a^-\rangle| \leq \|a\|_{C^*(K)}\|\varphi\|_{C(K)}$.

(a) For $\varphi \geq 0$ we get by the definition

$$\langle\varphi, a^+\rangle - \langle\varphi, a^-\rangle$$
$$= \sup\{\langle\psi, a\rangle : 0 \leq \psi \leq \varphi\} - \left[-\langle\varphi, a\rangle + \sup\{\langle\psi, a\rangle : 0 \leq \psi \leq \varphi\}\right]$$
$$= \langle\varphi, a\rangle.$$

(b) Using for a^+ the function $\psi \equiv 0$ and for a^- the function $\psi \equiv \varphi$, we get

$$\langle\varphi, a^+\rangle \geq 0 \quad \text{and} \quad \langle\varphi, a^-\rangle \geq 0 \quad \text{for} \quad \varphi \geq 0.$$

(c) For $\varphi_1, \varphi_2 \geq 0$ one has that

$$\langle\varphi_1+\varphi_2, a^+\rangle = \langle\varphi_1, a^+\rangle+\langle\varphi_2, a^+\rangle \quad \text{and} \quad \langle\varphi_1+\varphi_2, a^-\rangle = \langle\varphi_1, a^-\rangle+\langle\varphi_2, a^-\rangle.$$

As a is already linear we need to verify this property only for a^+, the property for a^- follows then from (a) and the property for a^+. Now, the inequality

$$\langle\varphi_1 + \varphi_2, a^+\rangle \geq \langle\varphi_1, a^+\rangle + \langle\varphi_2, a^+\rangle$$

follows by the definition. To verify the opposite inequality, assume $0 \leq \psi \leq \varphi_1 + \varphi_2$. We let

$$0 \leq \psi_1 := \min\{\psi, \varphi_1\} \leq \varphi_1 \quad \text{and} \quad 0 \leq \psi_2 := \psi - \min\{\psi, \varphi_1\} \leq \varphi_2.$$

Hence $\psi = \psi_1 + \psi_2 \leq \varphi_1 + \varphi_2$ and

$$\langle\psi, a^+\rangle = \langle\psi_1 + \psi_2, a^+\rangle \leq \langle\psi_1, a^+\rangle + \langle\psi_2, a^+\rangle \leq \langle\varphi_1, a^+\rangle + \langle\varphi_2, a^+\rangle,$$

where the first inequality follows by the special choice of ψ_1 and ψ_2.

(d) For $\lambda \geq 0$ and $\varphi \geq 0$ we observe that

$$\langle\lambda\varphi, a^+\rangle = \lambda\langle\varphi, a^+\rangle \quad \text{and} \quad \langle\lambda\varphi, a^-\rangle = \lambda\langle\varphi, a^-\rangle.$$

This property for a^+ follows directly from the definition. Having it for a^+ and a (a is linear by assumption), we use (a) to deduce it for a^-.

18.3 The Spaces $C(K)$

(e) Now we extend a^+ and a^- to all $\varphi \in C(K)$ by

$$\langle \varphi, a^+ \rangle := \langle \varphi^+, a^+ \rangle - \langle \varphi^-, a^+ \rangle,$$
$$\langle \varphi, a^- \rangle := \langle \varphi^+, a^- \rangle - \langle \varphi^-, a^- \rangle$$

where $\varphi^+ := \varphi \vee 0$ and $\varphi^- := (-\varphi) \vee 0$ so that $\varphi = \varphi^+ - \varphi^-$. Now we get that

$$\begin{aligned}
\langle \varphi, a^+ \rangle - \langle \varphi, a^- \rangle &= \langle \varphi^+, a^+ \rangle - \langle \varphi^-, a^+ \rangle - (\langle \varphi^+, a^- \rangle - \langle \varphi^-, a^- \rangle) \\
&= [\langle \varphi^+, a^+ \rangle - \langle \varphi^+, a^- \rangle] - [\langle \varphi^-, a^+ \rangle - \langle \varphi^-, a^- \rangle] \\
&= \langle \varphi^+, a \rangle - \langle \varphi^-, a \rangle \\
&= \langle \varphi, a \rangle.
\end{aligned}$$

(f) We verify that $a^+, a^- \in C^*(K)$. Because of (e) we only need to check $a^+ \in C^*(K)$. For general φ_1, φ_2 and $\varphi := \varphi_1 + \varphi_2$ we get using $\overline{\varphi}^\pm := \varphi_1^\pm + \varphi_2^\pm$ that

$$\begin{aligned}
\langle \varphi_1, a^+ \rangle + \langle \varphi_2, a^+ \rangle &= \langle \varphi_1^+, a^+ \rangle - \langle \varphi_1^-, a^+ \rangle + \langle \varphi_2^+, a^+ \rangle - \langle \varphi_2^-, a^+ \rangle \\
&= \langle \varphi_1^+ + \varphi_2^+, a^+ \rangle - \langle \varphi_1^- + \varphi_2^-, a^+ \rangle \\
&= \langle \overline{\varphi}^+, a^+ \rangle - \langle \overline{\varphi}^-, a^+ \rangle \\
&= \langle \varphi^+, a^+ \rangle - \langle \varphi^-, a^+ \rangle \\
&= \langle \varphi, a^+ \rangle
\end{aligned}$$

where the next to the last equality follows from

$$\overline{\varphi}^+ + \varphi^- = \varphi^+ + \overline{\varphi}^- \quad \text{and} \quad \langle \overline{\varphi}^+, a^+ \rangle + \langle \varphi^-, a^+ \rangle = \langle \varphi^+, a^+ \rangle + \langle \overline{\varphi}^-, a^+ \rangle.$$

For a general φ and $\lambda \geq 0$ we have

$$\langle \lambda \varphi, a^+ \rangle = \lambda \langle \varphi^+, a^+ \rangle - \lambda \langle \varphi^-, a^+ \rangle = \lambda \langle \varphi, a^+ \rangle.$$

For a general φ and $\lambda < 0$ we obtain similarly that

$$\langle \lambda \varphi, a^+ \rangle = \langle |\lambda| \varphi^-, a^+ \rangle - \langle |\lambda| \varphi^+, a^+ \rangle = -|\lambda| \langle \varphi, a^+ \rangle = \lambda \langle \varphi, a^+ \rangle.$$

We are left to show the continuity. Again we show it only for a^+, it follows from $a = a^+ - a^-$ for a^-. For a^+ we observe that

$$|\langle\varphi, a^+\rangle| \leqslant |\langle\varphi^+, a^+\rangle| + |\langle\varphi^-, a^+\rangle|$$
$$\leqslant \|a\|_{C^*(K)}[\|\varphi^+\|_{C(K)} + \|\varphi^-\|_{C(K)}]$$
$$\leqslant 2\|a\|_{C^*(K)}\|\varphi\|_{C(K)}.$$

(g) Now we check the minimality of our construction. For a^+ this can be seen as follows:

$$\langle\varphi, a^+\rangle = \sup\{\langle\psi, a\rangle : 0 \leqslant \psi \leqslant \varphi\}$$
$$= \sup\{\langle\psi, b^+\rangle - \langle\psi, b^-\rangle : 0 \leqslant \psi \leqslant \varphi\}$$
$$\leqslant \sup\{\langle\psi, b^+\rangle : 0 \leqslant \psi \leqslant \varphi\}$$
$$= \langle\varphi, b^+\rangle$$

where the last equality follows because $b^+ \geqslant 0$. Similarly,

$$\langle\varphi, a^-\rangle = \sup\{\langle\psi - \varphi, a\rangle : 0 \leqslant \psi \leqslant \varphi\}$$
$$= \sup\{\langle\psi - \varphi, b^+\rangle - \langle\psi - \varphi, b^-\rangle : 0 \leqslant \psi \leqslant \varphi\}$$
$$\leqslant \sup\{\langle\varphi - \psi, b^-\rangle : 0 \leqslant \psi \leqslant \varphi\}$$
$$= \langle\varphi, b^-\rangle.$$

$\square$

18.4 The Dual Space of $C(K)$

The dual space of $C(K)$ takes us back to measure theory. First we equip K with the Borel σ-algebra generated by the open sets. Then the space of all signed measures on the measurable space $(K, \mathcal{B}(K))$ is denoted by $\mathcal{M}(K)$. With this we get:

Theorem 18.4.1 *Let $[K, d]$ be a compact metric space. Then one has isometrically that*

$$C^*(K) = \mathcal{M}(K)$$

in the following sense: Each $\mu \in \mathcal{M}(K)$ defines an $a_\mu \in C^*(K)$ by

$$\langle f, a_\mu\rangle := \int_K f \, d\mu \quad \text{with} \quad \|a_\mu\|_{C^*(K)} = |\mu|_{\text{TV}},$$

and each $a \in C^*(K)$ can be represented by a $\mu \in \mathcal{M}(K)$ in this form.

18.4 The Dual Space of $C(K)$

Proof

(a) We show that $\mathcal{M}(K)$ is an isometric subspace of $C^*(K)$: By the Hahn-Jordan decomposition $\mu = \mu_J^+ - \mu_J^-$ we have that

$$\left| \int_K \varphi \, d\mu \right| = \left| \int_K \varphi \, d\mu_J^+ - \int_K \varphi \, d\mu_J^- \right|$$

$$\leq \|\varphi\|_{C(K)} (\mu_J^+(K) + \mu_J^-(K))$$

$$= \|\varphi\|_{C(K)} |\mu|_{\mathrm{TV}}.$$

Moreover, by Corollary 10.1.8,

$$\sup_{\|\varphi\|_{C(K)}=1} \left| \int_K \varphi \, d\mu \right| = |\mu|_{\mathrm{TV}}.$$

(b) Now we show that $\mathcal{M}(K) = C^*(K)$: Because of Proposition 18.3.2 we may assume that $a \geq 0$. As also the case $a = 0$ is trivial, we may assume without loss generality that $\langle 1, a \rangle = 1$. For an open set $G \subseteq K$ we define

$$\mu^{\mathrm{open}}(G) := \sup\{\langle \varphi, a \rangle : 0 \leq \varphi \leq 1, \overline{\{\varphi \neq 0\}} \subseteq G\}$$

and for any $A \subseteq K$,

$$\mu^*(A) := \inf\{\mu^{\mathrm{open}}(G) : A \subseteq G, G \text{ open}\}.$$

(b1) $\mu^*(A_1) \leq \mu^*(A_2)$ for $A_1 \subseteq A_2$ and $\mu^*(G) = \mu^{\mathrm{open}}(G)$ for any open G: The first relation follows from the definition of μ^*, the second one from the fact that for open sets $G_1 \subseteq G_2$ one has that $\mu^{\mathrm{open}}(G_1) \leq \mu^{\mathrm{open}}(G_2)$.

(b2) μ^* is an outer measure: The monotonicity and $\mu^*(\emptyset) = 0$ is due to (b1) (note that $\mu^{\mathrm{open}}(\emptyset) = 0$), so that we turn to the σ-sub-additivity. Assume set $A_1, A_2, \ldots \subseteq K$, we need to verify that

$$\mu^*\left(\bigcup_{i=1}^\infty A_i\right) \leq \sum_{i=1}^\infty \mu^*(A_i). \tag{18.3}$$

In fact, it is sufficient to show

$$\mu^*\left(\bigcup_{i=1}^\infty G_i\right) \leq \sum_{i=1}^\infty \mu^*(G_i) \tag{18.4}$$

for open sets G_i. To see this, we assume $\varepsilon > 0$ and find open sets $G_i \supseteq A_i$ such that $\mu^*(G_i) = \mu^{\text{open}}(G_i) \leqslant \frac{\varepsilon}{2^i} + \mu^*(A_i)$. Then $G := \bigcup_{i=1}^\infty G_i \supseteq \bigcup_{i=1}^\infty A_i$ is open. Assuming (18.4) implies $\mu^*(\bigcup_{i=1}^\infty A_i) \leqslant \mu^*(G) \leqslant \sum_{i=1}^\infty \mu^*(G_i) \leqslant \varepsilon + \sum_{i=1}^\infty \mu^*(A_i)$. By $\varepsilon \downarrow 0$ the inequality (18.3) follows.

By (b1) and the definition of μ^{open}, for $\varepsilon > 0$ we find φ_ε such that $0 \leqslant \varphi_\varepsilon \leqslant 1$ and $K_\varepsilon := \overline{\{\varphi_\varepsilon \neq 0\}} \subseteq G \subseteq \bigcup_{i=1}^\infty G_i$ and that

$$\mu^*(G) = \mu^{\text{open}}(G) \leqslant \varepsilon + \langle \varphi_\varepsilon, a \rangle.$$

As K_ε is compact, there is a finite cover

$$K_\varepsilon \subseteq \bigcup_{n=1}^N G_{i_n}.$$

By Theorem 4.1.8 we find continuous $0 \leqslant \varphi_{\varepsilon,n} \leqslant 1$ with $\overline{\{\varphi_{\varepsilon,n} \neq 0\}} \subseteq G_{i_n}$, $n = 1, \ldots, N$, such that

$$\sum_{n=1}^N \varphi_{\varepsilon,n} = 1 \text{ on } K_\varepsilon.$$

This implies that

$$\mu^*(G) \leqslant \varepsilon + \langle \varphi_\varepsilon, a \rangle \leqslant \varepsilon + \left\langle \sum_{n=1}^N \varphi_{\varepsilon,n}, a \right\rangle = \varepsilon + \sum_{n=1}^N \langle \varphi_{\varepsilon,n}, a \rangle$$

$$\leqslant \varepsilon + \sum_{n=1}^N \mu^{\text{open}}(G_{i_n}) = \varepsilon + \sum_{n=1}^N \mu^*(G_{i_n}).$$

(b3) All open sets G belong to $\mathcal{F}^{\mu^*}$ defined in Definition B.2.1: We check that for any B and open G one has

$$\mu^*(B) \geqslant \mu^*(B \cap G) + \mu^*(B \cap G^c).$$

Again, it is sufficient to assume that B is open as well. Given $\varepsilon > 0$ we choose $0 \leqslant \varphi_\varepsilon \leqslant 1$ such that

$$K_\varepsilon := \text{supp}(\varphi_\varepsilon) \subseteq B \cap G \quad \text{and} \quad \varepsilon + \langle \varphi_\varepsilon, a \rangle \geqslant \mu^{\text{open}}(B \cap G) = \mu^*(B \cap G).$$

Therefore $B \setminus K_\varepsilon \supseteq B \cap G^c$ is open and

$$\mu^*(B \setminus K_\varepsilon) \geqslant \mu^*(B \cap G^c).$$

18.4 The Dual Space of $C(K)$

We choose $0 \leqslant \varphi_\varepsilon^c \leqslant 1$ such that

$$\mathrm{supp}(\varphi_\varepsilon^c) \subseteq B \setminus K_\varepsilon \quad \text{and} \quad \varepsilon + \langle \varphi_\varepsilon^c, a \rangle \geqslant \mu^*(B \setminus K_\varepsilon)$$

and finish by

$$\mu^*(B) \geqslant \langle \varphi_\varepsilon + \varphi_\varepsilon^c, a \rangle$$
$$\geqslant \mu^*(B \cap G) - \varepsilon + \mu^*(B \setminus K_\varepsilon) - \varepsilon$$
$$\geqslant \mu^*(B \cap G) + \mu^*(B \cap G^c) - 2\varepsilon.$$

By $\varepsilon \downarrow 0$ we obtain the desired result.

By the above we obtain a probability measure μ on $\mathcal{B}(K)$.

(c) For the measure μ constructed in (b) under the conditions $a \geqslant 0$ and $\langle 1, a \rangle = 1$ we show that

$$\langle \varphi, a \rangle = \int_K \varphi \mathrm{d}\mu \quad \text{for} \quad \varphi \in C(K).$$

We only need to verify one inequality as by multiplication by -1 the other inequality follows since there is no restriction on $\varphi \in C(K)$. So we will verify

$$\langle \varphi, a \rangle \leqslant \int_K \varphi \mathrm{d}\mu.$$

There are $-\infty < A < B < \infty$ such that $\varphi : K \to [A, B]$. By a shift and positive scaling we can assume that $[A, B] = [0, 1]$. Choose $n \in \mathbb{N}$ and consider the partition $[0, 1] = \bigcup_{i=1}^n [a_i, b_i]$ with $b_i - a_i = 1/n$. Let $P_i := \varphi^{-1}([a_i, b_i])$. We fix $\varepsilon > 0$ and find open sets $G_i \supseteq P_i$ such that

$$\mu(G_i) \leqslant \mu(P_i) + \varepsilon \quad \text{and} \quad \varphi(x) \leqslant b_i + \varepsilon \quad \text{for } x \in G_i,$$

where we may choose $G_i := \{x \in K : \text{there exists } y \in P_i \text{ such that } d(x, y) < \varepsilon_i\}$ with a suitable chosen sequence $\varepsilon_i \downarrow 0$ (cf. Theorem 5.2.1). Because $K = \bigcup_{i=1}^n G_i$ we may use Theorem 4.1.8 to obtain a partition of unity $(h_i)_{i=1}^n$ of K, i.e.

$$1 = \sum_{i=1}^n h_i \quad \text{with} \quad \overline{\{h_i \neq 0\}} \subseteq G_i.$$

This implies that

$$\langle \varphi, a \rangle = \sum_{i=1}^{n} \langle \varphi h_i, a \rangle \leqslant \sum_{i=1}^{n} (b_i + \varepsilon) \langle h_i, a \rangle \leqslant \sum_{i=1}^{n} (b_i + \varepsilon) \mu(G_i)$$

$$\leqslant \sum_{i=1}^{n} (b_i + \varepsilon)(\mu(P_i) + \varepsilon) \leqslant \sum_{i=1}^{n} (a_i + \frac{1}{n} + \varepsilon)(\mu(P_i) + \varepsilon).$$

By $\varepsilon \downarrow 0$ we get that

$$\langle \varphi, a \rangle \leqslant \sum_{i=1}^{n} (a_i + \frac{1}{n}) \mu(P_i) = \sum_{i=1}^{n} a_i \mu(P_i) + \frac{1}{n} \leqslant \int_K \varphi d\mu + \frac{1}{n},$$

and by $n \to \infty$ that $\langle \varphi, a \rangle \leqslant \int_K \varphi d\mu$.

□

18.5 Exercises

Ex 1: We show that in Theorem 18.2.1 for $p = 1$ the condition σ-finite on the measure μ is needed. Let $\Omega := [0, 1]$ and $\mathcal{F} := \sigma(\{t\} : t \in [0, 1])$.

- Convince yourself, that $\mathcal{F}$ is the system of all sets $A \subseteq [0, 1]$ such that A or A^c is countable.
- Verify that a function $f : [0, 1] \to \mathbb{R}$ is $(\mathcal{F}, \mathcal{B}(\mathbb{R}))$-measurable if there is a partition $\bigcup_{i \in I} \Omega_i = [0, 1]$, where I is countable and $\emptyset \neq \Omega_i \in \mathcal{F}$, such that $f(t) = \sum_{i \in I} \alpha_i \mathbb{1}_{\Omega_i}(t)$.
- Let μ be the counting measure on $([0, 1], \mathcal{F})$, i.e. $\mu(A) := \#A$. Given an f as above we get that $f \in \mathcal{L}_1([0, 1], \mathcal{F}, \mu)$ if $\sum_{i \in I} |\alpha_i| \# \Omega_i < \infty$. As the only null sets are the empty sets, one has that the equivalence classes $[f]$ consist only of the function f itself. Verify that $f \in \mathcal{L}_1([0, 1], \mathcal{F}, \mu)$ if and only if $f(t) \neq 0$ only for countably many $t \in [0, 1]$ and that

$$\sum_{t \in [0,1]} |f(t)| := \sum_{\{t \in [0,1]: f(t) \neq 0\}} |f(t)| < \infty.$$

- We define the map $a : L_1([0, 1], \mathcal{F}, \mu) \to \mathbb{R}$ by

$$\langle f, a \rangle := \sum_{t \in [0,1]} t f(t).$$

Verify that a is linear and that $|\langle f, a \rangle| \leqslant \|f\|_{L_1([0,1], \mathcal{F}, \mu)}$ for all $f \in \mathcal{L}_1([0, 1], \mathcal{F}, \mu)$.

- Assume that there is an $(\mathcal{F}, \mathcal{B}(\mathbb{R}))$-measurable $g : [0, 1] \to \mathbb{R}$ with $\sup_{t \in [0,1]} |g(t)| < \infty$ such that

$$\langle f, a \rangle := \int_{[0,1]} f(t)g(t)\mathrm{d}\mu(t).$$

Check that the only candidate would be $g(t) = t$, but this candidate is not $(\mathcal{F}, \mathcal{B}(\mathbb{R}))$-measurable.

Ex 2: Assume a σ-finite measure space $(\Omega, \mathcal{F}, \mu)$ and $f \in \mathcal{L}_0(\Omega, \mathcal{F}, \mu)$. Prove that

$$\|f\|_{L_\infty(\mu)} = \sup_{\mu(A)>0} \frac{1}{\mu(A)} \left| \int_A f \mathrm{d}\mu \right|.$$

Hint: Denote the right-hand side by $|f|_{L_\infty(\mu)}$. Then check that $|f|_{L_\infty(\mu)} \leq \|f\|_{L_\infty(\mu)}$ follows directly from the definition. For the opposite inequality you can use the level-sets $\{\omega \in \Omega : |f(\omega)| \geq c\}$ for $c \geq 0$.

Ex 3: Here we show again, this time by Theorem 18.4.1, that the Riemann integral yields to the Lebesgue measure. Let $K = [0, 1]$ and consider the functional $\langle f, a \rangle := \int_0^1 f(t)\mathrm{d}t$. Show that $a \in C^*([0, 1])$. By Theorem 18.4.1 there exists a $\mu \in \mathcal{M}([0, 1])$ such that $\langle f, a \rangle = \int_{[0,1]} f \mathrm{d}\mu$ for all $f \in C([0, 1])$. Show that μ must be the Lebesgue measure.

18.6 Comments

For this chapter we also used Rudin [165], Werner [197], and Dudley [50], which we also recommend for further reading. For $p = 2$ Theorem 18.2.1 is the Riesz–Fréchet representation theorem named after F. Riesz [153] and Fréchet [65, 66]. Theorem 18.2.1 for $p \neq 2$ is due to F. Riesz, see [154, §11] and the duality $(L_1([a, b]))^* = L_\infty([a, b])$ due to Steinhaus [180]. Theorem 18.2.1 holds true for general measure spaces if $p \in (1, \infty)$, see McShane [129], J. Schwartz [171], and Behrends [7, Satz IV.2.4]. This cannot be extended to $p = 1$ as we have seen in Exercise 1, which we learned from Werner [197]. An earlier example is given in McShane [129] and attributed to a discussion with T.A. Botts and V.L. KLee. For further results about the dual space of $L_1(\Omega, \mathcal{F}, \mu)$ for measure spaces that are not σ-finite the reader is referred to [7, Section VI.3]. Theorem 18.4.1 for $K = [0, 1]$ is due to F. Riesz [155, 156], the general version Theorem 18.4.1 due to Kakutani, see [96].

Chapter 19
Banach Function Spaces

Banach function spaces form a class of Banach spaces consisting of functions defined on measure spaces. Prominent examples are Orlicz and Lorentz spaces. The first part of this chapter introduces the nonincreasing rearrangement of a random variable and the corresponding maximal function. It is shown that both are related to each other by the Hardy-Littlewood maximal theorem. The maximal function is also related to the K-functional from interpolation theory. Interpolation of Banach spaces is a method from functional analysis which is used to construct a new Banach space as an intermediate space from two other Banach spaces. After having dealt with these basic relations, Banach function spaces are introduced, and Lorentz and Orlicz spaces are discussed. The chapter concludes with a class of Lorentz and Orlicz spaces that describe different exponential tail behaviours of random variables.

19.1 Nonincreasing Rearrangement and Maximal Functions

Given a random variable $f : \Omega \to \mathbb{R}$ on some probability space $(\Omega, \mathcal{F}, \mathbb{P})$ there are the notions of the nonincreasing rearrangement f^* and the maximal function f^{**}. The functions f^* and f^{**} are used to describe function spaces in analysis and have their place in probability theory.

Let us start with the definition of the nonincreasing rearrangement and the maximal function:

Definition 19.1.1 The **nonincreasing rearrangement** $f^* : (0, 1] \to [0, \infty)$ of a random variable $f \in \mathcal{L}_0(\Omega, \mathcal{F}, \mathbb{P})$ is given by

$$f^*(t) := \inf\{u \geqslant 0 : \mathbb{P}(|f| > u) \leqslant t\}.$$

Moreover, for $t \in (0, 1]$ we define $f^{**} : (0, 1] \to [0, \infty]$ by

Fig. 19.1 Nonincreasing rearrangement and maximal function

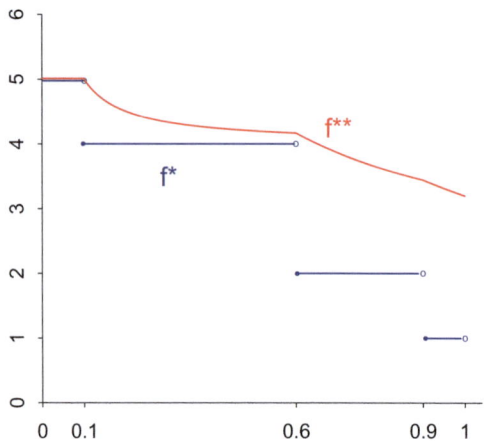

$$f^{**}(t) := \frac{1}{t} \int_{(0,t]} f^*(s) \mathrm{d}\lambda(s).$$

The function f^{**} is called **maximal function**.

Remark 19.1.2 Let $F_{|f|}$ be the distribution function of $|f|$. Then $f^*(1-t) = \inf\{u \geq 0 : F_{|f|}(u) \geq t\} =: F_{|f|}^{\leftarrow}(t)$ is known as the *left-inverse, generalized inverse* or *quantile function* of $F_{|f|}$.

Example 19.1.3 We assume a probability space $(\Omega, \mathcal{F}, \mathbb{P})$, a measurable partition $\Omega = A_1 \cup A_2 \cup A_3 \cup A_4$ with $\mathbb{P}(A_1) = 5/10$, $\mathbb{P}(A_2) = 1/10$, $\mathbb{P}(A_3) = 1/10$, and $\mathbb{P}(A_4) = 3/10$, and a random variable

$$f(\omega) := 4\mathbb{1}_{A_1}(\omega) - 5\mathbb{1}_{A_2}(\omega) + \mathbb{1}_{A_3}(\omega) + 2\mathbb{1}_{A_4}(\omega).$$

Then we get:

$$f^*(t) = 5\mathbb{1}_{(0,1/10)}(t) + 4\mathbb{1}_{[1/10,6/10)}(t) + 2\mathbb{1}_{[6/10,9/10)}(t) + \mathbb{1}_{[9/10,1)}(t).$$

See the graph of f^* and f^{**} in Fig. 19.1.

In the following we need to assume at some places that the underlying probability space $(\Omega, \mathcal{F}, \mathbb{P})$ is atomless. But this is not a restriction for us because of the following fact:

Remark 19.1.4 Assume a probability space $(\Omega, \mathcal{F}, \mathbb{P})$, which is not necessarily atomless, and a random variable $f : \Omega \to \mathbb{R}$. As in Example 8.14.3 we extend this probability space to $(\Omega \times [0, 1], \mathcal{F} \otimes \mathcal{B}([0, 1]), \mathbb{P} \otimes \lambda)$, which is atomless, and extend f canonically to $\tilde{f} : \Omega \times [0, 1] \to \mathbb{R}$ by $\tilde{f}(\omega, u) := f(\omega)$. It is obvious that $\mathrm{law}(f) = \mathrm{law}(\tilde{f})$, so that $f^* = (\tilde{f})^*$ and $f^{**} = (\tilde{f})^{**}$ as the nonincreasing rearrangement and the maximal function of a random variable depend only on the law of the random variable.

19.1 Nonincreasing Rearrangement and Maximal Functions

Now let us start by summarizing some basic properties of the nonincreasing rearrangement and the maximal function:

Theorem 19.1.5

(1) f^* is nonincreasing and right-continuous with $f^*(1) = 0$.
(2) $|f|$ and f^* have the same law. In particular, for any $(\mathcal{B}([0,\infty)), \mathcal{B}([0,\infty)))$-measurable $\psi : [0,\infty) \to [0,\infty)$ we have

$$\int_{(0,1]} \psi(f^*(s)) d\lambda(s) = \int_\Omega \psi(|f(\omega)|) d\mathbb{P}(\omega).$$

(3) $(f+g)^*(s+t) \leq f^*(s) + g^*(t)$ for $s, t, s+t \in (0, 1]$.
(4) $f^* \leq f^{**}$.
(5) For $A \in \mathcal{F}$ with $\mathbb{P}(A) = t \in (0, 1]$ one has

$$\int_A |f| d\mathbb{P} \leq \int_{(0,t]} f^*(s) d\lambda(s).$$

If $(\Omega, \mathcal{F}, \mathbb{P})$ is atomless, then we have for $t \in (0, 1]$ that

$$f^{**}(t) = \sup_{A \in \mathcal{F}, \mathbb{P}(A) = t} \frac{1}{t} \int_A |f| d\mathbb{P}.$$

(6) $(f+g)^{**} \leq f^{**} + g^{**}$.
(7) If $|f_n| \uparrow |f|$ a.s., then $f_n^* \uparrow f^*$ and $f_n^{**} \uparrow f^{**}$.

Proof For $t \in (0, 1]$ we first observe that

$$\mathbb{P}(|f| > f^*(t)) \leq t \leq \mathbb{P}(|f| \geq f^*(t)). \tag{19.1}$$

These inequalities can be verified as follows: We take a sequence $c_n \downarrow f^*(t)$, $n \in \mathbb{N}$, so that $\mathbb{P}(|f| > c_n) \leq t$ for all $n \in \mathbb{N}$. By continuity from below of the measure we know that $\lim_{n \to \infty} \mathbb{P}(|f| > c_n) = \mathbb{P}(|f| > f^*(t))$, so that the left inequality follows. To verify the right-hand side, we may assume $f^*(t) > 0$, otherwise $\mathbb{P}(|f| \geq f^*(t)) = 1$ and there is nothing to prove. Now we choose a sequence $0 < c_n \uparrow f^*(t)$, so that $\mathbb{P}(|f| > c_n) > t$ for all $n \in \mathbb{N}$. Similarly as in the first case continuity from above of the measure implies $t \leq \lim_{n \to \infty} \mathbb{P}(|f| > c_n) = \mathbb{P}(|f| \geq f^*(t))$.

(1) The fact that f^* is nonincreasing follows directly from the definition as $\mathbb{P}(|f| > c)$ is nonincreasing in $c \geq 0$. To verify the right-continuity assume $t_n, t \in (0, 1]$ with $t_n \downarrow t$. This immediately implies that $f^*(t) \geq \lim_{n \to \infty} f^*(t_n)$.

To understand the opposite inequality we use (19.1) to get $\mathbb{P}(|f| > f^*(t_n)) \leq t_n$ so that $\lim_{n \to \infty} \mathbb{P}(|f| > f^*(t_n)) \leq t$ and

$$\mathbb{P}\left(|f| > \lim_{n \to \infty} f^*(t_n)\right) \leq \mathbb{P}\left(\bigcap_{n=1}^{\infty} \{|f| > f^*(t_n)\}\right) = \lim_{n \to \infty} \mathbb{P}(|f| > f^*(t_n)) \leq t.$$

Therefore we obtain that $f^*(t) \leq \lim_{n\to\infty} f^*(t_n)$.

(2) Assume $t \in (0, 1]$ and $c \geq 0$. Then $\mathbb{P}(|f| > c) \leq t$ implies that $f^*(t) \leq c$. Since f^* is nonincreasing we conclude from $f^*(t) \leq c$ that $\{f^* > c\} \subseteq (0, t)$ so that $\lambda(f^* > c) \leq t$. Hence $\mathbb{P}(|f| > c) = t$ yields to $\lambda(f^* > c) \leq t$. Since this holds for all $t \in (0, 1]$ we deduce

$$\lambda(f^* > c) \leq \mathbb{P}(|f| > c) \quad \text{for all} \quad c \geq 0.$$

Conversely, assume that

$$\mathbb{P}(|f| > c) \leq \lambda(f^* > c) \quad \text{for all} \quad c \geq 0$$

does not hold. Then there exists a $c \geq 0$ and a $t \in (0, 1]$ such that $\mathbb{P}(|f| > c) = t$ but $\lambda(f^* > c) \leq t - \varepsilon$ for some $\varepsilon \in (0, t)$. By the right continuity of f^* this implies $f^*(t - \varepsilon) \leq c$ and $\mathbb{P}(|f| > c) \leq t - \varepsilon$, which contradicts our assumption.

(3) Using (19.1) we observe

$$\mathbb{P}(|f + g| > f^*(s) + g^*(t)) \leq \mathbb{P}(|f| > f^*(s)) + \mathbb{P}(|g| > g^*(t)) \leq s + t.$$

Therefore, $f^*(s) + g^*(t) \geq (f + g)^*(s + t)$.

(4) follows directly from (1).

(5) We define the random variable $g(\omega) := f(\omega)\mathbf{1}_A(\omega)$ so that $|g| \leq |f|$ and therefore $g^* \leq f^*$. Moreover, $g^*(t) = 0$ because $\mathbb{P}(|g| > 0) \leq \mathbb{P}(A) = t$. Using item (2) we get

$$\int_A |f| d\mathbb{P} = \mathbb{E}g = \int_{(0,1]} g^*(s) d\lambda(s) = \int_{(0,t]} g^*(s) d\lambda(s) \leq \int_{(0,t]} f^*(s) d\lambda(s).$$

To prove the second part we let $t \in (0, 1]$. Using (19.1) and that the probability space is atomless, we find a measurable subset $D_t \subseteq \{|f| = f^*(t)\}$ such that $\mathbb{P}(A_t) = t$ for $A_t := \{|f| > f^*(t)\} \cup D_t$. Using again (2), for this set we get

$$\frac{1}{t} \int_{A_t} |f| d\mathbb{P} = \frac{1}{t} \left[\int_{\{|f| > f^*(t)\}} |f| d\mathbb{P} + \int_{D_t} |f| d\mathbb{P} \right]$$

$$= \frac{1}{t} \left[\int_{\{f^* > f^*(t)\}} f^*(s) d\lambda(s) + f^*(t)\mathbb{P}(D_t) \right]$$

$$= \frac{1}{t} \left[\int_{\{f^* > f^*(t)\}} f^*(s) d\lambda(s) + f^*(t)(t - \mathbb{P}(|f| > f^*(t))) \right]$$

$$= \frac{1}{t} \left[\int_{\{f^* > f^*(t)\}} f^*(s) d\lambda(s) + f^*(t)(t - \lambda(f^* > f^*(t))) \right]$$

$$= \frac{1}{t} \int_{(0,t]} f^*(s) d\lambda(s).$$

(6) Using Remark 19.1.4 we may assume without loss of generality that the underlying probability space is atomless as the maximal functions f^{**}, g^{**}, and $(f+g)^{**}$ do not change. So we can use item (5) to deduce

$$(f+g)^{**}(t) = \sup_{A\in\mathcal{F},\mathbb{P}(A)=t} \frac{1}{t}\int_A |f+g|\,\mathrm{d}\mathbb{P}$$
$$\leqslant \sup_{A\in\mathcal{F},\mathbb{P}(A)=t} \frac{1}{t}\int_A |f|\,\mathrm{d}\mathbb{P} + \sup_{A\in\mathcal{F},\mathbb{P}(A)=t} \frac{1}{t}\int_A |g|\,\mathrm{d}\mathbb{P}$$
$$= f^{**}(t) + g^{**}(t).$$

(7) We may assume that $|f_n| \uparrow |f|$ on Ω as the nonincreasing rearrangement and maximal function are invariant with respect to changes on null sets of the random variable one is starting from. The statement $f_n^{**} \uparrow f^{**}$ would follow directly from $f_n^* \uparrow f^*$ by monotone convergence. Therefore, we only need to show $f_n^*(t) \uparrow f^*(t)$ for $t \in (0,1)$. From the definition it is clear that $f_n^*(t)$ is nondecreasing in n and that

$$\lim_{n\to\infty} f_n^*(t) \leqslant f^*(t).$$

Assume $\varepsilon > 0$ such that $f_n^*(t) < f^*(t) - \varepsilon$ for all $n \in \mathbb{N}$. Then

$$\mathbb{P}(|f_n| > f^*(t) - \varepsilon) \leqslant t$$

and by $n \to \infty$ we would get that $\mathbb{P}(|f| > f^*(t) - \varepsilon) \leqslant t$ because of $\{|f| > f^*(t) - \varepsilon\} = \bigcup_{n=1}^{\infty} \{|f_n| > f^*(t) - \varepsilon\}$. But this is a contradiction to the definition of $f^*(t)$. □

The following Hardy-Littlewood maximal theorem (Theorem 19.1.7) says that the size of f and f^{**} are comparable in certain situations. To verify this statement we use Hardy's inequality:

Lemma 19.1.6 (Hardy's Inequality) *Assume a* $(\mathcal{B}([0,\infty)), \mathcal{B}([0,\infty)))$-*measurable* $\phi : [0,\infty) \to [0,\infty)$ *and* $-\infty < \gamma < 1$. *Then*

$$\int_{(0,\infty)} \left[t^\gamma \frac{1}{t}\int_{(0,t]} \phi(s)\,\mathrm{d}\lambda(s) \right]^p \frac{\mathrm{d}\lambda(t)}{t} \leqslant \frac{1}{(1-\gamma)^p} \int_{(0,\infty)} [t^\gamma \phi(t)]^p \frac{\mathrm{d}\lambda(t)}{t}$$

for $p \in [1,\infty)$, *and, if* $p = \infty$,

$$\sup_{t\in(0,\infty)} \left[t^\gamma \frac{1}{t}\int_{(0,t]} \phi(s)\,\mathrm{d}\lambda(s) \right] \leqslant \frac{1}{1-\gamma} \sup_{t\in(0,\infty)} [t^\gamma \phi(t)].$$

Proof In the case $p = 1$ we have equality which follows directly from interchanging the order of integration. For $p = \infty$ we set $c := \sup_{t \in (0,\infty)} t^\gamma \phi(t)$, assume $c < \infty$, and estimate the left-hand side by using $\phi(s) \leqslant cs^{-\gamma}$ for $s \in (0, \infty)$.

For $p \in (1, \infty)$ and $t \in (0, \infty)$ we get that

$$\frac{1}{t}\int_{(0,t]} \phi(s)d\lambda(s) \leqslant \left(\frac{1}{t}\int_{(0,t]}\left[s^{-\frac{\gamma}{p'}}\right]^{p'} d\lambda(s)\right)^{\frac{1}{p'}} \left(\frac{1}{t}\int_{(0,t]}\left[s^{\frac{\gamma}{p'}}\phi(s)\right]^{p} d\lambda(s)\right)^{\frac{1}{p}}$$

$$= \left(\frac{1}{t}\int_{(0,t]} s^{-\gamma} d\lambda(s)\right)^{\frac{1}{p'}} \left(\frac{1}{t}\int_{(0,t]} s^{\gamma(p-1)}|\phi(s)|^p d\lambda(s)\right)^{\frac{1}{p}}$$

with $1 = \frac{1}{p} + \frac{1}{p'}$ and since $\frac{p}{p'} = p(1 - \frac{1}{p}) = p - 1$. Consequently, by computing the first factor,

$$\frac{1}{t}\int_{(0,t]} \phi(s)d\lambda(s) \leqslant \left(\frac{1}{1-\gamma}\right)^{\frac{1}{p'}} t^{\frac{-\gamma}{p'}-\frac{1}{p}} \left(\int_{(0,t]} s^{\gamma(p-1)}|\phi(s)|^p d\lambda(s)\right)^{\frac{1}{p}}$$

and

$$\int_{(0,\infty)} \left[t^\gamma \frac{1}{t}\int_{(0,t]} \phi(s)d\lambda(s)\right]^p \frac{d\lambda(t)}{t}$$

$$\leqslant (1-\gamma)^{-\frac{p}{p'}} \int_{(0,\infty)} t^{\gamma-2} \int_{(0,t]} s^{\gamma(p-1)}|\phi(s)|^p d\lambda(s) d\lambda(t)$$

$$= (1-\gamma)^{1-p} \int_{(0,\infty)} s^{\gamma(p-1)}|\phi(s)|^p \int_{[s,\infty)} t^{\gamma-2} d\lambda(t) d\lambda(s)$$

$$= (1-\gamma)^{1-p} \int_{(0,\infty)} s^{\gamma(p-1)}|\phi(s)|^p \frac{1}{1-\gamma} s^{\gamma-1} d\lambda(s)$$

$$= (1-\gamma)^{-p} \int_{(0,\infty)} s^{\gamma p}|\phi(s)|^p \frac{d\lambda(s)}{s}.$$

□

Now we formulate and prove the Hardy-Littlewood maximal theorem:

19.1 Nonincreasing Rearrangement and Maximal Functions

Theorem 19.1.7 (Hardy[1]-Littlewood[2] maximal theorem) *For $p \in (1, \infty)$ one has that*

$$\left(\int_\Omega |f|^p d\mathbb{P}\right)^{\frac{1}{p}} \leq \left(\int_{(0,1]} |f^{**}(t)|^p d\lambda(t)\right)^{\frac{1}{p}} \leq \frac{p}{p-1}\left(\int_\Omega |f|^p d\mathbb{P}\right)^{\frac{1}{p}}.$$

Moreover, for $p = \infty$ one has $\|f\|_{L_\infty(\Omega)} = \sup_{t \in (0,1]} f^{**}(t)$.

Proof For $p \in (1, \infty]$ we observe that

$$\|f\|_{L_p} = \|f^*\|_{L_p((0,1])} \leq \|f^{**}\|_{L_p((0,1])}.$$

Let $\gamma := 1/p \in [0, 1)$ and $\phi(s) := f^*(s)$ for $s \in (0, 1]$ and $\phi(s) := 0$ for $s > 1$. Now the first part of Lemma 19.1.6 gives

$$\int_{(0,1]} |f^{**}(t)|^p d\lambda(t) \leq \int_{(0,\infty)} \left[t^\gamma \frac{1}{t} \int_{(0,t]} \phi(s) d\lambda(s)\right]^p \frac{d\lambda(t)}{t}$$

$$\leq \left(\frac{1}{1-\gamma}\right)^p \int_{(0,\infty)} |t^\gamma \phi(t)|^p \frac{d\lambda(t)}{t}$$

$$= \left(\frac{1}{1-\gamma}\right)^p \int_{(0,1]} |f^*(t)|^p d\lambda(t).$$

For $p = \infty$ we use the second part of Lemma 19.1.6. □

Next we show that the maximal function describes a certain decomposition property, which is formulated in terms of the so-called K-functional. The K-functional is a key ingredient of interpolation theory for Banach spaces. To introduce this functional we need the notion of an interpolation couple:

Definition 19.1.8 We say that a pair of Banach spaces (E_0, E_1) is an **interpolation couple** provided that there is another Banach space E with $E_0 \subseteq E$ and $E_1 \subseteq E$ and such that the following holds:

(1) There are $c_0, c_1 > 0$ such that for $i = 0, 1$ one has

$$\|x_i\|_E \leq c_i \|x_i\|_{E_i} \quad \text{for} \quad x_i \in E_i,$$

(2) For $i = 0, 1$ and $x_i \in E_i$ one has $\|x_i\|_{E_i} = 0$ if and only if $\|x_i\|_E = 0$.

---

[1] Godfrey Harold Hardy, 07/02/1877 (Cranleigh, Surrey, England)–01/12/1947 (Cambridge, England).

[2] John Edensor Littlewood, 09/06/1885 (Rochester, Kent, England)–06/09/1977 (Cambridge, England).

Now we can define the K-functional:

Definition 19.1.9 Let (E_0, E_1) be an interpolation couple of Banach spaces, $x \in E_0 + E_1$, and $t \in (0, \infty)$. Then we define the K-**functional** by

$$K(x, t; E_0, E_1) := \inf\{\|x_0\|_{E_0} + t\|x_1\|_{E_1} : x = x_0 + x_1\}.$$

Although the K-functional is—by its very definition—an abstract concept from Banach space theory, it is used in probability theory (often without referring to the name K-functional) when decomposing random variables into a sum of two random variables and handling each term in this sum differently.

Let us summarize some basic properties of this functional:

Proposition 19.1.10 *Let (E_0, E_1) be an interpolation couple of Banach spaces, $x \in E_0 + E_1$, and $t \in (0, \infty)$. Then one has the following:*

(1) *The function $(0, \infty) \ni t \mapsto K(x, t; E_0, E_1)$ is nondecreasing and concave, and therefore also continuous.*
(2) $K(x, t; E_0, E_1) = tK(x, \frac{1}{t}; E_1, E_0)$.
(3) $\min\{1, t\} K(x, 1; E_0, E_1) \leqslant K(x, t; E_0, E_1) \leqslant \max\{1, t\} K(x, 1; E_0, E_1)$.
(4) $K(x, t; E_0, E_1) \leqslant t\|x\|_{E_1}$ *if* $x \in E_1$, *and* $K(x, t; E_0, E_1) \leqslant \|x\|_{E_0}$ *if* $x \in E_0$.

The proof uses only the definition of the K-functional and is recommended as an exercise to get acquainted with this functional. The idea of the K-functional is that the behavior of the function $t \mapsto K(x, t; E_0, E_1)$ describes somehow the position of an element $x \in E_0 + E_1$ relative to E_0 and E_1. Coming back to probability the K-functional relates to the maximal function as follows:

Theorem 19.1.11 *For $f \in \mathcal{L}_1(\Omega, \mathcal{F}, \mathbb{P})$ it holds that*

$$K(f, t; L_1(\Omega), L_\infty(\Omega)) = \begin{cases} \int_{(0,t]} f^*(s) d\lambda(s) \text{ for } t \in (0, 1], \\ \int_{(0,1]} f^*(s) d\lambda(s) \text{ for } t > 1. \end{cases}$$

Proof

(a) For $t > 1$ we have, taking into the account that $\|g\|_{L_1(\Omega)} \leqslant \|g\|_{L_\infty(\Omega)}$, that

$$\begin{aligned} &K(f, t; L_1(\Omega), L_\infty(\Omega)) \\ &= \inf\{\|f_0\|_{L_1(\Omega)} + t\|f_1\|_{L_\infty(\Omega)} : f = f_0 + f_1\} \\ &= \inf\{\|f_0\|_{L_1(\Omega)} + t\|f_1\|_{L_\infty(\Omega)} : f = f_0 + f_1, f_1 \equiv 0\} \\ &= \|f\|_{L_1(\Omega)} \\ &= \int_{(0,1]} f^*(s) d\lambda(s). \end{aligned}$$

19.1 Nonincreasing Rearrangement and Maximal Functions

(b) Assume $0 < t \leqslant 1$ and let $f = f_0 + f_1$ with $f_0 \in L_1(\Omega)$ and $f_1 \in L_\infty(\Omega)$. Then, by Theorem 19.1.5(6),

$$\int_{(0,t]} f^*(s) d\lambda(s) \leqslant \int_{(0,t]} f_0^*(s) d\lambda(s)$$

$$+ \int_{(0,t]} f_1^*(s) d\lambda(s) \leqslant \|f_0\|_{L_1(\Omega)} + t\|f_1\|_{L_\infty(\Omega)}$$

so that

$$\int_{(0,t]} f^*(s) d\lambda(s) \leqslant K(f, t; L_1(\Omega), L_\infty(\Omega)).$$

To show the converse inequality, we have to decompose f in the right way: Let

$$E := \{\omega \in \Omega : |f(\omega)| > f^*(t)\},$$

$$f_0(\omega) := \max\{|f(\omega)| - f^*(t), 0\} \operatorname{sgn}(f(\omega)),$$

$$f_1(\omega) := \min\{|f(\omega)|, f^*(t)\} \operatorname{sgn}(f(\omega))$$

so that $f = f_0 + f_1$. By (19.1) we have that $\mathbb{P}(E) \leqslant t$. If $\mathbb{P}(E) = 0$, then $\mathbb{P}(|f| \leqslant f^*(t)) = 1$ and we choose the decomposition $f = 0 + f$ so that $\|f_0\|_{L_1(\Omega)} + t\|f_1\|_{L_\infty(\Omega)} \leqslant tf^*(t) \leqslant \int_0^t f^*(s) d\lambda(s)$. For $\mathbb{P}(E) > 0$ we apply Theorem 19.1.5(5) to get

$$\|f_0\|_{L_1(\Omega)} = \int_E |f(\omega)| d\mathbb{P}(\omega) - \mathbb{P}(E) f^*(t)$$

$$\leqslant \mathbb{P}(E) f^{**}(\mathbb{P}(E)) - \mathbb{P}(E) f^*(t)$$

$$\leqslant \int_{(0,\mathbb{P}(E)]} f^*(s) d\lambda(s) - \mathbb{P}(E) f^*(t).$$

This implies, using the definition of f_1,

$$\|f_0\|_{L_1(\Omega)} + t\|f_1\|_{L_\infty(\Omega)} \leqslant \int_{(0,\mathbb{P}(E)]} f^*(s) d\lambda(s) - \mathbb{P}(E) f^*(t) + tf^*(t)$$

$$= \int_{(0,\mathbb{P}(E)]} f^*(s) d\lambda(s) + f^*(t)(t - \mathbb{P}(E))$$

$$\leqslant \int_{(0,t]} f^*(s) d\lambda(s).$$

□

Next we will provide upper and lower bounds for $\left\|\sup_{i=1,\ldots,n} |f_i|\right\|_{L_p}$ for i.i.d. random variables $f_1, \ldots, f_n$. We will use this result later in Theorem 20.7.4 to provide a lower bound for the approximation of the Brownian motion. We start with defining for $p \in (0, \infty)$ the maximal function

$$f_p^{**}(t) := \left(\frac{1}{t} \int_{(0,t]} |f^*(s)|^p d\lambda(s) \right)^{\frac{1}{p}} \quad \text{for } t \in (0, 1]. \tag{19.2}$$

It is not really a new object as we have $f_1^{**} = f^{**}$ and

$$\sqrt[p]{(|f|^p)^{**}(t)} = f_p^{**}(t) \quad \text{for } t \in (0, 1],$$

see Exercise 5.

Theorem 19.1.12 *For $p \in (0, \infty)$ and i.i.d. random variables $f_1, \ldots, f_n : \Omega \to \mathbb{R}$ on a probability space $(\Omega, \mathcal{F}, \mathbb{P})$ one has that*

$$\frac{1}{\sqrt[p]{2}} (f_1)_p^{**}\left(\frac{1}{n}\right) \leq \left\| \sup_{i=1,\ldots,n} |f_i| \right\|_{L_p} \leq (f_1)_p^{**}\left(\frac{1}{n}\right).$$

For the proof of the above result we use the following lemma:

Lemma 19.1.13 *For i.i.d. random variables $f_1, \ldots, f_n \in \mathcal{L}_0(\Omega, \mathcal{F}, \mathbb{P})$ and $c \geq 0$ one has*

$$\mathbb{P}\left(\sup_{i=1,\ldots,n} |f_i| > c \right) \geq \frac{n\mathbb{P}(|f_1| > c)}{1 + n\mathbb{P}(|f_1| > c)}.$$

Proof By independence of $f_1, \ldots, f_n$ we have that

$$\mathbb{P}\left(\sup_{i=1,\ldots,n} |f_i| > c \right) = 1 - \mathbb{P}\left(\sup_{i=1,\ldots,n} |f_i| \leq c \right)$$

$$= 1 - \prod_{i=1}^{n} \mathbb{P}(|f_i| \leq c)$$

$$= 1 - (1 - \mathbb{P}(|f_1| > c))^n.$$

So, for $p \in [0, 1]$ we have to verify that

$$1 - (1 - p)^n \geq \frac{np}{1 + np} = 1 - \frac{1}{1 + np} \quad \text{or} \quad (1 - p)^n \leq \frac{1}{1 + np}.$$

This is the same as $1 + np \leq (1 - p)^{-n}$ for $p \in [0, 1)$. For $p = 0$ both sides coincide. Differentiating both sides with respect to p gives $n \leq n(1 - p)^{-n-1}$, which turns out to be true. □

19.1 Nonincreasing Rearrangement and Maximal Functions

Proof of Theorem 19.1.12 We only need to prove the statement for $p = 1$: Assume that we know the statement for $p = 1$. Then we get that

$$\frac{1}{2}(|f_1|^p)^{**}\left(\frac{1}{n}\right) \leq \left\|\sup_{i=1,\ldots,n}|f_i|^p\right\|_{L_1} \leq (|f_1|^p)^{**}\left(\frac{1}{n}\right).$$

Taking the p-th root we deduce

$$\frac{1}{\sqrt[p]{2}}(f_1)_p^{**}\left(\frac{1}{n}\right) \leq \left\|\sup_{i=1,\ldots,n}|f_i|\right\|_{L_p} \leq (f_1)_p^{**}\left(\frac{1}{n}\right).$$

(a) To show the right-hand side of the claim let $f = g + h$ whereas $g \in L_\infty(\Omega)$ and $h \in L_1(\Omega)$. We model the independent $f_1, \ldots, f_n$ using the product space $(\Omega^n, \mathcal{F}^n, \mathbb{P}^n) = (\bigtimes_{k=1}^n \Omega, \bigotimes_{k=1}^n \mathcal{F}, \bigotimes_{k=1}^n \mathbb{P})$ and setting $f_k(\omega_1, \ldots, \omega_n) := f(\omega_k)$ for $k = 1, \ldots, n$. Then

$$\left\|\sup_{1 \leq i \leq n}|f(\omega_i)|\right\|_{L_1(\Omega^n)} \leq \|g\|_\infty + \left\|\sup_{1 \leq i \leq n}|h(\omega_i)|\right\|_{L_1(\Omega^n)}$$

$$\leq \|g\|_\infty + \left\|\sum_{i=1}^n |h(\omega_i)|\right\|_{L_1(\Omega^n)}$$

$$= \|g\|_\infty + n\|h\|_{L_1(\Omega)}.$$

Since this holds for all decompositions $f = g + h$ and using Theorem 19.1.11 we derive

$$\left\|\sup_{1 \leq i \leq n}|f(\omega_i)|\right\|_{L_1(\Omega^n)} \leq K(f, n; L_\infty, L_1)$$

$$= nK\left(f, \frac{1}{n}; L_1, L_\infty\right)$$

$$= n\int_{(0,1/n]} f^*(s)\,d\lambda(s)$$

$$= f^{**}\left(\frac{1}{n}\right).$$

(b) Let $M_n := \sup_{i=1,\ldots,n}|f_i|$. For $0 < t \leq \frac{1}{2}$ it follows from Lemma 19.1.13 that

$$M_n^*(t) = \inf\left\{c \geq 0 : \mathbb{P}\left(\sup_{i=1,\ldots,n}|f_i| > c\right) \leq t\right\}$$

$$\geq \inf\left\{c > 0 : \frac{n\mathbb{P}(|f_i| > c)}{1 + n\mathbb{P}(|f_i| > c)} \leq t\right\}$$

$$= \inf\left\{c > 0 : \mathbb{P}(|f_1| > c) \leq \frac{t}{n(1-t)}\right\} = (f_1)^*\left(\frac{t}{n(1-t)}\right).$$

Setting $s := \frac{t}{1-t} \in (0, 1]$ yields to

$$M_n^*\left(\frac{s}{2}\right) \geq M_n^*\left(\frac{s}{1+s}\right) \geq (f_1)^*\left(\frac{s}{n}\right).$$

Consequently,

$$(f_1)^{**}\left(\frac{1}{n}\right) = n\int_{(0,\frac{1}{n}]} (f_1)^*(s)\, d\lambda(s)$$

$$= \int_{(0,1]} (f_1)^*\left(\frac{s}{n}\right) d\lambda(s)$$

$$\leq \int_{(0,1]} (M_n)^*\left(\frac{s}{2}\right) d\lambda(s)$$

$$= 2\int_{(0,\frac{1}{2}]} (M_n)^*(s)\, d\lambda(s)$$

$$\leq 2\int_{(0,1]} (M_n)^*(s)\, d\lambda(s)$$

$$= 2\|M_n\|_{L_1(\Omega)}.$$

$\square$

We have the following examples:

Example 19.1.14 (Weibull[3] Random Variables) Let $p \in (0, \infty)$, $\alpha > 0$, and let $f \in \mathcal{L}_0(\Omega, \mathcal{F}, \mathbb{P})$ be a random variable such that

$$\mathbb{P}(|f| > c) = e^{-c^\alpha} \quad \text{for} \quad c \geq 0.$$

Then

$$f^*(t) = \sqrt[\alpha]{\log\frac{1}{t}} \quad \text{for} \quad t \in (0, 1],$$

---

[3] Ernst Hjalmar Waloddi Weibull, 18/06/1887 (Vittskövle, Sweden)–12/10/1979 (Annecy, France).

19.1 Nonincreasing Rearrangement and Maximal Functions

and for $c_{(19.3)} := \left(\int_{(0,1]} \left| 1 + \frac{\log \frac{1}{r}}{\log 2} \right|^{\frac{p}{\alpha}} d\lambda(r) \right)^{\frac{1}{p}}$ one has

$$\sqrt[\alpha]{\log \frac{1}{t}} \leq f_p^{**}(t) \leq c_{(19.3)} \sqrt[\alpha]{\log \frac{1}{t}} \quad \text{for} \quad t \in \left(0, \frac{1}{2}\right]. \tag{19.3}$$

Proof We get for the nonincreasing rearrangement that

$$f^*(t) = \inf\{c \geq 0 : \mathbb{P}(|f| > c) \leq t\} = \inf\{c \geq 0 : e^{-c^\alpha} \leq t\} = \sqrt[\alpha]{\log \frac{1}{t}}$$

and

$$\sqrt[\alpha]{\log \frac{1}{t}} = f^*(t) \leq f_p^{**}(t).$$

For the opposite inequality we let $\beta := \frac{\alpha}{p} > 0$ and $t \in (0, 1/2]$, and observe

$$\frac{(f_p^{**}(t))^p}{\sqrt[\beta]{\log \frac{1}{t}}} = \frac{\frac{1}{t} \int_{(0,t]} \sqrt[\beta]{\log \frac{1}{s}} \, d\lambda(s)}{\sqrt[\beta]{\log \frac{1}{t}}} = \frac{\int_{(0,1]} \sqrt[\beta]{\log \frac{1}{tr}} \, d\lambda(r)}{\sqrt[\beta]{\log \frac{1}{t}}}$$

$$= \int_{(0,1]} \sqrt[\beta]{1 + \frac{\log \frac{1}{r}}{\log \frac{1}{t}}} \, d\lambda(r)$$

$$\leq \int_{(0,1]} \sqrt[\beta]{1 + \frac{\log \frac{1}{r}}{\log 2}} \, d\lambda(r) < \infty.$$

$\square$

Example 19.1.15 (Gaussian Random Variables) Let $p \in (0, \infty)$ and $f \in \mathcal{L}_0(\Omega, \mathcal{F}, \mathbb{P})$ be a random variable such that $\text{law}(f) \sim N(0, 1)$. Then

$$f^*(t) \leq \sqrt{2 \log \left(\frac{1}{t}\right)} \quad \text{for} \quad t \in (0, 1],$$

and there is a constant $c_{(19.4)} = c_{(19.4)}(p) > 0$ such that

$$\frac{1}{c_{(19.4)}} \sqrt{\log \frac{1}{t}} \leq f_p^{**}(t) \leq c_{(19.3)} \sqrt{2 \log \frac{1}{t}} \quad \text{for} \quad t \in \left(0, \frac{1}{2}\right], \tag{19.4}$$

where $c_{(19.3)} > 0$ is the constant from Example 19.1.14 with parameters $(\alpha, p) := (2, p)$.

Proof Here we use Mill's ratio, Lemma 15.3.5. For $x > 0$ we have that

$$\mathbb{P}(f > c) \leq \frac{1}{2} e^{-\frac{c^2}{2}} \quad \text{for} \quad c > 0.$$

By symmetry this implies that

$$\mathbb{P}(|f| > c) \leq e^{-\frac{c^2}{2}} \quad \text{for} \quad c \geq 0.$$

So we obtain

$$f^*(t) = \inf\{c \geq 0 : \mathbb{P}(|f| > c) \leq t\} \leq \inf\{c \geq 0 : e^{-\frac{c^2}{2}} \leq t\} = \sqrt{2 \log\left(\frac{1}{t}\right)}.$$

Example 19.1.14 with $\alpha = 2$ gives that

$$f_p^{**}(t) \leq \sqrt{2} c_{(19.3)} \sqrt{\log \frac{1}{t}} \quad \text{for} \quad t \in \left(0, \frac{1}{2}\right].$$

Now we turn to the lower bound: Here Mill's ratio, Lemma 15.3.5, gives for $c \geq 0$ that

$$\frac{1}{\sqrt{2\pi}} \frac{c}{1+c^2} e^{-\frac{c^2}{2}} \leq \mathbb{P}(f > c).$$

Again, by symmetry,

$$\sqrt{\frac{2}{\pi}} \frac{c}{1+c^2} e^{-\frac{c^2}{2}} \leq \mathbb{P}(|f| > c).$$

We choose $c_0 > 0$ such that $\mathbb{P}(|f| > c_0) = \frac{1}{2}$. For $t \in (0, 1/2]$ we get that

$$f^*(t) = \inf\{c \geq 0 : \mathbb{P}(|f| > c) \leq t\}$$
$$= \inf\{c \geq c_0 : \mathbb{P}(|f| > c) \leq t\}$$
$$\geq \inf\left\{c \geq c_0 : \sqrt{\frac{2}{\pi}} \frac{c}{1+c^2} e^{-\frac{c^2}{2}} \leq t\right\}.$$

Choose $0 < A < 2$ such that

$$\sqrt{\frac{2}{\pi}} \frac{c}{1+c^2} e^{-\frac{c^2}{2}} \geq e^{-\frac{c^2}{A}} \quad \text{for} \quad c \geq c_0.$$

19.1 Nonincreasing Rearrangement and Maximal Functions

Then

$$\inf\left\{c \geq c_0 : \sqrt{\frac{2}{\pi}} \frac{c}{1+c^2} e^{-\frac{c^2}{2}} \leq t\right\} \geq \inf\left\{c \geq c_0 : e^{-\frac{c^2}{A}} \leq t\right\}$$

$$\geq \inf\left\{c \geq 0 : e^{-\frac{c^2}{A}} \leq t\right\}$$

$$= \sqrt{A}\sqrt{\log\left(\frac{1}{t}\right)}.$$

This implies for $t \in (0, 1/2]$ that

$$f_p^{**}(t) \geq f^*(t) \geq \sqrt{A}\sqrt{\log\left(\frac{1}{t}\right)}.$$

□

Example 19.1.16 (Pareto[4] Type Random Variables) Let $\kappa > 0$, $0 < p < \alpha$, and let $f \in \mathcal{L}_0(\Omega, \mathcal{F}, \mathbb{P})$ be a random variable such that

$$\mathbb{P}(|f| > c) = \left(\frac{\kappa}{\kappa + c}\right)^\alpha \quad \text{for} \quad c \geq 0.$$

Then

$$f^*(t) = \kappa\left(t^{-\frac{1}{\alpha}} - 1\right) \quad \text{for} \quad t \in (0, 1],$$

and there is a constant $c_{(19.5)} = c_{(19.5)}(\alpha, p) \geq 1$ such that

$$\frac{1}{c_{(19.5)}}\sqrt[\alpha]{\frac{1}{t}} \leq f_p^{**}(t) \leq c_{(19.5)}\sqrt[\alpha]{\frac{1}{t}} \quad \text{for} \quad t \in \left(0, \frac{1}{2}\right]. \tag{19.5}$$

Proof First we observe that

$$f^*(t) = \inf\{c \geq 0 : \mathbb{P}(|f| > c) \leq t\}$$

$$= \inf\left\{c \geq 0 : \left(\frac{\kappa}{\kappa + c}\right)^\alpha \leq t\right\}$$

$$= \kappa\left(t^{-\frac{1}{\alpha}} - 1\right).$$

---

[4] Wilfried Fritz Pareto, 15/07/1848 (Paris, France)–19/08/1923 (Céligny, Switzerland).

Then we obtain for the maximal function that

$$f_p^{**}(t) = \left(\frac{1}{t}\int_{(0,t]} \left|\kappa\left(s^{-\frac{1}{\alpha}}-1\right)\right|^p d\lambda(s)\right)^{\frac{1}{p}}$$

$$\leq \left(\frac{1}{t}\frac{\kappa^p}{1-\frac{p}{\alpha}}t^{1-\frac{p}{\alpha}}\right)^{\frac{1}{p}}$$

$$= \kappa \sqrt[p]{\frac{1}{1-\frac{p}{\alpha}}}\, t^{-\frac{1}{\alpha}}.$$

On the other side we have for $\varepsilon := 1 - 2^{-\frac{1}{\alpha}}$ and $t \in \left(0, \frac{1}{2}\right]$ that

$$f_p^{**}(t) = \left(\frac{1}{t}\int_{(0,t]} \left|\kappa\left(s^{-\frac{1}{\alpha}}-1\right)\right|^p d\lambda(s)\right)^{\frac{1}{p}}$$

$$\geq \kappa\varepsilon \left(\frac{1}{t}\int_{(0,t]} \left|\left(s^{-\frac{1}{\alpha}}\right)\right|^p d\lambda(s)\right)^{\frac{1}{p}}$$

$$= \kappa\varepsilon \sqrt[p]{\frac{\alpha}{\alpha-p}}\, t^{-\frac{1}{\alpha}}.$$

□

19.2 Banach Function Spaces

We have already introduced the L_p-spaces to describe certain types of convergence of random variables or to obtain upper bounds for the tail-behavior of a random variable by the Markov-inequality, i.e. to measure how fast $\mathbb{P}(|f| > c)$ converges to zero if $0 < c \uparrow \infty$. For the understanding of certain phenomena and for stating correct assumptions the scale of L_p-spaces is not always sufficient. There are, among others, two more general classes of spaces, the class of Lorentz spaces and the class of Orlicz spaces. Both generalize the L_p-spaces. Instead of treating them separately, there is a more abstract class, the class of Banach function spaces. So we start to investigate this class first, and afterwards we show that Orlicz and Lorentz spaces belong to this class. The definition of a Banach function norm and the associated Banach function spaces is as follows:

Definition 19.2.1 (Banach Function Space) Let $(\Omega, \mathcal{F}, \mathbb{P})$ be a probability space, a map $\rho : \mathcal{L}_0(\Omega, \mathcal{F}, \mathbb{P}) \to [0, \infty]$ is called **Banach function norm** provided that for $f, g, f_n \in \mathcal{L}_0(\Omega, \mathcal{F}, \mathbb{P})$ and $a \in \mathbb{R}, a \neq 0$, one has

19.2 Banach Function Spaces

(FS1) $\rho(f) = \rho(|f|)$,
(FS2) $\rho(f) = 0$ if and only if $f = 0$ a.s.,
(FS3) $\rho(af) = |a|\rho(f)$,
(FS4) $\rho(f+g) \leq \rho(f) + \rho(g)$,
(FS5) $\rho(f) \leq \rho(g)$ if $0 \leq f \leq g$ a.s.,
(FS6) $\rho(f_n) \uparrow \rho(f)$ if $0 \leq f_n \uparrow f$ a.s.,
(FS7) $\frac{1}{c}\|f\|_{L_1} \leq \rho(f) \leq c\|f\|_{L_\infty}$.

Moreover, we let

$$\mathcal{M}_\rho(\Omega, \mathcal{F}, \mathbb{P}) := \{f \in \mathcal{L}_0(\Omega, \mathcal{F}, \mathbb{P}) : \rho(f) < \infty\},$$
$$M_\rho(\Omega, \mathcal{F}, \mathbb{P}) := \{[f] \in L_0(\Omega, \mathcal{F}, \mathbb{P}) : \rho([f]) := \rho(f) < \infty\}.$$

The space $[M_\rho(\Omega, \mathcal{F}, \mathbb{P}), \rho]$ is called **Banach function space**.

The axiom (FS2) says that we use the standard equivalence classes $f \sim g$ if and only if $\mathbb{P}(f = g) = 1$. Moreover, (FS7) implies that

$$L_\infty(\Omega, \mathcal{F}, \mathbb{P}) \subseteq M_\rho(\Omega, \mathcal{F}, \mathbb{P}) \subseteq L_1(\Omega, \mathcal{F}, \mathbb{P}).$$

As we can guess by the name, Banach function space, the above axioms ensure that we obtain a Banach space:

Theorem 19.2.2 *The Banach function space $[M_\rho(\Omega, \mathcal{F}, \mathbb{P}), \rho]$ is a Banach space.*

Proof

(a) The space $[M_\rho(\Omega, \mathcal{F}, \mathbb{P}), \rho]$ is a normed space: (FS2) gives $\rho(f) = 0$ if and only if $[f] = 0$. The conditions (FS2)–(FS4) yield

$$\rho(af) = |a|\rho(f) \quad \text{and} \quad \rho(f+g) \leq \rho(f) + \rho(g)$$

for $f, g \in \mathcal{M}_\rho(\Omega, \mathcal{F}, \mathbb{P})$ and $a \in \mathbb{R}$.

(b) To show that $[M_\rho(\Omega, \mathcal{F}, \mathbb{P}), \rho]$ is complete we use Lemma 4.2.5: We assume that $M := \sum_{n=1}^\infty \rho(f_n) < \infty$ and let

$$S_N := |f_1| + \cdots + |f_N| \quad \text{and} \quad S_\infty := \lim_{N \to \infty} S_N.$$

The sequence $(S_N)_{N \in \mathbb{N}}$ is nonnegative and nondecreasing with $\rho(S_N) \leq \rho(S_{N+1}) \leq M$ by (FS5) and (FS4). Let $A := \{S_\infty = \infty\}$. Then $\mathbb{P}(A) > 0$ would yield $\|S_N\|_{L_1} \uparrow \infty$ so that $\rho(S_N) \uparrow \infty$ by (FS7), which is a contradiction. Therefore, $\mathbb{P}(A) = 0$ and we set $g := S_\infty \mathbb{1}_{A^c} \in \mathcal{L}_0(\Omega, \mathcal{F}, \mathbb{P})$ and get $\rho(g) \leq M$ by (FS6). We also have that

$$h(\omega) := \sum_{n=1}^\infty f_n(\omega) \mathbb{1}_{A^c}(\omega) \in \mathbb{R}$$

exists and $|h| \leq g$ so that $\rho(h) = \rho(|h|) \leq \rho(g) \leq M$ by (FS1) and (FS5). Finally, using (FS2), (FS1), (FS5), (FS6), and (FS4) we obtain the desired convergence by

$$\rho\left(h - \sum_{n=1}^{N} f_n\right) = \rho\left(h - \sum_{n=1}^{N} f_n \mathbb{1}_{A^c}\right)$$
$$= \rho\left(\sum_{n=N+1}^{\infty} f_n \mathbb{1}_{A^c}\right)$$
$$\leq \rho\left(\sum_{n=N+1}^{\infty} |f_n|\right)$$
$$= \sup_{M \to \infty, M > N} \rho\left(\sum_{n=N+1}^{M} |f_n|\right)$$
$$\leq \sup_{M \to \infty, M > N} \sum_{n=N+1}^{M} \rho(|f_n|)$$
$$= \sum_{n=N+1}^{\infty} \rho(|f_n|) \to 0 \text{ as } N \to \infty.$$

□

Next we define the Lorentz spaces. They are formally based on the maximal function f^{**} and not on the nonincreasing rearrangement f^*. The reason for this is the triangle inequality

$$(f + g)^{**}(t) \leq f^{**}(t) + g^{**}(t),$$

which we do not have to our disposal for the nonincreasing rearrangement f^*. The definition of the Lorentz spaces inspects the behavior of f^{**} as follows:

Definition 19.2.3 (Lorentz[5] Space) Let $(\Omega, \mathcal{F}, \mathbb{P})$ be a probability space, $w : [0, 1] \to [0, 1]$ be a strictly increasing bijection, and $p \in [1, \infty]$. Then we let $f \in \mathcal{L}_p^w(\Omega, \mathcal{F}, \mathbb{P})$ provided that

$$\|f\|_{L_p^w} := \|wf^{**}\|_{L_p((0,1])} < \infty.$$

The space of the corresponding equivalence classes is denoted by $L_p^w(\Omega, \mathcal{F}, \mathbb{P})$.

---

[5] George Gunter Lorentz, 25/02/1910 (St. Petersburg, Russia)–01/01/2006 (Chico, California, USA).

19.2 Banach Function Spaces

Theorem 19.2.4 *For $p \in [1, \infty]$ the Lorentz space $[L_p^w(\Omega, \mathcal{F}, \mathbb{P}), \|\cdot\|_{L_p^w}]$ is a Banach function space.*

Proof (FS1) is evident as the definition of the nonincreasing rearrangement of f relies only on $|f|$.

(FS2) We have $\|wf^{**}\|_{L_p((0,1])} = 0$ if and only if $f^{**} = 0$ on $(0, 1]$ (here we use that $w > 0$ on $(0, 1]$). But $f^{**} = 0$ on $(0, 1]$ is equivalent to $f = 0$ a.s.

(FS3)–(FS5) follow directly from the corresponding properties of the L_p-spaces and Theorem 19.1.5.

(FS6) If $0 \leq f_n \uparrow f$ a.s., then $0 \leq f_n^{**} \uparrow f^{**}$ by Theorem 19.1.5(7). For $p \in [1, \infty)$ we conclude by monotone convergence (Theorem 8.2.2), where for $p = \infty$ we additionally use Proposition 9.3.2(2) to exploit monotone convergence as for $p < \infty$.

(FS7) follows from

$$\|f\|_{L_p^w} = \|wf^{**}\|_{L_p((0,1])} \geq f^{**}(1)\|w\|_{L_p((0,1])} = \|f\|_{L_1}\|w\|_{L_p((0,1])}$$

and from

$$\|f\|_{L_p^w} = \|wf^{**}\|_{L_p((0,1])} \leq \|f\|_{L_\infty}\|w\|_{L_p((0,1])}.$$

$\square$

Now we turn to the class of Orlicz spaces:

Definition 19.2.5 (Orlicz[6] Spaces) Let $(\Omega, \mathcal{F}, \mathbb{P})$ be a probability space and $\Phi : [0, \infty) \to [0, \infty)$ be a Young function, i.e. Φ is a nondecreasing and convex bijection. Then $f \in \mathcal{L}^\Phi(\Omega, \mathcal{F}, \mathbb{P})$ provided that

$$\|f\|_{L^\Phi} := \inf\left\{c > 0 : \mathbb{E}\Phi\left(\frac{|f|}{c}\right) \leq 1\right\}$$

with $\inf \emptyset := \infty$. The space of the corresponding equivalence classes is denoted by $L^\Phi(\Omega, \mathcal{F}, \mathbb{P})$.

Theorem 19.2.6 *The Orlicz space $[L^\Phi(\Omega, \mathcal{F}, \mathbb{P}), \|\cdot\|_{L^\Phi}]$ is a Banach function space.*

Proof

(a) The property (FS1) follows by definition, (FS2) can be seen as follows: If $f = 0$ a.s., then $\mathbb{E}\Phi\left(\frac{|f|}{c}\right) = 0$ for all $c > 0$ as $\Phi(0) = 0$ so that $\|f\|_{L^\Phi} = 0$. Conversely, $\|f\|_{L^\Phi} = 0$ implies that $\mathbb{E}\Phi\left(\frac{|f|}{c}\right) \leq 1$ for all $c > 0$. Assuming

---

[6] Władysław Orlicz, 24/05/1903 (Okocim, now Poland)–09/08/1990 (Poznań, Poland).

$\mathbb{P}(|f|>0)>0$ implies the existence of some $\gamma>0$ such that $\mathbb{P}(|f|>\gamma)>0$. Hence
$$\mathbb{E}\Phi\left(\frac{|f|}{c}\right) \geqslant \mathbb{E}\mathbb{1}_{\{|f|>\gamma\}}\Phi\left(\frac{\gamma}{c}\right) = \Phi\left(\frac{\gamma}{c}\right)\mathbb{P}(|f|>\gamma) \to \infty \quad \text{if} \quad c\downarrow 0.$$

But this is a contradiction so that $\mathbb{P}(|f|>0)=0$.

(b) Properties (FS3) and (FS5) are easy to see, let us turn to (FS4). Assume that $\mathbb{E}\Phi\left(\frac{|f|}{c}\right) \leqslant 1$ and $\mathbb{E}\Phi\left(\frac{|g|}{d}\right) \leqslant 1$ for $c,d>0$. Then by the convexity of Φ we get that

$$\begin{aligned}
\mathbb{E}\Phi\left(\frac{|f+g|}{c+d}\right) &\leqslant \mathbb{E}\Phi\left(\frac{|f|+|g|}{c+d}\right) \\
&= \mathbb{E}\Phi\left(\frac{c}{c+d}\frac{|f|}{c} + \frac{d}{c+d}\frac{|g|}{d}\right) \\
&\leqslant \frac{c}{c+d}\mathbb{E}\Phi\left(\frac{|f|}{c}\right) + \frac{d}{c+d}\mathbb{E}\Phi\left(\frac{|g|}{d}\right) \\
&\leqslant 1
\end{aligned}$$

so that $\|f+g\|_{L^\Phi} \leqslant \|f\|_{L^\Phi} + \|g\|_{L^\Phi}$.

(c) Now we prove (FS6): By changing f_n, f on null sets we may assume $f_n \uparrow f$ on Ω. It is obvious that we have
$$\lim_{n\to\infty}\|f_n\|_{L^\Phi} \leqslant \|f\|_{L^\Phi}. \tag{19.6}$$

If $\|f\|_{L^\Phi}=0$ or $\|f_n\|_{L^\Phi}=\infty$ for some $n\in\mathbb{N}$, then we have equality in (19.6) so that we exclude these cases for the further proof. For $\|f\|_{L^\Phi}<\infty$ and $\varepsilon\in(0,\|f\|_{L^\Phi})$ let us assume that
$$\lim_{n\to\infty}\|f_n\|_{L^\Phi} + \varepsilon < \|f\|_{L^\Phi}.$$

Then
$$\mathbb{E}\Phi\left(\frac{f_n}{\|f\|_{L^\Phi}-\varepsilon}\right) \leqslant 1$$

and, by monotone convergence,
$$\mathbb{E}\Phi\left(\frac{f}{\|f\|_{L^\Phi}-\varepsilon}\right) \leqslant 1.$$

But the latter is a contradiction to the definition of $\|f\|_{L^\Phi}$. Now assume that $\|f\|_{L^\Phi} = \infty$. This implies

$$\mathbb{E}\Phi\left(\frac{f}{c}\right) = \infty \quad \text{for all} \quad c > 0. \tag{19.7}$$

In fact, if $\mathbb{E}\Phi\left(\frac{f}{c_0}\right) < \infty$ for some $c_0 > 0$, then we would find by dominated convergence some $c > 0$ such that $\mathbb{E}\Phi\left(\frac{f}{c}\right) \leqslant 1$. Relation (19.7) yields

$$\lim_{n\to\infty} \mathbb{E}\Phi\left(\frac{|f_n|}{c}\right) = \infty \quad \text{for all} \quad c > 0$$

and $\lim_{n\to\infty} \|f_n\|_{L^\Phi} \geqslant c$ for all $c > 0$, so that $\lim_{n\to\infty} \|f_n\|_{L^\Phi} = \infty$.

(d) Finally, we prove (FS7): If $|f| \leqslant M$ a.s., then

$$\|f\|_{L^\Phi} \leqslant \inf\left\{c > 0 : \Phi\left(\frac{M}{c}\right) \leqslant 1\right\}$$
$$= M \inf\left\{\frac{c}{M} > 0 : \Phi\left(\frac{M}{c}\right) \leqslant 1\right\}$$
$$= M \inf\left\{c > 0 : \Phi\left(\frac{1}{c}\right) \leqslant 1\right\}$$
$$= M \frac{1}{\Phi^{-1}(1)},$$

so that $\|f\|_{L^\Phi} \leqslant \frac{1}{\Phi^{-1}(1)} \|f\|_{L^\infty}$. On the other hand, if $\mathbb{E}|f| < \infty$, then the convexity of Φ allows to apply Jensen's inequality (Theorem 8.12.3) to deduce

$$\|f\|_{L^\Phi} = \inf\left\{c > 0 : \mathbb{E}\Phi\left(\frac{|f|}{c}\right) \leqslant 1\right\}$$
$$\geqslant \inf\left\{c > 0 : \Phi\left(\frac{\mathbb{E}|f|}{c}\right) \leqslant 1\right\}$$
$$= \frac{1}{\Phi^{-1}(1)} \mathbb{E}|f|.$$

□

In the next step we compare Lorentz and Orlicz spaces in the case the distribution has an exponential tail. To do so, we let

$$\Phi_r(u) := e^{|u|^r} - 1 \quad \text{and}$$

$$w_r(t) := \sqrt[r]{\frac{\log 2}{\log 2 - \log t}} \quad \text{for} \quad r \in [1, \infty) \text{ and } t \in (0, 1]. \tag{19.8}$$

Proposition 19.2.7 *If $r \in [1, \infty)$, then one has that*

$$\sup_{p \in [1,\infty)} \frac{\|f\|_{L_p}}{\sqrt[r]{p}} \sim \sup_{t \in (0,1]} w_r(t) f^*(t) \sim \|f\|_{L_\infty^{w_r}} \sim \|f\|_{L^{\Phi_r}} \qquad (19.9)$$

in the sense when A and B are two of the four items, then one has that $A \sim_c B$, where $c = c_{(19.9),r} \geq 1$ depends at most on r.

The expression $A = \sup_{t \in (0,1]} w_r(t) f^*(t)$ describes the exponential tail behaviour of the random variable f. Indeed, using Exercise 3 the condition $A := \sup_{t \in (0,1]} w_r(t) f^*(t) \in (0, \infty)$ can be expressed as

$$\mathbb{P}(|f| > Ac) \leq 2^{1-c^r} \quad \text{for all} \quad c \geq 1.$$

Proof of Proposition 19.2.7 To shorten the notation we use $|f|_{w_r} := \sup_{t \in (0,1]} w_r(t) f^*(t)$ and

$$A_1 := \sup_{p \in [1,\infty)} \frac{\|f\|_{L_p}}{\sqrt[r]{p}}, \quad A_2 := \sup_{t \in (0,1]} w_r(t) f^*(t), \quad A_3 := \|f\|_{L_\infty^{w_r}}, \quad A_4 := \|f\|_{L^{\Phi_r}}.$$

(a) $A_2 \leq A_3$: Because of $f^* \leq f^{**}$ one has $|f|_{w_r} \leq \|f\|_{L_\infty^{w_r}}$.
(b) $A_3 \leq a_r A_2$: we observe that

$$\|f\|_{L_\infty^{w_r}} = \sup_{t \in (0,1]} \left\{ w_r(t) \frac{1}{t} \int_{(0,t]} f^*(s) d\lambda(s) \right\}$$

$$= \sup_{t \in (0,1]} \left\{ w_r(t) \frac{1}{t} \int_{(0,t]} \frac{w_r(s) f^*(s)}{w_r(s)} d\lambda(s) \right\}$$

$$\leq \sup_{t \in (0,1]} \left\{ \frac{1}{t} \int_{(0,t]} \frac{w_r(t)}{w_r(s)} d\lambda(s) \right\} |f|_{w_r}$$

$$= \sup_{t \in (0,1]} \int_{(0,1]} \sqrt[r]{1 + \frac{\log \frac{1}{s}}{\log \frac{2}{t}}} d\lambda(s) |f|_{w_r}$$

$$\leq \int_{(0,1]} \sqrt[r]{1 + \frac{\log \frac{1}{s}}{\log 2}} d\lambda(s) |f|_{w_r}.$$

(c) $A_1 \leq b_r A_2$: For $p \in [1, \infty)$ we get that

$$\mathbb{E}|f|^p = \int_{(0,\infty)} p \mathbb{P}(|f| > u) u^{p-1} d\lambda(u)$$

$$= \int_{(0,\infty)} p \lambda(f^* > u) u^{p-1} d\lambda(u)$$

19.2 Banach Function Spaces

$$\leq \int_{(0,\infty)} p\lambda\left(\frac{|f|_{w_r}}{w_r} > u\right) u^{p-1} d\lambda(u)$$

$$= |f|_{w_r}^p \int_{(0,\infty)} p\lambda\left(\frac{1}{w_r} > u\right) u^{p-1} d\lambda(u)$$

$$= |f|_{w_r}^p \int_{(0,1]} \frac{1}{|w_r(t)|^p} d\lambda(t),$$

where the first and last equality follow from Corollary 8.10.3 and Exercise 7 from Chap. 8. Hence $\|f\|_{L_p} \leq \left\|\frac{1}{w_r}\right\|_{L_p((0,1])} |f|_{w_r} \leq b_r \sqrt[r]{p} |f|_{w_r}$.

(d) $A_2 \leq eA_1$: Assuming some $c > 0$ for which one has $\|f\|_{L_p} \leq c\sqrt[r]{p}$ for all $p \in [1,\infty)$, we get for $u > 0$ that

$$\mathbb{P}(|f| > u) \leq \frac{1}{u^p} \|f\|_{L_p}^p \leq \frac{1}{u^p} c^p p^{\frac{p}{r}}.$$

For $u \geq ce$ we put $p := \left(\frac{u}{ce}\right)^r$ so that $p \geq 1$ and

$$\mathbb{P}(|f| > u) \leq e^{-p} = e^{-\left(\frac{u}{ce}\right)^r}.$$

Consequently, for $t \leq 1/e$ we get

$$f^*(t) \leq ce \sqrt[r]{\log \frac{1}{t}}$$

which implies that

$$f^*(t) \leq ce \frac{1}{w_r(t)} \quad \text{for} \quad t \in (0,1].$$

(e) $A_4 \leq c_r A_2$: We have that

$$\|f\|_{L^{\Phi_r}} = \inf\left\{c > 0 : \mathbb{E}\Phi_r\left(\frac{|f|}{c}\right) \leq 1\right\}$$

$$\leq \inf\left\{c > 0 : \int_{(0,1]} \Phi_r\left(\frac{1/w_r(t)|f|_{w_r}}{c}\right) d\lambda(t) \leq 1\right\}$$

$$= |f|_{w_r} \left\|\frac{1}{w_r}\right\|_{L^{\Phi_r}((0,1])} =: c_r |f|_{w_r}$$

with

$$c_r := \frac{1}{\sqrt[r]{\log 2}} \inf\left\{c > 0 : \int_{(0,1]} \left(\frac{2}{t}\right)^{1/c^r} d\lambda(t) \leq 2\right\} < \infty.$$

(f) $A_2 \leq \sqrt[r]{\log 2} A_4$: Finally, assume that $\|f\|_{L^{\Phi_r}} = 1$. Then $\mathbb{E}\Phi_r(|f|) \leq 1$ and, for $u > 0$ such that $e^{(u^r)} \geq 2$, we get

$$\mathbb{P}(|f| > u) = \mathbb{P}(\Phi_r(|f|) > \Phi_r(u)) \leq \frac{1}{\Phi_r(u)} = \frac{1}{e^{(u^r)} - 1} \leq 2e^{-u^r}, \quad (19.10)$$

where for $e^{(u^r)} < 2$ and $u \geq 0$ we have $\mathbb{P}(|f| > u) \leq 1 \leq 2e^{-u^r}$ as well. This implies for $t \in (0, 1]$ by setting $t = 2e^{-u^r}$, that

$$f^*(t) \leq \sqrt[r]{\log \frac{2}{t}} = \sqrt[r]{\log 2} \sqrt[r]{\frac{\log 2 - \log t}{\log 2}} = \frac{\sqrt[r]{\log 2}}{w_r(t)}.$$

$\square$

Example 19.2.8 Assume a probability space $(\Omega, \mathcal{F}, \mathbb{P})$ and independent Rademacher variables $r_1, \ldots, r_n : \Omega \to \{-1, 1\}$ with $\mathbb{P}(r_i = \pm 1) = 1/2$, then from Hoeffding's inequality (Theorem 8.12.10) we get for $u \geq 0$ and $a = (a_1, \ldots, a_n) \in \mathbb{R}^n$, $a \neq 0$, with the Euclidean norm $|a| = (\sum_{i=1}^n |a_i|^2)^{\frac{1}{2}}$ that

$$\mathbb{P}\left(\frac{|a_1 r_1 + \cdots + a_n r_n|}{|a|} \geq u\right) \leq 2e^{-\frac{u^2}{2}}.$$

Choosing $u := \sqrt{2 \log \frac{2}{t}} > 0$ for $t \in (0, 1]$ this inequality turns into

$$\mathbb{P}\left(\frac{|a_1 r_1 + \cdots + a_n r_n|}{|a|} \geq \sqrt{2 \log \frac{2}{t}}\right) \leq t$$

and

$$\left(\frac{|a_1 r_1 + \cdots + a_n r_n|}{|a|}\right)^*(t) \leq \frac{\sqrt{2 \log 2}}{w_2(t)}$$

where w_2 was defined in (19.8). By Proposition 19.2.7 ($r = 2$) this implies

$$\|a_1 r_1 + \cdots + a_n r_n\|_{L_p} \leq [\sqrt{2 \log 2} c_{(19.9),2}] \sqrt{p} |a| =: c_{(19.11)} \sqrt{p} |a| \quad (19.11)$$

for $p \in [2, \infty)$. Assume on the same probability space $(\Omega, \mathcal{F}, \mathbb{P})$ i.i.d. random variables $f_1, f_2, \ldots \in L^{\Phi_2}$ that are symmetric, i.e. the law of f_i coincides with the law of $-f_i$. Using (19.11), Corollary 18.2.2, and for the last inequality again Proposition 19.2.7 ($r = 2$), we conclude for $p \in [2, \infty)$ that

$$\left\|\frac{1}{n}\sum_{i=1}^{n} f_i\right\|_{L_p} = \left\|\frac{1}{n}\sum_{i=1}^{n} r_i f_i\right\|_{L_p} \leq c_{(19.11)}\frac{\sqrt{p}}{n}\left\|\left(\sum_{i=1}^{n} |f_i|^2\right)^{\frac{1}{2}}\right\|_{L_p}$$

$$\leq c_{(19.11)}\frac{\sqrt{p}}{n}\left(\sum_{i=1}^{n} \|f_i\|_{L_p}^2\right)^{\frac{1}{2}} = c_{(19.11)}\frac{\sqrt{p}}{\sqrt{n}}\|f_1\|_{L_p}$$

$$\leq c_{(19.11)} c_{(19.9),2}\frac{p}{\sqrt{n}}\|f_1\|_{L^{\Phi_2}}.$$

For $p \in [1, 2)$ we have that

$$\left\|\frac{1}{n}\sum_{i=1}^{n} f_i\right\|_{L_p} \leq \left\|\frac{1}{n}\sum_{i=1}^{n} f_i\right\|_{L_2} \leq 2c_{(19.11)} c_{(19.9),2}\frac{p}{\sqrt{n}}\|f_1\|_{L^{\Phi_2}}.$$

Using Proposition 19.2.7 for $r = 1$ we obtain under the assumption $f_1 \in L^{\Phi_2}$ that

$$\left\|\frac{1}{n}\sum_{i=1}^{n} f_i\right\|_{L^{\Phi_1}} \leq 2c_{(19.11)} c_{(19.9),2}\, c_{(19.9),1}\frac{1}{\sqrt{n}}\|f_1\|_{L^{\Phi_2}} =: c_{(19.12)}\frac{1}{\sqrt{n}}\|f_1\|_{L^{\Phi_2}}. \tag{19.12}$$

Now we apply this estimate to get information about the accuracy of the Monte-Carlo method. Assume a random variable $g : \Omega \to \mathbb{R}$ such that we know that g has 'good tails', i.e. $g \in L^{\Phi_2}$, that g is symmetric w.r.t. its mean, i.e. the laws of $g - \mathbb{E}g$ and $-g + \mathbb{E}g$ coincide, and such that w.l.o.g. $\|g - \mathbb{E}g\|_{L^{\Phi_2}} = 1$. To estimate $m := \mathbb{E}g$ we take independent copies $g_1, g_2, \ldots : \Omega \to \mathbb{R}$ of g and obtain from the above computation together with (19.10) the accuracy of the Monte-Carlo method

$$\mathbb{P}\left(\left|\frac{g_1 + \cdots + g_n}{n} - m\right| > u\right) \leq 2e^{-\frac{\sqrt{n}u}{c_{(19.12)}}} \quad \text{for} \quad u \geq 0.$$

19.3 Exercises

Ex 1: For $0 < p < 1$ and a quasi-normed space $[E, \|\cdot\|_E]$ define

$$\|x\|_{E,p} := \inf\left\{\left(\sum_{i=1}^{n} \|x_i\|_E^p\right)^{\frac{1}{p}} : x = x_1 + \cdots + x_n, n \in \mathbb{N}\right\}.$$

Prove that $\|x+y\|_{E,p}^P \leq \|x\|_{E,p}^P + \|y\|_{E,p}^P$ for all $x, y \in E$.

Ex 2: Let $(\Omega, \mathcal{F}, \mathbb{P})$ be a probability space and define for $f \in \mathcal{L}_0(\Omega, \mathcal{F})$ the quantity

$$\|f\|_{L_1^{weak}} := \sup_{\varepsilon > 0} \varepsilon \mathbb{P}(|f| \geq \varepsilon).$$

Prove that

$$\|f + g\|_{L_1^{weak}} \leq 2[\|f\|_{L_1^{weak}} + \|g\|_{L_1^{weak}}].$$

Next, assume that $\|\cdot\|$ is a norm on $L_1^{weak}((0,1], \mathcal{B}((0,1]), \lambda)$ such that

$$\frac{1}{c}\|f\| \leq \|f\|_{L_1^{weak}} \leq c\|f\| \quad \text{for all} \quad f \in L_1^{weak}((0,1], \mathcal{B}((0,1]), \lambda)$$

for some $c \geq 1$.

(a) Prove that this would imply

$$\|f_1 + \cdots + f_n\|_{L_1^{weak}} \leq c^2[\|f_1\|_{L_1^{weak}} + \cdots + \|f_n\|_{L_1^{weak}}]$$

for all $n \in \mathbb{N}$ and $f_1, \ldots, f_n \in L_1^{weak}((0,1], \mathcal{B}((0,1]), \lambda)$.

(b) Consider $f_1(t) := 1/t$ and let $f_2, \ldots, f_n$ be period shifts by $1/n$, i.e. $f_{i+1}(t) := f_i(S_n t)$ for $i = 1, \ldots, n-1$, where

$$S_n(t) := \begin{cases} t + \frac{1}{n} & : t + \frac{1}{n} \leq 1 \\ t + \frac{1}{n} - 1 & : t + \frac{1}{n} > 1 \end{cases}.$$

Observe that $\|f_i\|_{L_1^{weak}} = 1$ for $i = 1, \ldots, n$.

(c) Prove that the inequality

$$\|f_1 + \cdots + f_n\|_{L_1^{weak}} \leq c^2 n$$

does not hold uniformly in n. This implies that there is no such norm $\|\cdot\|$ on $L_1^{weak}((0,1], \mathcal{B}((0,1]), \lambda)$.

Ex 3: Assume a strictly decreasing continuous bijection $\psi : [1, \infty) \to (0, 1]$. Given a probability space $(\Omega, \mathcal{F}, \mathbb{P})$ and a random variable $f : \Omega \to \mathbb{R}$, prove that the following assertions are equivalent:

(a) $\mathbb{P}(|f| > c) \leq \psi(c)$ for all $c \geq 1$.
(b) $f^*(t) \leq \psi^{-1}(t)$ for all $t \in (0, 1]$.

Ex 4: Verify Proposition 19.1.10.

Ex 5: Assume a probability space $(\Omega, \mathcal{F}, \mathbb{P})$ and $p \in (0, \infty)$. Prove for $t \in (0, 1]$ that $(f^*)^p(t) = (|f|^p)^*(t)$ and deduce that $\sqrt[p]{(|f|^p)^{**}(t)} = f_p^{**}(t)$, where f_p^{**} is defined in (19.2).

19.4 Comments

Section 19.1: The nonincreasing rearrangement is treated in the book of Hardy, Littlewood, and Pólya [79] with references to Steiner [179] and Schwarz [170]. Theorem 19.1.7 is due to Hardy and Littlewood [78]. The K-functional from Definition 19.1.9 was introduced by Peetre [145]. For a general study of the nonincreasing rearrangement and maximal function the reader is referred to [9]. The relation of the nonincreasing rearrangement f^* to the left-inverse, generalized inverse, or quantile function of the distribution function of $|f|$ is explained in [53]. Theorem 19.1.12 can be found in [70]. Lemma 19.1.13 goes back to [71](proof of Lemma 3.2) and can be also found in [95](Lemma 3).

Section 19.2: Banach function spaces have been rigorously introduced and investigated in the doctoral thesis of Luxemburg [126], for a recent survey see [125]. Orlicz spaces have been defined by Orlicz in [140]. Lorentz spaces have been defined by Lorentz in [123, 124]. Proposition 19.2.7 formulizes results appearing in different forms and contexts in the literature, for instance in the theory of high dimensional probability, see for example [192, Chapter 2].

Chapter 20
Probability in Banach Spaces

Random variables with values in a Banach space generalize $\mathbb{R}^d$-valued random variables and have their fixed place in probability theory and applications. There are parts of the theory of Banach space valued random variables where the statements are essentially the same as for $\mathbb{R}^d$-valued random variables, but there are important differences as well. Here we deal with both aspects.

First we show Pettis' measurability theorem about the equivalence of certain measurability concepts. Then the Bochner, Pettis, and Dunford integrals are introduced, which generalize the Lebesgue integral to different degrees. After that the strong law of large numbers and the Itô-Nisio theorem for Banach space valued random variables are shown. The chapter concludes with the construction of the Brownian motion as a $C([0, 1])$-valued random variable using the results of Ciesielski about Hölder functions. Here Banach function spaces are used to describe the modulus of continuity.

20.1 Banach Space Valued Measurable Maps

Given a Banach space E, where we exclude in this chapter the case $E = \{0\}$ to avoid degenerated situations. The Borel σ-algebra $\mathcal{B}(E)$ is defined to be the smallest σ-algebra generated by the norm-open sets in E. This is the σ-algebra from Definition 5.1.1. Moreover, given $x \in E$ and $\varepsilon > 0$ we denote by

$$U_\varepsilon(x) := \{y \in E : \|y - x\|_E < \varepsilon\}$$

the open ball around x with radius ε. We need an extension of the notion of a simple function:

Definition 20.1.1 Given a measurable space $(\Omega, \mathcal{F})$, we say that $f : \Omega \to E$ is a σ-**simple function** provided that f is measurable and the set $\{f(\omega) : \omega \in \Omega\}$ is countable. The representation

$$f(\omega) = \sum_{i \in I} x_i \mathbb{1}_{A_i}(\omega) \quad \text{for} \quad \omega \in \Omega$$

is called **canonical** if I is countable, $\Omega = \bigcup_{i \in I} A_i$ with $\emptyset \neq A_i \in \mathcal{F}$ is a partition, and $x_i \neq x_j$ if $i \neq j$.

The definition of σ-simple is an adaptation of the definition of a simple function from Definition 6.1.1. Here again σ stands for countable. In order to approximate a measurable function by a σ-simple function we introduce the map

$$\Phi_\varepsilon^D : E \to E.$$

Definition 20.1.2 Given a dense set $D = \{d_l : l \in \mathbb{N}\} \subseteq E$ and $\varepsilon > 0$, we define

$$B_{\varepsilon,1} := U_\varepsilon(d_1) \quad \text{and} \quad B_{\varepsilon,l} := U_\varepsilon(d_l) \setminus (B_{\varepsilon,1} \cup \cdots \cup B_{\varepsilon,l-1}) \quad \text{for} \quad l \geq 2.$$

If $B_{\varepsilon,l} \neq \emptyset$, then we choose and fix an element $x_{\varepsilon,l} \in B_{\varepsilon,l}$ and define

$$\Phi_\varepsilon^D(x) := \sum_{\{l \in \mathbb{N} : B_{\varepsilon,l} \neq \emptyset\}} x_{\varepsilon,l} \mathbb{1}_{B_{\varepsilon,l}}(x).$$

By construction we obtain a $(\mathcal{B}(E), \mathcal{B}(E))$-measurable function $\Phi_\varepsilon^D : E \to E$ such that

$$\|\Phi_\varepsilon^D(x) - x\|_E \leq 2\varepsilon \quad \text{for all} \quad x \in E.$$

Our first statement connects the strong and the weak measurability to each other:

Theorem 20.1.3 (Pettis[1] Measurability Theorem) *For a separable Banach space E, a countable norming subset $A \subseteq B_{E^*}$, a measurable space $(\Omega, \mathcal{F})$, and a map $f : \Omega \to E$ the following assertions are equivalent:*

(1) **Strong measurability:** *f is $(\mathcal{F}, \mathcal{B}(E))$-measurable.*
(2) **Pointwise approximation:** *There are σ-simple $f_n : \Omega \to E$, $n \in \mathbb{N}$, with*

$$\lim_{n \to \infty} f_n(\omega) = f(\omega) \quad \text{for all} \quad \omega \in \Omega.$$

(3) **Weak measurability I:** *The map $\omega \mapsto \langle f(\omega), a \rangle$ is $(\mathcal{F}, \mathcal{B}(\mathbb{R}))$-measurable for all $a \in E^*$.*
(4) **Weak measurability II:** *The map $\omega \mapsto \langle f(\omega), a \rangle$ is $(\mathcal{F}, \mathcal{B}(\mathbb{R}))$-measurable for all $a \in A$.*

---

[1] Billy James Pettis, 1913–14/04/1979 (USA).

20.1 Banach Space Valued Measurable Maps

In the case the conditions are satisfied, the σ-simple $f_n : \Omega \to E$ can be chosen such that $\lim_{n\to\infty}[\sup_{\omega\in\Omega} \|f(\omega) - f_n(\omega)\|_E] = 0$.

Proof (1) $\Rightarrow$ (2) For a dense set $D = \{d_l : l \in \mathbb{N}\} \subseteq E$ we choose $f_n(\omega) := \Phi_{\frac{1}{n}}^D(f(\omega))$ so that $\|f_n(\omega) - f(\omega)\|_E \leqslant 2/n$ for all $\omega \in \Omega$.
(2) $\Rightarrow$ (3) As each $\langle f_n, a\rangle$ is obviously measurable and $|\langle f_n, a\rangle - \langle f, a\rangle| \leqslant \|a\|_{E^*}(2/n)$ we may conclude by Proposition 6.1.5.
(3) $\Rightarrow$ (4) is obvious.
(4) $\Rightarrow$ (1) Given $x \in E$ and $\varepsilon > 0$, we have to show that $f^{-1}(U_\varepsilon(x)) \in \mathcal{F}$. For this we write

$$U_\varepsilon(x) = \bigcup_{n=1}^{\infty} \bigcap_{a\in A} \{y \in E : |\langle y, a\rangle - \langle x, a\rangle| \leqslant 1 - \varepsilon_n\} \tag{20.1}$$

for a fixed sequence $0 < \varepsilon_n \uparrow \varepsilon$. Now (4) implies that $f^{-1}(\{y \in E : |\langle y, a\rangle - \langle x, a\rangle| \leqslant 1 - \varepsilon_n\}) \in \mathcal{F}$ so that $f^{-1}(U_\varepsilon(x)) \in \mathcal{F}$. By Proposition 5.1.3 and Lemma 6.2.3 assertion (1) follows. □

As a corollary we extend properties known for real-valued measurable maps to E-valued measurable maps if E is separable:

Corollary 20.1.4 *Assume a separable Banach space E, a measurable space $(\Omega, \mathcal{F})$, and measurable maps $f, g, f_1, f_2, \ldots : \Omega \to E$.*

(1) *If $h(\omega) := \lim_{n\to\infty} f_n(\omega)$ converges in E for all $\omega \in \Omega$, then $h : \Omega \to E$ is measurable.*
(2) *$f + g : \Omega \to E$ is a measurable map.*

Proof Both statements follow from Theorem 20.1.3(3) as the statements are true for real-valued random variables. For example the assumption in (1) implies that $\langle h(\omega), a\rangle = \lim_{n\to\infty}\langle f_n(\omega), a\rangle$ for all $a \in E^*$, so that the map $\omega \mapsto \langle h(\omega), a\rangle$ is measurable for all $a \in \mathbb{R}$. Hence Theorem 20.1.3(3) implies that $h : \Omega \to E$ is measurable as well. □

Next we show that weak uniqueness of probability measures implies their strong uniqueness:

Theorem 20.1.5 *Assume a separable Banach space E and a norming subset $A \subseteq B_{E^*}$. Define $\mathrm{Lin}(A)$ to be the set of finite linear combinations of elements from A. Then, for probability measures μ, ν on $\mathcal{B}(E)$, the following assertions are equivalent:*

(1) **Strong uniqueness:** $\mu = \nu$.
(2) **Weak uniqueness I:** *For all $a \in \mathrm{Lin}(A)$ one has $\mu_a = \nu_a$, where $\mu_a, \nu_a \in \mathcal{M}_1(\mathbb{R})$ are the laws of μ, ν with respect to the map $a : E \to \mathbb{R} \in E^*$.*
(3) **Weak uniqueness II:** *For all $a \in \mathrm{Lin}(A)$ one has*

$$\int_E e^{i\langle x,a\rangle}\,d\mu(x) = \int_E e^{i\langle x,a\rangle}\,d\nu(x).$$

Proof (1) $\Rightarrow$ (2) is obvious, so we prove (2) $\Rightarrow$ (1): As Π-system we take the family of all cylinder-sets

$$\{x \in E : \langle x, a_1\rangle \in B_1, \ldots, \langle x, a_n\rangle \in B_n\}$$

with $n \in \mathbb{N}$, $B_1, \ldots, B_n \in \mathcal{B}(\mathbb{R})$, and $a_1, \ldots, a_n \in A$. Since for $c_1, \ldots, c_n \in \mathbb{R}$ and $a_1, \ldots, a_n \in A$ we have that $\sum_{i=1}^n c_i a_i \in \text{Lin}(A)$, by assumption all linear combinations $\sum_{i=1}^n c_i \langle x, a_i\rangle$, have the same distribution under μ and ν. This implies

$$\int_E e^{i\sum_{i=1}^n c_i \langle x,a_i\rangle}\,d\mu(x) = \int_E e^{i\sum_{i=1}^n c_i \langle x,a_i\rangle}\,d\nu(x).$$

Interpreted as a function of $(c_1, \ldots, c_n) \in \mathbb{R}^n$ we get by Theorem 11.5.5 that $E \ni x \mapsto (\langle x, a_1\rangle, \ldots, \langle x, a_n\rangle) \in \mathbb{R}^n$ has the same distribution under μ and ν. It follows from (20.1) that $U_\varepsilon(x) \in \sigma(\Pi)$. As each open set of a separable Banach space can be represented as a union of countable many open balls according to Proposition 4.1.4, we deduce that $\mathcal{B}(E) \subseteq \sigma(\Pi)$. And as the cylinder sets belong to $\mathcal{B}(E)$, we have that $\sigma(\Pi) = \mathcal{B}(E)$ and finally $\mu = \nu$ by Theorem 3.2.4.
(2) $\Leftrightarrow$ (3) follows from Theorem 11.5.5 as well. □

Now we return to Corollary 20.1.4 and show in Theorem 20.1.6 that for non-separable Banach spaces item (2) fails to be true in general. For example, the space $E := \{f : \mathbb{R} \to \mathbb{R} : \|f\| := \sup_{x \in \mathbb{R}} |f(x)| < \infty\}$ is a space where Theorem 20.1.6 applies to.

Theorem 20.1.6 *Let E be a Banach space with $\#E \geq \#2^{\mathbb{R}}$. Define $\Omega := E \times E$ and $\mathcal{F} := \mathcal{B}(E) \otimes \mathcal{B}(E)$, and let $f, g : \Omega \to E$ be the $(\mathcal{F}, \mathcal{B}(E))$-measurable maps defined by*

$$f((x, y)) := x \quad \text{and} \quad g((x, y)) := y.$$

Then $f - g$ is not $\mathcal{F}$-measurable.

For the proof we need two lemmas and start with a basic observation from measure theory:

Lemma 20.1.7 *Let Ω be a nonempty set and $\mathcal{G} \subseteq 2^{\Omega}$ be a nonempty family of subsets of Ω. Then*

$$\mathcal{F} := \bigcup_{\mathcal{C} \subseteq \mathcal{G} \text{ is countable}} \sigma(\mathcal{C})$$

is a σ-algebra and $\mathcal{F} = \sigma(\mathcal{G})$.

Proof Let $(A_i)_{i\in\mathbb{N}} \subseteq \mathcal{F}$. So for each $i \in \mathbb{N}$ there are countable $\mathcal{C}_i \subseteq \mathcal{G}$ and $G_{i,n} \in \mathcal{C}_i$, $i, n \in \mathbb{N}$, such that $A_i \in \sigma(G_{i,n} : n \in \mathbb{N}) \subseteq \sigma(G_{j,n} : j, n \in \mathbb{N})$. But $\{G_{j,n} : j, n \in \mathbb{N}\} \subseteq \mathcal{G}$ is countable again, so that $\bigcup_{i=1}^{\infty} A_i \in \mathcal{F}$ by definition of $\mathcal{F}$. If $A \in \mathcal{F}$, then $A^c \in \mathcal{F}$ is immediate. Moreover, $\emptyset, \Omega \in \mathcal{F}$ is obvious. As for any $G \in \mathcal{G}$ we can choose the system $\mathcal{C} = \{G\}$, we get $\mathcal{G} \subseteq \mathcal{F}$ and $\sigma(\mathcal{G}) \subseteq \mathcal{F}$. Finally, $\sigma(\mathcal{C}) \subseteq \sigma(\mathcal{G})$ whenever $\mathcal{C} \subseteq \mathcal{G}$, so that $\mathcal{F} \subseteq \sigma(\mathcal{G})$ holds as well. □

From this lemma we get:

Lemma 20.1.8 *Let Ω be a nonempty set and $\mathcal{G} \subseteq 2^{\Omega}$ be a nonempty system of subsets of Ω. Then for any $A \in \sigma(\mathcal{G})$ there exists $\mathcal{C} \subseteq \mathcal{G}$ such that $\#\mathcal{C} \leqslant \#\mathbb{N}$ and $A \in \sigma(\mathcal{C})$.*

Proof of Theorem 20.1.6 First we observe that

$$\{f - g = 0\} = \{f = g\} = \{(x, x) : x \in E\} =: \Delta_E.$$

Let us assume that $\Delta_E \in \mathcal{B}(E) \otimes \mathcal{B}(E)$. We let $\mathcal{G} \subseteq 2^{\Omega}$ be the system of $A_1 \times A_2$ with $A_1, A_2 \in \mathcal{B}(E)$. By Lemma 20.1.8 there exists a countable family $\{A_{1,n} \times A_{2,n} : n \in \mathbb{N}\} \subseteq \mathcal{G}$ such that

$$\Delta_E \in \sigma(A_{1,n} \times A_{2,n} : n \in \mathbb{N}).$$

For $\{A_n : n \in \mathbb{N}\} := \{A_{1,n} : n \in \mathbb{N}\} \cup \{A_{2,n} : n \in \mathbb{N}\}$ we get that

$$\Delta_E \in \sigma(A_n : n \in \mathbb{N}) \otimes \sigma(A_n : n \in \mathbb{N}),$$

and for all $x \in E$ that

$$\{x\} = \{y \in E : (x, y) \in \Delta_E\} \in \mathcal{E} := \sigma(A_n : n \in \mathbb{N}).$$

Hence, $\#\mathcal{E} \geqslant \#E \geqslant \#2^{\mathbb{R}}$. However, by construction, we also have $\#\mathcal{E} \leqslant \#\mathbb{R}$ by Theorem B.3.1. Hence we obtain $\#2^{\mathbb{R}} \leqslant \#\mathcal{E} \leqslant \#\mathbb{R}$ and $\#2^{\mathbb{R}} \leqslant \#\mathbb{R}$—which is a contradiction to Cantor's Theorem. □

20.2 The Bochner Integral

In this section we introduce the Bochner[2] integral which is one option to generalize the Lebesgue integral from Sect. 8.1 to random variables with values in a separable Banach space E. We start with the Bochner-Lebesgue spaces:

---

[2] Salomon Bochner, 20/08/1899 (Podgorze, near Kraków, now Poland)–02/05/1982 (Houston, Texas, USA).

Definition 20.2.1 (Bochner-Lebesgue Spaces) For $p \in (0, \infty]$ and a separable Banach space E we let

$$\mathcal{L}_p(\Omega, \mathcal{F}, \mathbb{P}; E) := \{f : \Omega \to E \text{ measurable with } \|f\|_{L_p(E)} := \|\|f\|_E\|_{L_p} < \infty\}.$$

The equivalence relation on $\mathcal{L}_p(\Omega, \mathcal{F}, \mathbb{P}; E)$ is given by $f \sim g$ if $\mathbb{P}(f = g) = 1$. The corresponding space of equivalence classes $[f]$ is denoted by $L_p(\Omega, \mathcal{F}, \mathbb{P}; E)$ and equipped with $\|[f]\|_{L_p(E)} := \|f\|_{L_p(E)}$.

We obtain a general version of Theorem 9.3.5:

Theorem 20.2.2 *For a separable Banach space E the space*

$$\left[L_p(\Omega, \mathcal{F}, \mathbb{P}; E), \|\cdot\|_{L_p(E)}\right]$$

is a Banach space if $p \in [1, \infty]$, a p-normed Banach space if $p \in (0, 1)$, and a quasi-Banach space if $p \in (0, \infty]$.

Proof The properties of a norm, a p-norm, and a quasi-norm, respectively, are verified by Minkowski's inequality (Theorem 8.12.9). In particular, we have that $\|[f]\|_{L_p(E)} = 0$ if and only if $\mathbb{P}(\|f\|_E = 0) = 1$ which is equivalent to $\mathbb{P}(f = 0) = 1$.

Completeness for $p \in (0, \infty)$: We let $q := \min\{p, 1\}$ and assume a Cauchy-sequence $(f_n)_{n=1}^\infty$ with respect to $\|\cdot\|_{L_p(E)}$ so that for all $\varepsilon, \eta > 0$ there is an $n(\varepsilon, \eta) \in \mathbb{N}$ such that

$$\mathbb{P}(\|f_n - f_m\|_E^q \geq \varepsilon) = \mathbb{P}(\|f_n - f_m\|_E^p \geq \varepsilon^{\frac{p}{q}}) \leq \varepsilon^{-\frac{p}{q}} \mathbb{E}\|f_n - f_m\|_E^p \leq \eta$$

for $n \geq m \geq n(\varepsilon, \eta)$. As in the proof of Theorem 9.2.4(2) we find a subsequence $1 \leq n_1 < n_2 < \cdots$ such that

$$\mathbb{P}(\Omega_0) = 1 \quad \text{with} \quad \Omega_0 := \left\{\omega \in \Omega : \sum_{j=1}^\infty \|f_{n_{j+1}}(\omega) - f_{n_j}(\omega)\|_E^q < \infty\right\}.$$

So we may define

$$f(\omega) := \begin{cases} f_{n_1}(\omega) + \sum_{j=1}^\infty (f_{n_{j+1}}(\omega) - f_{n_j}(\omega)) & : \omega \in \Omega_0 \\ 0 & : \omega \notin \Omega_0 \end{cases},$$

where we use that E is a complete metric space under the metric $d_{E,q}(x, y) := \|x - y\|_E^q$ (see Theorem 4.2.2(4)). Then, by Fatou's lemma,

$$\mathbb{E}\|f - f_m\|_E^p = \mathbb{E}\lim_{k \to \infty} \|f_{n_k} - f_m\|_E^p \leq \liminf_{k \to \infty} \mathbb{E}\|f_{n_k} - f_m\|_E^p \leq \varepsilon^p \quad \text{for} \quad m \geq m(\varepsilon).$$

20.2 The Bochner Integral

Completeness for $p = \infty$: For all $L \in \mathbb{N}$ there is an $n(L) \in \mathbb{N}$ such that one has ess $\sup_{\omega \in \Omega} \|f_n(\omega) - f_m(\omega)\|_E < \frac{1}{L}$ for $n \geq m \geq n(L)$. Therefore there is a null set $N(n, m, L) \in \mathcal{F}$ such that

$$\sup_{\omega \notin N(n,m,L)} \|f_n(\omega) - f_m(\omega)\|_E < \frac{1}{L} \quad \text{for all} \quad n \geq m \geq n(L).$$

For the null set $N := \bigcup_{n \geq m \geq n(L), L \in \mathbb{N}} N(n, m, L)$ and $\omega \in \Omega \setminus N$ this implies

$$\|f_n(\omega) - f_m(\omega)\|_E < \frac{1}{L} \quad \text{for all} \quad n \geq m \geq n(L) \quad \text{and} \quad L \in \mathbb{N}.$$

Again, as a Banach space is complete, we may define $f(\omega) := \lim_{n \to \infty} f_n(\omega)$ for $\omega \notin N$, and $f(\omega) = 0$ otherwise. For $m \geq n(L)$ and $\omega \notin N$ this implies that

$$\|f(\omega) - f_m(\omega)\|_E \leq \frac{1}{L} \quad \text{so that} \quad \lim_{m \to \infty} \|f - f_m\|_{L_\infty(E)} = 0.$$

$\square$

With the next two lemmas we prepare the definition of the Bochner integral:

Lemma 20.2.3 *For $f \in \mathcal{L}_1(\Omega, \mathcal{F}, \mathbb{P}; E)$ there are σ-simple functions $f_n : \Omega \to E$ such that $\lim_{n \to \infty} \|f - f_n\|_{L_1(E)} = 0$.*

Proof In Theorem 20.1.3 we constructed σ-simple functions $f_n : \Omega \to E$ such that $\sup_{\omega \in \Omega} \|f(\omega) - f_n(\omega)\|_E \to 0$ as $n \to \infty$ which implies $\|f - f_n\|_{L_1(E)} \leq \sup_{\omega \in \Omega} \|f(\omega) - f_n(\omega)\|_E \to 0$ as $n \to \infty$. $\square$

Lemma 20.2.4

(1) For a σ-simple function $f : \Omega \to E$ with canonical representation $f = \sum_{i \in I} x_i \mathbb{1}_{A_i}$ and $\int_\Omega \|f\|_E d\mathbb{P} < \infty$ one has $\int_\Omega \|f\|_E d\mathbb{P} = \sum_{i \in I} \|x_i\|_E \mathbb{P}(A_i)$, so that the sum

$$\mathbb{E}f := \sum_{i \in I} x_i \mathbb{P}(A_i) \in E$$

converges in E, does not depend on the order of summation, and one has

$$\|\mathbb{E}f\|_E \leq \mathbb{E}\|f\|_E.$$

(2) For $f \in \mathcal{L}_1(\Omega, \mathcal{F}, \mathbb{P}; E)$ assume two sequences of σ-simple functions $(f_n)_{n=1}^\infty$ and $(f_n')_{n=1}^\infty$ with

$$\lim_{n \to \infty} \|f - f_n\|_{L_1(E)} = \lim_{n \to \infty} \|f - f_n'\|_{L_1(E)} = 0.$$

Then $\lim_{n\to\infty} \mathbb{E} f_n \in E$ and $\lim_{n\to\infty} \mathbb{E} f'_n \in E$ exist and coincide. Denoting this limit by $\mathbb{E} f$, we have

$$\|\mathbb{E} f\|_E \leqslant \mathbb{E}\|f\|_E.$$

Proof
(1) As the existence of $\mathbb{E} f$ is obvious, we only show the inequality which follows from

$$\|\mathbb{E} f\|_E = \left\|\sum_{i\in I} x_i \mathbb{P}(A_i)\right\|_E \leqslant \sum_{i\in I} \|x_i\|_E \mathbb{P}(A_i) = \mathbb{E}\|f\|_E < \infty.$$

(2) The sequence $(f_n)_{n=1}^\infty \subseteq L_1(\Omega, \mathcal{F}, \mathbb{P}; E)$ is a Cauchy sequence and we have that

$$\|\mathbb{E} f_n - \mathbb{E} f_m\|_E \leqslant \|f_n - f_m\|_{L_1(E)}$$

by (1). Therefore $(\mathbb{E} f_n)_{n=1}^\infty \subseteq E$ is a Cauchy sequence with a limit $x := \lim_{n\to\infty} \mathbb{E} f_n$. Similarly, we have $x' := \lim_{n\to\infty} \mathbb{E} f'_n$. Finally,

$$\|x - x'\|_E = \lim_{n\to\infty} \|\mathbb{E} f_n - \mathbb{E} f'_n\|_E \leqslant \limsup_{n\to\infty} \mathbb{E}\|f_n - f'_n\|_E$$

$$\leqslant \limsup_{n\to\infty} \mathbb{E}\|f_n - f\|_E + \limsup_{n\to\infty} \mathbb{E}\|f - f'_n\|_E = 0.$$

Moreover, we have that

$$\|\mathbb{E} f\|_E = \lim_{n\to\infty} \|\mathbb{E} f_n\|_E \leqslant \limsup_{n\to\infty} \mathbb{E}\|f_n\|_E$$

$$\leqslant \mathbb{E}\|f\|_E + \limsup_{n\to\infty} \mathbb{E}\|f_n - f\|_E = \mathbb{E}\|f\|_E.$$

□

The above consideration yields to the introduction of the Bochner integral:

Definition 20.2.5 (Bochner Integral) For $f \in \mathcal{L}_1(\Omega, \mathcal{F}, \mathbb{P}; E)$ and σ-simple functions $f_n : \Omega \to E$ such that $\lim_{n\to\infty} \|f - f_n\|_{L_1(E)} = 0$ we let

$$\mathbb{E} f := \lim_{n\to\infty} \mathbb{E} f_n.$$

Proposition 20.2.6 *For $f \in \mathcal{L}_1(\Omega, \mathcal{F}, \mathbb{P})$ the Bochner and the Lebesgue integral coincide.*

Proof We find σ-simple $f_n : \Omega \to \mathbb{R}$ such that on $\{f \geqslant 0\}$ one has $0 \leqslant f_n \uparrow f$ and on $\{f \leqslant 0\}$ one has $0 \leqslant f_n \downarrow f$. By dominated convergence we get for the

Lebesgue integral

$$\lim_n \mathbb{E} f_n = \mathbb{E} f$$

which is also the definition of the Bochner integral. □

20.3 The Dunford and the Pettis Integral

The Bochner integral is defined as a limit in the Banach space E by Definition 20.2.5. Knowing the existence, it can be characterized by linear functionals $a \in E^*$:

Proposition 20.3.1 *Let E be a separable Banach space, $f \in \mathcal{L}_1(\Omega, \mathcal{F}, \mathbb{P}; E)$, and $x \in E$. Then the following assertions are equivalent:*

(1) $x = \mathbb{E} f$.
(2) *For all $a \in E^*$ one has $\mathbb{E}\langle f, a \rangle = \langle x, a \rangle$.*

Proof

(a) First we prove that

$$\langle \mathbb{E} f, a \rangle = \mathbb{E}\langle f, a \rangle \quad \text{for all} \quad a \in E^*.$$

For this we find a sequence of σ-simple $f_n : \Omega \to E$ such that

$$\lim_{n \to \infty} \sup_{\omega \in \Omega} \|f_n(\omega) - f(\omega)\|_E = 0 \quad \text{and hence} \quad \lim_{n \to \infty} \mathbb{E}\|f_n - f\|_E = 0.$$

This implies

$$|\mathbb{E}\langle f_n, a \rangle - \mathbb{E}\langle f, a \rangle| \leq \mathbb{E}|\langle f_n, a \rangle - \langle f, a \rangle| \leq \|a\|_{E^*} \mathbb{E}\|f_n - f\|_E \to 0 \text{ as } n \to \infty$$

and

$$\langle \mathbb{E} f, a \rangle = \langle \lim_{n \to \infty} \mathbb{E} f_n, a \rangle = \lim_{n \to \infty} \langle \mathbb{E} f_n, a \rangle = \lim_{n \to \infty} \mathbb{E}\langle f_n, a \rangle = \mathbb{E}\langle f, a \rangle. \tag{20.2}$$

Now (1) $\Rightarrow$ (2) follows immediately. Conversely, assume (2). By (20.2) we get that

$$\langle x - \mathbb{E} f, a \rangle = 0 \quad \text{for all} \quad a \in E^*.$$

The Hahn-Banach theorem (Theorem 4.4.2) states that there is an $a_0 \in B_{E^*}$ such that $\langle x - \mathbb{E} f, a_0 \rangle = \|x - \mathbb{E} f\|_E$. Now since for all $a \in E^*$ we have $\langle x - \mathbb{E} f, a \rangle = 0$, we conclude that $\|x - \mathbb{E} f\|_E = 0$ and hence $x = \mathbb{E} f$.

□

The above proposition enables different notions of integrability of Banach space valued random variables, where we are in the convenient setting that the Banach spaces are assumed to be separable so that strong and weak measurability coincide.

Definition 20.3.2 (Dunford[3] Integrability) A random variable $f : \Omega \to E$ is **Dunford integrable** (or weakly integrable) provided that for all $a \in E^*$ one has that $\mathbb{E}|\langle f, a \rangle| < \infty$.

Dunford integrability implies a formally stronger property:

Lemma 20.3.3 *The following assertions are equivalent:*

(1) *f is Dunford integrable.*
(2) *f is Dunford integrable and there is a constant $c \geqslant 0$ such that*

$$\mathbb{E}|\langle f, a \rangle| \leqslant c \|a\|_{E^*} \quad \text{for all} \quad a \in E^*.$$

Under these conditions there exits a unique $x^{**} \in E^{**}$ such that

$$\mathbb{E}\langle f, a \rangle = \langle a, x^{**} \rangle \quad \text{for all} \quad a \in E^*,$$

where $E^{**} := (E^*)^*$ stands for the bidual of E.

Proof We only need to prove (1) $\Rightarrow$ (2).

(a) We define the linear map $T : E^* \to L_1(\Omega, \mathcal{F}, \mathbb{P})$ with $Ta := \langle f, a \rangle$. To show that this map is closed we assume $a_n \to a$ in E^* and $Ta_n \to g$ in $L_1(\Omega, \mathcal{F}, \mathbb{P})$. Then we get that

$$\lim_{n \to \infty} (Ta_n)(\omega) = \lim_{n \to \infty} \langle f(\omega), a_n \rangle = \langle f(\omega), a \rangle = (Ta)(\omega) \quad \text{for all} \quad \omega \in \Omega.$$

Therefore, $Ta = g$ a.s. so that the map is closed and therefore continuous by the closed graph theorem (Theorem 4.4.4).

(b) From (a), by the continuity of T, we see that

$$|\mathbb{E}\langle f, a \rangle| \leqslant c \|a\|_{E^*}$$

which is assertion (2). In other words, we obtain a continuous linear map $E^* \ni a \mapsto \mathbb{E}\langle f, a \rangle \in \mathbb{R}$, i.e. an element from E^{**}.

□

Definition 20.3.4 (Dunford Integral) The element $x^{**} \in E^{**}$ is called **Dunford integral** and we write $x^{**} = \mathbb{E}f$.

---

[3] Nelson Dunford, 12/12/1906 (St. Louis, Missouri, USA)–07/09/1986 (Sarasota, Florida, USA).

20.3 The Dunford and the Pettis Integral

Directly by definition of the Dunford integral it follows that for $A \in \mathcal{F}$ and a Dunford integrable random variable $f : \Omega \to E$ the map $f\mathbb{1}_A$ is Dunford integrable as well. Next we define the Pettis integral:

Definition 20.3.5 A Dunford integrable random variable $f : \Omega \to E$ is **Pettis integrable** provided that for all $A \in \mathcal{F}$ the Dunford integral $\mathbb{E}f\mathbb{1}_A$ belongs to E for all $A \in \mathcal{F}$.

The role of the indicator function can be exemplary explained as follows: Assume that f is Dunford integrable, not Pettis integrable, but symmetric. In this case one would always get that $\mathbb{E}f = 0 \in E$ although there are A with $\mathbb{E}f\mathbb{1}_A \notin E$.

With the above definitions we have the relations

$$\text{Bochner integrable} \implies \text{Pettis integrable} \implies \text{Dunford integrable},$$

where the first relation is Proposition 20.3.1. The following example shows that these notions are different from each other in general:

Example 20.3.6 In Exercise 1 at the end of this chapter the Banach spaces c_0, ℓ_1 and ℓ_∞ are defined and the task is to show that assuming $E := c_0$ then $E^* = \ell_1$ and $E^{**} = \ell_\infty$. As probability space we choose $(\Omega, \mathcal{F}, \mathbb{P}) = (\mathbb{N}, 2^{\mathbb{N}}, \mathbb{P})$ with

$$\mathbb{P} := \sum_{n \in \mathbb{N}} p_n \delta_n, \quad \text{where} \quad p_n \in (0, 1) \quad \text{and} \quad \sum_{n \in \mathbb{N}} p_n = 1.$$

Further we assume $0 < \xi_1 \leqslant \xi_2 \leqslant \cdots$ and define

$$x_n := (\xi_1, \ldots, \xi_n, 0, 0, , \ldots) \in c_0.$$

As random variable $f : \mathbb{N} \to c_0$ we use $f(n) := x_n$ (recall that for $\mathcal{F} = 2^{\mathbb{N}}$ any map $f : \mathbb{N} \to c_0$ is Borel measurable). First we verify the following table:

	Integral	Condition	Value of the integral	
(A)	Bochner	$\sum_{n=1}^{\infty} p_n \xi_n < \infty$	$\mathbb{E}f = \left(\xi_k (\sum_{n \geqslant k} p_n)\right)_k$	$\in c_0$
(B)	Pettis	$\lim_{k \to \infty} \xi_k (\sum_{n=k}^{\infty} p_n) = 0$	$\mathbb{E}f\mathbb{1}_A = \left(\xi_k (\sum_{\substack{n \in A \\ n \geqslant k}} p_n)\right)_k$	$\in c_0$
(C)	Dunford	$\sup_{k \in \mathbb{N}} \xi_k (\sum_{n=k}^{\infty} p_n) < \infty$	$\mathbb{E}f = \left(\xi_k (\sum_{n \geqslant k} p_n)\right)_k$	$\in \ell_\infty$

(C) The random variable f is Dunford integrable if and only if there is a constant $c \geqslant 0$ such that

$$\left| \sum_{n=1}^{\infty} p_n \left(\sum_{k=1}^{n} \xi_k \alpha_k \right) \right| = |\mathbb{E}\langle f, a \rangle| \leqslant c \|a\|_{\ell_1} = c \sum_{k=1}^{\infty} |\alpha_k| \quad \text{for} \quad a = (\alpha_k)_{k=1}^{\infty} \in \ell_1.$$

One sees that the left hand side gets larger if all α_k are nonnegative or nonpositive. So we assume that $\alpha_k \geqslant 0$ and get the condition

$$\sum_{k=1}^{\infty} \alpha_k \left[\xi_k \left(\sum_{n=k}^{\infty} p_n\right)\right] \leqslant c \sum_{k=1}^{\infty} \alpha_k \quad \text{for} \quad a = (\alpha_k)_{k=1}^{\infty} \in \ell_1.$$

But this holds if and only if

$$\sup_{k \in \mathbb{N}} \left[\xi_k \left(\sum_{n=k}^{\infty} p_n\right)\right] \leqslant c. \tag{20.3}$$

The Dunford integral $\mathbb{E}f = (\eta_k)_{k=1}^{\infty} \in \ell_\infty$ is characterized by

$$\sum_{k=1}^{\infty} \alpha_k \left[\xi_k \left(\sum_{n=k}^{\infty} p_n\right)\right] = \sum_{k=1}^{\infty} \alpha_k \eta_k \quad \text{for all} \quad a = (\alpha_k)_{k=1}^{\infty} \in \ell_1$$

which gives

$$\eta_k = \xi_k \sum_{n=k}^{\infty} p_n.$$

(B) To verify the Pettis integrability we assume (20.3) and a nonempty set $A \in \mathcal{F}$. Then

$$\int_A \langle f, a \rangle \mathrm{d}\mathbb{P} = \sum_{n \in A} p_n \left(\sum_{k=1}^{n} \xi_k \alpha_k\right) = \sum_{k=1}^{\infty} \alpha_k \left[\xi_k \left(\sum_{n \in A, n \geqslant k} p_n\right)\right]$$

so that the Dunford integrals equals

$$\mathbb{E}f \mathbb{1}_A = \left(\xi_k \left(\sum_{n \in A, n \geqslant k} p_n\right)\right)_{k=1}^{\infty}.$$

The condition that $\mathbb{E}f \mathbb{1}_A \in c_0$ for all $A \in \mathcal{F}$ is simply $\lim_{k \to \infty} \eta_k = 0$.
(A) We have that f is Bochner integrable if and only if

$$\mathbb{E}\|f\|_{c_0} = \sum_{n=1}^{\infty} p_n \xi_n < \infty.$$

Now we look for some examples to see that conditions (A), (B), and (C) are different. For this purpose we let $p_n := 1/n - 1/(n+1)$ so that

$$\sum_{n=k}^{\infty} p_n = \frac{1}{k}.$$

For $\xi_k := k$ the Dunford integral equals

$$(1, 1, 1, \ldots) \in \ell_\infty \setminus c_0,$$

so that we do not have Pettis integrability. If $\xi_k := k/\log(k+1)$, which satisfies $0 < \xi_1 \leq \xi_2 \leq \cdots$, then

$$\lim_{k \to \infty} \xi_k \sum_{n=k}^{\infty} p_n = \lim_{k \to \infty} \frac{k}{\log(k+1)} \frac{1}{k} = 0.$$

On the other hand

$$\sum_{n=1}^{\infty} p_n \xi_n = \sum_{n=1}^{\infty} \left(\frac{1}{n} - \frac{1}{n+1} \right) \frac{n}{\log(n+1)} = \sum_{n=1}^{\infty} \frac{1}{(n+1)\log(n+1)} = \infty.$$

So we have an example that the Pettis integral exists, but not the Bochner integral.

20.4 Strong Law of Large Numbers in Separable Banach Spaces

We proved the strong law of large numbers (SLLN) for real-valued random variables. By considering the problem coordinate-wise one can extend the SLLN easily to $\mathbb{R}^d$-valued random variables. We will go one step ahead and prove the SLLN for random variables with values in a separable Banach space. Given a separable Banach space E, the independence of random variables $f_1, f_2, \ldots : \Omega \to E$ is defined as before, i.e. for all $n \in \mathbb{N}$ and all $B_1, \ldots, B_n \in \mathcal{B}(E)$ one has

$$\mathbb{P}(f_1 \in B_1, \ldots, f_n \in B_n) = \mathbb{P}(f_1 \in B_1) \cdots \mathbb{P}(f_n \in B_n).$$

The Banach space valued SLLN reads as follows:

Theorem 20.4.1 *Let E be a separable Banach space and let $f_1, f_2, \ldots : \Omega \to E$ be i.i.d. random variables with $\mathbb{E}\|f_1\|_E < \infty$. Then one has*

$$\lim_{n \to \infty} \frac{1}{n}(f_1 + \cdots + f_n) = \mathbb{E} f_1 \text{ a.s.}$$

Proof

(a) Assume that $f_1, f_2, \ldots : \Omega \to \{x_i : i \in I\} \subseteq E$ where the x_i are pairwise distinct and $I = \mathbb{N}$ or $I = \{1, \ldots, L\}$. If I is finite, then the real-valued SLLN implies that

$$\lim_{n \to \infty} \left\| \frac{1}{n}(f_1 + \cdots + f_n) - \mathbb{E} f_1 \right\|_E = 0 \text{ a.s.}$$

Now we assume that $I = \mathbb{N}$. For $L \in \mathbb{N}$ define $f_n^L := x_i$ if $f_n = x_i$ and $i \leqslant L$, otherwise we set f_n^L to zero. We have that $\mathbb{E} \|f_n^L\|_E \leqslant \mathbb{E} \|f_n\|_E$. Again by the real-valued SLLN we deduce that

$$\lim_{n \to \infty} \left\| \frac{1}{n}(f_1^L + \cdots + f_n^L) - \mathbb{E} f_1^L \right\|_E = 0 \quad \text{a.s.}$$

For $D_n^L := f_n - f_n^L$ the real-valued SLLN also implies

$$\lim_{n \to \infty} \frac{1}{n}(\|D_1^L\|_E + \cdots + \|D_n^L\|_E) = \mathbb{E} \|D_1^L\|_E \quad \text{a.s.}$$

Because $\lim_{L \to \infty} \|D_1^L(\omega)\|_E = 0$ for all $\omega \in \Omega$ and $\|D_1^L(\omega)\|_E \leqslant \|f_1(\omega)\|_E$ we get by dominated convergence that

$$\lim_{L \to \infty} \mathbb{E} \|D_1^L\|_E = 0.$$

Given $\varepsilon > 0$ we choose $L(\varepsilon) \in \mathbb{N}$ such that

$$\mathbb{E} \|D_1^{L(\varepsilon)}\|_E \leqslant \frac{\varepsilon}{4}.$$

Now we observe that

$$\left\| \frac{f_1 + \cdots + f_n}{n} - \mathbb{E} f_1 \right\|_E \leqslant \left\| \frac{f_1^{L(\varepsilon)} + \cdots + f_n^{L(\varepsilon)}}{n} - \mathbb{E} f_1^{L(\varepsilon)} \right\|_E$$

$$+ \frac{\|D_1^{L(\varepsilon)}\|_E + \cdots + \|D_n^{L(\varepsilon)}\|_E}{n} + \|\mathbb{E} f_1^{L(\varepsilon)} - \mathbb{E} f_1\|_E.$$

Since for $I = \{1, \ldots, L(\varepsilon)\}$ the SLLN holds there is a set $\Omega_{L(\varepsilon)}$ of measure one such that for all $\omega \in \Omega_{L(\varepsilon)}$ there is an $n(\omega, \varepsilon) \in \mathbb{N}$ such that for $n \geqslant n(\omega, \varepsilon)$ one has that

$$\left\| \frac{f_1^{L(\varepsilon)}(\omega) + \cdots + f_n^{L(\varepsilon)}(\omega)}{n} - \mathbb{E} f_1^{L(\varepsilon)} \right\|_E \leqslant \frac{\varepsilon}{4},$$

20.4 SLLN in Separable Banach Spaces

$$\frac{\|D_1^{L(\varepsilon)}(\omega)\|_E + \cdots + \|D_n^{L(\varepsilon)}(\omega)\|_E}{n} \leqslant \mathbb{E}\|D_1^{L(\varepsilon)}\|_E + \frac{\varepsilon}{4}.$$

Moreover,

$$\|\mathbb{E}f_1^{L(\varepsilon)} - \mathbb{E}f_1\|_E \leqslant \mathbb{E}\|D_1^{L(\varepsilon)}\|_E.$$

Combining the estimates gives for $n \geqslant n(\omega, \varepsilon)$ and $\omega \in \Omega_{L(\varepsilon)}$ that

$$\left\|\frac{f_1(\omega) + \cdots + f_n(\omega)}{n} - \mathbb{E}f_1\right\|_E \leqslant \frac{\varepsilon}{2} + 2\mathbb{E}\|D_1^{L(\varepsilon)}\|_E \leqslant \varepsilon.$$

For $\varepsilon_N := \frac{1}{N}$, $N \in \mathbb{N}$, and $\Omega_0 := \bigcap_{N=1}^{\infty} \Omega_{L(\frac{1}{N})}$ we get $\mathbb{P}(\Omega_0) = 1$ and

$$\lim_{n \to \infty} \left\|\frac{f_1(\omega) + \cdots + f_n(\omega)}{n} - \mathbb{E}f_1\right\|_E = 0 \quad \text{for} \quad \omega \in \Omega_0.$$

(b) By part (a) we have for some dense countable set $D \subseteq E$ and $\Phi_{\frac{1}{N}}^D$, $N \in \mathbb{N}$, from Definition 20.1.2 that

$$\lim_{n \to \infty} \left\|\frac{\Phi_{\frac{1}{N}}^D(f_1)(\omega) + \cdots + \Phi_{\frac{1}{N}}^D(f_n)(\omega)}{n} - \mathbb{E}\Phi_{\frac{1}{N}}^D(f_1)\right\|_E = 0 \quad \text{a.s.}$$

Moreover, we have

$$\|\mathbb{E}f_1 - \mathbb{E}\Phi_{\frac{1}{N}}^D(f_1)\|_E \leqslant \frac{2}{N}$$

and

$$\left\|\frac{f_1(\omega) + \cdots + f_n(\omega)}{n} - \frac{\Phi_{\frac{1}{N}}^D(f_1)(\omega) + \cdots + \Phi_{\frac{1}{N}}^D(f_n)(\omega)}{n}\right\|_E \leqslant \frac{2}{N}.$$

Therefore,

$$\limsup_{n \to \infty} \left\|\frac{f_1 + \cdots + f_n}{n} - \mathbb{E}f_1\right\|_E \leqslant \frac{4}{N} \quad \text{a.s.}$$

Since this is true for all $N \in \mathbb{N}$, the theorem follows.

$\square$

20.5 Maximal Inequalities

By maximal inequalities one aims to estimate the distribution of certain suprema of random variables. Maximal inequalities have various applications. In this section we prove two maximal inequalities which belong to the family of Lévy-Ottaviani-inequalities. We shall use them in the proofs of the Itô-Nisio theorem and of the law of iterated logarithm.

Theorem 20.5.1 *Let $a : E \to \mathbb{R}$ be measurable and subadditive, i.e. $a(x + y) \leq a(x) + a(y)$ for all $x, y \in E$. Assume independent random variables $f_1, \ldots, f_N : \Omega \to E$, $S_n := f_1 + \cdots + f_n$, $n = 1, \ldots, N$, and $s, t \geq 0$. Then one has that*

$$\mathbb{P}\left(\max_{n=1,\ldots,N} a(S_n) > s + t\right) \leq \frac{\mathbb{P}(a(S_N) > t)}{1 - \max_{n=1,\ldots,N} \mathbb{P}(a(S_n - S_N) > s)}$$

if $\max_{n=1,\ldots,N} \mathbb{P}(a(S_n - S_N) > s) < 1$.

Proof We use

$$\left\{\max_{n=1,\ldots,N} a(S_n) > s + t\right\} = \bigcup_{j=1}^{N} \left\{a(S_j) > s + t, \max_{n=1,\ldots,j-1} a(S_n) \leq s + t\right\}$$

where we interpret $\{a(S_1) > s+t, \max_{n=1,\ldots,1-1} a(S_n) \leq s+t\}$ as $\{a(S_1) > s+t\}$ and note that the sets are disjoint. This implies that

$$\left[\min_{n=1,\ldots,N} \mathbb{P}(a(S_n - S_N) \leq s)\right] \mathbb{P}\left(\max_{n=1,\ldots,N} a(S_n) > s + t\right)$$

$$\leq \sum_{j=1}^{N} \left[\mathbb{P}\left(a(S_j) > s + t, \max_{n=1,\ldots,j-1} a(S_n) \leq s + t\right) \mathbb{P}(a(S_j - S_N) \leq s)\right]$$

$$= \sum_{j=1}^{N} \mathbb{P}\left(a(S_j) > s + t, a(S_j - S_N) \leq s, \max_{n=1,\ldots,j-1} a(S_n) \leq s + t\right)$$

$$\leq \mathbb{P}(a(S_N) > t)$$

where we use the independence of $\{a(S_j) > s + t, \max_{n=1,\ldots,j-1} a(S_n) \leq s + t\}$ and $\{a(S_j - S_N) \leq s\}$. For the last inequality we exploit the subadditivity $a(S_j) - a(S_j - S_N) \leq a(S_N)$ which implies

$$\{a(S_j) > s + t, a(S_j - S_N) \leq s\} \subseteq \{a(S_N) > t\}.$$

□

20.5 Maximal Inequalities

One can deduce from Theorem 20.5.1 the following inequality (20.4), but with the constant 3 instead of 2 on the right-hand side. To get the better constant 2 we use the adaptation of Szewczak:

Theorem 20.5.2 *Let $a : E \to [0, \infty)$ be measurable, subadditive ($a(x + y) \leq a(x) + a(y)$ for all $x, y \in E$), and symmetric ($a(x) = a(-x)$ for all $x \in E$). Assume independent random variables $f_1, \ldots, f_N : \Omega \to E$, $S_n := f_1 + \cdots + f_n$, $n = 1, \ldots, N$, and $t \geq 0$. Then one has that*

$$\mathbb{P}\left(\max_{n=1,\ldots,N} a(S_n) > 3t\right) \leq 2 \max_{n=1,\ldots,N} \mathbb{P}(a(S_n) > t). \tag{20.4}$$

Proof For $n = 1, \ldots, N - 1$ we let

$$C_n := \{a(S_N - S_n) > s + t, a(S_N - S_{n+1}) \leq s + t, \ldots, a(S_N - S_{N-1}) \leq s + t\}.$$

Using that the sets C_n are pairwise disjoint, we get

$$\mathbb{P}\left(\max_{n=1,\ldots,N} a(S_n) > s + 2t\right)$$

$$\leq \mathbb{P}(a(S_N) > t) + \mathbb{P}\left(\max_{n=1,\ldots,N-1} a(S_n) > s + 2t, a(S_N) \leq t\right)$$

$$\leq \mathbb{P}(a(S_N) > t) + \mathbb{P}\left(\max_{n=1,\ldots,N-1} a(S_N - S_n) > s + t, a(S_N) \leq t\right)$$

$$= \mathbb{P}(a(S_N) > t) + \sum_{n=1}^{N-1} \mathbb{P}(C_n \cap \{a(S_N) \leq t\})$$

$$\leq \mathbb{P}(a(S_N) > t) + \sum_{n=1}^{N-1} \mathbb{P}(C_n \cap \{a(S_n) > s\})$$

$$= \mathbb{P}(a(S_N) > t) + \sum_{n=1}^{N-1} \mathbb{P}(C_n)\mathbb{P}(a(S_n) > s)$$

$$\leq \mathbb{P}(a(S_N) > t) + \max_{n=1,\ldots,N-1} \mathbb{P}(a(S_n) > s).$$

For the second and third inequality we used that a is subadditive and symmetric to get $a(S_N - S_n) \geq a(S_n) - a(S_N) > s + 2t - t = s + t$ and $a(S_n) \geq a(S_N - S_n) - a(S_N) > s + t - t = s$, respectively. Then, for example, the second inequality follows from

$$\left\{\max_{n=1,\ldots,N-1} a(S_n) > s + 2t, a(S_N) \leqslant t\right\}$$
$$\subseteq \left\{\max_{n=1,\ldots,N-1} a(S_N - S_n) > s + t, a(S_N) \leqslant t\right\},$$

and the third inequality can be shown similarly.

Finally, choosing $s = t$ yields to

$$\mathbb{P}\left(\max_{n=1,\ldots,N} a(S_n) > 3t\right) \leqslant 2 \max_{n=1,\ldots,N} \mathbb{P}(a(S_n) > t).$$

□

20.6 The Itô-Nisio Theorem

The following theorem due to Itô[4] and Nisio[5] generalizes results of P. Lévy[6] obtained for real-valued random variables to Banach space valued random variables:

Theorem 20.6.1 *For a separable Banach space E, a norming subset $A \subseteq B_{E^*}$, independent random variables $f_1, f_2, \ldots : \Omega \to E$, $S_n := f_1 + \cdots + f_n$, and a random variable $S : \Omega \to E$ we consider the following conditions:*

(1) $S_n \xrightarrow{a.s.} S$.
(2) $\langle S_n, a \rangle \xrightarrow{a.s.} \langle S, a \rangle$ for all $a \in A$.
(3) $S_n \xrightarrow{\mathbb{P}} S$.
(4) $\langle S_n, a \rangle \xrightarrow{\mathbb{P}} \langle S, a \rangle$ for all $a \in A$.
(5) For all $a \in A$ one has $\lim_{n \to \infty} \mathbb{E} e^{i\langle S_n, a \rangle} = \mathbb{E} e^{i\langle S, a \rangle}$.

Then one has the implications

$$(1) \iff (3) \implies (2) \iff (4) \iff (5).$$

If the random variables f_n are symmetric, then one has (4) $\Rightarrow$ (3).

Proof The implications (1) $\Rightarrow$ (3), (2) $\Rightarrow$ (4) $\Rightarrow$ (5), (1) $\Rightarrow$ (2), and (3) $\Rightarrow$ (4) are obvious. The equivalence (2) $\Leftrightarrow$ (4) is a special case of (1) $\Leftrightarrow$ (3) (by thinking of $E = \mathbb{R}$). The remaining implications are verified as follows:

(3) $\Rightarrow$ (1) As in Proposition 7.4.1 we need to verify

---

[4] Kiyosi Itô, 07/09/1915 (Hokusei-cho, now Inabe, Mie Prefecture, Japan)–10/11/2008 (Kyoto, Japan).
[5] Makiko Nisio, born 16/01/1931 (Kobe, Japan).
[6] Paul Pierre Lévy, 15/09/1886 (Paris, France)–15/12/1971 (Paris, France).

20.6 The Itô-Nisio Theorem

$$\lim_{n\to\infty} \mathbb{P}\left(\sup_{k\geq n} \|S_k - S_n\|_E > \varepsilon\right) = 0 \quad \text{for all} \quad \varepsilon > 0.$$

By Theorem 20.5.2 for $a(x) := \|x\|_E$ we have that

$$\mathbb{P}\left(\sup_{k\geq n} \|S_k - S_n\|_E > 3\varepsilon\right)$$

$$= \lim_{\substack{N\to\infty \\ N>n}} \mathbb{P}\left(\sup_{N\geq k\geq n} \|S_k - S_n\|_E > 3\varepsilon\right)$$

$$\leq 2 \lim_{\substack{N\to\infty \\ N>n}} \max_{k=n,\ldots,N} \mathbb{P}(\|S_k - S_n\|_E > \varepsilon)$$

$$\leq 2 \lim_{\substack{N\to\infty \\ N>n}} \max_{k=n,\ldots,N} \left[\mathbb{P}\left(\|S - S_n\|_E > \frac{\varepsilon}{2}\right) + \mathbb{P}\left(\|S - S_k\|_E > \frac{\varepsilon}{2}\right)\right]$$

$$\leq 4 \sup_{k\geq n} \mathbb{P}\left(\|S - S_k\|_E > \frac{\varepsilon}{2}\right) \xrightarrow{n\to\infty} 0.$$

(5) $\Rightarrow$ (4) We fix $a \in A$ and assume that (4) does not hold. Then we find an $\varepsilon \in (0, 1)$ and $1 \leq n_1 < m_1 < n_2 < m_2 < \cdots$ such that

$$\mathbb{P}(|\langle S_{m_k}, a\rangle - \langle S_{n_k}, a\rangle| > \varepsilon) > \varepsilon. \tag{20.5}$$

Assumption (5), Theorem 11.2.6(1), and Lemma 12.2.1 imply that the laws of $\langle S_{m_k}, a\rangle$ are tight, i.e. for all $\eta \in (0, 1)$ there is a compact set $K^\eta \subseteq \mathbb{R}$ such that $\mathbb{P}(\langle S_{m_k}, a\rangle \in K^\eta) \geq 1 - \frac{\eta}{2}$. Hence

$$\mathbb{P}(\langle S_{m_k}, a\rangle - \langle S_{n_k}, a\rangle \notin K^\eta - K^\eta) \leq \mathbb{P}(\langle S_{m_k}, a\rangle \notin K^\eta) + \mathbb{P}(\langle S_{n_k}, a\rangle \notin K^\eta) \leq \eta,$$

where $K^\eta - K^\eta := \{x - y : x, y \in K^\eta\}$. Therefore by Prokhorov's Theorem 12.1.5 there is a further subsequence such that the laws of $(\langle S_{m_{k_l}}, a\rangle - \langle S_{n_{k_l}}, a\rangle)_{l=1}^\infty$ converge weakly to a measure $\mu \in \mathcal{M}_1^+(\mathbb{R})$. Hence we have that the law of $\langle S_{m_{k_l}}, a\rangle - \langle S_{n_{k_l}}, a\rangle + \langle S_{n_{k_l}}, a\rangle$ converges weakly to $\mu * \text{law}(\langle S, a\rangle)$ and also to $\text{law}(\langle S, a\rangle)$. This implies that $\mu = \delta_0$ which is a contradiction to (20.5).

(4) $\Rightarrow$ (3) First we observe that the laws of S and $S - 2S_n$ coincide by the symmetry of the f_n as we have

$$\mathbb{E}e^{i\langle S,a\rangle} = \mathbb{E}e^{i\langle S-S_n,a\rangle}\mathbb{E}e^{i\langle S_n,a\rangle} = \mathbb{E}e^{i\langle S-S_n,a\rangle}\mathbb{E}e^{i\langle -S_n,a\rangle} = \mathbb{E}e^{i\langle S-2S_n,a\rangle}$$

for all $a \in E^*$. Here we use in the first equality that $\langle S - S_n, a \rangle$ and $\langle S_n, a \rangle$ are independent which follows from (4) and the independence of $\langle S_m - S_n, a \rangle$ and $\langle S_n, a \rangle$ for $m > n$. Therefore,

$$\mathbb{P}\left(S_n \notin \frac{1}{2}(K - K)\right) \leqslant \mathbb{P}(S \notin K) + \mathbb{P}(2S_n - S \notin -K) = 2\mathbb{P}(S \notin K)$$

for any compact set $K \subseteq E$. So the sequence $(S_n)_{n=1}^\infty$ is tight as S is tight and $K - K$ is compact. This implies that $(S_n - S)_{n=1}^\infty$ is tight. Suppose that (3) does not hold. This implies the existence of an $\varepsilon \in (0, 1)$, $\eta > 0$, and a subsequence $(n_k)_{k=1}^\infty$ such that

$$\mathbb{P}(\|S_{n_k} - S\|_E > \eta) > \varepsilon.$$

Now, as $(S_n - S)_{n=1}^\infty$ is tight, we find a compact set K such that

$$\mathbb{P}(S_n - S \in K) \geqslant 1 - \frac{\varepsilon}{2}.$$

Hence $\mathbb{P}(\|S_{n_k} - S\|_E > \eta, S_{n_k} - S \in K) > \frac{\varepsilon}{2}$. We find a finite cover of K of open balls of radius $\frac{\eta}{2}$. This implies the existence of one closed ball B not containing 0 of radius $\frac{\eta}{2}$ and an $\delta > 0$ such that

$$\mathbb{P}(S_{n_{k_j}} - S \in K \cap B) \geqslant \delta$$

for some further subsequence $1 \leqslant k_1 < k_2 < \cdots$. The separation Theorem 4.4.3 applied to $A := \{0\}$ and B as defined before implies the existence of an $a \in E^*$ such that $\langle y, a \rangle \geqslant 1$ for all $y \in B$. Hence for $\omega \in \{S_{n_{k_j}} - S \in K \cap B\}$ we have $\langle (S_{n_{k_j}} - S)(\omega), a \rangle \geqslant 1$ so that (4) does not hold. $\square$

20.7 Ciesielski's Theorem and the Wiener Process

Definition 20.7.1 A process $W = (W_t)_{t \in [0,1]}$ on a probability space $(\Omega, \mathcal{F}, \mathbb{P})$ is called a Wiener process provided that the following holds:

(1) The map $W : \Omega \to C([0, 1])$ is a random variable.
(2) For all $N \in \mathbb{N}$ and $0 \leqslant t_1 < \cdots < t_N \leqslant 1$ the random variable $(W_{t_1}, \ldots, W_{t_N}) : \Omega \to \mathbb{R}^N$ is Gaussian.
(3) For $s, t \in [0, 1]$ one has that $\mathbb{E} W_s W_t = \min\{s, t\}$.

The image measure of W on $(C([0, 1]), \mathcal{B}(C([0, 1])))$ is called the Wiener measure.

20.7 Ciesielski's Theorem and the Wiener Process

There are several constructions of the Wiener[7] measure on the separable Banach space $C([0, 1])$. We present an approach going back to Ciesielski[8] using Schauder[9] functions. For this we use the definition of Schauder functions from Appendix A.9: First one defines the Haar[10] functions $h_0, h_{pm} : [0, 1] \to \mathbb{R}$ by $h_0 :\equiv 1$ and

$$h_{pm}(t) := \begin{cases} 1 & : t \in \left[\frac{m-1}{2^p}, \frac{2m-1}{2^{p+1}}\right) \\ -1 & : t \in \left[\frac{2m-1}{2^{p+1}}, \frac{m}{2^p}\right) \\ 0 & : \text{otherwise} \end{cases}$$

for $p = 0, 1, 2, \ldots$ and $m = 1, \ldots, 2^p$ and obtains the Schauder functions

$$s_0(t) := t \quad \text{and} \quad s_{pm}(t) := \int_{[0,t]} h_{pm}(s) d\lambda(s)$$

by integration. The idea is to express the Wiener process as an expansion

$$W_t(\omega) = g_0(\omega) s_0(t) + \sum_{p=0}^{\infty} \sum_{m=1}^{2^p} g_{pm}(\omega) 2^{\frac{p}{2}} s_{pm}(t),$$

where $\{g_0, g_{pm} : p \in \mathbb{N}_0, m = 1, \ldots, 2^p\}$ is a family of independent standard Gaussian random variables. Before we formulate our main statement we give a meaning to this expansion. For this we use the following general observation:

Lemma 20.7.2 *Let $\psi : [1, \infty) \to [1, \infty)$ be a continuous and strictly increasing bijection such that there is a constant $b \geq 1$ with $3\psi(x) \leq \psi(bx)$ for $x \geq 1$. Assume random variables $f_1, f_2, \ldots : \Omega \to \mathbb{R}$ on a probability space $(\Omega, \mathcal{F}, \mathbb{P})$ satisfying*

$$\mathbb{P}(|f_i| > x) \leq e^{1-\psi(x)} \quad \text{for} \quad x \geq 1.$$

Then, for $w(i) := \psi^{-1}(1 + \log i)$, $i \in \mathbb{N}$, and $c_{(20.6)} := e^{-2} \sum_{i=1}^{\infty} \frac{1}{i^2} < \infty$, one has that

$$\mathbb{P}\left(\frac{1}{b} \sup_{i \in \mathbb{N}} \frac{|f_i|}{w(i)} > x\right) \leq c_{(20.6)} e^{1-\psi(x)} \quad \text{for} \quad x \geq 1. \tag{20.6}$$

---

[7] Norbert Wiener, 26/11/1894 (Columbia, Missouri, USA)–18/03/1964 (Stockholm, Sweden).
[8] Zbigniew Ciesielski, 01/10/1934 (Gdynia, Poland)–05/10/2020 (Sopot, Poland).
[9] Juliusz Pawel Schauder, 21/09/1899 (Lemberg, Galicia, Austrian Empire, now Lviv, Ukraine) September 1943 (Lwów, Poland, now Lviv, Ukraine).
[10] Alfréd Haar, 11/10/1885 (Budapest, Hungary)–16/03/1933 (Szeged, Hungary).

Proof For $x, y \geq 1$ we observe $\psi(x) + 2\psi(y) \leq 3\psi(xy) \leq \psi(bxy)$ and deduce, for $x \geq 1$,

$$\mathbb{P}\left(\frac{1}{b}\sup_{i\in\mathbb{N}}\frac{|f_i|}{w(i)} > x\right) \leq \sum_{i=1}^{\infty}\mathbb{P}\Big(|f_i| > bw(i)x\Big) \leq \sum_{i=1}^{\infty} e^{1-\psi(bw(i)x)}$$

$$\leq \sum_{i=1}^{\infty} e^{1-2\psi(w(i))-\psi(x)} = \left[\sum_{i=1}^{\infty} e^{-2\psi(w(i))}\right] e^{1-\psi(x)}$$

$$= e^{-2}\left[\sum_{i=1}^{\infty}\frac{1}{i^2}\right] e^{1-\psi(x)}.$$

$\square$

We apply Lemma 20.7.2 to Gaussian random variables. To do so, let g be a random variable with law $\mathcal{N}(0, 1)$. Mill's ratio (Lemma 15.3.5) implies the estimate

$$\mathbb{P}(|g| > x) \leq \sqrt{\frac{2}{\pi}}\frac{1}{xe^{\frac{x^2}{2}}} \quad \text{for} \quad x > 0.$$

For $x \geq 1$ and $\psi(x) = x^2$ this implies

$$\mathbb{P}\left(\frac{|g|}{\sqrt{2}} > x\right) = \mathbb{P}(|g| > \sqrt{2}x) \leq \sqrt{\frac{2}{\pi}}\frac{1}{\sqrt{2}xe^{\frac{2x^2}{2}}} \leq e^{-\psi(x)} \leq e^{1-\psi(x)}.$$

For the weights and the constant $b \geq 1$ we obtain $w(i) = \sqrt{1+\log i}$ and $b = \sqrt{3}$. Therefore,

$$\mathbb{P}\left(\frac{1}{\sqrt{6}}\sup_{i\in\mathbb{N}}\frac{|g_i|}{\sqrt{1+\log i}} \geq x\right) \leq c_{(20.6)}e^{1-x^2} \quad \text{for} \quad x \geq 1.$$

For our family of independent standard Gaussian random variables $\{g_0, g_{pm} : p \in \mathbb{N}_0, m = 1, \ldots, 2^p\}$ this implies that

$$\mathbb{P}\left(\frac{1}{\sqrt{6}}\frac{1}{\sqrt{\log 2}}\sup_{\substack{p\geq 0 \\ m=1,\ldots,2^p}}\frac{|g_{pm}|}{\sqrt{1+p}} \geq x\right) \leq c_{(20.6)}e^{1-x^2} \quad \text{for} \quad x \geq 1.$$

If

$$C(\omega) := \sup_{\substack{p\geq 0 \\ m=1,\ldots,2^p}}\frac{|g_{pm}(\omega)|}{\sqrt{1+p}},$$

20.7 Ciesielski's Theorem and the Wiener Process

then $C : \Omega \to [0, \infty]$ is almost surely finite and $C \in L^{\Phi_2}$ by Proposition 19.2.7. So for $\Omega_0 := \{\omega \in \Omega : C(\omega) < \infty\}$ we get $\mathbb{P}(\Omega_0) = 1$ and may define

$$W_t(\omega) := \begin{cases} g_0(\omega)s_0(t) + \sum_{p=0}^{\infty} \sum_{m=1}^{2^p} \left[g_{pm}(\omega) 2^{\frac{p}{2}} \right] s_{pm}(t) & : \omega \in \Omega_0 \\ 0 & : \omega \notin \Omega_0 \end{cases}.$$

In the notation of Appendix A.9 we have the random coefficients

$$a_0(\omega) := g_0(\omega) \quad \text{and} \quad a_{pm}(\omega) := g_{pm}(\omega) 2^{\frac{p}{2}}, \tag{20.7}$$

so that

$$|a_{pm}(\omega)| \leqslant C(\omega) 2^{\frac{p}{2}} \sqrt{1+p} =: C(\omega) w(p) \quad \text{for} \quad \omega \in \Omega_0. \tag{20.8}$$

Theorem 20.7.3

(1) *The map $W : \Omega \to C([0, 1])$ is a random variable.*
(2) *For all $N \in \mathbb{N}$ and $0 \leqslant t_1 < \cdots < t_N \leqslant 1$ the random variable $(W_{t_1}, \ldots, W_{t_N}) : \Omega \to \mathbb{R}^N$ is Gaussian.*
(3) *For $s, t \in [0, 1]$ one has that $\mathbb{E} W_s W_t = \min\{s, t\}$.*
(4) *For all $s, t \in [0, 1]$ and $\omega \in \Omega_0$ one has*

$$|W_t(\omega) - W_s(\omega)| \leqslant D(\omega) \sqrt{|t-s| \left(1 + \log_2 \frac{2}{|t-s|}\right)}$$

with $D := |g_0| + C\sqrt{2}[\underline{\kappa} + \overline{\kappa}] \in L^{\Phi_2}$, where $\underline{\kappa}, \overline{\kappa} > 0$ are taken from Corollary A.9.6.

Proof
(1) For $n \in \mathbb{N}$ we define

$$W_t^n(\omega) := \begin{cases} g_0(\omega)s_0(t) + \sum_{p=0}^{n} \sum_{m=1}^{2^p} \left[g_{pm}(\omega) 2^{\frac{p}{2}} \right] s_{pm}(t) & : \omega \in \Omega_0 \\ 0 & : \omega \notin \Omega_0 \end{cases}$$

and obtain a map $W^n : \Omega \to C([0, 1])$. For each $t \in [0, 1]$ the map $\omega \mapsto \langle W^n(\omega), \delta_t \rangle$ is a random variable. As the functionals $\delta_t : f \mapsto f(t)$ are norming we can apply the Pettis-measurability Theorem 20.1.3 to obtain that the map W^n is a random variable. Moreover, from Appendix A.9 we get that $|s_{pm}(t)| \leqslant 2^{-(p+1)}$ and

$$\|W(\omega) - W^n(\omega)\|_{C([0,1])} \leqslant \sum_{p=n+1}^{\infty} C(\omega) 2^{\frac{p}{2}} \sqrt{1+p} \, 2^{-(p+1)}$$

$$= \frac{C(\omega)}{2} \sum_{p=n+1}^{\infty} 2^{-\frac{p}{2}}\sqrt{1+p} \stackrel{n\to\infty}{\to} 0.$$

Therefore $W : \Omega \to C([0, 1])$ is a random variable by Corollary 20.1.4(1).
(2) We fix real numbers $b_1, \ldots, b_N \in \mathbb{R}$. Obviously, $\sum_{i=1}^{N} b_i W_{t_i}^n$ is Gaussian. Then

$$\sqrt{\mathbb{E}\left(\sum_{i=1}^{N} b_i W_{t_i}\right)^2} = \sqrt{\sum_{p=0}^{\infty}\sum_{m=1}^{2^p}\left|\sum_{i=1}^{N} b_i 2^{\frac{p}{2}} s_{pm}(t_i)\right|^2}$$

$$\leqslant \sum_{i=1}^{N} |b_i| \sqrt{\sum_{p=0}^{\infty}\sum_{m=1}^{2^p} \left|2^{\frac{p}{2}} s_{pm}(t_i)\right|^2}$$

$$\leqslant \sum_{i=1}^{N} |b_i| \sqrt{\sum_{p=0}^{\infty} 2^p 2^{-2(p+1)}} < \infty.$$

Therefore, since $\sum_{i=1}^{N} b_i W_{t_i}^n$ converges in $L_2(\Omega, \mathcal{F}, \mathbb{P})$ for $n \to \infty$ the map

$$\omega \mapsto \sum_{i=1}^{N} b_i \left[g_0(\omega) s_0(t_i) + \sum_{p=0}^{\infty}\sum_{m=1}^{2^p} g_{pm}(\omega) 2^{\frac{p}{2}} s_{pm}(t_i) \right]$$

is a Gaussian random variable as well. As this is true for all $b_1, \ldots, b_N \in \mathbb{R}$ the map $(W_{t_1}, \ldots, W_{t_N}) : \Omega \to \mathbb{R}^N$ is Gaussian and (2) is verified.
(3) This follows from

$$\mathbb{E} W_s W_t = s_0(s) s_0(t) + \sum_{p=0}^{\infty}\sum_{m=1}^{2^p} 2^p s_{pm}(s) s_{pm}(t)$$

$$= \langle \mathbb{1}_{[0,s)}, h_0 \rangle \langle \mathbb{1}_{[0,t)}, h_0 \rangle + \sum_{p=0}^{\infty}\sum_{m=1}^{2^p} \langle \mathbb{1}_{[0,s)}, 2^{\frac{p}{2}} h_{pm} \rangle \langle \mathbb{1}_{[0,t)}, 2^{\frac{p}{2}} h_{pm} \rangle$$

$$= \int_{[0,1]} \mathbb{1}_{[0,s)}(u) \mathbb{1}_{[0,t)}(u) d\lambda(u)$$

$$= \min\{s, t\},$$

where $\langle f, g \rangle := \int_{[0,1]} f(s) g(s) d\lambda(s)$ is the scalar product in $L_2([0, 1])$.
(4) On Ω_0 we have that

$$W_t(\omega) = g_0(\omega) s_0(t) + \sum_{p=0}^{\infty}\sum_{m=1}^{2^p} \left[g_{pm}(\omega) 2^{\frac{p}{2}} \right] s_{pm}(t)$$

20.7 Ciesielski's Theorem and the Wiener Process

$$= a_0(\omega)s_0(t) + \sum_{p=0}^{\infty} \sum_{m=1}^{2^p} a_{pm}(\omega) s_{pm}(t)$$

with a_0 and a_{pm} as in (20.7) with (20.8). Now Corollary A.9.6 implies

$$|W_t(\omega) - W_s(\omega)| \leq |g_0(\omega)||t-s| + C(\omega)\sqrt{2}[\underline{\kappa} + \overline{\kappa}]\sqrt{|t-s|\left(1 + \log_2 \frac{2}{|t-s|}\right)}$$

$$\leq \left[|g_0(\omega)| + C(\omega)\sqrt{2}[\underline{\kappa} + \overline{\kappa}]\right]\sqrt{|t-s|\left(1 + \log_2 \frac{2}{|t-s|}\right)}.$$

$\square$

We conclude by Theorem 20.7.3 and Definition 20.7.1 that the process $W = (W_t)_{t\in[0,1]}$ is a Wiener process.

Finally, we turn to an approximation property of the Wiener process to illustrate Theorem 19.1.12: Assume the Wiener process $W = (W_t)_{t\in[0,1]}$ and another process $A = (A_t)_{t\in[0,1]}$ with $A_0 \equiv 0$ which is *piecewise constant* on all intervals $\left(\frac{k-1}{n}, \frac{k}{n}\right]$, $k = 1, \ldots, n$. Then we have the following estimate:

Theorem 20.7.4 *For $p \in (0, \infty)$ and $n \in \mathbb{N}$ one has that*

$$\left\|\sup_{t\in[0,1]} |W_t - A_t|\right\|_{L_p} \geq c_p \sqrt{\frac{1 + \log n}{n}},$$

where $c_p > 0$ depends on p only.

Proof Define the intermediate points $t_k := \frac{k}{n} - \frac{1}{2n}$ for $k = 1, \ldots, n$. Then we get that

$$\left\|\sup_{t\in[0,1]} |W_t - A_t|\right\|_{L_p} \geq \left\|\sup_{k=1,\ldots,n} \left\{|W_{t_k} - A_{t_k}| \vee |W_{t_k + \frac{1}{2n}} - A_{t_k}|\right\}\right\|_{L_p}$$

$$\geq \frac{1}{2} \left\|\sup_{k=1,\ldots,n} \left|W_{t_k} - W_{t_k + \frac{1}{2n}}\right|\right\|_{L_p}$$

$$= \frac{1}{2}\sqrt{\frac{1}{2n}} \left\|\sup_{k=1,\ldots,n} |g_k|\right\|_{L_p}$$

$$\geq \frac{1}{2}\sqrt{\frac{1}{2n}} \frac{1}{\sqrt[p]{2}} (g_1)_p^{**}\left(\frac{1}{n}\right),$$

where $g_1, \ldots, g_n$ are independent standard Gaussian random variables and where we use Theorem 19.1.12 in the last estimate. Now we finish with Example 19.1.15. □

20.8 Exercises

Ex 1: Random variables with values in Lorentz sequence spaces

Let $p \in [1, \infty)$ and define the sequence spaces

$$\ell_p := \left\{ x = (x_n)_{n \in \mathbb{N}} \subseteq \mathbb{R} : \|x\|_{\ell_p} := \left(\sum_{n=1}^{\infty} |x_n|^p \right)^{\frac{1}{p}} < \infty \right\},$$

$$\ell_\infty := \left\{ x = (x_n)_{n \in \mathbb{N}} \subseteq \mathbb{R} : \|x\|_{\ell_\infty} := \sup_{n \in \mathbb{N}} |x_n| < \infty \right\},$$

$$c_0 := \left\{ x = (x_n)_{n \in \mathbb{N}} \subseteq \mathbb{R} : \text{ with } \lim_{n \to \infty} |x_n| = 0 \right\}$$

with $\|x\|_{c_0} := \sup_{n \in \mathbb{N}} |x_n|$.

(a) Prove that ℓ_p with $p \in [1, \infty)$ and c_0 are separable Banach spaces.
(b) Prove that ℓ_∞ is a non-separable Banach space.
(c) Prove that $c_0^* = \ell_1$ and $\ell_1^* = \ell_\infty$, where the identification is defined as $\langle (x_n)_{n=1}^\infty, (a_n)_{n=1}^\infty \rangle := \sum_{n=1}^\infty x_n a_n$ in both cases.

Hints:

- To show that ℓ_p with $p \in [1, \infty]$ is a Banach space one can use Theorem 18.1.2.
- For $c_0^* = \ell_1$ show first that $\ell_1 \subseteq c_0^*$ is an isometric subspace. After that check that all linear and continuous functionals $a : c_0 \to \mathbb{R}$ are of this form by using the evaluations $a_n := \langle e_n, a \rangle$, where $e_n \in c_0$ is the n-th unit vector, and using the fact that the space of all finite linear combinations of unit vectors is dense in c_0.
- The relation $\ell_1^* = \ell_\infty$ can be shown directly or by using Theorem 18.2.1.

(d) Assume a measurable space $(\Omega, \mathcal{F})$ and a map $f : \Omega \to \ell_p$ with $p \in [1, \infty)$ or $f : \Omega \to c_0$, where we write $f(\omega) = (f_n(\omega))_{n \in \mathbb{N}}$ in terms of the coordinate functionals. Prove that f is measurable if and only if all $f_n : \Omega \to \mathbb{R}$ are measurable.

Ex 2: Uncountable products of probability spaces

Let I be an uncountable index-set and assume for each $t \in I$ a probability space $(\Omega_t, \mathcal{F}_t, \mathbb{P}_t)$. With this we define:

20.8 Exercises

- $\Omega := \bigtimes_{t \in I} \Omega_t := \{(\omega_t)_{t \in I} : \omega_t \in \Omega_t\}$.
- For $(t_n)_{n \in \mathbb{N}} \subseteq I$ with $t_n \neq t_m$ for $n \neq m$ and $B \in \bigotimes_{n \in \mathbb{N}} \mathcal{F}_{t_n}$ we let

$$A := \{(\omega_t)_{t \in I} \in \Omega : (\omega_{t_n})_{n \in \mathbb{N}} \in B\}.$$

The set B is called basis of the cylinder set A. Note, that the basis B is not unique in general.
- We let $\mathcal{F}$ be the collection of all cylinder sets A, where one varies over $(t_n)_{n \in \mathbb{N}} \subseteq I$ and over $B \in \bigotimes_{n \in \mathbb{N}} \mathcal{F}_{t_n}$. Prove that $\mathcal{F}$ is a σ-algebra.
- For $A \in \mathcal{F}$ with a basis B we define

$$\mathbb{P}(A) := \left(\otimes_{n \in \mathbb{N}} \mathbb{P}_{t_n}\right)(B).$$

Prove that $\mathbb{P}(A)$ does not depend on the choice of the basis B and that $(\Omega, \mathcal{F}, \mathbb{P})$ is a probability space.

As for countable products we write $(\Omega, \mathcal{F}, \mathbb{P}) := \bigotimes_{t \in I}(\Omega_t, \mathcal{F}_t, \mathbb{P}_t)$.

Ex 3: Not all spaces $L_2(\Omega, \mathcal{F}, \mathbb{P})$ are separable
Take the construction from Exercise 2 with $I = [0, 1]$ and $(\Omega_t, \mathcal{F}_t, \mathbb{P}_t) = (\{-1, 1\}, 2^{\{-1,1\}}, \frac{1}{2}(\delta_{-1} + \delta_1))$. Choose the uncountable family of functions $(f_t)_{t \in [0,1]} \subseteq L_2(\mathbb{P})$ with $f_t(\omega) := \omega_t$ and investigate $\|f_s - f_t\|_{L_2(\mathbb{P})}$.

Ex 4: An identification that is often used
Let (M, Σ, μ) be σ-finite with $\mu(M) > 0$ and assume that $L_2(M, \Sigma, \mu)$ is separable. By this assumption there is an orthogonal basis $(h_n)_{n \in J} \subseteq L_2(M, \Sigma, \mu)$ with $J = \mathbb{N}$ or $J = \{1, \ldots, N\}$ for some $N \in \mathbb{N}$. For $F \in L_2(\Omega \times M, \mathcal{F} \otimes \Sigma, \mathbb{P} \otimes \mu)$ and $n \in J$ we define

$$\Omega_0 := \left\{\omega \in \Omega : \int_M |F(\omega, m)|^2 d\mu(m) < \infty\right\},$$

$$f_n(\omega) := 1_{\Omega_0}(\omega) \int_M F(\omega, m) h_n(m) d\mu(m).$$

Verify $\mathbb{P}(\Omega_0) = 1$ and that by monotone convergence we have

$$\sum_{n \in J} \|f_n\|_{L_2(\mathbb{P})}^2 = \int_{\Omega_0} \left[\int_M |F(\omega, m)|^2 d\mu(m)\right] d\mathbb{P}(\omega) = \|F\|_{L_2(\mathbb{P} \otimes \mu)}^2.$$

Hence we constructed an isometric embedding

$$J : L_2(\Omega \times M, \mathcal{F} \otimes \Sigma, \mathbb{P} \otimes \mu) \to L_2(\Omega, \mathcal{F}, \mathbb{P}; \ell_2(J)) \quad \text{by} \quad J(F) := (f_n)_{n \in J},$$

where $\ell_2(J) := \{x = (x_n)_{n \in J} : \|x\|_{\ell_2(J)}^2 := \sum_{n \in J} |x_n|^2 < \infty\}$. Prove that this map is surjective, i.e. we can identify $L_2(\Omega \times M, \mathcal{F} \otimes \Sigma, \mathbb{P} \otimes \mu)$ and $L_2(\Omega, \mathcal{F}, \mathbb{P}; \ell_2(J))$.

Ex 5: Let $E = C([0, 1])$. We know that E is separable. Given a probability space $(\Omega, \mathcal{F}, \mathbb{P})$ prove that $f : \Omega \to C([0, 1])$ is a random variable if and only if $f_t : \Omega \to \mathbb{R}$ is a random variable for all $t \in [0, 1]$, where $f_t(\omega) := (f(\omega))(t)$. Furthermore, for $f \in L_1(\Omega, \mathcal{F}, \mathbb{P}; C([0, 1]))$ show that $\mathbb{E}f = (\mathbb{E}f_t)_{t \in [0,1]}$.

20.9 Comments

General literature regarding probability in Banach spaces includes Bogachev [23, Chapter 4], Diestel and Uhl [46], Kwapień and Woyczyński [112], Ledoux and Talagrand [118], Li and Queffélec [120], van Neerven [190], Vakhania, Tarieladze and Chobanyan [189].

Section 20.1: Theorem 20.1.3 is due to Pettis [146]. Theorem 20.1.6 goes back to Nedoma [134] (see also [127, Theorem 2.16] and [189, Lemma 1.1, page 9]). We are grateful to M.S. Müller, F. Delbaen, and M. Riedle for transmitting the argument for the proof of Theorem 20.1.6 and the corresponding references. Lemma 20.1.7 is formulated in [77, Theorem 1.5.D] in the framework of σ-rings.

Section 20.2: The integral as given in Definition 20.2.5 was introduced by Bochner [20].

Section 20.3: The Dunford integral goes back to Dunford [51], the Pettis integral to Pettis [146]. An example that Dunford and Pettis integrability do not coincide in c_0 is given in [46, page 53]. Our example builds on this. In fact, the example covers the only possible situation in our setting: It is shown by Diestel [45] that the Dunford and Pettis integral coincide if the space E does not contain a copy of c_0 (see also [46, page 54]). For further reading about the Pettis integral we refer to Diestel and Uhl [46], Talagrand [186], and Musial [143, Chapter 12]. Regarding general integration theory the reader is referred to Dunford and Schwartz [52, Chapter III].

Section 20.4 follows the presentation in [142].

Section 20.5: Theorem 20.5.2 is due to Szewczak [183]. Earlier versions with larger constants have been proven by Etemadi [59, Theorem 1] and by Kwapień and Woyczyński in [112, Proposition 1.1.1]. A survey of Kühn and Schilling about maximal inequalities and their applications can be found in [109].

Section 20.6: The Itô-Nisio theorem is proven in [92] and further develops results of P. Lévy [119, Chapitre 6, Théorème 44] about real-valued random variables. Our proof of the Itô-Nisio theorem also uses the presentations in Kwapień and Woyczyński [112, Section 2.1] and van Neerven [190, Section 2.5].

Section 20.7: Ciesielski investigated in [40] Hölder conditions for realizations of Gaussian processes. This application to stochastic processes was based on his earlier work about the Schauder basis in $C([0, 1])$ in [38, 39]. Lemma 20.7.2, which is used to prove distributional properties of the modulus of continuity is taken from [70]. Regarding Theorem 20.7.4, there is a corresponding upper bound in [44].

Chapter 21
Law of Iterated Logarithm

In this chapter the law of iterated logarithm is proven using the Berry-Esseen theorem about the speed of convergence in the central limit theorem.

21.1 The Law of Iterated Logarithm

We did already study the strong law of large numbers in Theorem 13.1.1 which is a fundamental and far-reaching limit theorem. It states that for independent identically distributed and integrable random variables of mean zero one has

$$\lim_{n \to \infty} \frac{f_1 + \cdots + f_n}{n} = 0 \text{ a.s.}$$

If we change the re-normalization from n to $\sqrt{n}$ and suppose $\mathbb{E} f_n^2 = 1$, then we get another important theorem, the central limit Theorem 15.1.1, which states

$$\lim_{n \to \infty} \mathbb{P}\left(\frac{f_1 + \cdots + f_n}{\sqrt{n}} \leqslant c \right) = \int_{(-\infty, c]} \frac{1}{\sqrt{2\pi}} e^{-\frac{x^2}{2}} d\lambda(x) \quad \text{for} \quad c \in \mathbb{R}.$$

Looking at these two statements the next question is to understand the path-wise behaviour of $f_1(\omega) + \cdots + f_n(\omega)$ as $n \to \infty$, i.e. to find a function $\phi : \mathbb{N} \to (0, \infty)$ such that for any $\varepsilon > 0$ one has

$$-(1+\varepsilon)\phi(n) \leqslant S_n \leqslant (1+\varepsilon)\phi(n) \quad \text{with} \quad S_n := f_1 + \cdots + f_n$$

© The Author(s), under exclusive license to Springer Nature Switzerland AG 2025
H. Geiss, S. Geiss, *Measure, Probability and Functional Analysis*, Universitext,
https://doi.org/10.1007/978-3-031-84067-8_21

asymptotically for large n (in an appropriate sense). And in fact, it is possible to find such a function. This is the subject of the *Law of Iterated Logarithm* (LIL) in the form of Hartman[1]-Wintner:[2]

Theorem 21.1.1 (Law of Iterated Logarithm) *Let $f_1, f_2, \ldots : \Omega \to \mathbb{R}$ be an i.i.d. sequence of random variables such that $\mathbb{E} f_n^2 = 1$ and $\mathbb{E} f_n = 0$. Then one has that*

$$\mathbb{P}\left(\limsup_{\substack{n \to \infty \\ n \geqslant 3}} \frac{f_1 + \cdots + f_n}{\sqrt{2n \log \log n}} = 1\right) = 1.$$

To shorten the notation in this chapter we let

$$\phi(n) := \sqrt{2n \log \log n} \quad \text{for} \quad n \geqslant 3.$$

If $\mathbb{E} f_n^2 = \sigma^2$ for some $\sigma \in (0, \infty)$ and $\mathbb{E} f_n = 0$, then $(f_n/\sigma)_{n=1}^\infty$ and $(-f_n/\sigma)_{n=1}^\infty$ satisfy both the assumptions of Theorem 21.1.1 and we get, as the intersection of two sets of measure one is of measure one, that

$$\mathbb{P}\left(\limsup_{\substack{n \to \infty \\ n \geqslant 3}} \frac{f_1 + \cdots + f_n}{\phi(n)} = \sigma, \liminf_{\substack{n \to \infty \\ n \geqslant 3}} \frac{f_1 + \cdots + f_n}{\phi(n)} = -\sigma\right) = 1.$$

The LIL can be reformulated in terms of the two conditions

$$\mathbb{P}\left(\#\{n \geqslant 3 : S_n \geqslant (1 - \varepsilon)\sigma\phi(n)\} = \infty\right) = 1,$$

$$\mathbb{P}\left(\#\{n \geqslant 3 : S_n \geqslant (1 + \varepsilon)\sigma\phi(n)\} = \infty\right) = 0$$

for all $\varepsilon \in (0, 1)$. Therefore, $-\sigma\phi$ and $\sigma\phi$ are the correct lower and upper asymptotic bounds for the random walk $(S_n)_{n=1}^\infty$.

21.2 Berry-Esseen Meets the LIL

We prove the LIL under the slightly stronger condition that $\mathbb{E}|f_1|^{2+\alpha} < \infty$ for some $\alpha > 0$ in order to use the Berry-Esseen Theorem 15.2.1. The key observation to be able to deduce the LIL from the Berry-Esseen theorem is the following statement:

---

[1] Philip Hartman, 16/05/1915 (Baltimore, Maryland, USA)–28/08/2015.
[2] Aurel Friedrich Wintner, 08/04/1903 (Budapest, Hungary)–15/01/1958 (Baltimore, Maryland, USA).

21.2 Berry-Esseen Meets the LIL

Proposition 21.2.1 *Assume i.i.d. random variables $f_1, f_2, \ldots : \Omega \to \mathbb{R}$ such that $\mathbb{E} f_n^2 = 1$ and $\mathbb{E} f_n = 0$, and $\phi(n) = \sqrt{2n \log \log n}$ for $n \geq 3$. For $S_n := f_1 + \ldots + f_n$ suppose the upper and lower bounds*

(1) $\mathbb{P}(S_n > \varepsilon \phi(n)) \leq \overline{C}_{p,\varepsilon} (\log n)^{-p}$ *for all $\varepsilon > 1$ and $p \in (1, \varepsilon^2)$,*
(2) $\mathbb{P}(S_n > \varepsilon \phi(n)) \geq \underline{C}_{p,\varepsilon} (\log n)^{-p}$ *for all $\varepsilon > 0$ and $p \in (\varepsilon^2, 1)$,*

where $\overline{C}_{p,\varepsilon}, \underline{C}_{p,\varepsilon} > 0$ and $n \geq n_0 \geq 3$. Let assumption (1) also hold for the sequence $(-f_n)_{n=1}^\infty$ with a possibly different constant $\overline{C}_{p,\varepsilon} > 0$. Then one has

$$\mathbb{P}\left(\limsup_{\substack{n \to \infty \\ n \geq 3}} \frac{S_n}{\phi(n)} = 1\right) = 1.$$

Proof Assume the given upper and lower bounds. In the following we will use (1) to show that

$$\mathbb{P}\left(\limsup_{\substack{n \to \infty \\ n \geq 3}} \frac{S_n}{\phi(n)} \leq 1\right) = 1 \tag{21.1}$$

and in the second part we will use (1) for the sequence $(-f_n)_{n=1}^\infty$ and (2) for our initial sequence $(f_n)_{n=1}^\infty$ to deduce that

$$\mathbb{P}\left(\limsup_{\substack{n \to \infty \\ n \geq 3}} \frac{S_n}{\phi(n)} \geq 1\right) = 1. \tag{21.2}$$

Combining these results will lead us to the LIL.

To start, for $a > 1$ we define

$$n_k = n_k(a) := \lceil a^k \rceil \quad \text{for} \quad k \in \mathbb{N} \text{ with } k \geq k_0 = k_0(a) \in \mathbb{N},$$

where $k_0(a)$ is chosen such that $n_{k_0}(a) \geq 3$, and $\lceil x \rceil$ stands for the smallest natural number greater than or equal to x when $x > 0$.

Equation (21.1) For $\gamma > 1$ we choose $\theta \in (0, \gamma)$ and $a > 1$ such that

$$1 < \frac{\gamma - \theta}{\sqrt{a}} =: \gamma' < \gamma.$$

To be able to use Theorem 20.5.1 we observe that by Chebyshev's inequality (Corollary 8.12.7) we have, for $k > k_0$,

$$\max_{n\in[1,n_k]} \mathbb{P}(S_n > \theta\phi(n_{k-1})) \leq \max_{n\in[1,n_k]} \frac{\mathbb{E}S_n^2}{(\theta\phi(n_{k-1}))^2}$$

$$= \frac{\mathbb{E}S_{n_k}^2}{(\theta\phi(n_{k-1}))^2}$$

$$= \frac{n_k}{\theta^2 2n_{k-1}\log\log n_{k-1}}$$

$$\to 0 \text{ as } k \to \infty.$$

Therefore we may choose $k_1 > k_0$ such that

$$\max_{n\in[1,n_k]} \mathbb{P}(S_n > \theta\phi(n_{k-1})) \leq \frac{1}{2} \quad \text{for} \quad k \geq k_1.$$

Now we use Theorem 20.5.1 for $E = \mathbb{R}$ and $a(x) := x$ and $S_n = S_{n_{k-1}} + f_{n_{k-1}+1} + \ldots + f_n$ to estimate each term in the following sum

$$\sum_{k=k_1}^\infty \mathbb{P}\left(\max_{n\in[n_{k-1},n_k]} S_n > \gamma\phi(n_{k-1})\right)$$

$$\leq \sum_{k=k_1}^\infty \mathbb{P}\left(S_{n_k} > (\gamma-\theta)\phi(n_{k-1})\right) \frac{1}{\min_{n\in[n_{k-1},n_k]} \mathbb{P}(S_n - S_{n_k} \leq \theta\phi(n_{k-1}))}$$

$$\leq 2\sum_{k=k_1}^\infty \mathbb{P}\left(S_{n_k} > (\gamma-\theta)\phi(n_{k-1})\right)$$

$$= 2\sum_{k=k_1}^\infty \mathbb{P}\left(S_{n_k} > \left((\gamma-\theta)\frac{\phi(n_{k-1})}{\phi(n_k)}\right)\phi(n_k)\right). \tag{21.3}$$

Here we use $\min_{n\in[n_{k-1},n_k]} \mathbb{P}(S_n - S_{n_k} \leq \theta\phi(n_{k-1})) \geq \frac{1}{2}$ being equivalent to

$$\max_{n\in[n_{k-1},n_k]} \mathbb{P}(S_n - S_{n_k} > \theta\phi(n_{k-1})) \leq \frac{1}{2},$$

and which holds for $k \geq k_1$. By a calculation we also have that

$$\lim_{\substack{k\to\infty \\ k\geq k_1}} (\gamma-\theta)\frac{\phi(n_{k-1})}{\phi(n_k)} = \frac{\gamma-\theta}{\sqrt{a}} = \gamma' > 1.$$

Moreover, for all $p > 1$ one has

21.2 Berry-Esseen Meets the LIL

$$\sum_{k=k_0}^{\infty}\left(\frac{1}{\log n_k}\right)^p = \sum_{k=k_0}^{\infty}\left(\frac{1}{\log\lceil a^k\rceil}\right)^p < \infty.$$

Since for any $\varepsilon \in (1, \gamma')$ one has

$$(\gamma - \theta)\frac{\phi(n_{k-1})}{\phi(n_k)} \geq \varepsilon \quad \text{for} \quad k \geq k_2(\varepsilon) \geq k_1$$

we can use assumption (1) for $p \in (1, \varepsilon^2)$ to deduce from (21.3) that

$$\sum_{k=1}^{\infty} \mathbb{P}\left(\max_{n\in[n_{k-1},n_k]} S_n > \gamma\phi(n_{k-1})\right) < \infty.$$

By $\phi(n_{k-1}) \leq \phi(n)$ for $n \in [n_{k-1}, n_k]$ and the Lemma of Borel-Cantelli we conclude that

$$\mathbb{P}\left(\#\left\{k \geq k_1 : \max_{n\in[n_{k-1},n_k]} \frac{S_n}{\phi(n)} > \gamma\right\} = \infty\right) = 0$$

so that $\mathbb{P}\left(\limsup_{\substack{n\to\infty \\ n\geq 3}} \frac{S_n}{\phi(n)} \leq \gamma\right) = 1$. As this is true for all $\gamma > 1$ we verified (21.1).

Equation (21.2) We recall that (21.1) holds for the sequence $(-f_n)_{n=1}^{\infty}$. For the following we let $\gamma \in (0, 1)$, and choose $\varepsilon, p > 0$ and $a \in (1, \infty)$ such that

$$\gamma\sqrt{\frac{a}{a-1}} < \varepsilon < \sqrt{p} < 1. \tag{21.4}$$

Again we use $(n_k)_{k=k_0}^{\infty} = (n_k(a))_{k=k_0(a)}^{\infty}$ and get

$$1 = \mathbb{P}\left(\limsup_{\substack{n\to\infty \\ n\geq 3}} \frac{-S_n}{\phi(n)} \leq 1\right)$$

$$\leq \mathbb{P}\left(\limsup_{\substack{k\to\infty \\ k\geq k_0}} \frac{-S_{n_k}}{\phi(n_k)} \leq 1\right)$$

$$= \mathbb{P}\left(\limsup_{\substack{k\to\infty \\ k\geq k_0}} \frac{-S_{n_k}}{\phi(n_{k+1})}\frac{\phi(n_{k+1})}{\phi(n_k)} \leq 1\right)$$

$$= \mathbb{P}\left(\limsup_{\substack{k\to\infty \\ k\geq k_0}} \frac{-S_{n_k}}{\phi(n_{k+1})} \leq \frac{1}{\sqrt{a}}\right) \leq 1$$

because $\frac{\phi(n_{k+1})}{\phi(n_k)} \longrightarrow \sqrt{a}$ as $k \to \infty$. Therefore we can conclude that

$$\mathbb{P}\left(\liminf_{\substack{k\to\infty \\ k\geq k_0}} \frac{S_{n_k}}{\phi(n_{k+1})} \geq -\frac{1}{\sqrt{a}}\right) = 1. \tag{21.5}$$

Keeping (21.5) in mind for later use, we look at $Y_k := S_{n_{k+1}} - S_{n_k}$ for $k \geq k_0$. These are independent random variables, with $Y_k \stackrel{d}{=} S_{n_{k+1}-n_k}$ (where $S_0 := 0$), and we get for $n_{k+1} - n_k \geq 3$, which is satisfied for all $k \geq k_3$ for some $k_3 \geq k_0$, that

$$\mathbb{P}(Y_k > \gamma\phi(n_{k+1})) = \mathbb{P}\left(S_{n_{k+1}-n_k} > \gamma\phi(n_{k+1})\right)$$
$$= \mathbb{P}\left(S_{n_{k+1}-n_k} > \gamma \frac{\phi(n_{k+1})}{\phi(n_{k+1}-n_k)}\phi(n_{k+1}-n_k)\right).$$

We observe that $\gamma \frac{\phi(n_{k+1})}{\phi(n_{k+1}-n_k)} \to \gamma\sqrt{\frac{a}{a-1}} < \varepsilon$ as $k \to \infty$. We use assumption (2) for $\varepsilon^2 < p < 1$ to conclude that

$$\sum_{k=k_3}^{\infty} \mathbb{P}(Y_k > \gamma\phi(n_{k+1})) = \infty$$

as a consequence of $\gamma\sqrt{\frac{a}{a-1}} < \varepsilon$ and $\sum_{k=k_3}^{\infty}\left(\frac{1}{\log(n_{k+1}-n_k)}\right)^p = \infty$ as $p < 1$. Now the Lemma of Borel-Cantelli gives us

$$\mathbb{P}\left(\limsup_{\substack{k\to\infty \\ k\geq k_0}} \frac{Y_k}{\phi(n_{k+1})} \geq \gamma\right) = 1.$$

We recall (21.5) and that

$$\limsup_{\substack{k\to\infty \\ k\geq k_0}}\left(\frac{S_{n_k}}{\phi(n_{k+1})} + \frac{Y_k}{\phi(n_{k+1})}\right) \geq \liminf_{\substack{k\to\infty \\ k\geq k_0}} \frac{S_{n_k}}{\phi(n_{k+1})} + \limsup_{\substack{k\to\infty \\ k\geq k_0}} \frac{Y_k}{\phi(n_{k+1})}$$

on the set

$$\left\{\liminf_{\substack{k\to\infty \\ k\geq k_0}} \frac{S_{n_k}}{\phi(n_{k+1})} \geq -\frac{1}{\sqrt{a}},\ \limsup_{\substack{k\to\infty \\ k\geq k_0}} \frac{Y_k}{\phi(n_{k+1})} \geq \gamma\right\}$$

and deduce that

21.2 Berry-Esseen Meets the LIL

$$\mathbb{P}\left(\limsup_{\substack{n\to\infty\\n\geqslant 3}} \frac{S_n}{\phi(n)} \geqslant \gamma - \frac{1}{\sqrt{a}}\right)$$

$$\geqslant \mathbb{P}\left(\limsup_{\substack{k\to\infty\\k\geqslant k_0}} \frac{S_{n_{k+1}}}{\phi(n_{k+1})} \geqslant \gamma - \frac{1}{\sqrt{a}}\right)$$

$$= \mathbb{P}\left(\limsup_{\substack{k\to\infty\\k\geqslant k_0}} \left(\frac{S_{n_k}}{\phi(n_{k+1})} + \frac{Y_k}{\phi(n_{k+1})}\right) \geqslant -\frac{1}{\sqrt{a}} + \gamma\right) = 1.$$

Since (21.4) remains to be true if we replace a by any larger value than the originally chosen one we may let $a \uparrow \infty$ and get that

$$\mathbb{P}\left(\limsup_{\substack{n\to\infty\\n\geqslant 3}} \frac{S_n}{\phi(n)} \geqslant \gamma\right) = 1.$$

As $\gamma \in (0, 1)$ was chosen arbitrarily, we finally get

$$\mathbb{P}\left(\limsup_{\substack{n\to\infty\\n\geqslant 3}} \frac{S_n}{\phi(n)} \geqslant 1\right) = 1$$

and (21.2) follows. □

Now we are in a position to deduce the law of iterated logarithm:

Proof of Theorem 21.1.1 We prove the statement under the additional assumption that $\mathbb{E}|f_1|^{2+\alpha} < \infty$ for some $\alpha > 0$ which is required by the Berry-Esseen theorem (Theorem 15.2.1). We apply the Berry-Esseen theorem to verify the assumptions in Theorem 21.2.1:

Condition (1) By Theorem 15.2.1 there exists a constant $c > 0$ such that for all $\varepsilon > 1$ and for all $n \geqslant 3$ we get

$$\mathbb{P}(S_n > \varepsilon\sqrt{2n \log \log n}) = \mathbb{P}\left(\frac{S_n}{\sqrt{n}} > \varepsilon\sqrt{2 \log \log n}\right)$$

$$\leqslant c n^{-\alpha/2} + \int_{\varepsilon\sqrt{2\log\log n}}^{\infty} e^{-\frac{x^2}{2}} \frac{d\lambda(x)}{\sqrt{2\pi}}$$

$$\leqslant c n^{-\alpha/2} + \frac{1}{2}(\log n)^{-\varepsilon^2}$$

$$\leqslant \overline{C}_{p,\varepsilon,\alpha}(\log n)^{-p}$$

for $p \in (1, \varepsilon^2]$, where we used Mills' ratio (Lemma 15.3.5) for the second inequality.

Condition (2) Similarly, for $\varepsilon > 0$ and $n \geq 3$ we get

$$\mathbb{P}(S_n > \varepsilon\sqrt{2n \log \log n}) = \mathbb{P}\left(\frac{S_n}{\sqrt{n}} > \varepsilon\sqrt{2 \log \log n}\right)$$

$$\geq \int_{\varepsilon\sqrt{2\log\log n}}^{\infty} e^{-\frac{x^2}{2}} \frac{d\lambda(x)}{\sqrt{2\pi}} - cn^{-\alpha/2}$$

$$\geq \frac{1}{\sqrt{2\pi}} \frac{\varepsilon\sqrt{2\log\log n}}{1 + \varepsilon^2 2\log\log n} (\log n)^{-\varepsilon^2} - cn^{-\alpha/2}$$

where we again used Mills' ratio. Hence for $p \in (\varepsilon^2, 1)$ there is an $n_0 \geq 3$ and a $\underline{C}_{p,\varepsilon,\alpha} > 0$ such that

$$\mathbb{P}(S_n \geq \varepsilon\sqrt{2n \log \log n}) \geq \underline{C}_{p,\varepsilon,\alpha} (\log n)^{-p} \quad \text{for} \quad n \geq n_0.$$

$\square$

21.3 Exercises

Ex 1: Experimenting with R: Run the R-program from Appendix C.4 where the LIL is demonstrated with a symmetric random walk and experiment with the parameters.

21.4 Comments

Section 21.1: Khinchin [101] proved the LIL in the case that f_n is the indicator function of an event E appearing with probability $p \in (0, 1)$, i.e. $f_n \in \{0, 1\}$. This was extended by Kolmogorov [104] to sequences of bounded independent random variables that do not necessarily have the same distribution, and developed further by Hartman and Wintner [80]. Our form of the LIL is covered by the version of Hartman and Wintner. Stout [181] proved a version for stationary ergodic martingale differences and Kuelbs [110] proved Kolmogorov's variant of the LIL for Banach space valued random variables. Acosta [26] gave an alternative proof for the versions of Hartman-Wintner and Kuelbs.

Section 21.2: We follow the approach from [191].

Chapter 22
An Application to Non-life Insurance

In this chapter the strong law of large numbers, Wald's identity, and results about Banach function spaces are applied to non-life insurance mathematics: it is shown that if the net-profit condition is not satisfied, ruin happens with probability one. Finally, estimates about expected maximal waiting times between consecutive claims are given.

22.1 The Sparre-Anderson Model

We start by modelling the process that describes the premium income from the customers and the payment of claims to the customers. Usually this is done by using stochastic processes in continuous time. However, in order to check only the probability of ruin, processes with a discrete time parameter can be used.

Let us start with the model: We assume a probability space $(\Omega, \mathcal{F}, \mathbb{P})$, and independent and integrable random variables $W_1, X_1, W_2, X_2, \ldots : \Omega \to (0, \infty)$, where $W_1, W_2, \ldots$ and $X_1, X_2, \ldots$ are identically distributed, respectively. Furthermore, we assume constants $c, \kappa > 0$, and interpret these objects as follows:

- The constant $c > 0$ is the capital of the insurer at time $t = 0$.
- The constant $\kappa > 0$ is the premium rate the insurer charges over time. It is assumed that the number of customers is large, so that the premium in-flow is constant over time.
- The i.i.d. random variables $W_1, W_2, \ldots : \Omega \to (0, \infty)$ are the waiting times between two claims and there is *no constant* $w > 0$ *such that* $\mathbb{P}(W_k = w) = 1$.
- The i.i.d. random variables $X_1, X_2, \ldots : \Omega \to (0, \infty)$ are the claims sizes.

The balance of the insurer after the n-th claim is modelled as

$$U_n := c + \sum_{k=1}^{n}(\kappa W_k - X_k).$$

The event of ruin with initial capital c is

$$Ruin := \{\omega \in \Omega : \text{there exists an } n \in \mathbb{N} \text{ such that } U_n(\omega) < 0\}.$$

The above describes the Sparre-Anderson model. In case the waiting times are exponentially distributed, it is called the Cramér-Lundberg model. To avoid an a.s. ruin, the following condition is used:

Definition 22.1.1 The **net-profit condition** NPC holds provided that

$$\mathbb{E}X_1 < \kappa \mathbb{E}W_1.$$

This NPC condition means that the expected premium income is larger than the expected size of a claim.

22.2 Necessity of the NPC Condition and Expected Waiting Times

We show that one has ruin with probability one in the absence of the NPC condition:

Theorem 22.2.1 *If* $\mathbb{E}X_1 \geq \kappa \mathbb{E}W_1$, *then* $\mathbb{P}(\text{Ruin}) = 1$.

Proof We set $f_n := X_n - \kappa W_n$, so that $m := \mathbb{E}f_1 = \mathbb{E}X_1 - \kappa \mathbb{E}W_1$, and put $S_n := f_1 + \cdots + f_n$ for $n \in \mathbb{N}$.
(a) If $\mathbb{E}f_1 > 0$, then the SLLN of Kolmogorov (Theorem 13.1.1) implies that

$$\lim_{n \to \infty} \frac{S_n}{n} = m > 0 \quad \text{a.s.,}$$

so that $\lim_{n \to \infty} S_n = \infty$ a.s. and $\mathbb{P}(\text{Ruin}) = 1$.
(b) Assume $\mathbb{E}f_1 = 0$. There are $\kappa^{\pm} > 0$ such that $\mathbb{P}(f_1 > \kappa^+) > 0$ and $\mathbb{P}(f_1 < -\kappa^-) > 0$. To verify this, assume $f_1 = 0$ a.s., i.e. $X_1 = \kappa W_1$ a.s. for some $\kappa > 0$. This would imply

$$\mathbb{E}(X_1 - \mathbb{E}X_1)^2 = \mathbb{E}[(\kappa W_1 - \mathbb{E}X_1)(X_1 - \mathbb{E}X_1)] = \mathbb{E}(\kappa W_1 - \mathbb{E}X_1)\mathbb{E}(X_1 - \mathbb{E}X_1)$$
$$= 0,$$

where we used the independence of X_1 and W_1. Hence $X_1 = \mathbb{E}X_1$ a.s. and $W_1 = (1/\kappa)\mathbb{E}X_1$ a.s. But this case is excluded by our assumption on the sequence $W_1, W_2, \ldots$

22.2 NPC and Expected Waiting Times

Now, for $\varepsilon \in (0, \kappa^+)$ we let

$$A_\varepsilon := \left\{ \limsup_{n\to\infty} S_n > \varepsilon \right\}.$$

Then $\mathbb{P}(A_\varepsilon) \in \{0, 1\}$ by the 0-1 law of Hewitt and Savage (Theorem 7.7.4) as for any finite permutation of the summands $f_1, f_2, \ldots$ the event A_ε does not change.
(b1) Assume that $\mathbb{P}(A_\varepsilon) > 0$, so that $\mathbb{P}(A_\varepsilon) = 1$. Below in step (b2) we will show that indeed there exists an $\varepsilon > 0$ with $\mathbb{P}(A_\varepsilon) > 0$. For $L \geq 2$ with $L \in \mathbb{N}$ we obtain

$$\mathbb{P}\left(\limsup_{n\to\infty} S_n > L\varepsilon\right)$$

$$\geq \mathbb{P}\left(\{f_1 + \cdots + f_{L-1} > (L-1)\varepsilon\} \cap \left\{\limsup_{n\to\infty, n \geq L}(S_n - f_1 - \cdots - f_{L-1}) > \varepsilon\right\}\right)$$

$$= \mathbb{P}(f_1 + \cdots + f_{L-1} > (L-1)\varepsilon) > 0,$$

so that $\mathbb{P}(\limsup_{n\to\infty} S_n > L\varepsilon) = 1$ by the 0-1 law of Hewitt and Savage. Here we exploited the independence of $f_1 + \cdots + f_{L-1}$ and $S_n - f_1 - \cdots - f_{L-1}$, the assumption that

$$\mathbb{P}\left(\limsup_{n\to\infty, n\geq L}(S_n - f_1 - \cdots - f_{L-1}) > \varepsilon\right) = \mathbb{P}(A_\varepsilon) = 1,$$

and that $\mathbb{P}(f_1 + \cdots + f_{L-1} > (L-1)\varepsilon) > 0$ as a consequence of $\varepsilon \in (0, \kappa^+)$. Now we have $\mathbb{P}(\limsup_{n\to\infty} S_n > L\varepsilon) = 1$ for all $L \in \mathbb{N}$ so that $\mathbb{P}(\limsup_{n\to\infty} S_n = \infty) = 1$, which proves our claim.
(b2) To conclude the proof we have to show that it is not possible that $\mathbb{P}(A_{\frac{1}{n}}) = 0$ for all $n \in \mathbb{N}$. In fact, if this would be the case, then $\mathbb{P}(\limsup_{n\to\infty} S_n \leq 0) = 1$. So let us assume that

$$\mathbb{P}\left(\limsup_{n\to\infty} S_n \leq 0\right) = 1.$$

Now we repeat the argument from (b1). For $L \in \mathbb{N}$ we get

$$\mathbb{P}\left(\limsup_{n\to\infty} S_n < -L\kappa^-\right)$$

$$\geq \mathbb{P}\left(\{f_1 + \cdots + f_L < L\kappa^-\} \cap \left\{\limsup_{n\to\infty, n\geq L+1}(S_n - f_1 - \cdots - f_L) \leq 0\right\}\right)$$

$$> 0,$$

so that $\mathbb{P}(\limsup_{n\to\infty} S_n < -L\kappa^-) = 1$ for all $L \in \mathbb{N}$ and $\mathbb{P}(\lim_{n\to\infty} S_n = -\infty) = \mathbb{P}(\limsup_{n\to\infty} S_n = -\infty) = 1$. For $\omega \in \Omega$ with $\lim_{n\to\infty} S_n(\omega) = -\infty$, there is a finite maximum of the sequence $S_1(\omega), S_2(\omega), S_2(\omega), \ldots$ Define the σ-algebras $\mathcal{F}_0 := \{\emptyset, \Omega\}$ and $\mathcal{F}_n := \sigma(f_1, \ldots, f_n)$ for $n \in \mathbb{N}$, and the random time

$$\tau(\omega) := \inf\{n \geqslant 1 : S_n(\omega) \leqslant 0\} \quad \text{with} \quad \inf \emptyset := \infty.$$

It is easy to see that τ is a stopping time in the sense of Definition 8.13.1. We get

$$1 = \sum_{n=1}^{\infty} \mathbb{P}(\{S_1 < S_n, \ldots, S_{n-1} < S_n\} \cap \{S_n \geqslant S_{n+1}, S_n \geqslant S_{n+2}, \ldots\})$$

$$= \sum_{n=1}^{\infty} \mathbb{P}(S_1 < S_n, \ldots, S_{n-1} < S_n) \mathbb{P}(S_n \geqslant S_{n+1}, S_n \geqslant S_{n+2}, \ldots)$$

$$= \mathbb{P}\left(\sup_{n \in \mathbb{N}} S_n \leqslant 0\right)$$

$$\sum_{n=1}^{\infty} \mathbb{P}(f_n < f_1 + \cdots + f_n, \ldots, f_2 + \cdots + f_n < f_1 + \cdots + f_n)$$

$$= \mathbb{P}\left(\sup_{n \in \mathbb{N}} S_n \leqslant 0\right) \sum_{n=1}^{\infty} \mathbb{P}(f_1 + \cdots + f_{n-1} > 0, \ldots, f_1 > 0)$$

$$= \mathbb{P}\left(\sup_{n \in \mathbb{N}} S_n \leqslant 0\right) \sum_{n=1}^{\infty} \mathbb{P}(\tau \geqslant n)$$

$$= \mathbb{P}\left(\sup_{n \in \mathbb{N}} S_n \leqslant 0\right) \sum_{n=1}^{\infty} n\, \mathbb{P}(\tau = n)$$

$$= \mathbb{P}\left(\sup_{n \in \mathbb{N}} S_n \leqslant 0\right) \mathbb{E}\tau.$$

This implies that $\mathbb{E}\tau < \infty$ since we can argue that $\mathbb{P}\left(\sup_{n\in\mathbb{N}} S_n \leqslant 0\right)$ can not be zero. This follows from the first two lines of the above computation because $\mathbb{P}\left(\sup_{n\in\mathbb{N}} S_n \leqslant 0\right) = \mathbb{P}(S_n \geqslant S_{n+1}, S_n \geqslant S_{n+2}, \ldots) = 0$ would lead to a contradiction. Using Wald's identity Theorem 8.13.2 this gives

$$\mathbb{E}S_\tau = \mathbb{E}\left(\sum_{n=1}^{\tau} f_n\right) = \mathbb{E}\tau\, \mathbb{E}f_1 = 0.$$

As $\mathbb{P}(S_\tau \leqslant 0) = 1$ this implies $S_\tau = 0$ a.s. But this is a contradiction because $\mathbb{P}(S_\tau < 0) \geqslant \mathbb{P}(f_1 < 0) > 0$. □

As another application of the previous chapters we estimate the expected maximal waiting time between the first n claims, i.e.

$$\mathbb{E}\sup_{i=1,\ldots,n} W_i.$$

Applying Example 19.1.14 for the one-sided Weibull distribution (Weibull(α)), in particular for the exponential distribution (Weibull(1)), and Example 19.1.16 for the one-sided Pareto distribution (Pareto(α, κ)) we get the following table:

Name of the model	Distribution of waiting times W_i	Expected maximal waiting time between n arrivals
Cramér-Lundberg	$\mathrm{Exp}(\lambda), \lambda > 0$	$\log(n+1)$
Sparre-Anderson	Pareto(α, κ), $\alpha > 1, \kappa > 0$	$n^{\frac{1}{\alpha}}$
Sparre-Anderson	Weibull(α), $\alpha > 0$	$\sqrt[\alpha]{\log(n+1)}$

22.3 Comments

Regarding Non-Life Insurance mathematics the reader is referred to the monographs of Rolski et al. [162] and of Mikosch [130]. Step (b1) of the proof of Theorem 22.2.1 is based on [162, pp. 233-235] with the modification that we use the 0-1 law of Hewitt and Savage.

Appendix A
Analysis

This chapter recalls the basic definitions and statements, which are used in the book. Moreover, identities for oscillatory integrals, the Riesz–Thorin theorem about interpolation, and Ciesielski's theorems about representations of Hölder continuous functions are proven.

A.1 Complex Numbers

We shall identify the complex numbers $\mathbb{C}$ with $\mathbb{R}^2$, i.e.

$$\mathbb{C} \cong \mathbb{R}^2 = \{(x, y) : x, y \in \mathbb{R}\}$$

and write

$$\mathbb{C} \ni z = x + iy \cong (x, y) \in \mathbb{R}^2,$$

where $x = \mathrm{Re}(z)$ is the **real part** of z and $y = \mathrm{Im}(z)$ is the **imaginary part** of z. For $z_1 = (x_1, y_1)$ and $z_2 = (x_2, y_2)$ the addition and multiplication of the complex numbers are given by

$$z_1 + z_2 := (x_1 + x_2, y_1 + y_2) = (x_1 + x_2) + i(y_1 + y_2),$$
$$z_1 z_2 := (x_1 x_2 - y_1 y_2, x_1 y_2 + x_2 y_1) = (x_1 x_2 - y_1 y_2) + i(x_1 y_2 + x_2 y_1).$$

The above definition yields to $i^2 = (-1, 0) \cong -1$ which formally leads to

$$(x_1 + iy_1)(x_2 + iy_2) = x_1 x_2 + i^2 y_1 y_2 + i(x_1 y_2 + x_2 y_1)$$
$$= (x_1 x_2 - y_1 y_2) + i(x_1 y_2 + x_2 y_1).$$

© The Author(s), under exclusive license to Springer Nature Switzerland AG 2025
H. Geiss, S. Geiss, *Measure, Probability and Functional Analysis*, Universitext,
https://doi.org/10.1007/978-3-031-84067-8

The **modulus of a complex number** is defined by

$$|(x, y)| = |x + iy| := \sqrt{x^2 + y^2}.$$

If $z = (x, y) \in \mathbb{C}$, then the **conjugate complex number** is $\bar{z} := (x, -y) = x - iy$. We have that $z\bar{z} = (x^2 + y^2, 0) \cong x^2 + y^2$. The **polar coordinates** of $z = (x, y) \in \mathbb{C}$ are given as $(r, \varphi) \in [0, \infty) \times [0, 2\pi)$ with

$$x = r \cos \varphi \quad \text{and} \quad y = r \sin \varphi.$$

Note that $r = |z|$ and that φ is not unique whenever $r = 0$. For $z \in \mathbb{C}$ we define its **exponential** by

$$e^z := \sum_{n=0}^{\infty} \frac{z^n}{n!},$$

where e^z is the unique complex number such that $\lim_{N \to \infty} \left| e^z - \sum_{n=0}^{N} \frac{z^n}{n!} \right| = 0$. The following properties can be proved directly by the above definition:

Proposition A.1.1

(1) For all $z_1, z_2 \in \mathbb{C}$ one has $e^{z_1+z_2} = e^{z_1} e^{z_2}$.
(2) Euler[1]'s formula: One has $e^{ix} = \cos x + i \sin x$ for $x \in \mathbb{R}$.
(3) For $x \in \mathbb{R}$ and $N \in \mathbb{N}_0$ one has

$$\left| e^{ix} - \left[1 + \sum_{n=1}^{N} \frac{(ix)^n}{n!} \right] \right| \leq \frac{|x|^{N+1}}{(N+1)!}$$

with $\sum_{n=1}^{0} := 0$.

A.2 Limit Inferior and Limit Superior

First we recall the notion of the limit inferior and limit superior of sequences of real numbers, concepts that are used in probability theory:

Definition A.2.1 ($\liminf a_n$ AND $\limsup a_n$) For $a_1, a_2, \ldots \in \mathbb{R}$ we let

$$\liminf_{n \to \infty} a_n := \lim_{n \to \infty} \inf_{k \geq n} a_k \in [-\infty, \infty],$$

---

[1] Leonhard Euler 15/04/1707 (Basel, Switzerland)–18/09/1783 (St Petersburg, Russia), Swiss mathematician.

$$\limsup_{n\to\infty} a_n := \lim_{n\to\infty} \sup_{k\geq n} a_k \in [-\infty, \infty],$$

be the **limit inferior** and the **limit superior** of the sequence $(a_n)_{n=1}^\infty$, respectively.

Remark A.2.2

(1) The $\liminf_{n\to\infty} a_n$ does exist because $\inf_{k\geq n} a_k$ is a nondecreasing sequence and the limit of a monotone sequence does exist. Similarly, $\limsup_{n\to\infty} a_n$ does exist.
(2) The value $\liminf_{n\to\infty} a_n$ is the smallest $c \in [-\infty, \infty]$ such that there is a subsequence $n_1 < n_2 < n_3 < \cdots$ with $\lim_{k\to\infty} a_{n_k} = c$. In other words, it is the smallest accumulation point of the sequence $(a_n)_{n=1}^\infty$.
(3) Similarly, the value $\limsup_{n\to\infty} a_n$ is the largest $c \in [-\infty, \infty]$ such that there is a subsequence $n_1 < n_2 < n_3 < \cdots$ with $\lim_{k\to\infty} a_{n_k} = c$. In other words, it is the largest accumulation point of the sequence $(a_n)_{n=1}^\infty$.
(4) By definition one has that

$$\liminf_{n\to\infty}(-a_n) = -\limsup_{n\to\infty} a_n, \tag{A.1}$$

$$-\infty \leq \liminf_{n\to\infty} a_n \leq \limsup_{n\to\infty} a_n \leq \infty. \tag{A.2}$$

(5) It holds that $\liminf_{n\to\infty} a_n = \limsup_{n\to\infty} a_n = a \in \mathbb{R}$ if and only if $\lim_{n\to\infty} a_n = a$.

A.3 Theorem of Stone and Weierstrass

There are different versions of the Stone[2]-Weierstrass[3] theorem. For the version we use, we recall that $C_0(\mathbb{R}^d; \mathbb{C})$ is the space of all continuous functions $\varphi : \mathbb{R}^d \to \mathbb{C}$ such that $\lim_{n\to\infty} |\varphi(x)| = 0$.

Theorem A.3.1 (Stone and Weierstrass) *Assume that $\mathcal{A} \subseteq C_0(\mathbb{R}^d; \mathbb{C})$ satisfies the following properties:*

(1) $\mathcal{A}$ *is a linear space.*
(2) $\alpha_1, \alpha_2 \in \mathcal{A}$ *implies* $\alpha_1 \alpha_2 \in \mathcal{A}$.
(3) $\alpha \in \mathcal{A}$ *implies* $\bar{\alpha} \in \mathcal{A}$.
(4) *For all $x_0 \in \mathbb{R}^d$ there is an $\alpha \in \mathcal{A}$ such that $\alpha(x_0) \neq 0$.*
(5) *For all $x_0 \neq x_1$ there is an $\alpha \in \mathcal{A}$ such that $\alpha(x_0) \neq \alpha(x_1)$.*

---

[2] Marshall Harvey Stone, 08/04/1903 (New York, USA)–09/01/1989 (Madras, now Chennai, India).
[3] Karl Theodor Wilhelm Weierstrass, 31/10/1815 (Ostenfelde, Westphalia (now Germany))–19/02/1879 (Berlin, Germany).

Then for $\varphi \in C_0(\mathbb{R}^d; \mathbb{C})$ there exist $\alpha_n \in \mathcal{A}$ such that

$$\lim_{n \to \infty} \sup_{x \in \mathbb{R}^d} |\alpha_n(x) - \varphi(x)| = 0.$$

A proof can be found in [42, Corollary V.8.3]. To match the assumption in the reference, one considers the closure of $\mathcal{A}$ with respect to the metric $d(\varphi_1, \varphi_2) := \sup_{x \in \mathbb{R}^d} |\varphi_1(x) - \varphi_2(x)|$.

A.4 Oscillatory Integrals and Dirichlet Integral

The following techniques are also known as, or related to, *Feynman*[4]*'s trick*, see the explanations in [73]. For our purposes two oscillatory integrals are needed:

Lemma A.4.1 *For $y \in \mathbb{R}$ one has that*

$$\int_{[0,\infty)} \cos(yx) e^{-x} d\lambda(x) = \frac{1}{1+y^2}, \tag{A.3}$$

$$\int_{[0,\infty)} \sin(yx) e^{-x} d\lambda(x) = \frac{y}{1+y^2}. \tag{A.4}$$

Proof By applying partial integration twice we have

$$\int_{[0,\infty)} \cos(yx) e^{-x} d\lambda(x)$$

$$= \cos(yx)(-e^{-x})|_{x=0}^{x=\infty} - \int_{[0,\infty)} (-y \sin(yx))(-e^{-x}) d\lambda(x)$$

$$= 1 - y \int_{[0,\infty)} \sin(yx) e^{-x} d\lambda(x)$$

$$= 1 - y \left[\sin(yx)(-e^{-x})|_{x=0}^{x=\infty} - \int_{[0,\infty)} y \cos(yx)(-e^{-x}) d\lambda(x) \right]$$

$$= 1 - y^2 \int_{[0,\infty)} \cos(yx) e^{-x} d\lambda(x)$$

so that $\int_{[0,\infty)} \cos(yx) e^{-x} d\lambda(x) = \frac{1}{1+y^2}$ which is (A.3). Inserting (A.3) into the above computation yields (A.4) as well. □

---

[4] Richard Phillips Feynman, 11/05/1918 (New York, USA)–15/02/1988 (Los Angeles, California, USA).

A Analysis

With the above in our hands we can compute the Dirichlet integral:

Lemma A.4.2 *One has* $\lim_{c \to \infty, c > 0} \int_{(0,c]} \frac{\sin(y)}{y} d\lambda(y) = \frac{\pi}{2}$.

Proof We have that $\lim_{\substack{c \to \infty \\ c > 0}} \int_{(0,c]} \frac{\sin(y)}{y} d\lambda(y)$ converges to some value in $[0, \infty]$ because

$$\int_{(2\pi(k-1), 2\pi k]} \frac{\sin(y)}{y} d\lambda(y) \geq 0 \quad \text{for} \quad k \geq 1,$$

and

$$\lim_{k \to \infty, k \geq 1} \int_{(2\pi(k-1), 2\pi k]} \left| \frac{\sin(y)}{y} \right| d\lambda(y) = 0.$$

Furthermore, by Lemma A.4.1 we have by a change of variables that

$$\int_{[0,\infty)} \sin(x) e^{-xy} d\lambda(x) = \frac{1}{1 + y^2} \quad \text{for} \quad y > 0.$$

Hence, by Fubini's theorem we get that

$$\lim_{\substack{k \to \infty \\ k \geq 1}} \int_{(0, 2\pi k]} \frac{\sin(y)}{y} d\lambda(y)$$

$$= \lim_{\substack{k \to \infty \\ k \geq 1}} \int_{(0, 2\pi k]} \left[\int_{[0,\infty)} e^{-xy} d\lambda(x) \right] \sin(y) d\lambda(y)$$

$$= \lim_{\substack{k \to \infty \\ k \geq 1}} \int_{[0,\infty)} \int_{(0, 2\pi k]} e^{-xy} \sin(y) d\lambda(y) d\lambda(x)$$

$$= \int_{[0,\infty)} \left[\lim_{\substack{k \to \infty \\ k \geq 1}} \int_{(0, 2\pi k]} e^{-xy} \sin(y) d\lambda(y) \right] d\lambda(x)$$

$$= \int_{[0,\infty)} \frac{1}{1 + x^2} d\lambda(x) = \frac{\pi}{2}$$

where the limit can moved into the integral because of monotone convergence as one has

$$\int_{(2\pi(k-1), 2\pi k]} e^{-xy} \sin(y) d\lambda(y) \geq 0 \quad \text{for} \quad k \geq 1.$$

□

A.5 Schur's Product Theorem

A matrix $A = (a_{kl})_{k,l=1}^{n} \in \mathbb{C}^{n \times n}$ is positive semidefinite provided that

$$\sum_{k,l=1}^{n} a_{kl} \lambda_l \overline{\lambda_k} \geq 0 \quad \text{for all} \quad \lambda_1, \ldots, \lambda_n \in \mathbb{C}.$$

A positive semidefinite matrix *over* $\mathbb{C}$ is automatically a Hermitian[5] matrix, i.e. $a_{kl} = \overline{a_{lk}}$ which is easy to see by checking the case $n = 2$ (which is sufficient for this). Now one has the following Schur[6] product theorem [169, Theorem VII]:[7]

Theorem A.5.1 *Let $A = (a_{kl})_{k,l=1}^{n} \in \mathbb{C}^{n \times n}$ and $B = (b_{kl})_{k,l=1}^{n} \in \mathbb{C}^{n \times n}$ be positive semidefinite. Then the Hadamard[8] product $C := (a_{kl} b_{kl})_{k,l=1}^{n}$ is positive semidefinite.*

A.6 The Axiom of Choice and the Lemma of Zorn-Kuratowski

The Axiom of choice is part of the set theory and was formulated by Zermelo[9]:

Axiom A.6.1 (Axiom of choice) Let I be a nonempty set and $(M_i)_{i \in I}$ be a collection of nonempty sets M_i. Then there exists a family $(m_i)_{i \in I}$ with $m_i \in M_i$.

Definition A.6.2 A nonempty set S with a relation $\leq$ is called **partially ordered** if for certain pairs $(x, y) \in S \times S$ one has the order $x \leq y$ and the following axioms are satisfied:

(1) For all $x \in S$ one has $x \leq x$.
(2) If $x \leq y$ and $y \leq x$, then $x = y$.
(3) If $x \leq y$ and $y \leq z$, then $x \leq z$.

Definition A.6.3 Assume a nonempty partially ordered set $(S, \leq)$.

(1) An element $m \in S$ is called **maximal element** provided that $m \leq x$ implies $m = x$.

---

[5] Charles Hermite, 24/12/1822 (Dieuze, Lorraine, France)–14/01/1901 (Paris, France).
[6] Issai Schur, 10/01/1875 (Mogilev, Russian Empire, now Belarus)–10/01/1941 (Tel Aviv, British Mandate, now Israel).
[7] Schur wrote 'In enger Beziehung zu dem Satze V steht folgender Satz, der trotz seiner Einfachheit nicht bekannt zu sein scheint'.
[8] Jacques Salomon Hadamard, 08/12/1865 (Versailles, France)–17/10/1963 (Paris, France).
[9] Ernst Friedrich Ferdinand Zermelo, 27/07/1871 (Berlin, Germany)–21/05/1953 (Freiburg im Breisgau, Germany).

(2) A nonempty $A \subseteq S$ is called **totally ordered** if for all $x, y \in A$ one has either $x \leqslant y$ or $y \leqslant x$.
(3) An element $b \in S$ is called an **upper bound** of a nonempty $A \subseteq S$ if $x \leqslant b$ for all $x \in A$.

The following statement from set-theory is due to Zorn[10] and Kuratowski[11]:

Lemma A.6.4 *Let S be a nonempty partially ordered set such that each nonempty totally ordered set has an upper bound in S. Then S contains a maximal element.*

The lemma is usually named after Zorn [203] and goes also back to Kuratowski [111].

A.7 Equivalence Classes

Definition A.7.1 Let $M \neq \emptyset$ be an arbitrary set. We say that a relation $x \sim y$ is **an equivalence class relation** on M, provided that

(1) $x \sim x$ for all $x \in M$ (reflexivity),
(2) if $x \sim y$, then $y \sim x$ (symmetry),
(3) if $x \sim y$ and $y \sim z$, then $x \sim z$ (transitivity).

The following fundamental theorem is easy to prove:

Theorem A.7.2 *Let $M \neq \emptyset$ be equipped with an equivalence class relation $x \sim y$ and define $M_x := \{y \in M : y \sim x\}$. Then the following holds:*

(1) $M = \bigcup_{x \in M} M_x$.
(2) *One has either $M_x = M_y$ or $M_x \cap M_y = \emptyset$.*
(3) *One has $x \sim y$ if and only if there is some $z \in M$ such that $x, y \in M_z$.*

Definition A.7.3 The sets M_x are called **equivalence classes**, an element $y \in M_x$ is called **representative**. We also use the notation $M_x = [y]$ if $y \in M_x$.

A.8 Riesz–Thorin Theorem

The Riesz[12]–Thorin[13] theorem is a special case of a class of interpolation theorems. Its early form goes back to M. Riesz [157] and was developed further by G.O. Thorin (which was a student of M. Riesz) [188]. Interpolation techniques have their origin

---

[10] Max August Zorn, 06/06/1906 (Krefeld, Germany)–09/03/1993 (Bloomington, Indiana, USA).
[11] Kazimierz Kuratowski, 02/02/1896 (Warsaw, Poland)–18/06/1980 (Warsaw, Poland).
[12] Marcel Riesz, 16/11/1886 (Györ, Hungary)–04/09/1969 (Lund, Sweden).
[13] G. Olof Thorin, 23/02/1912 (Halmstad, Sweden)–14/02/2004 (Danderyd, Sweden).

in functional analysis [9, 11] and became a powerful tool in probability as well. The following Riesz–Thorin theorem and its proof is standard in the corresponding literature, see for instance [9, Chapter 4.1] where a variant for real L_p spaces, instead of complex ones, can be found as well.

We assume two σ-finite measure spaces $(M, \mathcal{M}, \mu)$ and $(N, \mathcal{N}, \nu)$ and let $S(M, \mathcal{M}, \mu; \mathbb{C})$ be the linear space of all equivalence classes of simple functions $f = \sum_{i=1}^{m} \alpha_i \mathbb{1}_{A_i}$ with $\alpha_i \in \mathbb{C}$, $A_i \in \mathcal{M}$, and $\mu(A_i) < \infty$, and assume a linear operator

$$T : S(M, \mathcal{M}, \mu; \mathbb{C}) \to L_0(N, \mathcal{N}, \nu; \mathbb{C})$$

such that

$$\int_B |T(\mathbb{1}_A)| d\nu < \infty$$

for all $A \in \mathcal{M}$ and $B \in \mathcal{N}$ with $\mu(A) < \infty$ and $\nu(B) < \infty$.

Theorem A.8.1 (Riesz–Thorin) *Let $p_0, p_1, q_0, q_1 \in [1, \infty]$ and $\theta \in (0, 1)$ such that for*

$$\frac{1}{p} = \frac{1-\theta}{p_0} + \frac{\theta}{p_1} \quad \text{and} \quad \frac{1}{q} = \frac{1-\theta}{q_0} + \frac{\theta}{q_1}$$

one has $p, q \in (1, \infty)$. If there are $c_0, c_1 \geq 0$ such that

$$\|Tf\|_{L_{q_i}} \leq c_i \|f\|_{L_{p_i}} \quad \text{for } f \in S(M, \mathcal{M}, \mu; \mathbb{C}) \text{ and } i = 0, 1,$$

then one has

$$\|Tf\|_{L_q} \leq c_0^{1-\theta} c_1^{\theta} \|f\|_{L_p} \quad \text{for all} \quad f \in S(M, \mathcal{M}, \mu; \mathbb{C}).$$

To prove the Riesz–Thorin Theorem, we use the three-tine theorem of Hadamard:

Theorem A.8.2 (Hadamard) *Let $S := \{\theta + it : 0 \leq \theta \leq 1, t \in \mathbb{R}\}$ and $H : S \to \mathbb{C}$ be a bounded function holomorphic in the interior of S and continuous on S. Then, for $\theta \in (0, 1)$,*

$$\sup_{t \in \mathbb{R}} |H(\theta + it)| \leq \sup_{t \in \mathbb{R}} |H(it)|^{1-\theta} \sup_{t \in \mathbb{R}} |H(1 + it)|^{\theta}.$$

Proof We choose $c_0, c_1 > 0$ such that

$$\sup_{t \in \mathbb{R}} |H(it)| \leq c_0 \quad \text{and} \quad \sup_{t \in \mathbb{R}} |H(1 + it)| \leq c_1.$$

A Analysis

For $\varepsilon > 0$ and $\lambda \in \mathbb{R}$ we define the function

$$H_\varepsilon(z) := \exp(\varepsilon z^2 + \lambda z) H(z).$$

For $t \in \mathbb{R}$ we get

$$|H_\varepsilon(it)| = |\exp(-\varepsilon t^2 + i\lambda t) H(it)| \leq |H(it)| \leq c_0 \quad (A.5)$$

and

$$\begin{aligned}|H_\varepsilon(1+it)| &= |\exp(\varepsilon(1+2it-t^2) + \lambda(1+it)) H(1+it)| \\ &= |\exp(\varepsilon(1-t^2) + \lambda) H(1+it)| \\ &\leq \exp(\varepsilon + \lambda) c_1.\end{aligned} \quad (A.6)$$

Moreover, we have

$$\begin{aligned}|H_\varepsilon(\theta+it)| &= \exp(\varepsilon \theta^2 - \varepsilon t^2 + \lambda \theta) |H(\theta+it)| \\ &\leq \exp(\varepsilon(1-t^2) + |\lambda|) \sup_{z \in S} |H(z)| \to 0\end{aligned} \quad (A.7)$$

as $t \uparrow \infty$. Because H_ε is a bounded function holomorphic in the interior of S and continuous on S satisfying (A.5), (A.6), and (A.7), by the maximum principle we deduce for $\theta \in (0, 1)$ that

$$|H_\varepsilon(\theta+it)| \leq \max\{c_0, c_1 \exp(\varepsilon + \lambda)\}$$

or

$$\begin{aligned}|H(\theta+it)| &\leq \exp(-\varepsilon(\theta^2 - t^2) - \lambda \theta) \max\{c_0, c_1 \exp(\varepsilon + \lambda)\} \\ &= \exp(-\varepsilon(\theta^2 - t^2)) \max\{\exp(-\lambda \theta) c_0, \exp(\lambda(1-\theta) + \varepsilon) c_1\}.\end{aligned}$$

By $\varepsilon \downarrow 0$ we get that

$$|H(\theta+it)| \leq \max\{\exp(-\lambda \theta) c_0, \exp(\lambda(1-\theta)) c_1\}$$

and the choice of $\lambda \in \mathbb{R}$ such that $\exp(\lambda) = \frac{c_0}{c_1}$ yields

$$\sup_{t \in \mathbb{R}} |H(\theta+it)| \leq c_0^{1-\theta} c_1^\theta.$$

By taking the infimum over all admissible $c_0, c_1 > 0$ the assertion follows. $\square$

Proof of Theorem A.8.1 Let $1 = \frac{1}{q} + \frac{1}{q'} = \frac{1}{q_i} + \frac{1}{q'_i}$ with $q', q'_i \in [1, \infty]$. By density it suffices to prove

$$\left| \int_N (Tf) b \, d\nu \right| \leq c_0^{1-\theta} c_1^{\theta} \quad \text{for} \quad f \in S(M, \mathcal{M}, \mu; \mathbb{C}) \quad \text{and} \quad b \in S(N, \mathcal{N}, \nu; \mathbb{C})$$

with $\|f\|_{L_p} = \|b\|_{L_{q'}} = 1$. We define $F(z) : M \to \mathbb{C}$ and $B(z) : N \to \mathbb{C}$ by

$$F(z) := |f|^{(1-z)\frac{p}{p_0} + z\frac{p}{p_1}} \frac{f}{|f|} \mathbb{1}_{\{f \neq 0\}} \quad \text{and} \quad B(z) := |b|^{(1-z)\frac{q'}{q'_0} + z\frac{q'}{q'_1}} \frac{b}{|b|} \mathbb{1}_{\{b \neq 0\}}$$

so that by assumption $F(\theta) = f$ and $B(\theta) = b$. With this definition we get

$$\|F(it)\|_{L_{p_0}} = \||f|^{\frac{p}{p_0}}\|_{L_{p_0}} = 1,$$

$$\|F(1+it)\|_{L_{p_1}} = \||f|^{\frac{p}{p_1}}\|_{L_{p_1}} = 1,$$

$$\|B(it)\|_{L_{q'_0}} = \||b|^{\frac{q'}{q'_0}}\|_{L_{q'_0}} = 1,$$

$$\|B(1+it)\|_{L_{q'_1}} = \||b|^{\frac{q'}{q'_1}}\|_{L_{q'_1}} = 1.$$

Furthermore, let

$$H(z) := \int_N (TF(z)) B(z) \, d\nu.$$

More precisely, if $f = \sum_{i=1}^m \alpha_i \mathbb{1}_{A_i}$ and $b = \sum_{j=1}^n \beta_j \mathbb{1}_{B_j}$ with $\alpha_i, \beta_j \neq 0$, then

$$H(z) = \sum_{i=1}^m \sum_{j=1}^n |\alpha_i|^{(1-z)\frac{p}{p_0} + z\frac{p}{p_1}} |\beta_j|^{(1-z)\frac{q'}{q'_0} + z\frac{q'}{q'_1}} \frac{\alpha_i}{|\alpha_i|} \frac{\beta_j}{|\alpha_j|} \int_{B_j} (T\mathbb{1}_{A_i}) \, d\nu.$$

With this definition we get that

$$|H(it)| \leq c_0 \|F(it)\|_{L_{p_0}} \|B(it)\|_{L_{(q_0)'}} \leq c_0,$$

$$|H(1+it)| \leq c_1 \|F(1+it)\|_{L_{p_1}} \|B(1+it)\|_{L_{(q_1)'}} \leq c_1.$$

Consequently, $F(\theta) = f$ and $B(\theta) = b$ and Theorem A.8.2 imply that

$$\left| \int_N (Tf) b \, d\nu \right| = \left| \int_N (TF(\theta)) B(\theta) \, d\nu \right| = |H(\theta)| \leq \sup_{t \in \mathbb{R}} |H(\theta + it)| \leq c_0^{1-\theta} c_1^{\theta}.$$

$\square$

A.9 Ciesielski's Theorems

In this section we present results developed by Ciesielski in a series of articles [37–39] that describe the modulus of continuity of continuous functions by means of series expansions consisting of Schauder functions. We consider here the easiest case, Hölder continuous functions defined on the interval $[0, 1]$. An extension to higher dimensions and its connections to Schauder bases can be found in [41].

We start by recalling the definition of the Haar functions $h_0, h_{pm} : [0, 1] \to \mathbb{R}$: We set $h_0 :\equiv 1$ and

$$h_{pm}(t) := \begin{cases} 1 & : t \in \left[\frac{m-1}{2^p}, \frac{2m-1}{2^{p+1}}\right) \\ -1 & : t \in \left[\frac{2m-1}{2^{p+1}}, \frac{m}{2^p}\right) \\ 0 & : \text{otherwise} \end{cases}$$

for $p = 0, 1, 2, \ldots$ and $m = 1, \ldots, 2^p$. We get that

$$\|h_0\|_{L_\infty([0,1])} = \|h_{pm}\|_{L_\infty([0,1])} = 1.$$

By integrating the Haar functions we obtain the Schauder functions

$$s_0(t) := t \quad \text{and} \quad s_{pm}(t) := \int_{[0,t]} h_{pm}(s) d\lambda(s)$$

with $\|s_0\|_{L_\infty([0,1])} = 1$ and $\|s_{pm}\|_{L_\infty([0,1])} = 2^{-(p+1)}$. Moreover, given a continues function $f : [0, 1] \to \mathbb{R}$ with $f(0) = 0$ we let

$$a_0 := f(1) - f(0) = f(1),$$
$$a_{pm} := -2^p \left[f(L_{pm}) - 2f(M_{pm}) + f(R_{pm}) \right],$$
$$A_p(f) := \sup_{m=1,\ldots,2^p} |a_{pm}|,$$

with $[L_{pm}, R_{pm}] := \overline{\{t \in [0, 1] : h_{pm}(t) \neq 0\}}$ and $M_{pm} := \frac{1}{2}\left[L_{pm} + R_{pm}\right]$.

The classical Ciesielski theorem we want to present is the equivalence obtained from Theorem A.9.1 and Corollary A.9.5 which gives the characterization of the Hölder continuous functions in term of expansions into Schauder functions. Corollary A.9.6 is another version of Corollary A.9.5 which is used to describe the modulus of continuity of the Brownian motion.

For $f : [0, 1] \to \mathbb{R}$ and $\theta \in (0, 1)$ we will use the Hölder norm

$$|f|_\theta = \sup_{x \neq y} \frac{|f(x) - f(y)|}{|x - y|^\theta}.$$

Theorem A.9.1 (Ciesielski I) *Let $\theta \in (0, 1)$ and let $f : [0, 1] \to \mathbb{R}$ be θ-Hölder continuous with $f(0) = 0$. Then one has*

(1) $|a_0| \leq |f|_\theta$ and $\sup_{p \geq 0} 2^{(\theta-1)p} A_p(f) \leq 2^{1-\theta} |f|_\theta$,
(2) $\sum_{p=0}^\infty \sum_{m=1}^{2^p} |a_{pm}| s_{pm}(t) < \infty$ for $t \in [0, 1]$,
(3) $f(t) = a_0 s_0(t) + \sum_{p=0}^\infty \sum_{m=1}^{2^p} a_{pm} s_{pm}(t)$ for $t \in [0, 1]$.

Proof
(1) By the Hölder continuity of f we obtain that $|a_0| \leq |f|_\theta$ and

$$|a_{pm}| \leq 2^p |f|_\theta [|M_{pm} - L_{pm}|^\theta + |R_{pm} - M_{pm}|^\theta] \leq 2 \cdot 2^p |f|_\theta 2^{-(p+1)\theta}$$
$$= 2^{(p+1)(1-\theta)} |f|_\theta.$$

(2) follows from (1) because the functions $s_{p1}, \ldots, s_{p2^p}$ have disjoint supports so that

$$\sum_{p=0}^\infty \sum_{m=1}^{2^p} |a_{pm}| s_{pm}(t) \leq \sum_{p=0}^\infty \left[2^{(p+1)(1-\theta)} |f|_\theta\right] 2^{-(p+1)} = |f|_\theta \sum_{p=0}^\infty 2^{-(p+1)\theta} < \infty.$$

(3) For $P \in \mathbb{N}_0$ we consider the dyadic grid points $t_k^P := \frac{k}{2^P}$ for $k = 0, \ldots, 2^P$. First we show that

$$f(t_k^P) = a_0 s_0(t_k^P) + \sum_{p=0}^\infty \sum_{m=1}^{2^p} a_{pm} s_{pm}(t_k^P) \quad \text{for} \quad P \in \mathbb{N}_0 \text{ and } k = 0, \ldots, 2^P. \tag{A.8}$$

$P = 0$: As for $t \in \{0, 1\}$ the above right-hand side equals $a_0 s_0(t)$ we only need to check $f(t) = a_0 s_0(t) = a_0 t$ for $t \in \{0, 1\}$. But this follows from the definition of a_0.

$P \in \mathbb{N}$: Here we have that $\sum_{p=P}^\infty \sum_{m=1}^{2^p} a_{pm} s_{pm}(t_k^P) = 0$ since by definition $s_{pm}(t_k^P) = 0$ for all $p \geq P$ and $m = 1, \ldots, 2^p$. Therefore we only need to check that

$$f(t_k^P) = a_0 s_0(t_k^P) + \sum_{p=0}^{P-1} \sum_{m=1}^{2^p} a_{pm} s_{pm}(t_k^P).$$

A Analysis

Letting

$$f_P(t) := a_0 \int_{[0,t]} h_0(s)\,\mathrm{d}\lambda(s) + \sum_{p=0}^{P-1}\sum_{m=1}^{2^p} a_{pm} \int_{[0,t]} h_{pm}(s)\,\mathrm{d}\lambda(s)$$

$$= \int_{[0,t]} \left[a_0 h_0(s) + \sum_{p=0}^{P-1}\sum_{m=1}^{2^p} a_{pm} h_{pm}(s) \right] \mathrm{d}\lambda(s),$$

for all $t \in [0,1] \setminus \{t_k^P : k = 0, \ldots, 2^P\}$, we have that

$$(f_P)'(t) = a_0 h_0(t) + \sum_{p=0}^{P-1}\sum_{m=1}^{2^p} a_{pm} h_{pm}(t)$$

and set $(f_P)'(t_k^P) := 0$. With this we get that

$$\int_{[0,1]} (f_P)'(t) h_0(t)\,\mathrm{d}\lambda(t) = a_0 \quad \text{and} \quad \int_{[0,1]} (f_P)'(t) h_{pm}(t)\,\mathrm{d}\lambda(t) = a_{pm} 2^{-P}.$$

Using the continuity of f_P and the definitions of h_0 and h_{pm} we compute

$$a_0 = f_P(1) - f_P(0) \quad \text{and} \quad a_{pm} = -2^p \left[f_P(L_{pm}) - 2 f_P(M_{pm}) + f_P(R_{pm}) \right],$$

which means $f(1) - f(0) = f_P(1) - f_P(0)$ and

$$f(L_{pm}) - 2f(M_{pm}) + f(R_{pm}) = f_P(L_{pm}) - 2f_P(M_{pm}) + f_P(R_{pm}).$$

This implies that $f_P(t_k^P) = f(t_k^P)$ for $k = 0, \ldots, 2^P$.

So we proved (A.8) and can conclude: Because f and

$$t \mapsto a_0 s_0(t) + \sum_{p=0}^{\infty}\sum_{m=1}^{2^p} a_{pm} s_{pm}(t)$$

are continuous functions on $[0,1]$, the relation (A.8) implies item (3). □

To prove the converse to Theorem A.9.1 in Corollary A.9.5 below we need some preparations and start with a lemma that provides even a more general variant of a converse to Theorem A.9.1:

Lemma A.9.2 *Assume a nondecreasing function $w : [0,\infty) \to (0,\infty)$ with*

$$\sum_{p=0}^{P} w(p) \leq \underline{\kappa} w(P) \quad \text{and} \quad \sum_{p=P+1}^{\infty} 2^{-P} w(p) \leq \overline{\kappa} 2^{-P} w(P) \tag{A.9}$$

for some $\underline{\kappa}, \overline{\kappa} > 0$ and $P \in \mathbb{N}_0$. Let $f : [0, 1] \to \mathbb{R}$ be continuous and such that for each $t \in [0, 1]$ the sum $f(t) = a_0 s_0(t) + \sum_{p=0}^{\infty} \sum_{m=1}^{2^p} a_{pm} s_{pm}(t)$ converges absolutely. Assume a constant $c > 0$ such that

$$A_p(f) = \sup_{m=1,\ldots,2^p} |a_{pm}| \leq cw(p) \quad \text{for} \quad p \in \mathbb{N}_0.$$

Then one has for $s, t \in [0, 1]$ with $s \neq t$ that

$$|f(t) - f(s)| \leq |t - s| \left[|a_0| + c \left[\underline{\kappa} + \overline{\kappa} \right] w \left(\log_2 \frac{2}{|t - s|} \right) \right].$$

Proof Given $P \in \mathbb{N}$ we write

$$f(t) = a_0 s_0(t) + \sum_{p=0}^{\infty} \sum_{m=1}^{2^p} a_{pm} s_{pm}(t)$$

$$= \left[a_0 s_0(t) + \sum_{p=0}^{P} \sum_{m=1}^{2^p} a_{pm} s_{pm}(t) \right] + \left[\sum_{p=P+1}^{\infty} \sum_{m=1}^{2^p} a_{pm} s_{pm}(t) \right]$$

$$=: f_1^P(t) + f_\infty^P(t).$$

We may assume $s \neq t$ and choose $P \in \mathbb{N}$ such that $2^{-P} < |t - s| \leq 2^{1-P}$. Then we get

$$|f(t) - f(s)| \leq |f_1^P(t) - f_1^P(s)| + |f_\infty^P(t) - f_\infty^P(s)|$$

$$\leq |f_1^P|_1 |t - s| + 2 \sup_{u \in [0,1]} |f_\infty^P(u)|$$

$$\leq \left[|a_0| + \sum_{p=0}^{P} A_p \right] |t - s| + 2 \sum_{p=P+1}^{\infty} A_p 2^{-p-1}$$

$$= \left[|a_0| + \sum_{p=0}^{P} A_p \right] |t - s| + \sum_{p=P+1}^{\infty} 2^{-p} A_p$$

$$\leq \left[|a_0| + c \sum_{p=0}^{P} w(p) \right] |t - s| + c \sum_{p=P+1}^{\infty} 2^{-p} w(p)$$

$$\leq \left[|a_0| + c\underline{\kappa} w(P) \right] |t - s| + c\overline{\kappa} 2^{-P} w(P)$$

A Analysis

$$\leqslant |t-s|\Big[|a_0| + c\big[\underline{\kappa} + \overline{\kappa}\big] w(P)\Big]$$

$$\leqslant |t-s|\Big[|a_0| + c\big[\underline{\kappa} + \overline{\kappa}\big] w\Big(\log_2 \frac{2}{|t-s|}\Big)\Big].$$

□

Now we give two examples for the condition (A.9):

Example A.9.3 For $w(x) := 2^{\eta x}$ with $\eta \in (0,1)$ the condition (A.9) is satisfied with $\underline{\kappa} = \frac{2^{\eta}}{2^{\eta}-1}$ and $\overline{\kappa} = \frac{1}{2^{1-\eta}-1}$. Indeed, one has the estimates

$$\sum_{p=0}^{P} w(p) = \sum_{p=0}^{P} 2^{\eta p} = \frac{2^{\eta(P+1)} - 1}{2^{\eta} - 1} \leqslant \frac{2^{\eta}}{2^{\eta}-1} w(P)$$

and

$$\sum_{p=P+1}^{\infty} 2^{-p} w(p) = \sum_{p=P+1}^{\infty} 2^{-p} 2^{\eta p} = \frac{1}{2^{1-\eta}-1} 2^{-P} w(P).$$

Example A.9.4 For $w(x) := 2^{\frac{x}{2}} \sqrt{x+1}$ the condition (A.9) is satisfied with $\underline{\kappa} = \frac{\sqrt{2}}{\sqrt{2}-1}$ and $\overline{\kappa} = \sum_{p=1}^{\infty} 2^{-\frac{p}{2}} \sqrt{p+1}$. Similarly to Example A.9.3 one has the estimates

$$\sum_{p=0}^{P} w(p) \leqslant \sqrt{P+1} \sum_{p=0}^{P} 2^{\frac{p}{2}} \leqslant \sqrt{P+1} \left[\frac{\sqrt{2}}{\sqrt{2}-1}\right] 2^{\frac{P}{2}}$$

and

$$\sum_{p=P+1}^{\infty} 2^{-p} w(p) = \sum_{p=P+1}^{\infty} 2^{-p} 2^{\frac{p}{2}} \sqrt{p+1}$$

$$= 2^{-\frac{P}{2}} \sqrt{P+1} \sum_{p=1}^{\infty} 2^{-\frac{p}{2}} \sqrt{\frac{p+P+1}{P+1}}$$

$$\leqslant \left[\sum_{p=1}^{\infty} 2^{-\frac{p}{2}} \sqrt{p+1}\right] 2^{-\frac{P}{2}} \sqrt{P+1}.$$

Combining the two above examples with Lemma A.9.2 yields to two statements regarding a converse implication to Theorem A.9.1.

Corollary A.9.5 (Ciesielski II) *If $\theta \in (0, 1)$ and $f : [0, 1] \to \mathbb{R}$ is a continuous function such that $f(0) = 0$ and $A_p(f) \leqslant c2^{(1-\theta)p}$ for $p \in \mathbb{N}_0$ and for some $c \geqslant 0$, then we have*

$$|f(t) - f(s)| \leqslant |a_0||t - s| + c[\underline{\kappa} + \overline{\kappa}]2^{1-\theta}|t - s|^\theta,$$

where $\underline{\kappa} = \frac{2^{1-\theta}}{2^{1-\theta}-1}$ and $\overline{\kappa} = \frac{1}{2^\theta - 1}$.

Proof For $x \geqslant 0$ we let $w(x) := 2^{(1-\theta)x}$. Then we use Lemma A.9.2 together with Example A.9.3 and

$$w\left(\log_2 \frac{2}{|t-s|}\right) = \left(\frac{2}{|t-s|}\right)^{1-\theta}$$

to derive for $s \neq t$ that

$$|f(t) - f(s)| \leqslant |t - s||a_0| + |t - s|c[\underline{\kappa} + \overline{\kappa}]\left(\frac{2}{|t-s|}\right)^{1-\theta}.$$

$\square$

Corollary A.9.6 (Ciesielski III) *If $f : [0, 1] \to \mathbb{R}$ is a continuous function such that $f(0) = 0$ and*

$$A_p(f) \leqslant c2^{\frac{p}{2}}\sqrt{1+p} \quad \text{for } p \in \mathbb{N}_0 \text{ and for some } c \geqslant 0,$$

then we have

$$|f(t) - f(s)| \leqslant |a_0||t - s| + c\sqrt{2}[\underline{\kappa} + \overline{\kappa}]\sqrt{|t-s|\left(1 + \log_2 \frac{2}{|t-s|}\right)}$$

with $\underline{\kappa} = \frac{\sqrt{2}}{\sqrt{2}-1}$ and $\overline{\kappa} = \sum_{p=1}^\infty 2^{-\frac{p}{2}}\sqrt{p+1}$.

Proof For $x \geqslant 0$ we let $w(x) := \sqrt{1+x}\, 2^{\frac{x}{2}}$. Then we use Lemma A.9.2 together with Example A.9.4 and

$$w\left(\log_2 \frac{2}{|t-s|}\right) = \sqrt{1 + \log_2 \frac{2}{|t-s|}}\, 2^{\frac{\log_2 \frac{2}{|t-s|}}{2}} = \sqrt{1 + \log_2 \frac{2}{|t-s|}}\sqrt{\frac{2}{|t-s|}}$$

to derive

$$|f(t) - f(s)| \leqslant |a_0||t - s| + c\sqrt{2}[\underline{\kappa} + \overline{\kappa}]\sqrt{|t-s|\left(1 + \log_2 \frac{2}{|t-s|}\right)}.$$

$\square$

Appendix B
Measure and Integration Theory

In this chapter the monotone class theorems for sets and functions and the π-λ-theorem are proven. The chapter continues with the proof of the extension theorem based on outer measures due to Carathéodory and concludes with the Lebesgue σ-algebra including two constructions of non-Lebesgue measurable sets. So we prove statements that we already used, but also present supplementary results that are recommended for those who want to go deeper into probability theory. Besides the particular references mentioned, we also used the monographs of Shiryaev [173], Schilling [168], and Tao [187].

B.1 Monotone Class Theorems and π-λ-Theorem

Definition B.1.1 (λ-System) Given $\Omega \neq \emptyset$, a collection $\mathcal{L}$ of subsets of Ω is called λ**-system** or **Dynkin**[1]**-system** if

(1) $\Omega \in \mathcal{L}$,
(2) $A, B \in \mathcal{L}$ and $A \subseteq B$ imply $B \setminus A \in \mathcal{L}$,
(3) $A_1, A_2, \dots \in \mathcal{L}$ and $A_i \subseteq A_{i+1}, i = 1, 2, \dots$ imply $\bigcup_{i=1}^\infty A_i \in \mathcal{L}$.

Definition B.1.2 (π-System) Given $\Omega \neq \emptyset$, a nonempty collection $\mathcal{P}$ of subsets of Ω is called π**-system** provided that

$$A \cap B \in \mathcal{P} \quad \text{for all} \quad A, B \in \mathcal{P}.$$

Any algebra is a π-system, but a π-system is not an algebra in general, take for example the π-system $\{(a, b) : -\infty < a < b < \infty\} \cup \{\emptyset\}$.

---

[1] Eugene Borisovich Dynkin, 11/05/1924 (Leningrad, USSR, now St Petersburg, Russia)–14/11/2014 (Ithaca, New York, USA).

© The Author(s), under exclusive license to Springer Nature Switzerland AG 2025
H. Geiss, S. Geiss, *Measure, Probability and Functional Analysis*, Universitext,
https://doi.org/10.1007/978-3-031-84067-8

Definition B.1.3 Given $\Omega \neq \emptyset$, a nonempty collection $\mathcal{M}$ of subsets of Ω is called **μ-system** or **monotone class** provided that

(1) $\bigcup_{i=1}^{\infty} A_i \in \mathcal{M}$ for all $A_1, A_2, \ldots \in \mathcal{M}$ with $A_1 \subseteq A_2 \subseteq A_3 \subseteq \cdots$,
(2) $\bigcap_{i=1}^{\infty} A_i \in \mathcal{M}$ for all $A_1, A_2, \ldots \in \mathcal{M}$ with $A_1 \supseteq A_2 \supseteq A_3 \supseteq \cdots$.

Now let us collect some operations and conditions on a nonempty collection $\mathcal{G} \subseteq 2^{\Omega}$, where $\Omega \neq \emptyset$ is an arbitrary set:

(C1) $\Omega \in \mathcal{G}$.
(C2) $\emptyset \in \mathcal{G}$.
(C3) $A \in \mathcal{G}$ implies $A^c \in \mathcal{G}$.
(C4) $A, B \in \mathcal{G}$ with $A \subseteq B$ implies $B \setminus A \in \mathcal{G}$.
(C5) $A, B \in \mathcal{G}$ implies $A \cup B \in \mathcal{G}$.
(C6) $A, B \in \mathcal{G}$ implies $A \cap B \in \mathcal{G}$.
(C7) $A_1, A_2, \cdots \in \mathcal{G}$ and $A_i \subseteq A_{i+1}, i = 1, 2, \ldots$ imply $\bigcup_{i=1}^{\infty} A_n \in \mathcal{G}$.
(C8) $A_1, A_2, \cdots \in \mathcal{G}$ and $A_i \supseteq A_{i+1}, i = 1, 2, \ldots$ imply $\bigcap_{i=1}^{\infty} A_n \in \mathcal{G}$.
(C9) $A_1, A_2, \cdots \in \mathcal{G}$ implies $\bigcup_{i=1}^{\infty} A_n \in \mathcal{G}$.
(C10) $A_1, A_2, \cdots \in \mathcal{G}$ implies $\bigcap_{i=1}^{\infty} A_n \in \mathcal{G}$.

With this notation we obtain the following table:

Definition	Associated with
Algebra	$C1, C2, C3, C5$
σ-algebra	$C1, C2, C3, C9$
λ-system	$C1, C4, C7$
μ-system	$C7, C8$
π-system	$C6$

Lemma B.1.4 Let $\Omega \neq \emptyset$, let $\mathcal{G}$ be a nonempty collection of subsets of Ω, and let $\emptyset \neq J \subseteq \{1, \ldots, 10\}$. Then there is a smallest collection $\rho(\mathcal{G}) \subseteq 2^{\Omega}$ such that

(1) $\mathcal{G} \subseteq \rho(\mathcal{G})$,
(2) $\rho(\mathcal{G})$ satisfies the conditions $(Cj)_{j \in J}$.

Proof The proof is a copy of the proof of Proposition 3.3.1. In fact, the collection 2^{Ω} satisfies $(C1), \ldots, (C10)$ and contains all sets from $\mathcal{G}$. So we let

$$\rho(\mathcal{G}) := \{A \subseteq \Omega : A \in \mathcal{F} \text{ for all } \mathcal{F} \in I\},$$

where I consists of all $\mathcal{F} \subseteq 2^{\Omega}$ that satisfy $(C_j)_{j \in J}$ and such that $\mathcal{G} \subseteq \mathcal{F}$. We have $I \neq \emptyset$ as $2^{\Omega} \in I$ and $\rho(\mathcal{G})$ is the collection of subsets of Ω we were looking for. □

Now let us start with Dynkin's π-λ-theorem:

Theorem B.1.5 (Dynkin's π-λ-Theorem) *For a π-system $\mathcal{P}$ one has*

$$\sigma(\mathcal{P}) = \lambda(\mathcal{P}),$$

where $\lambda(\mathcal{P})$ stands for the smallest λ-system which contains $\mathcal{P}$.

Proof As any σ-algebra is a λ-system, one has $\sigma(\mathcal{P}) \supseteq \lambda(\mathcal{P})$. So it remains to show that $\sigma(\mathcal{P}) \subseteq \lambda(\mathcal{P})$. For this it is sufficient to check that $\lambda(\mathcal{P})$ is a σ-algebra. Because an algebra that is a λ-system is a σ-algebra, we only need to verify that $\lambda(\mathcal{P})$ is an algebra. By properties (1) and (2) of Definition B.1.1 we have $\Omega, \emptyset \in \lambda(\mathcal{P})$ and that $A^c \in \lambda(\mathcal{P})$ whenever $A \in \lambda(\mathcal{P})$. Assume now $A, B \in \lambda(\mathcal{P})$. We wish to show that $A \cap B \in \lambda(\mathcal{P})$. We verify this in two steps:
Step 1: We fix $B \in \mathcal{P}$ and define

$$\mathcal{S}_B := \{A \in \lambda(\mathcal{P}) : A \cap B \in \lambda(\mathcal{P})\}.$$

From the definition of a π-system we see that $\mathcal{P} \subseteq \mathcal{S}_B$. Moreover, it is easy to check that $\mathcal{S}_B$ is a λ-system. Therefore $\mathcal{S}_B = \lambda(\mathcal{P})$, i.e. we checked that $A \cap B \in \lambda(\mathcal{P})$ for all $A \in \lambda(\mathcal{P})$ and $B \in \mathcal{P}$.
Step 2: By symmetry, we have $A \cap B \in \lambda(\mathcal{P})$ for all $A \in \mathcal{P}$ and $B \in \lambda(\mathcal{P})$ by Step 1. Now we fix $B \in \lambda(\mathcal{P})$ and recall the argument from Step 1. This gives finally that $A \cap B \in \lambda(\mathcal{P})$ for all $A, B \in \lambda(\mathcal{P})$ and the proof is complete. $\square$

The next two statements are monotone class theorems, one for sets, the other one for functions:

Theorem B.1.6 (Monotone Class Theorem for Sets) *For an algebra $\mathcal{A}$ one has*

$$\sigma(\mathcal{A}) = \mu(\mathcal{A}),$$

where $\mu(\mathcal{A})$ stands for the smallest monotone class which contains $\mathcal{A}$.

Proof The proof uses the same methodology as the proof of *Dynkin*'s π-λ-Theorem. As any σ-algebra is a μ-system, one has $\sigma(\mathcal{A}) \supseteq \mu(\mathcal{A})$. So it remains to show that $\sigma(\mathcal{A}) \subseteq \mu(\mathcal{A})$. For this it is sufficient to check that $\mu(\mathcal{A})$ is a σ-algebra and by (1) of Definition B.1.3 only that $\mu(\mathcal{A})$ is an algebra.
(a) Since $\mathcal{A} \subseteq \mu(\mathcal{A})$, we have that $\Omega, \emptyset \in \mu(\mathcal{A})$.
(b) We let

$$\mathcal{C} := \{A \in \mu(\mathcal{A}) : A^c \in \mu(\mathcal{A})\}.$$

We have that $\mathcal{A} \subseteq \mathcal{C}$ as $\mathcal{A}$ is an algebra and $\mathcal{A} \subseteq \mu(\mathcal{A})$, and one easily checks that $\mathcal{C}$ is a monotone class. Hence $\mathcal{C} = \mu(\mathcal{A})$ and $A \in \mu(\mathcal{A})$ if and only if $A^c \in \mu(\mathcal{A})$.
(c) Finally, we prove that $A \cup B \in \mu(\mathcal{A})$ whenever $A, B \in \mu(\mathcal{A})$ and do this again in two steps:
Step 1: We fix $B \in \mathcal{A}$ and consider

$$\mathcal{U}_B := \{A \in \mu(\mathcal{A}) : A \cup B \in \mu(\mathcal{A})\}.$$

As $\mathcal{A}$ is an algebra, $\mathcal{A} \subseteq \mathcal{U}_B$. Moreover, one can see that $\mathcal{U}_B$ is a monotone class. Therefore, $\mathcal{U}_B = \mu(\mathcal{A})$, or $A \cup B \in \mu(\mathcal{A})$ whenever $B \in \mathcal{A}$ and $A \in \mu(\mathcal{A})$.
Step 2: By symmetry it holds that $A \cup B \in \mu(\mathcal{A})$ whenever $A \in \mathcal{A}$ and $B \in \mu(\mathcal{A})$. Now we recall step 1 for $B \in \mu(\mathcal{A})$, which gives the desired result. □

The twin of Theorem B.1.6 is the monotone class theorem for functions:

Theorem B.1.7 (Monotone Class Theorem for Functions) *Let $\mathcal{P}$ be a π-system that contains Ω. Assume that for a set $\mathcal{H}$ of functions $f : \Omega \to \mathbb{R}$ the following holds:*

(1) $\mathbb{1}_A \in \mathcal{H}$ for $A \in \mathcal{P}$.
(2) $\mathcal{H}$ is a linear space.
(3) If $(f_n)_{n=1}^\infty \subseteq \mathcal{H}$ such that $0 \leqslant f_n \uparrow f$ and if f is bounded, then $f \in \mathcal{H}$.

Then $\mathcal{H}$ contains all bounded functions that are $\sigma(\mathcal{P})$-measurable.

Proof We follow [84, Theorem I.1.4] and define

$$\mathcal{L} := \{A \subseteq \Omega : \mathbb{1}_A \in \mathcal{H}\}.$$

First we verify that $\mathcal{L}$ is a λ-system: Assumptions $\Omega \in \mathcal{P}$ and (1) imply that $\Omega \in \mathcal{L}$. Assumption (2) yields for $A, B \in \mathcal{L}$ with $A \subseteq B$ that $B \setminus A \in \mathcal{L}$. Finally, (3) implies Definition B.1.1(3). Therefore $\mathcal{L}$ is a λ-system with $\mathcal{P} \subseteq \mathcal{L}$. Exploiting Theorem B.1.5 we have that

$$\sigma(\mathcal{P}) = \lambda(\mathcal{P}) \subseteq \lambda(\mathcal{L}) = \mathcal{L}.$$

Therefore we get that $\mathbb{1}_A \in \mathcal{H}$ for all $A \in \sigma(\mathcal{P})$. By (2) all $\sigma(\mathcal{P})$-measurable simple functions are in $\mathcal{H}$. If $f : \Omega \to [0, \infty)$ is a bounded $\sigma(\mathcal{P})$-measurable function and $0 \leqslant f_n(\omega) \uparrow f(\omega)$, $\omega \in \Omega$, is a sequence of $\sigma(\mathcal{P})$-measurable simple functions f_n, (3) implies that $f \in \mathcal{H}$. The case of general bounded f is handled by decomposing f into its positive and negative part. □

Proof of Theorem 3.2.4 Define the collection

$$\mathcal{L} := \{A \in \mathcal{F} : \mu(A) = \nu(A)\} \supseteq \mathcal{P}.$$

If we can prove that $\mathcal{L}$ is a λ-system, then Theorem B.1.5 implies that $\mathcal{F} = \sigma(\mathcal{P}) \subseteq \mathcal{L}$ and the statement follows. Property (1) of Definition B.1.1 follows by $\Omega \in \mathcal{P}$, property (2) by the finite additivity, and property (3) from the monotonicity of the measures μ and ν from below. □

B.2 Outer Measures

We start with the notion of an outer measure which goes back to Carathéodory[2].

Definition B.2.1 Let $\Omega \neq \emptyset$. A map $\mu : 2^\Omega \to [0, \infty]$ is called **outer measure** provided that

(1) $\mu(\emptyset) = 0$,
(2) $\mu(B_1) \leq \mu(B_2)$ for all $B_1 \subseteq B_2$,
(3) $\mu(\bigcup_{i=1}^\infty B_i) \leq \sum_{i=1}^\infty \mu(B_i)$ for all $B_i \subseteq \Omega$.

Moreover, we let

$$\mathcal{F}^\mu := \{A \subseteq \Omega : \mu(B) = \mu(B \cap A) + \mu(B \cap A^c) \text{ for all } B \subseteq \Omega\} \quad (B.1)$$

be the collection of μ-**measurable sets**.

By the following statement the collection $\mathcal{F}^\mu$ turns out to be the collection of sets $A \subseteq \Omega$, on which the outer measure μ behaves like a measure:

Theorem B.2.2 $(\Omega, \mathcal{F}^\mu, \mu)$ is a complete measure space.

Proof Before we start we remark that we also have that

$$\mu\left(\bigcup_{i=1}^n B_i\right) \leq \sum_{i=1}^n \mu(B_i)$$

by setting $\emptyset = B_{n+1} = B_{n+2} = \cdots$ and using $\mu(\emptyset) = 0$.

(a) Because $\mu(B \cap \emptyset) + \mu(B \cap \Omega) = \mu(B)$ we have that $\emptyset, \Omega \in \mathcal{F}^\mu$.
(b) By the symmetry of the definition of $\mathcal{F}^\mu$ with respect to A we also have that $A \in \mathcal{F}^\mu$ if and only if $A^c \in \mathcal{F}^\mu$.
(c) Now we assume $A_1, A_2 \in \mathcal{F}^\mu$ and prove that $A_1 \cap A_2 \in \mathcal{F}^\mu$: This follows from

$$\mu(B) = \mu(B \cap A_1) + \mu(B \cap A_1^c)$$
$$= \mu(B \cap A_1 \cap A_2) + \mu(B \cap A_1 \cap A_2^c) + \mu(B \cap A_1^c)$$
$$\geq \mu(B \cap A_1 \cap A_2) + \mu\left((B \cap A_1 \cap A_2^c) \cup (B \cap A_1^c)\right)$$

---

[2] Constantin Carathéodory, 13/09/1873 (Berlin, Germany)–02/02/1950 (Munich, Germany).

$$= \mu(B \cap A_1 \cap A_2) + \mu\bigl(B \cap (A_1 \cap A_2)^c\bigr)$$
$$\geq \mu(B).$$

Therefore, $\mathcal{F}^\mu$ is an algebra.

(d) Now let us assume that $A_1, A_2, \ldots \in \mathcal{F}^\mu$. By using that $\mathcal{F}^\mu$ is an algebra we may assume that the A_n are pairwise disjoint while proving that

$$\mu(B) \geq \mu\left(B \cap \left(\bigcup_{i=1}^\infty A_i\right)\right) + \mu\left(B \cap \left(\bigcup_{i=1}^\infty A_i\right)^c\right).$$

By the fact that $\mathcal{F}^\mu$ is an algebra we know that

$$\mu(B) = \mu\left(B \cap \left(\bigcup_{i=1}^n A_i\right)\right) + \mu\left(B \cap \left(\bigcup_{i=1}^n A_i\right)^c\right)$$
$$= \sum_{i=1}^n \mu(B \cap A_i) + \mu\left(B \cap \left(\bigcup_{i=1}^n A_i\right)^c\right)$$

where the second equality follows from a successive application of the definition of $\mathcal{F}^\mu$. By $n \to \infty$ this implies that

$$\mu(B) \geq \sum_{i=1}^\infty \mu(B \cap A_i) + \mu\left(B \cap \left(\bigcup_{i=1}^\infty A_i\right)^c\right)$$
$$\geq \mu\left(B \cap \left(\bigcup_{i=1}^\infty A_i\right)\right) + \mu\left(B \cap \left(\bigcup_{i=1}^\infty A_i\right)^c\right)$$
$$\geq \mu(B)$$

so that we are done.

(e) Now we prove that μ is a measure on $\mathcal{F}^\mu$. Assuming pairwise disjoint $A_1, A_2, \ldots \in \mathcal{F}^\mu$ we have that

$$\mu\left(\bigcup_{i=1}^\infty A_i\right) \leq \sum_{i=1}^\infty \mu(A_i)$$

by the definition of the outer measure. To check the converse inequality, we use step (d) for $B := \bigcup_{i=1}^\infty A_i$ and get

$$\mu\left(\bigcup_{i=1}^\infty A_i\right) \geq \sum_{i=1}^\infty \mu(A_i).$$

(f) To verify the completeness we let $A \in \mathcal{F}^\mu$ with $\mu(A) = 0$ and $N \subseteq A$. Then for $B \subseteq \Omega$ we obtain

$$\mu(B) \leqslant \mu(B \cap N) + \mu(B \cap N^c) \leqslant \mu(A) + \mu(B) = \mu(B)$$

so that $N \in \mathcal{F}^\mu$. □

Now we prove the extension theorem and the statement about outer regularity:

Proof of Theorems 3.2.1 and 3.2.2 We define

$$\mu_0^*(B) := \inf \left\{ \sum_{i=1}^\infty \mu_0(A_i) : B \subseteq \bigcup_{i=1}^\infty A_i, A_i \in \mathcal{A} \right\}. \quad (B.2)$$

(a) We have $\mu_0(\emptyset) = 0$: Take $A_1 := \Omega_1$ and $A_n := \emptyset$ for $n \geqslant 2$, we get that

$$\infty > \mu_0(\Omega_1) = \mu_0\left(\bigcup_{n=1}^\infty A_n\right) = \mu_0(\Omega_1) + \sum_{n=2}^\infty \mu_0(\emptyset)$$

so that necessarily $\mu_0(\emptyset) = 0$.

(b) Similar to (a) one shows finite additivity on $\mathcal{A}$, so that for $A_1, A_2 \in \mathcal{A}$ with $A_1 \subseteq A_2$ it follows that $\mu_0(A_2) = \mu_0(A_2 \setminus A_1) + \mu_0(A_1) \geqslant \mu_0(A_1)$.

(c) We check the properties (1), (2), and (3) for μ_0^* being an outer measure:
  (1) Because $\mu_0(\emptyset) = 0$, we have $\mu_0^*(\emptyset) = 0$.
  (2) Any covering of B_2 is a covering of B_1, so that $\mu_0^*(B_1) \leqslant \mu_0^*(B_2)$.
  (3) Let $\varepsilon > 0$ and find coverings

$$B_i \subseteq \bigcup_{j=1}^\infty A_{i,j} \text{ with } A_{i,j} \in \mathcal{A} \text{ and } \sum_{j=1}^\infty \mu_0(A_{i,j}) \leqslant \frac{\varepsilon}{2^i} + \mu_0^*(B_i),$$

where on both sides the value infinity is allowed. Then $\bigcup_{i=1}^\infty B_i \subseteq \bigcup_{i,j=1}^\infty A_{i,j}$ and

$$\mu_0^*\left(\bigcup_{i=1}^\infty B_i\right) \leqslant \sum_{i,j=1}^\infty \mu_0(A_{i,j}) \leqslant \sum_{i=1}^\infty \left(\frac{\varepsilon}{2^i} + \mu_0^*(B_i)\right) = \varepsilon + \sum_{i=1}^\infty \mu_0^*(B_i).$$

Letting $\varepsilon \downarrow 0$, the subadditivity follows.

(d) One has $\mathcal{A} \subseteq \mathcal{F}^{\mu_0^*}$ (see formula (B.1)), so that $\sigma(\mathcal{A}) \subseteq \mathcal{F}^{\mu_0^*}$: Indeed, let $A \in \mathcal{A}$ and $B \subseteq \Omega$. We have to show that

$$\mu_0^*(B) = \mu_0^*(B \cap A) + \mu_0^*(B \cap A^c).$$

Because $\mu_0^*(B) \leqslant \mu_0^*(B \cap A) + \mu_0^*(B \cap A^c)$ follows from the property that μ_0^* is an outer measure, it remains to check that $\mu_0^*(B) \geqslant \mu_0^*(B \cap A) + \mu_0^*(B \cap A^c)$. For $\varepsilon > 0$ choose an $\mathcal{A}$-cover

$$B \subseteq \bigcup_{i=1}^{\infty} A_i \quad \text{with} \quad \sum_{i=1}^{\infty} \mu_0(A_i) \leqslant \mu_0^*(B) + \varepsilon.$$

For $B \cap A$ and $B \cap A^c$ we obtain the $\mathcal{A}$-covers

$$B \cap A \subseteq \bigcup_{i=1}^{\infty}(A_i \cap A) \quad \text{and} \quad B \cap A^c \subseteq \bigcup_{i=1}^{\infty}(A_i \cap A^c),$$

and that

$$\mu_0^*(B \cap A) \leqslant \sum_{i=1}^{\infty} \mu_0(A_i \cap A) \quad \text{and} \quad \mu_0^*(B \cap A^c) \leqslant \sum_{i=1}^{\infty} \mu_0(A_i \cap A^c).$$

By assumption on μ_0 this gives that

$$\mu_0^*(B \cap A) + \mu_0^*(B \cap A^c) \leqslant \sum_{i=1}^{\infty} \left(\mu_0(A_i \cap A) + \mu_0(A_i \cap A^c) \right)$$

$$= \sum_{i=1}^{\infty} \mu_0(A_i)$$

$$\leqslant \mu_0^*(B) + \varepsilon.$$

Again, letting $\varepsilon \downarrow 0$ yields the assertion.

(e) One has $\mu_0 = \mu_0^*$ on $\mathcal{A}$: Let $A \in \mathcal{A}$. Then A is an $\mathcal{A}$-cover for A so that $\mu_0^*(A) \leqslant \mu_0(A)$. Now we check $\mu_0(A) \leqslant \mu_0^*(A)$, i.e. that for any $\mathcal{A}$-cover $A \subseteq \bigcup_{i=1}^{\infty} A_i$ one has

$$\mu_0(A) \leqslant \sum_{i=1}^{\infty} \mu_0(A_i).$$

This follows from

$$\mu_0(A) = \mu_0 \left(\bigcup_{i=1}^{\infty} (A \cap A_i) \right) = \sum_{i=1}^{\infty} \mu_0(A \cap A_i) \leqslant \sum_{i=1}^{\infty} \mu_0(A_i).$$

(f) It remains to show the uniqueness. For the case $\mu_0(\Omega) < \infty$ this follows immediately from Theorem 3.2.4. The general case can be checked as follows:

We consider the trace σ-algebras of $\sigma(\mathcal{A})$ on Ω_l, and get that μ and ν coincide on these trace σ-algebras. This implies that μ and ν coincide globally. $\square$

Theorem B.2.3 *Assume the setting and notation from Theorem 3.2.1 and define the outer measure μ_0^* as in* (B.2). *Then*

$$\overline{\mathcal{F}}^\mu = \mathcal{F}^{\mu_0^*},$$

where $\overline{\mathcal{F}}^\mu$ is the completion of $\mathcal{F}$ with respect to μ according to Sect. 2.4.

Proof Since it holds that $\mathcal{F} \subseteq \mathcal{F}^{\mu_0^*}$, the inclusion $\overline{\mathcal{F}}^\mu \subseteq \mathcal{F}^{\mu_0^*}$ follows from the completeness of $\mathcal{F}^{\mu_0^*}$ proved in Theorem B.2.2. So we only need to verify $\mathcal{F}^{\mu_0^*} \subseteq \overline{\mathcal{F}}^\mu$.

(a) Let $A \in \mathcal{F}^{\mu_0^*}$ with $\mu(A) < \infty$. For all $n \in \mathbb{N}$ we find $\mathcal{A}$-covers with

$$A \subseteq \bigcup_{i=1}^\infty A_{n,i} \quad \text{and} \quad \sum_{i=1}^\infty \mu(A_{n,i}) \leq \mu(A) + \frac{1}{n}.$$

For $\overline{A} := \bigcap_{n=1}^\infty \left(\bigcup_{i=1}^\infty A_{n,i}\right)$ we deduce

$$A \subseteq \overline{A} \in \mathcal{F} \quad \text{and} \quad \mu(A) = \mu(\overline{A}).$$

Therefore, $\mu(\overline{A} \setminus A) = 0$ where we exploit $\mu(A) < \infty$. Now we repeat the construction from above for $N := \overline{A} \setminus A \in \mathcal{F}^{\mu_0^*}$ and find $\overline{N} \in \mathcal{F}$ with $\overline{N} \supseteq N = \overline{A} \setminus A$ and $\mu(\overline{N}) = 0$. Hence $\overline{A} \setminus A \in \overline{\mathcal{F}}^\mu$ and therefore $A \in \overline{\mathcal{F}}^\mu$.

(b) Let $A \in \mathcal{F}^{\mu_0^*}$ be general and let $A_l := A \cap \Omega_l$ so that $\mu(A_l) < \infty$. Then $A_l \in \overline{\mathcal{F}}^\mu$ by step (a) and $A = \bigcup_{l=1}^\infty A_l \in \overline{\mathcal{F}}^\mu$. $\square$

B.3 Countably Generated σ-Algebras

The following statement is intuitive, but its proof is involved and beyond the scope of this book. We use Theorem B.3.1 twice: firstly, to distinguish between the Borel and the Lebesgue σ-algebras in Appendix B.4, and secondly to show in Theorem 20.1.6 that the sum of two Banach space valued random variables is not necessarily measurable.

Theorem B.3.1 *Let $\Omega \neq \emptyset$, let $\mathcal{G} \subseteq 2^\Omega$ be a countable collection of subsets, and set $\mathcal{F} := \sigma(\mathcal{G})$. Then one has that $\#\mathcal{F} \leq \#\mathbb{R}$.*

For a recent proof of Theorem B.3.1 we refer to [99, Theorem 4].

B.4 The Lebesgue σ-Algebra

The Lebesgue σ-algebra on $\mathbb{R}^d$ can be approached in different ways, either by the completion of the Borel σ-algebra or by the extension Theorem 3.2.1. We take the first route and link this approach to the extension theorem in a second step.

Definition B.4.1 The Lebesgue σ-algebra $\mathcal{L}(\mathbb{R}^d)$ is the completion of the Borel σ-algebra $\mathcal{B}(\mathbb{R}^d)$ with respect to the Lebesgue measure λ_d in the sense of Sect. 2.4.

Inspired by Sect. 3.3.3, where we constructed the Lebesgue measure on $\mathcal{B}(\mathbb{R})$, we assume the following:

(1) We let $\mathcal{A}$ be an algebra of subsets from $\mathbb{R}^d$ such that $\sigma(\mathcal{A}) = \mathcal{B}(\mathbb{R}^d)$.
(2) Using the algebra $\mathcal{A}$ we define the outer measure

$$\lambda_d^*(B) := \inf\left\{\sum_{i=1}^{\infty} \lambda_d(A_i) : B \subseteq \bigcup_{i=1}^{\infty} A_i, A_i \in \mathcal{A}\right\},$$

where λ_d is the Lebesgue measure on $\mathcal{B}(\mathbb{R}^d)$.

The main result of this section is the link of $\mathcal{L}(\mathbb{R}^d)$ to the construction behind our proof of Theorem 3.2.1:

Theorem B.4.2

(1) $\mathcal{L}(\mathbb{R}^d) = \mathcal{F}^{\lambda_d^*}$, where $\mathcal{F}^{\lambda_d^*}$ is defined in (B.1).
(2) For $B \in \mathcal{L}(\mathbb{R}^d)$ and $\varepsilon > 0$ there exists a closed set $F \subseteq \mathbb{R}^d$ and an open $G \subseteq \mathbb{R}^d$ such that

$$F \subseteq B \subseteq G \quad \text{and} \quad \lambda_d(G \setminus F) \leqslant \varepsilon.$$

(3) $\#\mathcal{B}(\mathbb{R}^d) = \#\mathbb{R}$.
(4) $\#\mathcal{L}(\mathbb{R}^d) = \#2^{\mathbb{R}}$.
(5) There exists a set $H \subseteq (0, 1]$, called a Vitali set, such that $H \notin \mathcal{L}(\mathbb{R})$.

Proof

(1) follows from Theorem B.2.3.
(2) We apply Theorem B.2.3 and Proposition 2.4.2 to get $A, C \in \mathcal{B}(\mathbb{R}^d)$ with $A \subseteq B \subseteq C$ and $\lambda_d(C \setminus A) = 0$. Then we use Theorem 5.2.4(2) to find a closed set $F \subseteq A$ with $\lambda_d(A \setminus F) < \varepsilon/2$ and to find an open set $G \supseteq C$ with $\lambda_d(G \setminus C) < \varepsilon/2$. This implies our assertion as $\lambda_d(G \setminus F) \leqslant \varepsilon$.
(3) The inequality $\#\mathcal{B}(\mathbb{R}^d) \leqslant \#\mathbb{R}$ follows from Theorem B.3.1 as we can generate $\mathcal{B}(\mathbb{R}^d)$ by a countable collection, for example open cuboids with rational vertices. On the other hand, taking sets $A_n := (n-1, n] \times \mathbb{R}^{d-1}$, $n \in \mathbb{N}$, the set of all possible unions of these sets A_n (the unions belong to $\mathcal{B}(\mathbb{R}^d)$) has the cardinality of $\mathbb{R}$.

(4) From $\mathcal{L}(\mathbb{R}^d) \subseteq 2^{(\mathbb{R}^d)}$ it follows that $\#\mathcal{L}(\mathbb{R}^d) \leqslant \#2^{(\mathbb{R}^d)} = \#2^{\mathbb{R}}$. Since the Cantor set $C \subseteq [0, 1]$, $C \in \mathcal{B}(\mathbb{R})$, has the cardinality of $\mathbb{R}$ (see Exercise 5 in Chap. 10) and $\lambda_d(C \times \mathbb{R}^{d-1}) = 0$, we obtain that $\#\mathcal{L}(\mathbb{R}^d) \geqslant \#2^{\mathbb{R}}$.

(5) Given $x, y \in [0, 1)$ and $A \subseteq [0, 1)$, we use the addition modulo one

$$x \oplus y := \begin{cases} x + y & \text{if } x + y \in [0, 1) \\ x + y - 1 & \text{if } x + y \in [1, 2) \end{cases},$$

$$A \oplus x := \{a \oplus x : a \in A\}$$

and define on $[0, 1)$ the equivalence relation

$$x \sim y \quad \text{if and only if} \quad x \oplus r = y \quad \text{for some rational} \quad r \in [0, 1);$$

it is fairly easy to see that $x \sim y$ defines an equivalence relation. Using the axiom of choice Axiom A.6.1 we let $H \subseteq [0, 1)$ be a set consisting of exactly one representative from each equivalence class. Then $H \oplus r_1$ and $H \oplus r_2$ are disjoint for rational numbers $r_1, r_2 \in [0, 1)$ with $r_1 \neq r_2$. Indeed, if they were not disjoint, then there would exist $h_1, h_2 \in H$ with $h_1 \oplus r_1 = h_2 \oplus r_2$. But this implies $h_1 \sim h_2$ and hence $h_1 = h_2$ and $r_1 = r_2$. So it follows that $[0, 1)$ is the countable union of pairwise disjoint sets

$$[0, 1) = \bigcup_{r \in [0,1) \text{ rational}} (H \oplus r).$$

Assume that $H \in \mathcal{L}(\mathbb{R})$. If we would know that

$$H \oplus r \in \mathcal{L}(\mathbb{R}) \quad \text{and} \quad \lambda(H \oplus r) = \lambda(H) \tag{B.3}$$

for all $r \in [0, 1) \cap \mathbb{Q}$, then we would get

$$1 = \lambda([0, 1)) = \sum_{r \in [0,1) \text{ rational}} \lambda(H \oplus r) = \sum_{r \in [0,1) \text{ rational}} \lambda(H),$$

which is apparently impossible. To verify (B.3) we fix a rational number $r \in [0, 1)$ and define the function $f_r : \mathbb{R} \to \mathbb{R}$, which is a periodic translation on $[0, 1)$, by

$$f_r(x) := \begin{cases} x & : x \notin [0, 1) \\ x - r & : x \in [r, 1) \\ x - r + 1 & : x \in [0, r) \end{cases}. \tag{B.4}$$

The function is $(\mathcal{B}(\mathbb{R}), \mathcal{B}(\mathbb{R}))$-measurable and preserves the Lebesgue measure, i.e. $\lambda^{(r)} = \lambda$ on $\mathcal{B}(\mathbb{R})$ for $\lambda^{(r)} := \lambda \circ f_r^{-1}$. The latter follows from $\lambda^{(r)}((a, b)) =$

$\lambda((a, b))$ for $-\infty < a \leqslant b < \infty$ so that Theorem 3.2.4 implies $\lambda^{(r)} = \lambda$ on $\mathcal{B}((-n, n))$ for $n \in \mathbb{N}$ and therefore $\lambda^{(r)} = \lambda$ on $\mathcal{B}(\mathbb{R})$. Because $\lambda(f_r^{-1}(N)) = 0$ if $\lambda(N) = 0$, $N \in \mathcal{B}(\mathbb{R})$, we get that $f_r : \mathbb{R} \to \mathbb{R}$ is also $(\mathcal{L}(\mathbb{R}), \mathcal{L}(\mathbb{R}))$-measurable and preserves the Lebesgue measure as well (here one can use Proposition 2.4.2). Finally, we remark that $H \oplus r = f_r^{-1}(H)$ so that $\lambda(H \oplus r) = \lambda(H)$. □

Remark B.4.3 The proof of Theorem B.4.2(5) also says that the *Lebesgue* measure λ on $\mathcal{L}(\mathbb{R})$ cannot be extended to a measure μ on $2^{\mathbb{R}}$ in a way such that the periodic translations f_r from (B.4) are still measure preserving, i.e. $\mu(f_r^{-1}(B)) = \mu(B)$ for all $B \subseteq \mathbb{R}$.

We add another construction of a non-measurable set in Theorem B.4.5 below. To do so we need the following proposition which is of independent interest:

Proposition B.4.4 *Let $B \in \mathcal{L}(\mathbb{R})$ such that $\lambda(B) > 0$. Then there is an $\varepsilon > 0$ such that $(-\varepsilon, \varepsilon) \subseteq B - B$.*

Proof By taking one of the sets $B \cap (n, n + 1]$, $n \in \mathbb{Z}$, we may assume that $B \subseteq (n, n + 1]$ and that $\lambda(B \cap (n, n + 1]) \in (0, 1]$. So we may replace the set B by some $B \subseteq (n, n + 1]$ with $\lambda(B) \in (0, 1]$. Applying Theorem B.4.2(2) together with Exercise (3b) from Chap. 3 we find $n < a_l < b_l < n+2$ such that $B \subseteq \bigcup_{l \in I} (a_l, b_l)$, where I is countable, the intervals (a_l, b_l) are pairwise disjoint, and

$$\frac{3}{4} \leqslant \frac{\lambda(B)}{\sum_{l \in I}(b_l - a_l)} \leqslant 1.$$

Consequently there is one interval $(a, b) = (a_l, b_l)$ such that

$$\frac{\lambda(B \cap (a, b))}{b - a} \geqslant \frac{3}{4}.$$

Similarly as above, by replacing B by $B \cap (a, b)$ we can assume that $B \subseteq (a, b)$ as

$$B - B \supseteq A - A \quad \text{whenever} \quad \emptyset \neq A \subseteq B.$$

If for $|x| < \varepsilon := (b - a)/2$ we can show that

$$(x + B) \cap B \neq \emptyset, \tag{B.5}$$

then there is an $y \in (x + B) \cap B$ which implies $y \in B$, $z := y - x \in B$, and $x = y - z \in B - B$. To verify (B.5) we consider a subset $A \subseteq B$ such that $\lambda(A) = \frac{3}{4}|b - a|$. To find such a set, we choose $B_0 \in \mathcal{B}(\mathbb{R})$ with $B_0 \subseteq B$ and $\lambda(B_0) = \lambda(B)$, and consider the sections $B_0 \cap (-\infty, u]$ with $u \in \mathbb{R}$ (cf. Example 8.14.3). Then we observe that

$$\lambda((x + A) \cap (a, b)^c) < \frac{1}{2}|b - a|.$$

This implies that

$$\lambda((x+A) \cap (a,b)) + \lambda(A) > \frac{3}{4}|b-a| - \frac{1}{2}|b-a| + \frac{3}{4}|b-a| = |b-a|$$

and $(x+B) \cap B \supseteq (x+A) \cap A \neq \emptyset$. □

Now we are able to provide the second construction of a non-measurable set using the axiom of choice:

Theorem B.4.5 *There exists a set $A \subseteq [0,1]$ such that A and $[0,1] \setminus A$ do not contain any set of $\mathcal{L}(\mathbb{R})$ of positive measure. In particular, the set A cannot be Lebesgue-measurable.*

Proof We fix $\alpha \in \mathbb{R} \setminus \mathbb{Q}$ and define the sets $A_x := x + \mathbb{Z} + \alpha\mathbb{Z}$. The sets A_x are either disjoint or coincide and we obviously have that $\mathbb{R} = \bigcup_{x \in \mathbb{R}} A_x$. By the axiom of choice we find a set $C \subseteq \mathbb{R}$ such that for distinct $x, y \in C$ the sets A_x and A_y are disjoint and such that $\mathbb{R} = \bigcup_{x \in C} A_x$. Further, we define

$$A_x^{\text{even}} := x + 2\mathbb{Z} + \alpha\mathbb{Z} \quad \text{and} \quad A_x^{\text{odd}} := A_x^{\text{even}} + 1 = x + 2\mathbb{Z} + 1 + \alpha\mathbb{Z},$$

so that we have the disjoint union $A_x = A_x^{\text{even}} \cup A_x^{\text{odd}}$. We let

$$A^{\text{even}} := \bigcup_{x \in C} A_x^{\text{even}} \quad \text{and} \quad A^{\text{odd}} := \bigcup_{x \in C} A_x^{\text{odd}}$$

and get a disjoint union $\mathbb{R} = A^{\text{even}} \cup A^{\text{odd}}$. Finally, we restrict these sets to $[0,1]$ by

$$A_{[0,1]}^{\text{even}} := A^{\text{even}} \cap [0,1] \quad \text{and} \quad A_{[0,1]}^{\text{odd}} := A^{\text{odd}} \cap [0,1].$$

Again, we have the disjoint union $[0,1] = A_{[0,1]}^{\text{even}} \cup A_{[0,1]}^{\text{odd}}$. We know that

$$[(C-C) + 2\mathbb{Z} + \alpha\mathbb{Z}] \cap [2\mathbb{Z} + \alpha\mathbb{Z} + 1] = \emptyset,$$
$$[(C-C) + 2\mathbb{Z} + \alpha\mathbb{Z} + 1] \cap [2\mathbb{Z} + \alpha\mathbb{Z}] = \emptyset,$$

and that the sets $2\mathbb{Z} + \alpha\mathbb{Z}$ and $2\mathbb{Z} + \alpha\mathbb{Z} + 1$ are dense. The latter can be seen as follows: If $2\mathbb{Z}+\alpha\mathbb{Z}$ is dense, so $2\mathbb{Z}+\alpha\mathbb{Z}+1$ is. Therefore, we only need to see that $2\mathbb{Z} + \alpha\mathbb{Z}$ is dense. For this we define

$$\delta := \inf\{x = 2a + \alpha b > 0 : a, b \in \mathbb{Z}\}.$$

If we can show that $\delta = 0$, then $2\mathbb{Z} + \alpha\mathbb{Z}$ is dense. We find $x_n = 2a_n + \alpha b_n \downarrow \delta$ so that $2(a_n - a_{n+1}) + \alpha(b_n - b_{n+1}) \downarrow 0$, whence $\delta = 0$.

Therefore $A^{\text{even}}_{[0,1]} - A^{\text{even}}_{[0,1]}$ and $A^{\text{odd}}_{[0,1]} - A^{\text{odd}}_{[0,1]}$ cannot contain any interval $(-\varepsilon, \varepsilon)$ for some $\varepsilon > 0$ and, by Proposition B.4.4, the sets $A^{\text{even}}_{[0,1]}$ and $A^{\text{odd}}_{[0,1]}$ cannot contain any Lebesgue-measurable subset of positive measure. $\square$

Corollary B.4.6 *There exists a separable metric space $[M, d]$ and a probability measure μ on $\mathcal{B}(M)$ such that $\mu(K) = 0$ for all compact $K \subseteq M$.*

Corollary B.4.6 shows that the property 'complete' cannot be omitted in Ulam's Theorem 5.2.7:

Proof of Corollary B.4.6 We take the set $A \subseteq [0, 1]$ obtained in Theorem B.4.5 and define the metric space $[M, d]$ as the set A equipped with the Euclidean distance. As $[0, 1]$ with the Euclidean distance is separable, $[M, d]$ is separable. We define the outer measure $\mu(C) := \inf\{\lambda(B) : C \subseteq B, B \in \mathcal{L}(\mathbb{R})\}$ on 2^A and obtain the σ-algebra $\mathcal{F}^\mu_A$, where μ becomes a measure. By Theorem B.4.2(1) it can be verified that $\mathcal{L}(\mathbb{R})|_A \subseteq \mathcal{F}^\mu_A$. This also implies that all open balls from $[M, d]$ belong to $\mathcal{L}(\mathbb{R})|_A$, whence to $\mathcal{F}^\mu_A$ so that $\mathcal{B}(M) \subseteq \mathcal{F}^\mu_A$. To compute $\mu(A)$ we assume a Lebesgue set $B \supseteq A$ and aim to minimise $\lambda(B)$. As $A \subseteq [0, 1]$ we may assume $B \subseteq [0, 1]$ as well. Then $[0, 1] \setminus B \subseteq [0, 1] \setminus A$. Hence $\lambda([0, 1] \setminus B) = 0$ by Theorem B.4.5 and $\lambda(B) = 1$. Therefore, $\mu(A) = 1$. Now we assume $K \subseteq A$ to be compact (within $[M, d]$). Because the natural embedding $J : M \hookrightarrow [0, 1]$ is continuous, K is compact in $[0, 1]$ as well and therefore $K \in \mathcal{L}(R)$. But then Theorem B.4.5 applies again and $\lambda(K) = 0$ and $\mu(K) = 0$. $\square$

Additional Remarks About the Literature The existence of a non-measurable set as in Theorem B.4.2(5) is due to Vitali in 1905 (see the concluding footnote from Lebesgue in [117]). Unaware about this existence van Vleck [194] constructed a set like in Theorem B.4.5 in 1908. With Proposition B.4.4 and Theorem B.4.5 we follow the presentation of Dudley [50, pages 107–108]. Regarding this and other counterexamples, the monograph of Kühn and Schilling [108] is recommended.

Appendix C
R-code

The following simple programs, written for the *The R Project for Statistical Computing*, https://www.r-project.org/, illustrate the strong law of large numbers (SLLN), the central limit theorem (CLT), and the law of iterated logarithm (LIL). The files can be downloaded from https://github.com/hannahgeiss/R-for-Probability.

C.1 monte-carlo.r

Computation of the integral $\int_{[0,1]} \frac{1}{2}(\cos(4\pi x) + 1)\, d\lambda(x)$ (which is obviously equal to $\frac{1}{2}$). The program shows a plot where the random points below the cosine curve appear and the approximate value of the integral.

```
K <- 1000          # Number of samples, outer loop
N <- 1000          # Number of independent random
                   # variables, inner loop
T <- 1:N           # Discretization of the x values
                   # when plotting the function
X <- 1:K           # Random x-values
Y <- 1:K           # Random y-values

for (i in 0:N){
        F[i] <- (cos(4*pi*i/N) + 1)/2
        T[i] <- i/N}
X <- runif(K, min = 0, max = 1)
Y <- runif(K, min = 0, max = 1)
plot(T,F,type="l",col="blue1",lwd=0.3)
title( "Monte Carlo method",
                        cex.main=1, font.main=1)
I <- 0
```

```
for (k in 1:K){
        if (Y[k] <=   (cos(4*pi* X[k]) + 1)/2)
{points(X[k],Y[k],type="p",col="blue1",lwd=1.1)
        I <- I+1}
}
I <- I/K
text(.7,1,"Integral equals")
text(.95,1,I)
```

C.2 monte-carlo_for_pi.r

Computation of π by computing the area of the ball with radius 1 around the origin in $\mathbb{R}^2$. The program shows a plot where the uniformly on $[-1, 1] \times [-1, 1]$ distributed points which are inside the ball are shown and counted and hence provides the approximate value for π.

```
N <- 10000   # Number of discretization points for
             # drawing the circle
K <- 1000    # Number of sample points

for (i in 0:N){
        X[i] <- cos(4*pi*i/N)
        Y[i] <- sin(4*pi*i/N)
        }

plot(X,Y,type="l",col="blue1",lwd=0.3)
title( "Monte Carlo method to compute Pi"  ,
                        cex.main=1, font.main=1)

A <- runif(K, min = -1, max = 1)
B <- runif(K, min = -1, max = 1)

I <- 0

for (k in 1:K){
        if (A[k]^2 + B[k]^2 <= 1)
    {points(A[k],B[k],type="p",col="blue1",lwd=1.1)
        I <- I+1}
    }

I <- 4 * I/K

text(.6,1,"Pi approximately = ")
text(.95,1,I)
```

C.3 slln_vs_clt.r

Simulation of the SLLN (parameter 0) and the CLT (parameter 1) by a symmetric random walk with histograms for the empirical distributions. The program sums up independent random variables which are equal to 1 or − 1 with probability 1/2 and scales them, either according to the SLLN or to the CLT. A sequence of the plotted random walk appears and finally the histogram of the final values of all random walks. To recognize the difference between SLLN and CLT notice that the histograms have different scales for the D axis.

```
K <- 1000       # Number of samples, outer loop
N <- 1000       # Number of independent random variables,
                # inner loop
D <- 1:K        # Sample vector
S <- 1:N        # Sum of random variables
W <- 1:N        # Renormalizes sum of random variables
T <- 1:N        # Number of step in the computation of the
                # sum

# Choice of the limit theorem
param <- 1 # 0 for SLLN
           # 1 for CLT
if (param == 0)
        {sigma <- 1/N}
if (param == 1)
        {sigma <- 1/sqrt(N)}

# The computation
for(k in 1:K){
N <- 1000
f <- 2*rbinom(n=N, size=1, prob=0.5) - 1
S[1] <- f[1]
for (i in 2:N){S[i] <- S[i-1] + f[i]}
for (i in 1:N){W[i] <- sigma * S[i]}
for (i in 1:N){T[i] <- i}
plot(T,W, type="l", col="blue1", lwd=0.3)
if (param == 0)
        title( "Strong_Law_of_Large_Numbers" ,
                                cex.main=1, font.main=1)
if (param == 1)
        title( "Central_Limit_Theorem" ,
                                cex.main=1, font.main=1)
D[k] <- W[N]
}

# Output histogram
if (param == 0)
    hist(D, main="Histogram_for_Strong_Law_of_Large_Numbers")
if (param == 1)
    hist(D, main="Histogram_for_Central_Limit_Theorem")
```

C.4 lil.r

Simulation of the LIL by a symmetric random walk. In the plot we see how the area inside the ' log log curve' is filled up by the paths of the random walk while only a few of the paths happen to be outside the curve.

```
# For a direct output to the screen comment out the commands
# pdf("lil.pdf") and dev.off().
# Otherwise your output will be lil.pdf.

pdf("lil.pdf")

N <- 1000              # Number of independent random
                       # variables, inner loop
K <- 1000              # Number of realizations
S <- 1:N               # Sum of random variables
T <- 1:N               # Number of steps in the
                       # computation
                       # of the sum
U <- 1:N               # Upper bound
L <- 1:N               # Lower bound
B <- sqrt(2*N*log(log(N)))  # Upper and lower bound for the
                            # graph

for (i in 3:N){L[i] <- - sqrt(2*i*log( log(i)))}
for (i in 3:N){U[i] <-   sqrt(2*i*log( log(i)))}
plot(T,U,type="l",col="red1",lwd=0.5,ylim=c(-B,B))
lines(T,L,type="l",col="red1",lwd=0.5)

for (k in 1:K){
f <- 2*rbinom(n=N,size=1,prob=0.5) - 1
S[1] <- f[1]
for (i in 2:N){S[i] <- S[i-1] + f[i]}
lines(T,S,type="l",col="blue1",lwd=0.1)
}

title( "Law of Iterated Logarithm" ,
                            cex.main=1, font.main=1)

dev.off()
```

Bibliography

1. Aleksandrov, A.D.: Additive set-functions in abstract spaces. Mat. Sb. **8**, 307–348 (1940); **9**, 563–628 (1941); **13**, 169–238 (1943)
2. Aoki, T.: Locally bounded linear topological spaces. Proc. Imp. Acad. Tokyo **18**(10) (1942)
3. Bauer, H.: Probability Theory. De Gruyter, Berlin (1995)
4. Bauer, H.: Measure and Integration Theory. De Gruyter, Berlin (2001)
5. Babenko, K.I.: An inequality in the theory of Fourier integrals. Izvestiya Akademii Nauk SSSR **25**, 531–542 (1961)
6. Beckner, W.: Inequalities in Fourier analysis. Ann. Math. **102**, 159–182 (1975)
7. Behrends, E.: Maß- und Integrationstheorie. Springer, Berlin (1987)
8. Behrends, E., Gritzmann, P., Ziegler, G.M.: Pi und Co. Springer, Berlin (2008)
9. Bennett, C., Sharpley, R.: Interpolation of Operators. Academic Press, London (1988)
10. Bentkus, V.Y.: Dependence of the Berry–Esseen estimate on the dimension. Lith. Math. J. **26**, 110–114 (1986)
11. Bergh, J., Löfström, J.: Interpolation Spaces. Springer, Berlin (1976)
12. Bernstein, S.N.: An attempt to axiomatize the foundations of the theory of probability. Soobshch. Khar'k. matem. ob-va, series 2 **15**, 209–274 (1917) (in Russian)
13. Berry, A.C.: The accuracy of the Gaussian approximation to the sum of independent variates. Trans. Am. Math. Soc. **49**, 122–136 (1941)
14. Bienaymé, M.: Considérations à l'appui de la découverte de Laplace sur la loi de probabilité dans la méthode des moindres carrés. J. Math. Pures Appl. 2e série **12**, 158–176 (1867)
15. Billingsley, P.: Weak Convergence of Measures: Applications in Probability. Philadelphia, SIAM (1971)
16. Billingsley, P.: Probability and Measure, 3rd edn. Wiley, Hoboken (1995)
17. Billingsley, P.: Convergence of Probability Measures, 2nd edn. Wiley, Hoboken (1999)
18. Billingsley, P.: Probability and Measure, anniversary edn. Wiley, Hoboken (2012)
19. Birkhoff, G.D.: Proof of the ergodic theorem. Proc. Nat. Acad. Sci. USA **17**, 656–660 (1931)
20. Bochner, S.: Integration von Funktionen, deren Werte die Elemente eines Vektorraumes sind. Fundamenta Mathematicae **20**, 262–276 (1933)
21. Bochner, S.: Monotone Funktionen, Stieltjessche Integrale, und harmonische Analyse. Math. Ann. **108**, 378–410 (1933)
22. Bogachev, V.I.: Measure Theory. Springer, Berlin (2007)
23. Bogachev, V.I.: Weak Convergence of Measures. Mathematical Surveys and Monographs, vol. 234. American Mathematical Society (2018)
24. Borel, É.: Les probabilités dénombrables et leurs applications arithmétiques. Supplemento di rend. circ. Mat. Palermo **27**, 247–271 (1909)

25. Breiman, L.: Probability. Addison-Wesley, Boston (1968)
26. de Acosta, A.: A new proof of the Hartman-Wintner law of the iterated logarithm. Ann. Probab. **11**(2), 270–276 (1983)
27. Comte de Buffon, Leclerc, G.L.: Essai d'arithmétique morale. Appendix to Histoire naturelle générale et particulière, 4 (1777)
28. Calderon, A.P., Zygmund, A.: On the existence of certain singular integrals. Acta Math. **88**, 85–139 (1952)
29. Cantelli, F.P.: Sulla probabilità come limite della frequenza. Accad. Lincei, Rendiconti, Cl. Sci. Fis. Mat. Nat. Ser. 5 **26**, 39–45 (1917)
30. Cantelli, F.P.: Su due applicazione di un teorema di G. Boole alla statistica matematica. Accad. Lincei, Rendiconti, Cl. Sci. Fis. Mat. Nat. Ser. 5 **26**, 295–302 (1917)
31. Carathéodory, C.: Vorlesungen über Reelle Funktionen, 2. Auflage. Springer, Berlin (1927)
32. Chatterjee, S., and Students: Stein's Method and Applications. Lecture Notes Stanford University (2007). https://souravchatterjee.su.domains//stat206Afall07.html
33. Chebyshev, P.L.: Des valeurs moyennes. J. Math. Pures Appl. 2e série **12**, 177–184 (1867)
34. Chen, L.H.Y.: Stein meets Malliavin in normal approximation. Acta Math Vietnam **40**, 205–230 (2015)
35. Chen, L.H.Y., Goldstein, L., Shao, Q.-M.: Normal Approximation by Stein's Method. Springer, Berlin (2011)
36. Chow, Y.S., Teicher, H.: Probability Theory: Independence Interchangeability Martingales. Springer, Berlin (1978)
37. Ciesielski, Z.: On Haar functions and on the Schauder basis of the space $C_{<0,1>}$. Bull. Acad. Polon. Sci. Sér. Sci. Math. Astr. Phys. **4**, 227–232 (1959)
38. Ciesielski, Z.: Some properties of Schauder basis of the space $C_{<0,1>}$. Bull. Acad. Polon. Sci. Sér. Sci. Math. Astr. Phys. **8**, 141–144 (1960)
39. Ciesielski, Z.: On the isomorphisms of the space H_α and m. Bull. Acad. Polon. Sci. Sér. Sci. Math. Astr. Phys. **8**, 217–222 (1960)
40. Ciesielski, Z.: Hölder conditions for realizations of Gaussian processes. Trans. Am. Math. Soc. **99**, 403–413 (1961)
41. Ciesielski, Z.: Fractal functions and Schauder bases. Comput. Math. Appl. **30**(3–6), 283–291 (1995)
42. Conway, J.B.: A Course in Functional Analysis. Springer, Berlin (1990)
43. Daniell, P.J.: Stieltjes derivatives. Bull. Am. Math. Soc. **26**, 444–448 (1920)
44. Djehiche, B., Geiss, H., Geiss, S., Labart, C., Nykänen, J.: Convergence rate for random walk approximations of mean field BSDEs (2024). arXiv 2409.14212
45. Diestel, J.: Applications of weak compactness and bases to vector measures and vectorial integration, Rev. Roumaine Math. Pures Appl. **18**, 211–224 (1973)
46. Diestel, J., Uhl, J.J.: Vector Measures. Mathematical Surveys, vol. 15. American Mathematical Society, Providence (1977)
47. Dieudonne, J.: Foundations of Modern Analysis. Academic Press, Cambridge (1969)
48. Dudley, R.M.: Distances of probability measures and random variables. Ann. Math. Stat. **39**, 1563–1572 (1968)
49. Dudley, R.M.: An extended Wichura theorem, definitions of Donsker class, and weighted empirical distributions. In: A. Beck et al. (eds.) Probability in Banach Spaces V (Proc. Conf. Medford, 1984). Lecture Notes in Mathematics, vol. 1153, pp. 141–178. Springer, Berlin (1985)
50. Dudley, R.M.: Real Analysis and Probability, 2nd edn. Cambridge University Press, Cambridge (2002)
51. Dunford, N.: Integration of vector-valued functions. Bull. Am. Math. Soc. **43**, 24 (1937) (abstract)
52. Dunford, N., Schwartz, J.T.: Linear Operators, Part I: General Theory, 4th edn. Wiley, Hoboken (1967)
53. Embrechts, P., Hofert, M.: A note on generalized inverses. Math. Meth. Oper. Res. **77**, 423–432 (2013)

54. Engelbert, H.-J., Schladitz, K.: On probability density functions which are their own characteristic functions. Theor. Probab. Appl. **40**(3), 694–698 (1996)
55. Esseen, C.G.: On the Liapounoff limit of error in the theory of probability. Ark. Math., Astr.o.Fysik **28A**(9), 1–19 (1942)
56. Esseen, C.G.: Fourier analysis of distribution functions. Acta Math. **77**, 1–125 (1945)
57. Etemadi, N.: An elementary proof of the strong law of large numbers. Z. Wahrsch. verw. Gebiete **55**, 119–122 (1981)
58. Etemadi, N.: On the law of large numbers for nonnegative random variables. J. Multivariate Anal. **13**, 187–193 (1983)
59. Etemadi, N.: On some classical results in probability theory. Sankhyä Indian J. Stat. (Series A) **47**, 215–221 (1985)
60. Fatou, P.J.L.: Séries trigonométriques et séries de Taylor. Acta Math. **30**, 335–400 (1906)
61. Feller, W.: An Introduction to Probability Theory and Its Applications I, 3rd edn. Wiley, Hoboken (1968)
62. Feller, W.: An Introduction to Probability Theory and Its Applications II, 2nd edn. Wiley, Hoboken (1971)
63. Fischer, H.: A History of the Central Limit Theorem. From Classical to Modern Probability Theory. Springer, Berlin (2011)
64. Folland, G.B.: A Course in Abstract Harmonic Analysis. Textbooks in Mathematics, vol. 29, 2nd edn. Chapman and Hall/CRC, Boca Raton (2016)
65. Fréchet, M.: Sur les ensembles de fonctions et les opérations linéaires. Les Comptes rendus de l'Académie des Sciences **144**, 1414–1416 (1907)
66. Fréchet, M.: Sur les opérations linéaires (troisième note). Trans. Am. Math. Soc. **8**(4), 433–446 (1907)
67. Fréchet, M.: Des familles et fonctions additives d'ensembles abstraits (Suite). Fundamenta Mathematicae **5**(1), 206–251 (1924)
68. Tonelli, L.: Sugli integrali multipli. Accad. Lincei, Rendiconti, Cl. Sci. Fis. Mat. Nat. Ser. 5 **16**, 608–614 (1907)
69. Garsia, A.M.: A simple proof of E. Hopf's maximal erogodic theorem. J. Math. Mech. **14**, 381–382 (1965)
70. Geiss, S.: BMO_ψ-spaces and applications to extrapolation theory. Stud. Math. **122**, 235–274 (1997)
71. Giné, E., Zinn, J.: Central limit theorems and weak laws of large numbers in certain Banach spaces, Z. Wahrsch. verw. Gebiete **62**, 323–354 (1983)
72. Gendenko, B.V.: The Theory of Probability, 6th edn. CRC Press, Boca Raton (1997)
73. Goldmaker, L.: Differentiation under the integral sign. https://web.williams.edu/Mathematics/lg5/Feynman.pdf
74. Grattan-Guinness, I.: Joseph Fourier, 1768–1830. Cambridge, MIT Press (1972)
75. Hahn, H.: Theorie der Reellen Funktionen. Springer, Berlin (1921)
76. Hahn, H.: Über die Multiplikation total-additiver Mengenfunktionen. Annali della Scuola Normale Superiore di Pisa, Classe di Scienze 2e série **2**(4), 429–452 (1933)
77. Halmos, P.R.: Measure Theory. Springer, Berlin (1950)
78. Hardy, G.H., Littlewood, J.E.: A maximal theorem with function-theoretic applications. Acta Math. **54**, 81–116 (1930)
79. Hardy, G.H., Littlewood, J.E., Pólya, G.: Inequalities. Cambridge University Press, Cambridge (1934)
80. Hartman, P., Wintner, A.: On the law of the iterated logarithm. Am. J. Math. **63**(1), 169–176 (1941)
81. Hausdorff, F.: Eine Ausdehnung des Parsevalschen Satzes über Fourierreihen. Math. Zeitschrift **16**, 163–169 (1923)
82. Hausdorff, F.: Grundzüge der Mengenlehre, 1st edn. Chelsea Publishing, Von Veit, Leipzig/Chelsea/New York (1965)
83. Herglotz, G.: Über Potenzreihen mit positiven Teil im Einheitskreis. Leipziger Berichte **63**, 501–511 (1911)

84. He, S., Wang, J., Yan, J.: Semimartingale Theory and Stochastic Calculus. Taylor & Francis, Milton Park (1992)
85. Hewitt, E., Savage, L.J.: Symmetric measures on Cartesian products. Trans. Am. Math. Soc. **80**, 470–501 (1955)
86. Hilbert, D.: Über eine Anwendung der Integralgleichungen auf ein Problem der Funktionentheorie. Verhandlungen des dritten internationalen Mathematiker Kongresses in Heidelberg 1904, pp. 233–240. Teubner, Leipzig (1905)
87. Hodges, S.L., Cam, L.: The Poisson approximation to the binomial distribution. Ann. Math. Stat. **31**, 737–740 (1960)
88. Hölder, O.: Über einen Mittelwertsatz. Nachrichten von der Königl. Gesellschaft der Wissenschaften und der Georg-August-Universität zu Göttingen, Göttingen, 38–47 (1889)
89. Hoeffding, W.: Probability inequalities for sums of bounded random variables. J. Am. Stat. Assoc. **58**, 13–30 (1963)
90. Hoffmann-Jørgensen, J.: The theory of analytic spaces. Various Publication Series 10. Aarhus Universitet, Matematisk Institut, Aarhus (1970)
91. Hoffmann-Jørgensen, J.: Probability of B-spaces. Lecture Notes Series, vol. 48. Aarhus. Aarhus Universitet, Matematisk Institut, Aarhus (1977)
92. Itô, K., Nisio, M.: On the convergence of sums of independent Banach space valued random variables. Osaka J. Math. **5**, 35–48 (1968)
93. Jacod, J., Protter, P.: Probability Essentials, 2nd edn. Springer, Berlin (2004)
94. Jensen, J.L.W.V.: Sur les fonctions convexes et les inégalités entre les valeurs moyennes. Acta Math. **30**, 175–193 (1906)
95. Johnson, W.B., Schechtman, G.: Sums of independent random variables in rearrangement invariant function spaces, Ann. Prob. **17**(2), 789–808 (1989)
96. Kakutani, S.: Concrete representation of abstract (M)-spaces. Ann. Math. **42**, 994–1024 (1941)
97. Kallenberg, O.: Foundations of Modern Probability, 3rd edn. Springer, Berlin (2021)
98. Kalton, N.J., Peck, N.T., Roberts, J.W.: An F-space Sampler. Lecture Notes, vol. 89. London Mathematical Society, London (1984)
99. Kánnai, Z.: An elementary proof that the Borel class of the reals has cardinality continuum. Acta Math. Hungar. **159**, 124–130 (2019)
100. Karatzas, I., Shreve, E.: Brownian motion and Stochastic Calculus, 2nd edn. Springer, Berlin (1999)
101. Khintchine, A.: Über einen Satz der Wahrscheinlichkeitsrechnung. Fundamenta Mathematicae **6**, 9–20 (1924)
102. Khintchine, A.: Zur mathematischen Begründung der statistischen Mechanik. Zeitschr. für angew. Math. Mech. **13**, 101–103 (1933)
103. Klenke, A.: Probability Theory. A Comprehensive Course. Springer, Berlin (2014)
104. Kolmogorov, A.: Über das Gesetz des iterierten Logarithmus. Math. Ann. **101**, 126–135 (1929)
105. Kolmogorov, A.N.: Grundbegriffe der Wahrscheinlichkeitsrechnung. Springer, Berlin (1933)
106. Kolmogorov, A.N.: Ein vereinfachter Beweis des Birkhoff-Khintchinchen Ergodensatzes. Recueil Math. **44**, 366–368 (1937)
107. Kress, R.: Linear Integral Equations. Springer, Berlin (1989)
108. Kühn, F., Schilling, R.: Counterexamples in Measure and Integration. Cambridge University Press, Cambridge (2021)
109. Kühn, F., Schilling, R.: Maximal inequalities and some applications. Probab. Surv. **20**, 382–485 (2023)
110. Kuelbs, J.: Kolmogorov's law of the iterated logarithm for Banach space valued random variables. Ill. J. Math. **21**(4), 784–800 (1977)
111. Kuratowski, C.: Une méthode d'élimination des nombres transfinis des raisonnements mathématiques. Fundamenta Mathematicae **3**, 76–108 (1922)
112. Kwapień, S., Woyczyński, W.A.: Random Series and Stochastic Integrals: Single and Multiple. Birkhäuser, Basel (1992)

113. Lamberton, D., Lapeyre, B.: Stochastic Calculus Applied to Finance. Chapman & Hall, Boca Raton (2008)
114. Cam, L.: An approximation theorem for the Poisson binomial distribution. Pacific J. Math. **10**, 1181–1197 (1960)
115. Cam, L.: On the distribution of sums of independent random variables, Bernoulli, Bayes, Laplace. In: Neyman, J., Le Cam, L.M. (eds.) Proceeding of an International Research Seminar, pp. 179–202. Springer, Berlin (1963)
116. Lebesgue, H.: Leçons sur l'intégration et la recherche des fonctions primitives. Gauthier-Villars, Paris (1904)
117. Lebesgue, H.: Contribution à l'étude des correspondances de M. Zermelo. Bulletin de la Société Math'ematique de France **35**, 202–212 (1907)
118. Ledoux, M., Talagrand, M.: Probability in Banach Spaces. Springer, Berlin (1991)
119. Lévy, P.: Théorie de l'addition des variables aléatoires. Gauthier-Villars, Paris (1937)
120. Li, D., Queffélec, H.: Introduction to Banach Spaces: Analysis and Probability I and II. Cambridge University Press, Cambridge (2018)
121. Loève, M.: Probability I, 4th edn. Springer, Berlin (1977)
122. Loève, M.: Probability II, 4th edn. Springer, Berlin (1978)
123. Lorentz, G.: Some new function spaces. Ann. Math. **51**, 37–55 (1950)
124. Lorentz, G.: On the theory of spaces Λ. Pacific J. Math. **1**, 411–429 (1951)
125. Lorist, E., Nieraeth, Z.: Banach function spaces done right. Indagationes Mathematicae **35**, 247–268 (2024)
126. Luxemburg, W.A.J.: Banach Function Spaces. TU Delft, Delft (1955)
127. Masani, P.: Measurability and Pettis integration in Hilbert spaces. Journal für die reine und angewandte Mathematik **297**, 92–135 (1978)
128. Massart, A.: Concentration Inequalities and Model Selection. Ecole d'Eté de Probabilités de Saint-Flour XXXIII, 2003. Lecture Notes in Mathematics. Springer, Berlin (1896)
129. McShane, E.J.: Linear functionals on certain Banach spaces. Proc. Am. Math. Soc. **1**, 402–408 (1950)
130. Mikosch, T.: Non-life Insurance Mathematics: An Introduction with the Poisson Process. Springer, Berlin (2009)
131. Minkowski, H.: Diophantische Approximationen. Springer, Berlin (1907)
132. Mises, R.V.: Grundlagen der Wahrscheinlichkeitsrechnung. Math. Zeitschrift **5**, 52–99 (1919)
133. Müller-Gronbach, T., Novak, E., Ritter, K.: Monte Carlo-Algorithmen. Springer, Berlin (2012)
134. Nedoma, J.: Note on generalized random variables. In: Transactions of the First Prague Conference on Information Theory, Statistical Decision Functions, Random Processes (Liblice, 1956), pp. 139–141. Publishing House of the Czechoslovak Academy of Sciences, Prague (1957)
135. Nagaev, S.V.: An estimate of the remainder term in the multidimensional central limit theorem. In: Proceedings of the Third Japan–USSR Symposium on Probability Theory (Tashkent, 1975). Lecture Notes in Mathematics, vol. 550, pp. 419–438. Springer, Berlin (1976)
136. Nikodym, O.M.: Sur une généralisation des mesures de M. J. Radon. Fundamenta Math. **15**, 131–179 (1930)
137. Nourdin, I., Peccati, G.: Stein's method on Wiener chaos. Probab. Theory Relat. Fields **145**(1–2), 75–118 (2009)
138. Nourdin, I., Peccati, G.: Normal Approximation with Malliavin Calculus: From Stein's Method to Universality. Cambridge Tracts in Mathematics, vol. 192. Cambridge University Press, Cambridge (2012)
139. Nualart, D., Peccati, G.: Central limit theorems for sequences of multiple stochastic integrals. Ann. Probab. **33**, 177–193 (2005)
140. Orlicz, W.: Über eine gewisse Klasse von Räumen vom Typus B. Bull. Int. Acad. Polon. Sci. A (8–9), 207–220 (1932)

141. Oxtoby, J.C., Ulam, S.M.: On the existence of a measure invariant under a transformation. Ann. Math. **40**, 560–566 (1939)
142. Padgett, W.J., Taylor, R.L.: Laws of Large Numbers for Normed Linear Spaces and Certain Frechet Spaces. Lecture Notes in Mathematics, vol. 360. Springer, Berlin (1973)
143. Pap, E.: Handbook of Measure Theory: In two volumes. North Holland, Amsterdam (2002)
144. Parthasarathy, K.R.: Probability Measures on Metric Spaces. Academic Press, Cambridge (1967)
145. Peetre, J.: A theory of interpolation of normed spaces. Notas de Matematica, Rio de Janeiro **39**, 1–86 (1963)
146. Pettis, B.J.: On integration in vector spaces. Trans. Am. Math. Soc. **44**, 277–304 (1938)
147. Plancherel, M.: Contribution à l'étude de la représentation d'une fonction arbitraire par des intégrales définies. Rendiconti del Circolo Matematico di Palermo (3), 289–335 (1910)
148. Pólya, G.: Remarks on characteristic functions. Proceedings of the Berkeley Symposium on Mathematical Statistics and Probability, pp. 115–123 (1949)
149. Prokhorov, Y.V.: Asymptotic behavior of the binomial distribution (in Russian). Uspekhi Mat. Nauk **8**(3), 135–142 (1953)
150. Prokhorov, Y.V.: Convergence of random processes and limit theorems in probability theory. Theor. Probab. Appl. **1**(2), 157–214 (1956)
151. Radon, J.: Theorie und Anwendungen der absolut additiven Mengenfunktionen. Sitzungsber. Akad. Wiss. Wien Abt. IIa **122**, 1295–1438 (1913)
152. Raič, M.: A multivariate Berry–Esseen theorem with explicit constants. Bernoulli **25**(4A), 2824–2853 (2019)
153. Riesz, F.: Sur une espèce de géométrie analytique des systèmes de fonctions sommables. Comptes rendus de l'Académie des Sciences **144**, 1409–1411 (1907)
154. Riesz, F.: Untersuchungen über Systeme integrierbarer Funktionen. Math. Ann. **69**, 449–497 (1910)
155. Riesz, F.: Sur les opérations fonctionnelles linéaires. Comptes rendus de l'Académie des Sciences **149**, 974–977 (1910)
156. Riesz, F.: Sur certains systèmes singuliers d'équations intégrales. Annales scientifiques de IÉ.N.S., 3e série **28**, 33–62 (1911)
157. Riesz, M.: Sur les maxima des formes bilinéaires et sur les fonctionnelles linéaires. Acta Math. **49**, 465–497 (1926)
158. Riesz, M.: Sur les fonctions conjuguées. Math. Zeitschrift **27**, 218–244 (1928)
159. Rogers, L.J.: An extension of a certain theorem in inequalities. Messenger Math. New Series, XVII **10**, 145–150 (1888)
160. Rolewicz, S.: On a certain class of linear metric spaces. Bull. Acad. Polon. Sci. **5**, 471–473 (1957)
161. Rolewicz, S.: Metric linear spaces. PWN-Polish Scientific Publishers, Warsaw (1984)
162. Rolski, T., Schmidli, H., Schmidt, V., Teugels, J.: Stochastic Processes for Insurance and Finance. John Wiley, Hoboken (1999)
163. Romano, J., Siegel, A.F.: Counterexamples in Probability and Statistics. Chapman & Hall, Boca Raton (1986)
164. Ross, N.: Fundamentals of Stein's method, Probab. Surv. **8**, 210–293 (2011)
165. Rudin, W.: Real and Complex Analysis, 3rd edn. Mc Graw Hill, New York (1987)
166. Rudin, W.: Functional Analysis. McGraw-Hill, New York (1989)
167. Schilling, R.L.: Wahrscheinlichkeit. De Gruyter, Berlin (2017)
168. Schilling, R.L.: Measures, Integrals, and Martingales, 2nd edn. Cambridge University Press, Cambridge (2017)
169. Schur, J.: Bemerkungen zur Theorie der beschränkten Bilinearformen mit unendlich vielen Veränderlichen. Journal für die reine und angewandte Mathematik **140**, 1–28 (1911)
170. Schwarz, H.A.: Beweis des Satzes, dass die Kugel kleinere Oberfläche besitzt, als jeder anderer Körper gleichen Volumens. Göttinger Nachrichten 1–13 (1884)
171. Schwartz, J.: A note on the space L_p^*. Proc. Am. Math. Soc. **2**, 270–275 (1951)

172. Schwartz, L.: Radon Measures on Arbitrary Topological Spaces and Cylindrical Measures. Oxford University Press, Oxford (1973)
173. Shiryaev, A.N.: Probability, 2nd edn. Springer, Berlin (1996)
174. Sierpinski, W.: Sur les fonctions d'ensemble additives et continues. Fundamenta Mathematicae **3**, 240–246 (1922)
175. Skorohod, A.V.: Limit theorems for stochastic processes. Theor. Probab. Appl. **1**, 261–290 (1956)
176. Steele, J.M.: Le Cam's inequality and Poisson approximations. Am. Math. Monthly **101**(1), 48–54 (1994)
177. Stein, C.: A bound for the error in the normal approximation to the distribution of a sum of dependent random variables. In: Proceedings of the Sixth Berkeley Symposium on Mathematical Statistics and Probability (Univ. California, Berkeley, Calif., 1970/1971). Probability Theory, vol. II, pp. 583–602. University California Press, Berkeley (1972)
178. Stein, C.: Approximate Computation of Expectations. IMS Lecture Notes Monograph Series 7. Institute of Mathematical Statistics, Hayward (1986)
179. Steiner, J.: Gesammelte Werke. Springer, Berlin (1881–1882)
180. Steinhaus, H.: Additive und stetige Funktionaloperationen. Math. Zeitschrift **5**, 186–221 (1919)
181. Stout, W.F.: The Hartman-Wintner law of the iterated logarithm for martingales. Ann. Math. Stat. **41**(6), 2158–2160 (1970)
182. Stoyanov, J.M.: Counterexamples in Probability. Dover Publications, Mineola (2013)
183. Szewczak, Z.S.: On the maximal Lévy–Ottaviani inequality for sums of independent and dependent random vectors. Bull. Polish Acad. Sci. (Math.) **61**, 155–160 (2013)
184. Tonelli, L.: Sull'integrazione per parti. Accad. Lincei, Rendiconti, Cl. Sci. Fis. Mat. Nat. Ser. 5 **18**, 246–253 (1909)
185. Urysohn, P.: Über die Mächtigkeit der zusammenhängenden Mengen. Math. Ann. **94**, 262–295 (1925)
186. Talagrand, M.: Pettis integral and measure theory. Mem. Am. Math. Soc. **307** (1984)
187. Tao, T.: An Introduction to Measure Theory. Graduate Studies in Mathematics, vol. 126. AMS, Providence (2011)
188. Thorin, G.O.: Convexity Theorems Generalizing those of M. Riesz and Hadamard with some Applications. Comm. Sem. Math. University of Lund, Lund (1948)
189. Vakhania, N.N., Tarieladze, V.I., Chobanyan, S.A.: Probability distributions in Banach spaces. Translated from the Russian. Kluwer, Dordrecht (1991)
190. van Neerven, J.: Stochastic Evolution Equations. ISEM Lecture Notes 2007/2008. TU Delft, Delft (2007)
191. Varadhan, S.R.S.: Probability Theory. Courant Lecture Notes. Courant Institute of Mathematical Sciences, New York (2001)
192. Vershynin, R.: High-Dimensional Probability. An Introduction with Applications in Data Science. Cambridge University Press, Cambridge (2018)
193. Vitali, G.: Sul problema della misura dei gruppi di punti di una retta. Bologna, Tip. Gamberini e Parmeggiani (1905)
194. Vleck, E.B.: On non-measurable sets of points with an example. Trans. Am. Math. Soc. **9**, 237–244 (1908)
195. Wald, A.: On cumulative sums of random variables. Ann. Math. Stat. **15**(3), 283–296 (1944)
196. Weil, A.: L'Intégration dans les Groupes Topologiques et ses Applications, 2nd edn. Hermann, Paris (1940)
197. Werner, D.: Funktionalanalysis, 8th edn. Springer, Berlin (2018)
198. Williams, D.: Probability with Martingales. Cambridge University Press, Cambridge (1991)
199. Williams, R.: Diffusions, Markov Processes and Martingales. Foundations, vol. 1, 2nd edn. Cambridge University Press, Cambridge (2000)
200. Wichura, M.J.: On the construction of almost uniformly convergent random variables with given weakly convergent image laws. Ann. Math. Stat. **41**, 284–291 (1970)

201. Yosida, K., Kakutani, S.: Birkhoff's ergodic theorem and the maximal ergodic theorem. Proc. Imp. Acad. **15**(6), 165–168 (1939)
202. Young, W.H.: On the determination of the summability of a function by means of its Fourier constants. Proc. Lond. Math. Soc. **12**, 71–88 (1913)
203. Zorn, M.: A remark on method in transfinite algebra. Bull. Am. Math. Soc. **41**, 667–670 (1935)

Index

Symbols
$(\mathcal{F}, \Sigma)$-measurable map, 71
K-functional, 334
μ-measurable sets, 417
σ-additivity, 11
σ-algebra, 8
   Borel σ-algebra on metric spaces, 57
   Borel σ-algebra on $\mathbb{R}$, 33
   generated by random variables, 85
   intersection, 31
   of invariant sets, 253
   Lebesgue σ-algebra, 422
   product σ-algebra, 88
   smallest σ-algebra containing $\mathcal{G}$, 32
   trace σ-algebra, 42

A
Algebra, 8
Axiom of choice, 402

B
Bayes' formula, 20
Buffon's needle problem, 249

C
Canonical representation, 68
Cauchy sequence, 46
   in probability, 157
Central limit theorem
   classical form, 259
   Lindeberg form, 260
Change of variables, 122
   affine, 124
Characteristic function, 200
Concavity, 137
Conditional expectation
   definition, 185
   existence, 185
Convergence
   almost sure convergence, 152
   convergence in probability, 157
   convergence in L_p, 162
   in distribution of random variables, 228
   weak convergence of measures, 228
Convexity, 137
Convolution
   of functions, 207, 208
   of measures, 205

D
Density, 37, 206
Distance
   Kantorovich-Rubinstein distance, 275
   Kolmogorov distance, 275
   Wasserstein distance, 275
Distribution, 11
   binomial distribution, 26
   Cauchy distribution, 297
   distribution of a random variable (75 (*see also* Random variables))
   exponential distribution, 39
   Gaussian distribution on $\mathbb{R}$, 38
   geometric distribution, 29
   normal distribution on $\mathbb{R}$, 38
   Pareto distribution, 341
   Poisson distribution, 28

uniform distribution, 42
Weibull distribution, 338
Distribution function
    cumulative distribution function of a random variable, 75
    cumulative distribution function on $\mathbb{R}$, 75
    of a random variable, 75
    on $\mathbb{R}$, 75
    on $\mathbb{R}^d$, 77

E
Elementary event, 7
Equivalence class, 403
Equivalence class relation, 403
Ergodic map, 253
Essential supremum, 162
    of a set of maps, 174
Euler's formula, 398
Event, 8

F
Filtration, 144
Finite permutation, 99
Fourier transform
    of a function, 204, 301
    inverse, 293, 295
    of a measure, 200
    multiplier, 308
    uniqueness, 212
Function
    σ-simple function, 355
    Dirichlet function, 121
    Haar function, 375, 407
    Schauder function, 375, 407
    simple function, 68

H
Hilbert transform, 309

I
Independence
    of a family of events, 16
    of a family of random variables, 84
    of a finite family of random variables, 84
    of σ-algebras, 85
    pairwise independence, 17
Inequality
    of Chebyshev, 140
    of Hölder, 138
    of Hölder, conditional, 188

    of Hadamard, 404
    of Hardy, 331
    of Hausdorff and Young, 307
    of Hoeffding, 142
    of Jensen, 137
    of Jensen, conditional, 188
    of Lévy and Ottaviani, 370, 371
    of Markov, 136
    maximal, 370, 371
    of Minkowski, 140
    of Minkowski, conditional, 188
Integral
    Bochner integral, 362
    Dirichlet integral, 401
    Dunford integrability, 364
    Dunford integral, 364
    Lebesgue integral for extended measurable maps, 114
    Lebesgue integral for measurable maps, 108
    Lebesgue-Stieltjes integral, 122
    Pettis integral, 365
    uniform integrability, 165
Interpolation couple, 333

L
Lemma
    of Borel and Cantelli, 17
    of Doob and Dynkin about factorization, 190
    of Fatou for random variables, 116
    of Fatou for sets, 16
    of Riemann and Lebesgue, 210
    of Stein, 278
    of Urysohn, 47
    of Zorn and Kuratowski, 403
Limit inferior, 398
Limit superior, 398
Lindeberg condition, 260

M
Matrix
    positive semidefinite, 217
    symmetric, 217
Maximal function, 327
Measurable map, 68
    extended, 114
    integrable, 108
Measurable space, 8
Measure
    σ-finite measure, 11
    absolute continuity, 176

Index

absolute moment of order $(l_1, \ldots, l_d)$, 213
Carathéodory's construction, 417
continuity from above, 13
continuity from below, 13
counting measure, 12
Dirac measure, 12
finite measure, 11
Gaussian
    degenerated, 218
    density on $\mathbb{R}$, 38
    nondegenerated, 218
    on $\mathbb{R}$, 38
Hahn-Jordan decomposition of a signed measure, 179
image measure, 75
inner regular on metric spaces, 59, 60
Lebesgue measure
    on $\mathbb{R}$, 37
    on $\mathbb{R}^d$, 94
measure, 11
moment of order $(l_1, \ldots, l_d)$, 213
outer measure, 30, 417
outer regularity, 30
outer regular on metric spaces, 59, 60
product measure, 93
push forward measure, 75
signed measure, 176
support of a measure, 63
tightness, 62
tightness of a set of measures, 228
total variation of a signed measure, 180
uniqueness, 31
Measure preserving map, 253
Measure space
    complete measure space, 14
    product of σ-finite spaces, 93
Metric space, 45
    closed set, 33, 46
    closure, 46
    compactness, 46
    completeness, 46
    continuity, 47
    dense set, 46
    Ky-Fan pseudo metric, 161
    metric, 45
    open set, 33, 46
    Polish, 46
    pre-compact set, 46
    separable, 46
Mill's ratio, 277
Moments
    computation using Fourier transform, 213
    n-th absolute moment, 123
    n-th moment, 123

Monotone class, 414

N
Net-profit condition, 392
Nonincreasing rearrangement, 327
Non-negative $a \in C^*(K)$, 317
Norm, 48
    Banach function norm, 342
    norming set, 54
    p-norm, 48
    quasi-norm, 48

P
Partition, 11
Partition of unity, 48
Polar coordinates, 131
Positive semidefinite
    function, 202
Probability
    conditional probability, 19, 185
    a posteriori probability, 21
    prior probability, 21
    probability measure, 11
    probability space, 11
        atomless, 145
        product of, 89
Process
    Wiener process, 374

R
Radon-Nikodym derivative, 181
Random variable, 68
    Bernoulli, 98
    continuous, 77
    discrete, 77
    distribution of a random variable, 75
    existence of independent random variables, 95
    expectation of a random variable, 108
    expected value of a random variable, 108
    Gaussian random variable, 221
    law of a random variable, 75
    standard Gaussian random variable, 221
    uncorrelated random variables, 222
    variance of a random variable, 135
Representative, 403

S
Set
    non Lebesgue measurable, 425

pre-image, 67
symmetric, 99
Vitali set, 422
Space
    $\mathcal{L}_0(\Omega, \mathcal{F}, \mathbb{P})$, 151, 157, 162
    $L_0(\Omega, \mathcal{F}, \mathbb{P})$, 162
    Banach function, 342
    Banach space, 50
    bidual space, 53
    Bochner-Lebesgue, 360
    dual space, 53
    Lorentz, 344
    normed, 48
    p-normed, 48
    p-normed Banach space, 50
    Orlicz, 345
    quasi-normed, 48
    quasi-normed Banach space, 50
Stirling's formula, 101
Stochastic basis, 144
Stopping time, 144
Symbols
    B_E, 53
    $C(K)$, 317
    $C(M)$, 180
    $C^*(K)$, 317
    $C_0(\mathbb{R}^d; \mathbb{C})$, 210
    $C_c(\mathbb{R}^d; \mathbb{C})$, 302
    E^*, 53
    E^{**}, 53
    $L^\Phi(\Omega, \mathcal{F}, \mathbb{P})$, 345
    $L_0(\Omega, \mathcal{F}, \mathbb{P})$, 161
    $L_p(\Omega, \mathcal{F}, \mu)$, 311
    $L_p(\Omega, \mathcal{F}, \mathbb{P})$, 164
    $L_p(\Omega, \mathcal{F}, \mathbb{P}; E)$, 360
    $L_p^w(\Omega, \mathcal{F}, \mathbb{P})$, 344
    $M_\rho(\Omega, \mathcal{F}, \mathbb{P})$, 342
    $[-c, c]_0$, 289
    $\bigotimes_{i \in I} \mathcal{F}_i$, 88
    $\check{\varphi}$, 291
    $\frac{d\nu}{d\mu}$, 181
    λ_d, 94
    $\langle x, a \rangle$, 53
    $\liminf_n A_n$, 15
    $\liminf_{n \to \infty} a_n$, 398
    $\limsup_n A_n$, 15
    $\limsup_{n \to \infty} a_n$, 398
    Lip(1), 275
    $\mu_1 * \cdots * \mu_n$, 205
    $\psi_1 * \psi_2$, 207, 208
    φ_f, 200
    $\widehat{\mu}$, 200
    $d_W(\cdot, \cdot)$, 275
    f^*, 327
    f^{**}, 327
    $f_n \xrightarrow{L_p} f$, 162
    $f_n \xrightarrow{\mathbb{P}} f$, 157
    $f_n \xrightarrow{a.s.} f$, 152
    f_p^{**}, 336
    $\mathbb{E}[f \mid \gamma = y]$, 191
    $\mathbb{E}[\mathbb{1}_A \mid \gamma = y]$, 191
    $\mathbb{E}[f \mid \mathcal{G}]$, 185
    $\mathbb{E}[\mathbb{1}_A \mid \mathcal{G}]$, 185
    $\mathcal{F}^\infty$, 86
    $\mathcal{L}^\Phi(\Omega, \mathcal{F}, \mathbb{P})$, 345
    $\mathcal{L}_p(\Omega, \mathcal{F}, \mu)$, 311
    $\mathcal{L}_p(\Omega, \mathcal{F}, \mu; E)$, 198
    $\mathcal{L}_p(\Omega, \mathcal{F}, \mathbb{P})$, 162
    $\mathcal{L}_p(\Omega, \mathcal{F}, \mathbb{P}; E)$, 360
    $\mathcal{L}_p^w(\Omega, \mathcal{F}, \mathbb{P})$, 344
    $\mathcal{M}(\mathbb{R}^d)$, 199
    $\mathcal{M}_1(\mathbb{R}^d)$, 199
    $\mathcal{M}_\rho(\Omega, \mathcal{F}, \mathbb{P})$, 342
    $\mathcal{U}_{[c,d]}$, 42
    ess sup$_\Omega f$, 162
    $\mid \nu \mid_{\text{TV}}$, 180
    Bin$_{n,p}$, 26
    Exp$_\lambda$, 39
    Geom$_p$, 29
    Pois$_\lambda$, 28
    $\sigma(f_i : i \in I)$, 85
    $\sigma(\mathcal{G})$, 32
    p_{m,σ^2}, 38
    $\mathcal{B}(\mathbb{R})$, 33
    $\mathcal{F}^\mu$, 417
    $\mathcal{L}(\mathbb{R}^d)$, 422
    $\mathcal{N}_{m,\sigma^2}$, 38
    i.i.d., 243
System
    Dynkin system, 413
    λ-system, 413
    μ-system, 414
    π-system, 30, 413

T

Theorem
    about independence and product of laws, 97
    about monotone convergence, 111, 115
    about normal numbers, 247
    about ruin probability, 392
    of Aoki and Rolewicz, 48
    of Berry and Esseen, 267
    of Birkhoff and Khinchin, ergodic theorem, 255
    of Bochner and Khinchin, 294

Index 443

of Ciesielski, 412
    characterization of Hölder functions, 408
    representation of Brownian motion, 377
closed graph theorem, 54
continuous mapping
    for convergence in probability, 160
    for the weak convergence, 229
of Dudley and Skorohod, 238
of Dynkin, $\pi - \lambda$-theorem, 415
extension theorem, 30
of Fubini, 128
of Hahn and Banach, 53
Hahn decomposition, 177
of Hardy and Littlewood, maximal theorem, 333
of Hewitt and Savage, 0-1 law, 99
inversion formula, 289, 290
of Itô and Nisio, 372
of Kakutani, 320
of Kolmogorov, 0-1 law, 86
law of iterated logarithm, 384
of Lévy's, continuity theorem, 228
of Lebesgue
    criterion for Riemann integrability, 119
    theorem about dominated convergence I, 117
    theorem of dominated convergence II, 167
maximal ergodic theorem, 254
monotone class
    for functions, 416
    for sets, 415
of Pettis, measurability theorem, 356
of Plancherel, 302
of Poisson, limit theorem, 284
of Polya, 295
portmanteau theorem, 227
of Prokhorov, 229
of Radon and Nikodym, 181
of Riesz and Fréchet, 313
of Riesz and Thorin, 404
of Schur, product theorem, 402
separation theorem, 54
of Sierpinski, 145
of Slutsky, 230
Stein's method, 276
of Stone and Weierstrass, 399
strong law of large numbers
    in Banach spaces, 367
    converse form, 244
    in Etemadi form, 243
    4th-moment condition, 156
    in Kolmogorov form, 244
of Szewczak, 371
of Tonelli, 125
of Ulam, 62
uniqueness of weak limit, 229
of Vitali, convergence theorem, 167
Wald's identity, 144
weak law of large numbers, 161

MIX
Papier aus verantwortungsvollen Quellen
Paper from responsible sources
FSC® C105338

If you have any concerns about our products,
you can contact us on
ProductSafety@springernature.com

In case Publisher is established outside the EU,
the EU authorized representative is:
**Springer Nature Customer Service Center GmbH
Europaplatz 3, 69115 Heidelberg, Germany**

Printed by Libri Plureos GmbH
in Hamburg, Germany